THE POLITICS OF THE OCEAN

The Politics

of the Ocean

By EDWARD WENK, JR.

UNIVERSITY OF WASHINGTON PRESS

Seattle and London

Library of Congress Cataloging in Publication Data

Wenk, Edward.
    The politics of the ocean.

    Includes bibliographical references.
    1. Oceanography and state—United States. I. Title.
GC58.W45   333.9′164′0973   72–5814
ISBN 0–295–95240–7

This book was published with the assistance of a grant from the Ford
Foundation.

Preface

In the historical pattern of alternating interest and neglect, the decade of the 1960s saw a reawakened and deeper consideration of the seas. For the first time, however, problems and opportunities that held serious portent for all citizens on the planet were examined by the President, by Congress, by officials of other lands, and by the United Nations. And now the issues concerned the public order of the oceans; their sharply intensified use for navigation and extraction of fish, energy, and mineral resources; conflicts in development along the ocean's rim; and pollution threats to the health of the ocean itself.

During this period, I held four policy advisory posts in the executive and legislative branches that dealt with these issues, the last one as Executive Secretary of the National Council on Marine Resources and Engineering Development that was created to advise and assist the Chief Executive directly. Upon my resignation in 1970, colleagues urged that I prepare a book drawing on observations from these unique vantage points. One group felt that pending marine decisions on both a national and international scale might be more sensitively and rationally approached in the future, if seen against the background of an integrated and systemic reflection of the subtle, causal relationships between political events of the past. A second group believed that a careful examination of marine affairs as a field unified by its geographical theater could elucidate the roles, dilemmas, processes, and bureaucratic behavior of governmental apparatus, and thus illustrate the de facto policy processes that generally defy scholarly classification. Perhaps such a case study could contribute to decision making in fields remote from marine affairs. Two different books seemed indicated.

v

Studying this duality of substance and process, I found that these twin aspects were so interwoven that the politics of the ocean viewed from either perspective had best be portrayed as a single theme. Nevertheless, the internal organization of the book was designed to aid readers who may seek to engage the text from a wide range of different interests.

Chapters 1 through 3 contain perspectives on recent policy development in ocean affairs—with a blend of scientific, technological, economic, legal, social, and political factors that couple the sea to affairs of men. Chapter 1 on political geography of the oceans sets the stage. Chapter 2 deals with the recent neglect of ocean affairs; the unprecedented political interest by the scientific community and the U.S. Congress; and legislative initiatives through the 1956–66 interval, leading to the seminal Marine Resources and Engineering Development Act. Chapter 3 deals with responses from 1966 through 1970 by Presidents Lyndon B. Johnson and Richard M. Nixon and their respective Vice-Presidents, who were charged with integrating the contributions of implementing agencies in the executive branch. Chapters 4 through 7 are anecdotal studies in selected cases of coastal use management, multinational ocean exploration, evolving law of the sea, and marine resource development. General discussion is resumed in Chapter 8 on organizational reform with the creation in 1970 of the National Oceanic and Atmospheric Administration. Chapters 9 and 10 are analytical, diagnostic, and to some extent prescriptive. From the record of marine affairs as evidence, questions are raised about the capacity of modern government to meet the stresses of the future, especially in dealing with transdepartmental problems and with perplexing issues that arise from interactions of science and technology with our society and its institutions. Chapter 10 on alternative marine futures sets forth the author's concerns with national and international issues and proposals for action. A chronology threading together key events in marine affairs is given in Appendix 32.

Some major elements in marine affairs were not implemented in relation to the 1966 Act, particularly concerning the merchant marine and naval activities, and these have not been treated in detail. Moreover, even with the topics covered, only a very small fraction of the total endeavor in and outside the federal government could be reported, namely, that which appeared to be influential at policy levels. One can only hope that omissions did not invalidate analysis and that the many unmentioned participants will understand the author's quandary in seeking both authenticity and brevity.

This book could not have been written without the assistance, advice, and felicitous cooperation of a large number of persons. First, there were those individuals intimately involved in the events portrayed in the book who, with perception and candor, kindly shared their recollections in personal interviews on transactions that bare official records rarely revealed: Robert B. Abel, David Z. Beckler, Enoch L. Dillon, John M. Drewry, Senator Hubert H. Humphrey, Congressman Alton A. Lennon, Senator Warren G. Magnuson, Dan Markel, Congressman George P. Miller, Congressman Charles A. Mosher, Robert Packard, Ambassador Arvid Pardo, Herman Pollack, Richard C. Vetter, and James H. Wakelin, Jr.

Second, there were those who waded through the lengthy rough drafts of individual chapters, checking facts and interpretations and providing incisive commentary that significantly balanced the author's necessarily limited perspective: Dayton L. Alverson, Henry A. Arnold, Vladimir Baum, William T. Burke, William D. Carey, James A. Crutchfield, Enoch L. Dillon, George A. Doumani, Robert G. Fleagle, R. Boyd Ladd, Bill L. Long, William H. Mansfield III, Dan Markel, Don K. Price, Glenn E. Schweitzer, George A. Shipman, Julius A. Stratton, and Dael Wolfle.

Third, there were those who assisted in the library research and mechanics of preparation: graduate students Hector J. Cyre, Jr., L. Ellen Levenseller, John G. Malinka, and Brian D. Wool. Especially, I am indebted to Robert H. Stockman, whose indefatigable research and persistence aided documentation and detailed proofing; to Marion Impola, whose editorial skills and scalpel helped pare redundancies and sharpen prose; and to Virginia V. Stringer, who carefully supervised preparation of the manuscript drafts.

Finally, there were three individuals who read the complete manuscript and whose advice and counsel helped enormously to pilot the author through a host of obstacles. Hugh L. Elsbree provided an insightful critique of the objectives, organization, and tone of the entire book, and of the author's interpretation of public process. Florence Broussard, who was my research assistant through almost this entire interval, provided an invaluable corrective to errant memory. And my wife Carolyn gave the final draft a careful reading, quite apart from having to suffer through nearly twelve years with a man living, sleeping, and energetically engaged in the hypnotic excitement of policy planning.

Many, many others served directly or indirectly as the author's teachers and must be recognized as significant influences on the final product. The author, however, must take the full responsibility for the accuracy of the text, and interpretations.

The Ford Foundation made possible the preparation of the book through a most generous grant, and for its confidence and support, the author is eternally grateful. Study of international institutions discussed in Chapter 10 was supported by the Secretariat of the United Nations. Support for student research assistance was also contributed from the College of Engineering of the University of Washington and the NOAA Sea Grant program.

Last, there must be acknowledgment to those elected and appointed servants of the people who animate this account. To those in the crucible of tough decisions who selflessly choose what they believe is right rather than what is expedient, this book is dedicated.

EDWARD WENK, JR.
Seattle, Washington

Contents

Appendixes

ABBREVIATIONS

ACDA	Arms Control and Disarmament Agency
AEC	Atomic Energy Commission
AID	Agency for International Development
ARPA	Advanced Research Projects Agency
ASO	American Society for Oceanography
BCF	Bureau of Commercial Fisheries
BOB	Bureau of the Budget
C&GS	Coast and Geodetic Survey
CEQ	Council on Environmental Quality
CIPME	Committee on International Policy in the Marine Environment (MSC)
CMUCZ	Committee on Multiple Use of the Coastal Zone (MSC)
CNEXO	Centre National pour l'Exploration des Océans
C of E	Army Corps of Engineers
COLD	Council of Oceanographic Laboratory Directors
COMSER	Commission on Marine Science, Engineering and Resources
COSPUP	Committee on Science and Public Policy (NAS)
CPR	Committee for Policy Review (MSC)
DOD	Department of Defense
DOT	Department of Transportation
ECOSOC	Economic and Social Council (UN)
ENDC	Eighteen Nation Disarmament Committee (UN)
ESSA	Environmental Science Services Administration
FAO	Food and Agriculture Organization (UN)
FCST	Federal Council for Science and Technology
FDA	Food and Drug Administration
FWPCA	Federal Water Pollution Control Administration
HEW	Department of Health, Education and Welfare
IAEA	International Atomic Energy Agency
ICO	Interagency Committee on Oceanography (FCST)
ICSU	International Council of Scientific Unions
IDOE	International Decade of Ocean Exploration

IGY	International Geophysical Year
IMCO	Intergovernmental Maritime Consultative Organization (UN)
IOC	Intergovernmental Oceanographic Commission (UNESCO)
LRS	Legislative Reference Service (Library of Congress)
MSC	Marine Sciences Council—National Council on Marine Resources and Engineering Development
MTS	Marine Technology Society
NACOA	National Advisory Committee on the Oceans and Atmosphere
NAE	National Academy of Engineering
NAS	National Academy of Sciences
NASA	National Aeronautics and Space Administration
NASCO	National Academy of Sciences' Committee on Oceanography
NEPA	National Environmental Policy Act
NOA	National Oceanography Association
NOAA	National Oceanic and Atmospheric Administration
NPC	National Petroleum Council
NSC	National Security Council
NSF	National Science Foundation
NSIA	National Security Industrial Association
OMB	Office of Management and Budget
ONR	Office of Naval Research
OST	Office of Science and Technology
PSAC	President's Science Advisory Committee
SCOR	Special (later Scientific) Committee on Oceanic Research (ICSU)
TENOC	Ten Years in Oceanography (U.S. Navy)
UNESCO	UN Educational, Scientific and Cultural Organization
UNGA	General Assembly of the United Nations
WHO	World Health Organization
WHOI	Woods Hole Oceanographic Institution
WMO	World Meteorological Organization

THE POLITICS OF THE OCEAN

Introduction

Most of our world is ocean. Recent color photographs of the Earth from space confirm that the predominant terrestrial feature is marine blue. All the human inhabitants of this planet reside on continental islands embedded in this vast ocean, and most of man's activities have been conspicuously if imperceptibly affected by that pervasive marine environment. While the role occupied by the oceans in the destiny of both man and nations can never be fully isolated from the remainder of the geopolitical theater, only recently has man consciously sought a tangible nexus with the sea, a marked transition from random exploration and uncritical exploitation of marine resources to a thoughtful appraisal of the potential treasures locked within the sea. Then followed a policy and a strategy to employ these resources to fulfill wide-ranging social goals and aspirations, and an exercise of stewardship of the marine environment as a common heritage of mankind.

Even our imagery about the sea has undergone sharp change. What is best remembered from both historical annals and fiction are gripping accounts of unequal contest between diminutive, fragile man in his puny cockleshells of seacraft, and the vast, tempestuous sea exerting irresistible forces amidst a howling gale. We need only recall grizzled mariners in oilskins from Joseph Conrad's *Typhoon;* Nicholas Monsarrat's *The Cruel Sea;* Herman Melville's *Moby Dick;* and Jack London's *The Cruise of the Snark.*

In contrast, another school of salt water authors considered the influence of that sublime environment on the human spirit. Rachel Carson with *The Sea Around Us* and Henry Beston in *The Outermost House*, both in the seductive traditions of Thoreau, related man to the tranquility and restorative qualities of a restless sea. And now the rapture of marine landscapes under the sea is reported eloquently by skindiving Jacques Cousteau, whose artful television series has brought to a mass audience all of the beauty and the poetry of the submarine world. In part from that medium, the term "oceanography" is now a household word stimulating almost voluptuous pleasure from playful schools of brilliantly colored fish amidst a twinkling blue-green background. In a technological age, visual excitement has been reinforced by colorful research submarines with owlish eyes and dexterous mechanized arms, at work on the previously inaccessible seabed.

While notions about oceanography have been rapidly changing, the word itself was coined only two-hundred years ago when the literature began to reflect careful observations of the then rare marine naturalist. Benjamin Franklin first published charts of the Gulf Stream in 1786. Nathaniel Bowditch structured thinking about the sea with his *New American Practical Navigator* in 1802. Charles Darwin reported systematically on marine life in his journals of the *Beagle* from 1831 to 1836. *The Physical Geography of the Sea* by Matthew Fontaine Maury, published in 1855, opened the door to technical hydrography, mapping, and charting.

Then came organization of scientific knowledge on marine life by Louis Agassiz around 1873, and the first systematic oceanographic expedition in the years 1872–76 by H. M. S. *Challenger*. From these roots a new group of marine sciences was born. They were to evolve, however, almost in isolation. As recently as 1932, a famous oceanographer wrote, "Physical oceanography was too frequently considered a branch of science having little importance except to those interested in the ocean itself. . . ."[1] The problem, however, was far more serious than simply a gap between marine science and the classical disciplines; it also was a fact that the science of the seas treated this blue planet as though it were uninhabited. Almost completely neglected were relationships to those human affairs, events, institutions, individuals and their transactions that coupled mankind as well as men to the ubiquitous oceans. Paradoxically, for a nation that was settled by sea and that enrolled a rich heritage of maritime accomplishments, the political as well as scientific communities proved almost indifferent to the influence of the sea on national affairs except in time of war.

The turning point for the United States occurred in 1966. Perhaps it occurred then for all nations. This book is an account of circumstances leading up to and following what may be the single most important event in enlisting the seas to the service of man—the enactment of the

Marine Resources and Engineering Development legislation and its subsequent implementation.

The awakening—most succinctly represented by the 1966 transition in terminology from "scientific oceanography" to "marine science affairs" —did not occur easily, spontaneously, or even visibly. No new scientific discoveries shook the world. Nor did any natural catastrophe threaten survival. The security of no nation was significantly menaced by the maritime power of another; no underwater Sputnik triggered a frantic U.S. response as did the 1957 space surprise of the Soviet Union. While the promise of sudden wealth was espoused by some enthusiasts, it did not serve as a spur. No powerful vested interest or special group of citizens clamored for action. Indeed, a cliché was spawned in 1966 that proved a marvelous symbol of political lethargy: "The fish don't vote."

With no threat and no effective lobby, political analysts would probably agree that for anything to happen at all was somewhat remarkable. To make things happen amidst such apathy meant bending events, persuading people, molding new understandings, even generating new institutions. But notions as to what could or should be done to couple the oceans more effectively to human needs and aspirations varied among individuals, among private and public institutions, and among nations. It is thus not surprising that progress was escorted by a familiar appurtenance to human affairs—by disagreements blossoming into conflict: by conflicts in goals, in motivation, in priorities, and in power. What may prove surprising, however, is that the slow, quiet collision of events and issues that was to force a political resolution was impelled more by logic than by crisis.

The past few years have witnessed numerous policy initiatives by three presidents and five congresses in regard to improving management of the coastal environment, meeting world hunger with food from the sea, initiating an International Decade of Ocean Exploration, negotiating an unprecedented legal regime and disarmament for the seabed, and creating a new federal agency. This account of these developments is based primarily on the public record and on personal interviews, but is supplemented with observations and perspective from direct involvement. Its objective is to explain not only what happened, but how and why. The determination and decisions, the anonymous acts, and the activity of a few men who exercised personal leadership to influence the outcome of the game both positively and negatively—these, when coupled with the responses of the other participants—constitute "The Politics of the Ocean."

CHAPTER 1

Political Geography
of the Blue Planet

AN UNSTEADY LOVE AFFAIR WITH THE SEA

Man has looked on the sea as a two-faced Janus: its mysteries fascinated, its hazards frightened. It has been a protective moat insulating peoples from aggression and isolating human endeavors, but it also has provided an inviting and almost hypnotic medium for the bold and imaginative, to exchange Polynesian as well as Christian culture, to foster peaceful transport and trade, and to extract fish and salt. Initially, the sea was of practical importance to man as a vehicle of national power. Naval warfare was first practiced 2,500 years ago by expansionist Phoenicians and kings of Crete and Athens in the Mediterranean. Fifteen hundred years later, Norsemen, Genoans, and Venetians exploited the sea for military purposes, followed by Portuguese, Dutch, and Spaniards; they in turn lost maritime adroitness and precedence to the bold mercantile British. Exploration of the sea fostered the world-wide march of empire. Vasco da Gama beat around the Cape of Good Hope; then Columbus traveled to North America; Magellan's unintentional circumnavigation of the globe was followed intentionally by Sir Francis Drake. Exploration was not of the sea, however, but of the lands beyond; not an end in itself, but a means toward what lay on the distant shore. While participants sought adventure, it was the hope of practical treasure that enticed explorer, commercial investor, and monarchal sponsor alike: the wealth to be confiscated, the inhabitants to be regimented, the political power of sovereignty to be embossed on new colonies. The sea was thus a highway, to be swiftly traversed and forgotten. In the course of those long, uncomfortable voyages, man inevitably made observations about the nature of the sea—from sargassum floating on the surface to porpoises and whales gamboling in the vessel's wake and fish caught beneath. Folklore filled in some gaps of knowledge, but systematic marine inquiry was almost completely neglected.

Clearly the sea was a vital factor in the exploration and development of America, but even our own national commitment proved fickle. Once released from the umbilical cord that tied the colonies to their logistical sources 3,000 miles away, settlers in America focused their attention on the interior. Even the more recent romantic interlude of majestic passenger vessels that offered maritime exposure to world travelers closed silently in 1956, when travelers by air to Europe outnumbered those by ship. Only a decade ago, five unrelated circumstances began to sprout and converge to set the stage for a spectacular change.

First, scientific oceanography began to generate deeper comprehension of what is in and under the sea. The theater of maritime operations depicted in figure 1 was perceived to cover some 140 million square miles of the earth's surface, three times the area of all the continents combined. Its depths were found to average 2½ miles, with the submarine landscape revealing shallow continental shelves and slopes, canyons and sweeping abyssal plains, volcanic seamounts, and flat-topped guyots. Its 350 million cubic miles of water carrying 165 million tons of solids per cubic mile constitute the earth's largest, albeit diluted, continuous ore body. The underwater extension of the continents forms a continuous shelf with an average depth of 100 fathoms, varying in breadth but equivalent to approximately one fourth the area of the continental land masses. Offshore the United States, the shelf occupies about 862,100 square miles.[1]

From systematic observation over the past hundred years, we have learned a great deal about these waters, their motions and their contents; about the dissolved minerals and gases; the energy generated by tides, waves, and surges; the great horizontal surface and submarine currents and the nutrient-loaded upwelling currents near the continents. We have charted the topography of the seabed, located mineral-rich, potato-shaped manganese nodules resting on the ocean floor, and pockets of oil and gas buried under it. Tens of thousands of different species of animal and plant life were observed to abound in the oceans, from microscopic bacteria and miniscule plankton to 100-ton whales—life that swims, drifts, or rides piggyback as a parasite. At the turbulent interface between ocean and atmosphere, energy injected by heat from the sun and matter were found to be exchanged, generating our global weather and climate essential to life on the planet. And along the shoreline were found the ecologically most complex and practically most important sectors of the marine environment.

(Areas of Selected Seas in Square Miles)

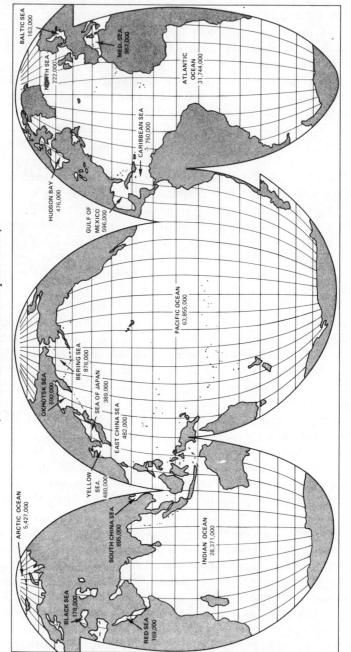

Fig. 1. The global sea. Source: *MSA*, 3d report (1969), p. 17

As a second development, man began to realize that he could now engage in activities in, on, and under the sea that were historically thwarted by its hostile and strenuous environment. The technology was ripening. Initially spun off from national security and space exploration, marine engineering nurtured deep-diving submersibles, radio-linked buoys, satellite observation platforms, compact nuclear power plants, and precise ocean navigation systems that made it feasible to reach, extract, and harvest resources that were once inaccessible.

Third, world populations, outracing their food supply and seeking higher living standards, became more heavily industrialized and displayed a voracious appetite for nutrition, energy, and raw materials. The oceans that lay relatively untapped beyond the coastline beckoned as the new frontier.

Fourth, as these populations grew, they concentrated more and more along the seacoast, following industries located there by compelling economics of low-cost sea transport of bulk materials. Their mounting waste was indiscriminately dumped into coastal waters to spoil a fragile and limited ecosystem. Simultaneously, urban-scarred populations sought tranquil sanctuary at the nearby sea; and gradually, the conscience of man awakened him to his role as protector of the coastal margin for future generations.

Finally, as international tensions persisted even under the ominous threat of nuclear holocaust, peoples everywhere intensified their quest for new pathways to live in peace. Longfellow had already remarked on "the dim, dark sea—that divides—yet unites mankind." Now the oceans were perceived as a new arena for peaceful cooperation that could contribute to world order and understanding.[2] Simultaneously, the number of coastal and land-locked nations granted independence and having interest in benefits to be derived from the sea increased sharply. So did the potential for conflict.

These were the emerging external circumstances that countered generations of neglect of our maritime legacy and promoted a maritime renaissance.

Poets and prophets had envisioned the importance to mankind of this undersea world that covers 71 percent of the earth's surface. But it was the scientific community in 1959 that called the attention of our government to these benefits and to the support needed for oceanographic research to realize them, citing the potential of an incalculable source of food, fuel, and minerals within reach of contemporary technological muscle. With this vast, rich frontier stretching out on all sides, the United States government subsequently recalled in a presidential document[3] the eloquent question of the British philosopher, Thomas Huxley, when he visited the western United States a hundred years ago. Said Huxley, "I cannot say that I am in the slightest degree impressed by your bigness or your material resources as such. The great issue about which hangs the terror of overhanging fate is, What are you going to do with these things?" What were we about to do with the vast ocean?

NEW MARITIME SOLUTIONS FOR OLD PROBLEMS

As the nation or more particularly some of its leaders pondered issues related to peace, economic vitality, health, safety, humane living conditions and a quality environment for all citizens, it became clearer that since our social and economic problems do not stop at the water's edge, neither should their solutions. To relate marine sciences to the affairs of men required study and understanding of the sea itself. But there was more: marine science affairs involved all the disciplines and techniques, specialized interests and institutions, that characterize man's activities on land—the entire political, economic, and social fabric of our nation. It is thus not surprising that marine activities evolved historically in many federal agencies, congressional committees, state, regional, and international organizations, numerous universities, and a wide variety of industries catering to a full array of citizen aspirations, social goals, and consumer needs.

This breadth of involvement, however, was misleading, for indeed the total activity was thinly spread. In the first instance, we knew too little about this zone of our planet. Our maps were crude; our geological samples were inadequate; our fishery statistics were based on the fish caught, not those uncaught. Of the millions of square miles of seabed beyond the shallows, man had seen directly less than one hundred. If the sea were barren, if it were remote, if it had little influence on our daily life, this poor comprehension would evoke little concern. But the oceans seemed a promising source of critically needed protein and energy. Aided by modern science and technology, means but not ends themselves, man could begin to understand, nurture, and develop his last frontier. The inherent human curiosity about the world around us— the desire to explore and the urge to inquire—formed the springboard for intelligent and effective use of the sea in a twentieth century, knowledge-based society. But scientific programs to realize the benefits sought required definition of public purposes.

Until the catalytic 1966 policy, this endeavor was a sometime thing. Federal investments for marine research were disproportionately small, less than 3 percent of the total public support of research and development, only half of what we were spending in the field of astronomy. In setting national research priorities, the destiny of marine exploration had been lost in the noise level of debate. Also neglected were issues on strengthening of capabilities for ocean environmental observation and prediction; provision of stable support for academic research institutions; coordination of ocean and atmospheric sciences and services; and mounting a world-wide effort to collect facts on the sea in a reasonably short time, supported by a far more effective exchange of data.

In short, to control and use the seas, man would need to predict marine phenomena; and to predict, he would have to comprehend. The science of the seas thus provided the arsenal for attacking old problems by new and creative employment of the marine environment: as a naval arena and a commercial highway; as a coastal gateway to urban development and wildlife sanctuaries; as a storehouse of nutritional protein, fuel, and minerals; and as a locus for fresh international cooperation. These opportunities led to a wide array of questions in public policy, many present but unanswered for decades.

National Security

Long coastlines have historically exposed our flanks on both sides of the continent, while the same ocean expanses have separated us from allies and adversaries alike. Naval power comprised a shield for defense, a springboard for attack, and a lifeline for support; and the simple display of the flag in distant waters has constituted a strong, long arm of diplomacy. As the twenty-first century approaches, the sea is being restored as the historical arena of geopolitical influence. Updating its missions, the Navy has extended its operations both to the air above and to the waters beneath to provide a strategic deterrent, to help prevent and contain limited wars, and to undergird collective security arrangements (some of which are depicted in fig. 2). To carry out these missions, the Navy designed submarines armed with nuclear missiles, strengthened capabilities for antisubmarine and amphibious operations, bombardment and air support for ground operations, protection for essential shipping, ocean surveillance and highly sophisticated intelligence. While the oceans offered to a surface force the tactical advantages of mobility in deployment and freedom to disperse and concentrate swiftly, the opacity of the sea served to conceal submarine operations so as to decrease their strategic vulnerability while improving the opportunity of tactical surprise.

No nation, however, has a monopoly on sea-based systems or on naval technology. With reduction of Britain's naval forces, the United States and the U.S.S.R. have emerged as the two major competitors for that primacy at sea that is a major factor in every ramification of cold war diplomacy and of arms limitations. The size and prestige of the U.S. and the Soviet navies, however, should not obscure the existence of other fleets. With the submarine likely to play an even greater future role, France and Britain have expanded their nuclear-armed undersea forces, Mainland China is constructing three nuclear-powered submarines, and a number of the developing nations are acquiring older submarines. *Jane's Fighting Ships*[4] catalogs nine countries with aircraft carriers, fifteen with cruisers, and more than thirty with destroyers. More than forty

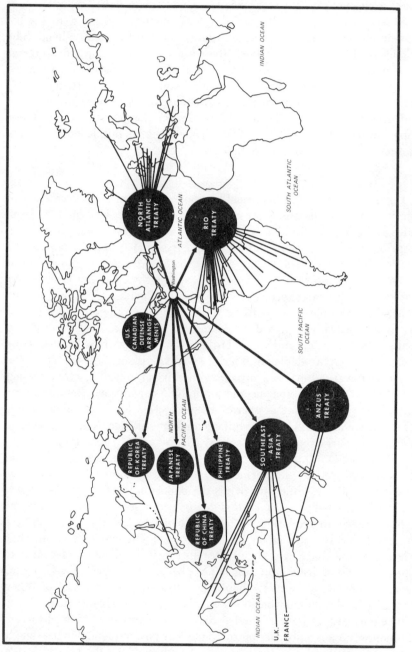

Fig. 2. U.S. participation in collective defense arrangements. Source: *MSA*, 3d report (1969), p. 41

new coastal nations have emerged since World War II, and their purchase of patrol boats and corvettes underscores the importance attached to a military presence on the seas. Today, the effectiveness of naval operations depends more upon technical sophistication, than on sheer size, of navigation systems, command and control systems, detection and tracking systems, and the weapons themselves. All require an aggressive, competent technological base. Equally important, a Navy must understand the fluid medium in which it operates.

By far the greatest portion of U.S. oceanic endeavors, public and private, has been devoted to national security—whether measured by annual expenditures or by sea power capabilities. Even with investments in oceanographic research, mapping, and exploration of the sea, the Navy's contribution has exceeded those of all other federal agencies combined, until 1971 representing as much as two thirds of the total. Much of this knowledge has been unclassified, however, and made available for general use.

With emphasis growing on the sea as a theater for strategic deterrents and possible siting for antiballistic missiles, requirements have grown for ocean science and naval engineering. One major issue thus arises on how, and how much, to nourish, extend, and utilize these capabilities. A second concerns how to increase world order and to inhibit use of the seas for military conflict.

The Merchant Marine

Of all the uses of the sea, maritime transport is the most indispensable to modern society, particularly because maritime trade routes form lifelines for the exchange of bulk materials such as ore, fossil fuel, and chemicals, and of manufactured goods vital to all industrial economies. Nations that trade with each other usually develop cultural, political, and economic ties, as suggested by the major trade routes to the United States in figure 3. Additionally, a nation's merchant marine can be profitable, earning foreign exchange and, as an economic counterpart to military seapower, enhancing national prestige.

While air transport is becoming increasingly important, especially for high-value, low-density cargo, the bulk of world-wide intercontinental commerce and manufactured goods still moves prosaically by sea. The cost per ton mile is roughly one-tenth that by rail, one-hundredth that by air. Consequently, over 95 percent of the world's trade goes by ship; it will probably always do so, and, as shown in figure 4, is projected to grow rapidly in the decades ahead.

In some countries, a collateral benefit is shipbuilding. By the end of 1970, enormous ships of 477,000 tons and upwards were being built in Japanese drydocks. Eight massive building ways were planned in Britain, France, Portugal, Italy, Germany, and Scandinavia. In con-

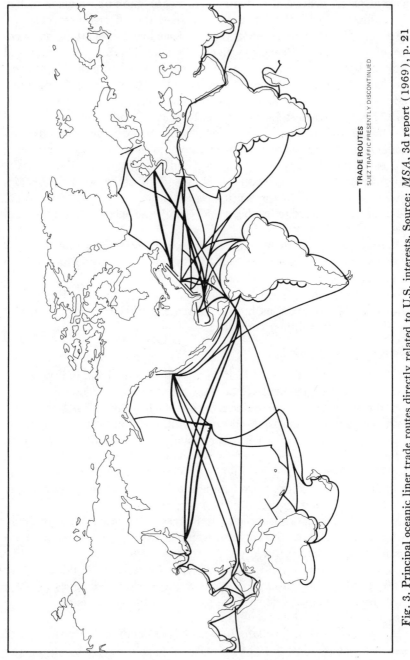

TRADE ROUTES
SUEZ TRAFFIC PRESENTLY DISCONTINUED

Fig. 3. Principal oceanic liner trade routes directly related to U.S. interests. Source: *MSA*, 3d report (1969), p. 21

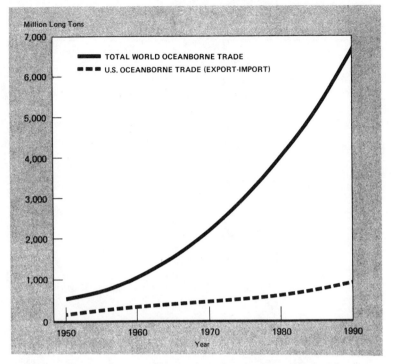

Fig. 4. Projections of ocean-borne trade. Source: *MSA*, 3d report (1969), p. 22

trast, the United States had only one yard capable of building ships of 300,000 tons. Because bulk carriers are rapidly increasing in size, the actual number of ships in the world fleet is expected to grow more slowly than in the past. In fact, with petroleum representing 60 percent of the bulk cargo, the number of supertankers larger than 200,000 DWT is expected to compose one-half of the total world tonnage capacity by 1973, with tankers in the 400,000 to 600,000 DWT category accounting for 10 percent of total world capacity. Likewise, the shipping of "break-bulk" cargo is embracing new techniques of "unitized" loads. By 1973, 23 percent of all U.S. flag cargos may move in container ships; by 1983, that figure should increase to over 40 percent.

As ocean traffic increases, so do its hazards. In 1971, there were 2,577 commercial vessel casualties in U.S. waters,[5] with unwelcome oil stains on sea and shore from normal tanker operations as well as from accidents.

Although waterborne U.S.–foreign cargo trade almost trebled between 1950 and 1969, the portion carried by U.S. flag vessels was dropping rapidly, decreasing from 39.3 to 4.9 percent in that same period. In contrast, the Soviet merchant fleet doubled in tonnage during the five-

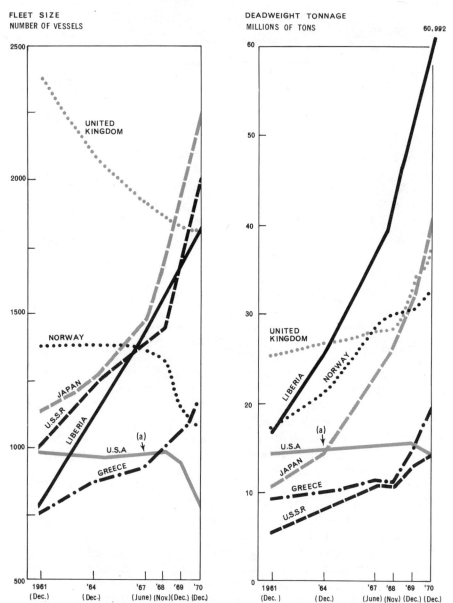

Fig. 5. Merchant fleets of the world (ocean-going steam and motor ships of 1,000 gross tons or more; excludes ships operating exclusively on inland waterways and special ships such as channel ships, icebreakers, military ships, and so forth). Sources: *MSA*, 3d report (1969), p. 23; *Merchant Fleets of the World as of Dec. 31, 1969*, and . . . *as of Dec. 31, 1970*, Department of Commerce, Maritime Administration (Washington: GPO, 1970)

[a] Only privately owned U.S. ships are included. U.S. government ships, excluding the reserve fleet, rose from 64 vessels in 1961 to 201 vessels in 1967, but this increase largely reflects activation of reserve ships for war duty.

year period 1961–65, and by 1970 grew another 50 percent, from 9.5 to 14 million deadweight tons. Within a single decade of sustained growth at a forced pace, the Soviet Union has leapt from obscurity in the realm of ocean shipping to take its place as the sixth leading maritime nation in the world (fig. 5). A similar contrast between U.S. and Soviet capabilities shows up in the fleets for oceanographic survey and fisheries research.[6] The tonnage of their ships and ours was equal in 1964, but by 1971 theirs was approximately 50 percent greater. So, presumably, will be their seagoing laboratory accommodations for scientific personnel. The Soviet leaders have thus begun to rely more deliberately and openly on the use of ships as an instrument of geopolitical influence, directing their maritime policy toward active utilization of their merchant fleet in the economic competition between East and West.[7]

This has presented the United States with a nagging unresolved issue about the role of the flag merchant marine in national security and economic development, and the steps needed to strengthen this maritime presence.

The Coastal Zone

The most intensively used and most accessible part of the ocean environment is the margin where the land meets the sea. Of equal significance, it is the band where people meet the sea. This coastal zone encompasses a broad variety of physical features. It is characterized by the dynamic interaction of land with wind, tides, currents, waves, storms; it supports a rich variety of flora and fauna. This region has great values for man. The unique natural beauty of the shore holds fascination for everyone. More tangibly, the coastal zones throughout the world have been the unplanned sites for concentration of industry and commerce, because the port cities are gateways for the raw materials that feed industry. The thirty coastal and Great Lake states contain more than 75 percent of America's population. Eighteen of the 21 million increase in population during the decade of the sixties located there. It is a locale for heavy industrial investment, for $40 billion annually of maritime trade, $1 billion in offshore oil, and over $700 million in fisheries. And because fuel, ore and bulk chemicals can be transported most economically by the sea, this gateway has sustained an ever growing frenzy of diversified industrial development. Practically all of the megalopoli projected to the year 2000 touch the sea. With the U.S. population expected to increase by about 60 percent, coastal populations are projected to concentrate as shown in figure 6.

Beyond their economic significance, the salubrious climate and aesthetic pleasures of our shores make them the recreational refuge for a

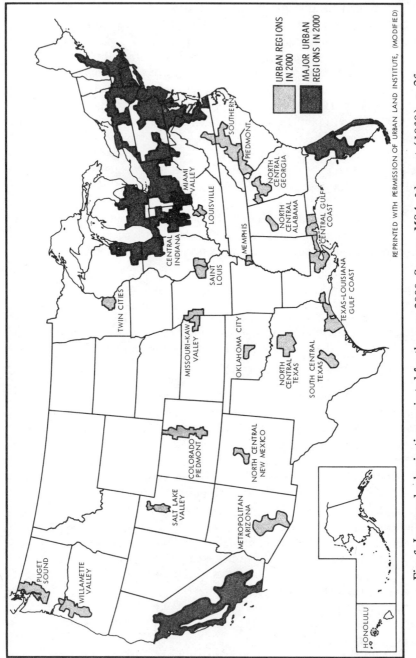

Fig. 6. Increased urbanization projected for the year 2000. Source: *MSA*, 3d report (1969), p. 26

REPRINTED WITH PERMISSION OF URBAN LAND INSTITUTE. (MODIFIED)

URBAN REGIONS
IN 2000

MAJOR URBAN
REGIONS IN 2000

PUGET
SOUND

WILLAMETTE
VALLEY

SALT LAKE
VALLEY

METROPOLITAN
ARIZONA

COLORADO
PIEDMONT

NORTH CENTRAL
NEW MEXICO

NORTH
CENTRAL
TEXAS

SOUTH CENTRAL
TEXAS

OKLAHOMA CITY

MISSOURI-KAW
VALLEY

TWIN CITIES

SAINT
LOUIS

CENTRAL INDIANA

MIAMI
VALLEY

LOUISVILLE

MEMPHIS

NORTH
CENTRAL
ALABAMA

NORTH
CENTRAL
GEORGIA

PIEDMONT

SOUTHERN

CENTRAL GULF
COAST

TEXAS-LOUISIANA
GULF COAST

HONOLULU

busy people; in 1964, 33 million turned annually to the sea to swim; 8 million to fish; 20 million to cruise and sail.[8]

Paradoxically, with this heightened pressure on the coastal areas, the resource itself is shrinking and subject to abuse. From urban centers and industry concentrated along the coast flow the subtle, potent contaminants of an affluent society: chemical waste from factories; heat from power plants; domestic waste and sewage from cities and towns; insecticides and fertilizers from land runoff; low-level radioactive waste from reactors, laboratories, and hospitals; and petroleum wastes from distributed sources. With their sheer bulk and chemical stability, waste products are no longer diluted, dispersed, or degraded.

Clearly this sector of the marine environment is "where the action is." But the action is characterized by an anarchy of utilization: everybody wants to do "his own thing." Inevitably, competition for a scarce resource begets conflict, and that is the nub of the coastal problem. In the early stages of a shoreline's development, scattered individual actions exert relatively innocuous impact on the ecology and on other potential users. But as pressure builds, the environment is modified. Each user influences his neighbor. Early entrenched users preempt later equally legitimate demands. For example, we have filled wetlands for building construction, exhausted sites for dredging spoil, wiped out breeding grounds for shad and other anadromous fish, removed waters from shellfish culture because of pollution, sharply reduced public access to choice beaches by limitations of private ownership. At the same time, the coastal waters and marine resources are legally a public trust.

Out of this mélange has emerged an unexpectedly compelling issue on how to foster prudent resource management for more balanced use and protection of coastal waters and land, for both public and private purposes.

A Source of Protein

By the year 2000, only one generation off, more than 6 billion people will be competing for the earth's food resources. The "green revolution" of enhanced agricultural productivity does not promise a full solution to this demand; moreover, agriculture is potentially vulnerable to drought and disease. Many authorities believe that more effective use of the living resources of the ocean can provide both better and less expensive food to help alleviate the world hunger problem and a wider variety of nutritious foods for affluent societies, while enhancing domestic economic opportunities by a more productive, competitive, fishing industry.

One half of the world's population lives in countries critically short of animal protein. Such protein malnutrition in preschool children can permanently retard mental development, blocking maturity to a healthy, productive adult. Fish are approximately 15 percent protein; in addition,

CATCH OF SEAFOODS
BILLIONS OF POUNDS

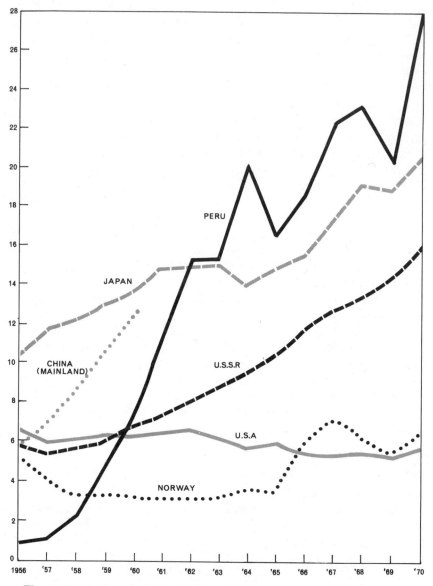

Fig. 7. Catch of seafoods by leading countries on live-weight basis. Sources: *MSA*, 4th report (1970), p. 85; *Yearbook of Fishery Statistics*, vol. 30 (Rome: FAO, 1970)

fish protein contains a favorable balance of different amino acids required in human diets but not commonly found in plant sources. Conservative estimates indicate that the world fish catch of wild stocks can be increased several times without depleting stocks. Aquaculture in bays, estuaries, and lagoons could double that estimate. Fish protein could potentially alleviate a significant fraction of present and future nutritional deficiencies.

Currently, five nations are catching approximately 50 percent of the total 69 million-ton world fish catch. Ironically, most nations do not go to sea to fish for fish; they fish for money. Thus the species sought are those of high market value, and those fishery products involved in world trade are primarily consumed by the more affluent who have less need for the protein essence.

In the United States, the use of fishery products for all purposes climbs steadily. The annual U.S. fish utilization, which peaked at 80 pounds per person in 1969, is among the highest in the world, although significant fractions are consumed as pet food. Yet in the United States many harvesting sectors of the fishing industry are in decline, with the U.S. fish catch remaining relatively constant for the past thirty years (fig. 7). The processing and marketing portions of the industry have, however, grown by an increase in U.S. fish imports and by expansion of U.S. fish processing firms into areas abroad to obtain raw materials.

Two policy issues have been generated by these facts. What steps are needed to realize the potential of marine protein in the war on hunger, including steps to build an economically viable enterprise? And what is the proper role of the domestic fishing industry and what steps are needed to sharpen its productivity?

Nonliving Resources

In the last thirty years, the industrial economy of the United States has consumed more minerals and fuels than the entire world appropriated in all previous history. Projected world and U.S. demands for energy to the year 2000, as well as the estimated sources from offshore oil and gas, are summarized in figure 8. The United States is the heaviest consumer of petroleum, relying on both imports and domestic production. Western Europe, another heavy consumer, produces less than 10 percent of the oil it consumes. Figure 9 portrays the patterns of international oil trade that have developed to satisfy geographical disparities in supply and demand.

As both developed and developing nations step up their industrial economies, competition for low-cost raw materials has intensified, and the reserves of crude oil on the continental shelves of the world have already been partially tapped to meet this demand. Reserves in North America are considered adequate to sustain offshore production for

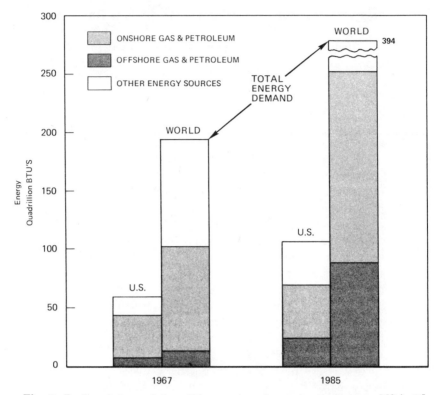

Fig. 8. Projected demand for offshore gas and petroleum. Source: *MSA*, 3d report (1969), p. 31

approximately thirty-three years on the basis of 1968 data, and recent discoveries off the coasts of Alaska, Washington, California, and the Atlantic and Gulf states suggest that additional deposits may soon be located. Oil and gas production, which constitute over 90 percent by value of minerals extracted from the marine environment, were limited in 1968 to water depths of 340 feet and to a distance of 70 miles from shore due to the additional costs of moving into deeper waters, but that too will change as economics of production entice exploration farther out to sea. In 1960 only three or four nations and about five private companies held major offshore petroleum interests. By 1970, as shown in figure 10, subsea exploration and development was underway or about to start off the coasts of 28 countries and exploratory surveys were being carried out on the continental shelves of another 50. Offshore deposits were responsible for 16 percent of the oil and 6 percent of the natural gas produced by the free world. Projections indicate that by 1980 a third of the oil production—four times the 1968 output of 6.5 million barrels per day—will come from beneath the ocean; the increase in gas production is expected to be comparable.

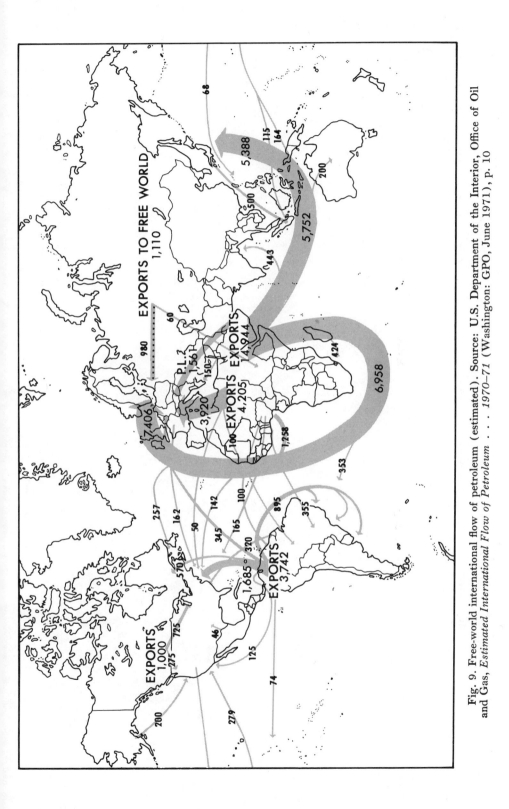

Fig. 9. Free-world international flow of petroleum (estimated). Source: U.S. Department of the Interior, Office of Oil and Gas, *Estimated International Flow of Petroleum . . . 1970–71* (Washington: GPO, June 1971), p. 10

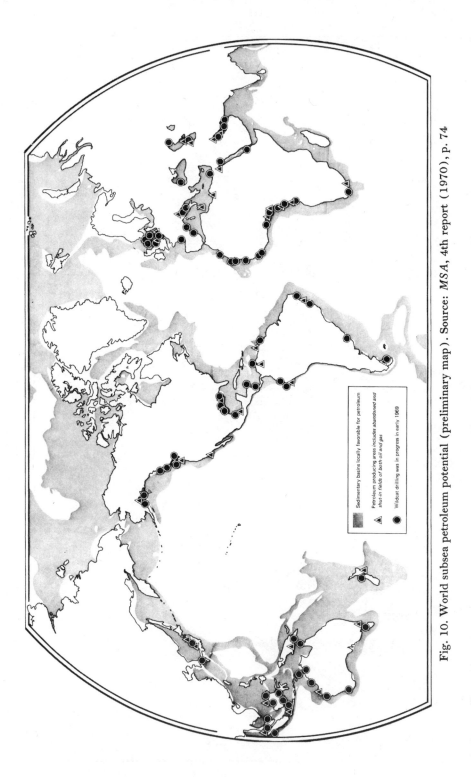

Fig. 10. World subsea petroleum potential (preliminary map). Source: *MSA*, 4th report (1970), p. 74

Sedimentary basins locally favorable for petroleum

Petroleum producing areas *includes abandoned and shut-in fields of both oil and gas*

Wildcat drilling was in progress in early 1969

In the United States, where some 80 percent of the most favorable onshore areas have been explored, in contrast with less than 10 percent of the offshore shelf areas, subsea petroleum development is increasing rapidly. Between 1946 and 1968 approximately 15,300 wells had been drilled offshore and it has been estimated that 3,000 to 5,000 wells will be drilled annually by 1980. Through 1968, over $13 billion has been invested by industry for exploration and development activities, which have resulted in production of $7 billion worth of oil and gas from the submerged wedge of sediments composing the U.S. continental shelf, mostly off Louisiana. By 1968, the industry had paid the federal government about $3.4 billion in bonuses to lease 6.5 million acres of the outer continental shelf. All but $35 million of this was paid for oil and gas leases, the remainder going for rights to salt and sulphur.[9]

The world-wide demand for nonfuel minerals is also increasing rapidly, and a fivefold expansion may occur before the year 2000. In the U.S., demand is expected to double by 1985 and perhaps to triple by 2000. Competition for the world's mineral resources will become even more intense as the developing countries join the list of industrial consumers.

The ocean floor and seabed are certain to be attractive as a source. Extensive deposits of phosphorites; metalliferous muds rich in copper, zinc, and other metals; manganese nodules that also contain cobalt, nickel, and copper; gold, tin, platinum, and other heavy minerals; coal, sulphur, iron ore and limestone, and sand and gravel, are all tapped modestly. More lie in wait on and beyond the continental shelf. Present knowledge of the extent of these resources and their distribution is limited, and the cost of extraction is high compared to alternative terrestrial sources. Yet the growing world demand for resources, the unexplored areas of the continental shelf having geological continuity with adjacent land areas that have yielded mineral wealth, and the scattered positive indications that resources exist, all direct attention to such nonliving resources under the sea.

In the exploitation of marine resources, as on land, private enterprise is expected to serve as entrepreneur and investor, responsible for extraction and marketing of resources. The petroleum and gas industry has already taken the initiative in exploration of continental shelf resources. But a very significant difference separates terrestrial from marine sources: the latter are public rather than private property. In the case of the United States, access to such resources is provided by leases, leading to a public/private partnership of significant proportions. In a fashion somewhat analogous to the development of the West, the government has assumed five key roles: in providing economic incentives that motivated private industry to move seaward; in establishing the administrative, legal, and financial framework to serve public and private interests; in providing such services as geological mapping and coastal weather predictions; in making available applicable military technology; and in protecting the environment through regulation of off-shore structures and drilling procedures.

Major issues have erupted, however, as to the appropriate federal and private roles in development of marine resources. What shall be the specific mode of federal assistance to an emerging marine industry, particularly in providing environmental services or sponsoring research and development in those areas where investment involves high initial cost, unusual risk, or long-deferred return on investments? And how may the marine environment best be protected against accidental oil spills?

LEVERAGE OF TECHNOLOGY

Man's victory over his environment has always been linked to his technological prowess, and his mastery of the sea is no exception. Until a brimming curiosity about what lay over the horizon stimulated invention of the first raft, then a canoe, then a sailing craft, aboriginal populations could only dig their toes in seashore sands and channel their migrations and settlements over land. But that once impassable barrier was overcome, and the distribution of human populations around the Pacific Basin is clear evidence that seamanship leaped across the geographical barrier. Modern seafarers such as Thor Heyerdahl have proved that voyages over 6,000-mile stretches of open ocean were technically feasible in crude vehicles, at least with favorable currents and winds. When the water craft was made steerable, marine technology was born. In every era, successive improvements gave man ever greater freedom to explore, to conquer, to settle, to trade. In the fifteenth century, larger ships and more precise navigation via the magnetic compass were immediately put to use by governments bent on fulfilling nationalistic ambitions through a quest for treasure and territorial claims.[10]

Practical goals coupled with technological opportunities have released man from his terrestrial prison. In the last two-hundred years, steam propulsion replaced sail, freeing the navigator from the caprice of the unreliable winds. Then came the iron ship whose strength released the shipbuilder and the skipper from size limitations imposed by wood. The same superior material properties made possible another dream, that of navigating underwater by submarine.

Liquid fuels, steam turbines, and diesel further improved performance. In 1954, nuclear power was married to the submersible in the U.S.S. *Nautilus*. Freed from requirements for a supply of air to run propulsion engines or to recharge batteries, the sleek man-made fish could now roam the entire ocean from surface to seabed. Because the oceans can effectively hide activities within their depths, submarine warfare gained ominous potency. Underwater surveillance becomes a nightmare as compared to surveillance of the surface and sky that can be patrolled at great distances by radar, a deadly game of blindman's buff in which the winning side is most likely to be that with the most acute hearing. Nuclear technologies were soon merged with missile and weapon technologies, spawning the sea-based Polaris deterrent on which U.S. defense now relies so heavily for security against a first strike.

Although nuclear submarines proved their versatility with North Pole traverses under the Arctic ice cap and a completely submerged circumnavigation of the globe, their long endurance was not matched with capabilities for deep diving. A bathyscaph, *Trieste*, made a descent in 1960 to the bottom of the deepest known trench, and the *Aluminaut*, completed in 1964, opened the technology of deeper operation of true submarines.[11]

In the last twenty years, the variety and sophistication of marine technologies have sharply accelerated to meet other nonmilitary uses of the sea, both to amplify man's capabilities to explore, and to nurture exploitation of marine resources. The oceanographer now happily extends his previously limited capabilities by such innovations as radio and satellite aids to navigation; more sensitive instruments to detect properties of the sea; computerized data storage and retrieval; automation in seismic and bathymetric surveys; unmanned buoys as remote-sensing stations with radio communications links; satellites to observe oceanic phenomena directly; techniques for retrieving cores from the seabed in great depths of water; and small research submarines that carry instruments, closed circuit TV cameras, or human observers to scout biomass and seabed.

The same revolution in capability facilitated exploitation of offshore gas and oil. At first, operations in shallow water involved simple extensions of petroleum drilling technology that had been well developed on land, pushed to sea by depletion of land reserves and the economics of the market place. Reinforced by incentives of a depletion allowance, new technologies were bred to extend the exploration to deeper water farther out on the continental shelf. Drilling from stationary surface platforms became commonplace, followed by semisubmerged and dynamically positioned floating rigs.[12] Completely submerged rigs promise freedom from surface storm and wave damage, a boon indeed, since hurricanes abound in many of the areas where oil and gas are in richest supply. New technologies also facilitated the search for and capture of fish: new fish-finding sonar gear, deepwater trawls. Refrigeration extended the range of operations.

The options are clearly open to exploit the seas with sophisticated technology. But what is feasible is not necessarily economical. Oil, gas, minerals, and protein can be found on land as well as at sea. In most respects, the consumer is neither interested nor aware of the original source of the raw material. The question of whether sources of supply are marine rather than terrestrial depends largely on the economics of exploitation—costs of exploration and development, richness and distribution of supply, proximity to convenient transportation and markets. New civilian technology must thus afford cost reduction and improved productivity in marine activities, as well as innovation.

Who pays the bill for such development cannot be answered simply when public objectives of full employment or environmental protection must be merged with private goals of profit. A delicate partnership of industry and government is required. In marine affairs, as in other national endeavors, that balance has yet to be found. As one result, the United States that was once a leader in most areas of marine technology is being sharply challenged. In commercial ship design and construction, U.S. post war leadership has yielded to initiatives by several nations of Western Europe and Japan. In fisheries, several nations have overtaken American leadership. In naval technology, the Soviet Union is a growing contender.

Elsewhere in marine technology, whether in tools for exploration and research or in hardware for offshore petroleum development, American leadership is preeminent.[13] But as the other nations awaken to their interests in the sea, they are taking steps to close this "technological gap." How this capability is nurtured and guided at home, especially at a time when American science and technology are underfunded and underutilized and the deficiency grows in balance of payments, opens still another major issue in marine policy.

THE PLURALITY OF GLOBAL INTERESTS

National Interests and International Cooperation

A number of nations, including Canada, France, Germany, Japan, the United Kingdom, and the Soviet Union, as well as the United States, have recently increased their priorities for ocean and underseas research. Their motivations are clear; they have consulted the future and expect dividends from more intensive use of the sea.

In terms of economic development: (1) fishery production could readily double in ten years and eventually increase by a factor of four without depleting stocks; (2) aquaculture is likely to expand; (3) extraction of offshore oil and gas could increase by the year 2000 by a factor of five; (4) revenues to governments from offshore rents and royalties could amount to over $50 billion during that interval; (5) world-wide ocean shipping will increase in the next thirty years by a

factor of four; that of the U.S. will at least double; (6) offshore platforms could be built as sites for nuclear power generation, for supertanker terminals, or for metropolitan jetports, to meet problems of
land-based sites.

In contributing to humanitarian concerns: (1) inexpensive fish
protein could counter, by the year 1980, 20 percent of the nutritional
deficiencies projected for the world's peoples, to limit the tragic cycle of
famine and despair; (2) ports and harbors that have become the festering sores of urban decay could be rehabilitated to become the most attractive portions of cities; (3) with urban populations along the shoreline continuing to increase, expanded seashore recreation could be made
more widely available and the natural legacy protected for public enjoyment; (4) bays and estuaries could be protected from pollution for
future generations.

In the world community: (1) world order could be improved by placing portions of the sea "off limits" to weapons of mass destruction or as
arenas of conflict; (2) international arrangements could assure that
living resources are harvested in an equitable manner while maintaining
their continued abundance, that mineral resources are extracted in a way
to benefit developing as well as advanced nations, and that pollution of
the sea from waste, oil spills and pesticides be reduced; (3) the concept
of the marine environment as a common heritage of mankind could be
employed to foster international cooperation and open new avenues for
international understanding.

Historically international, the oceans impartially wash the shorelines
of over one-hundred nations. Many of these countries have had rich
maritime histories, but even the new and developing nations, including
most landlocked ones, are interested in what the seas hold for rapid
strengthening of their economies. New ocean technologies extend the
maritime reach to resources in more distant seas or in deeper and previously inaccessible stretches of the seabed.

Mounting international interest and activity in the oceans were reflected in the results of a 1968 UN poll of its member states requesting
information on ocean activities. Of 58 nations responding to the questionnaire, roughly half of those touching the sea:

— 53 were engaged in basic or applied marine research;
— 42 had nautical charting programs;
— 22 operated position fixing systems or performed related naviga-
    tional services;
— 20 monitored or forecast physical ocean conditions;
— 12 monitored or forecast pollutants;
— 37 explored, monitored, and forecast fish stocks;
— 21 undertook submarine geological and geophysical surveys;
— 19 were concerned with modifications of coasts and channels.

As to using the sea:
- —55 engaged in fishing;
- —19 exploited offshore petroleum and gas resources;
- —14 used the ocean for recreation;
- —13 exploited minerals from ocean water and the seabed;
- —5 extracted sand and gravel from the sea;
- —1 utilized the ocean tides for power;
- —7 disposed of wastes in the ocean.[14]

The General Assembly of the United Nations opened discussion of ocean space in 1966, although numerous UN specialized agencies had long-standing concern: for example, 55 countries participate in the activities of the FAO Committee on Fisheries; 60 countries are members of the Intergovernmental Maritime Consultative Organization; and the membership of the UNESCO Intergovernmental Oceanographic Commission increased by 50 percent to more than 60 in less than eight years.

As man has moved into the deep oceans he has encountered new questions of jurisdiction over the seabed and has sought new legal principles to guide its development and use. As high seas fishing stocks become accessible to more nations and threatened with depletion, traditions of free access to common resources have partially given way to international agreements—and disputes. But the territorial imperative has led some coastal states to assert national jurisdiction over resources hundreds of miles seaward, eroding the traditional freedoms of the sea and opening the way to anarchy, rivalry, and conflict. As use of the ocean increases, the chances also multiply of disturbing the natural ecological balance of the seas, particularly through pollution which is not deterred by national boundaries. Debate at the UN has quickened on contrasting questions of marine area sovereignty and cooperation. Between the pressures on limited fishing stocks, disputes on seabed and freedom of navigation, and global concern for the environment as an integral ecosystem, motivation is heightened for concerted development of more and better information about the sea. Many nations share experience and capabilities in marine science. All potentially derive benefit from this collective capability, because the oceans are simply too vast for any single nation to map alone. Moreover, scientific study of the planet requires understanding of environmental attributes that transcend national boundaries.

Major issues commanding serious attention in the United States and the world community have centered around: increasing international collaboration in research and dissemination of results; strengthening international law to maintain freedom of the seas and to insure equitable opportunities for all nations to share in development of the wealth of the oceans with a minimum of conflict; taking steps to protect the ecological quality and balance of the seas; encouraging mutual restraint among nations.

In dealing with these future prospects, each nation is obliged to consider its own self interest. But will nations seek collective and harmonious relationships conducive to world order rather than short-term, individual ambitions that beget conflict?

Intimately and inextricably intertwined with the purely national considerations are international questions derived from the common property resources that lie beyond the limits of national sovereignty. What are these limits? And once they are defined, how should resources in the area of universal sovereignty be managed and their economic benefits dispersed?

KEY PARTICIPANTS AND KEY EVENTS

As the plot unfolds on these different uses of the sea and on unresolved issues, it becomes clear that both past and future events depend critically on who makes things happen. In a sense, the characters in the play have been writing their own dialogue. While folksong and fantasy would have the action in the hands of sailing men, fishermen or, more recently, oceanographers, a more pragmatic appraisal reveals a sophisticated cast embracing lawyers, bankers, naval strategists, diplomats and politicians. But few if any of these have influenced the action alone; each operated from some institutional podium.

Clearly a wide range of national and international interests was involved, matched by a diversity of institutions:

Twenty-nine bureaus and eleven departments and agencies within the executive branch of the U.S. government;

Thirty-three subcommittees of the two houses of Congress;

Thirty states bordering our seacoasts and Great Lakes;

Sixty-three universities engaged in marine science teaching and research;

In the commercial sector: chemical, mineral, oil, fishing, shipping, shipbuilding, dredging, construction and recreational industries as well as the high-technology sector of aerospace and electronics, and investments;

Maritime unions concerned with ship manning and cargo handling;

The general public with its interests in recreation and environmental conservancy.

Beyond purely domestic institutions, there were:

One hundred twelve maritime nations fronting on the sea;

More than fifty-five international organizations, including many specialized agencies of the UN;

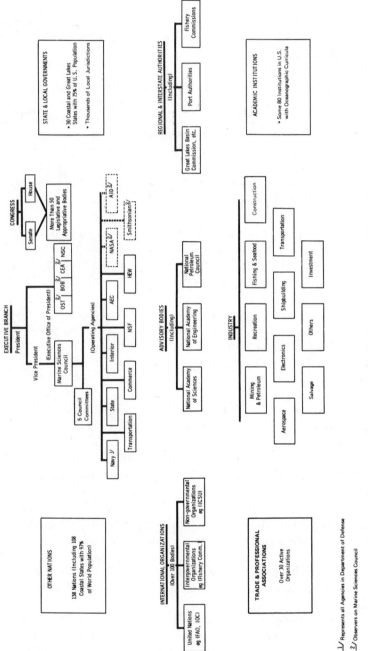

Fig. 11. Participants in the marine sciences. Source: *MSA*, 3d report (1969), p. 198

Note: This schematic representation of U.S. governmental bodies involved predates creation of EPA and the retirement of the Marine Sciences Council in 1971.

1/ Represents all Agencies in Department of Defense

2/ Observers on Marine Sciences Council

Domestic and international scientific unions providing for information exchange among scientists.

A catalog of these participants is diagramed in figure 11. What is not shown, however, are the many explicit and implicit relationships that wire them together. Moreover, in many cases the maritime activity has closer ties with its terrestrial counterpart than it does with sister water-related activities; petroleum is treated as a single commodity with minor reference to its seabed or dry-land origins. And so everything at sea relates a little to everything else.

In one sense, the different clientele and their interests in marine science affairs and the public and private institutions created to respond to these needs and aspirations form a "system." It may be portrayed much like a wiring diagram of a complex instrument such as a radio set. The issues represent the music flowing through the system.

All of these issues matured slowly, pushed by advocates, obstructed by opponents, modulated by external as well as internal events. In addition, each subsystem generated a separate rhythm and different harmony. With such diversity of interest, fragmentation, and random motivation, it is not surprising that America has lacked a coherent sense of purpose and of agreement on goals, priorities, strategies, and the necessary governmental apparatus. The challenge was how to manage such diversity. The primary single theme that emerged as the common concern of all was: What priority—measured by funds and by leadership—is this nation and its government to give to affairs of the sea?

The locus of action to deal with that question was for ten years to oscillate between the Congress and the office of the President. Then, in 1966 came the epochal Marine Resources and Engineering Development Act. For the first time in our history, the nation had a mandate to improve scientific understanding, to accelerate exploration, and to tap dormant marine resources to serve all mankind. In addition to a vigorous statement of purpose, the legislation was unequivocal about who was responsible for making things happen in a ponderous, atomized bureaucracy: the President of the United States. To assist him in carrying out the unprecedented charge, the legislation provided dual instruments of advice and coordination: a cabinet-level policy-planning council, chaired by the Vice President of the United States, and a presidentially appointed public advisory commission.

A number of simple indicators mark the transition since 1966 in man's relationship with the sea. A sharp increase in public interest has been revealed by the tripling of newspaper headlines in the *New York Times*.[15] Policy level involvement has similarly risen in the last six years: the President of the United States has made significantly more marine decisions, given more speeches, issued more directives than ever before. Marine-related budgets of the United States have grown. So have those in the U.S.S.R. and in many other countries. Debates on oceanic issues of seabed legal regimes and disarmament have appeared far more frequently on the agenda at the United Nations.

Two factors helped sectorial interests to discover each other and develop an appreciation of their common concerns: the need for scientific information about the marine environment, and recognition of the key role of the federal government in every sector of activity, especially in support of research, exploration, commercial incentives and regulation. Correspondingly, the tempo of debate among these interests has accelerated: between government and the scientific community; between governmental agencies over budget priorities and jurisdiction within the government; between federal, state, and local governments having various jurisdictions over coastal uses; between mineral and oil interests and sport and commercial fishermen; between industrialists and other users of the shoreline, including advocates of environmental conservancy.

Internationally, as well, nations are squabbling over fishing rights, freedoms of navigation and of research, the breadth of territorial seas, and seabed sovereignty. Economically advanced nations are debating with disadvantaged late starters over ways and means to distribute future benefits of marine resources held in common. Finally, there has been jostling between the bewildering array of rival international organizations concerned with ocean affairs.

As the different players interacted with each other, on their initiative or in response to other stimuli, some ripples of activity became conspicuous to all interests, often culminating in key events. Some of these beacons of progress signaled the end of a long hibernation, often influenced by other events of unusual significance having nothing to do with the oceans. In turn, such key events generated a cascade of consequences through different channels that were to flow together at a later time and place to make new history.

This account thus focuses on such key events to the extent that they could be detected, on their gestation and consequences, and on the train of decisions that they affected. A chronological summary is given in the appendix.

As these issues unroll in subsequent chapters and onionskins of subtleties are peeled back, the past uncertainties that were temporarily relieved after 1966 may again be threatening to return. The central theme continues to be our national commitment to the sea.

CHAPTER 2

Translating Scientific Discovery into Cultural Development

For man to use the ocean, he must understand it. Before he can fully accomplish tasks—to increase the yield of food, to extract fuel and minerals, to reconcile use and preservation of the coastal zone, to forecast the weather, or to enhance world security and peace—he must explore. But exploration and research are also ends in themselves. Every individual has a latent curiosity about the world around him, that collectivized and institutionalized as scientific discovery has now become a goal of society, especially of affluent societies. Indeed, exploration, science, innovative engineering, and invention are conspicuous and widely admired features of American culture.

Historically, study of the oceans was motivated by practical objectives of ship navigation: to arrive safely at a selected destination, and if possible on schedule. Early nautical charts portrayed coastlines, water depths, shoals, rip tides and rocks, coastal prominences and prevailing winds, although some portions were fabricated from pure rumor, imagination, fantasy, and hallucination. Many became works of art as elaborate marginal decorations occupied blank spaces of geographical unknowns. Whatever their shortcomings, these charts were treated as enormously valuable property, technological insurance against ship loss and a source of enlightenment that in the age of growing maritime competition conferred advantages of speed and reliability on their proprietors.

Advances from the charting of static shorelines to study of ocean dynamics were prompted by ship logs that yielded data on transit times and weather, sufficient to construct atlases of currents and trade winds. By this technique, Benjamin Franklin discovered the Gulf Stream and recommended sailing routes that gave American clipper ships yet a further competitive edge in speeding transatlantic freight. Matthew Fontaine Maury, a young and inquisitive naval officer, by devoting much of his career to this research, produced the first tables ever to be published, which earned him world-wide acclaim. But interest in this pragmatic marine research waned as sail gave way to the brute force of steam propulsion and schedules depended less on the vagaries of oceanographic phenomena. As Ralph Waldo Emerson so poignantly noted, man "no longer waits for favoring gales but by the means of steam he . . . carries the two and thirty winds in the boiler of his boat."

By the middle of the nineteenth century, most of the shorelines of the new world had been explored and charted, and the infantile empirical science—oceanology, as study of the seas was then widely denoted—dipped to a new low of activity. Paradoxically, this decline occurred while teaching and research in classical disciplines of science were beginning to flourish. The first scientific expedition of note to examine the sea rather than new lands that lay beyond was the British-sponsored voyage around the world of the *Challenger* in 1872–76. Out of this monumental tour de force came an encyclopedia of 50 volumes of data on marine life, bottom sediments, and physical characteristics of the water column. Inspired by this accomplishment, a number of minuscule oceanic research laboratories were founded by individual scientists in Europe, but little impact was felt in America. This lag in American oceanography became so apparent that in 1927 a special committee of the National Academy of Sciences (NAS) was convened to examine academic and research deficiencies in oceanography and to suggest remedies.[1] Out of that study came a report sufficiently persuasive to tickle the fancy of private foundations, which then were the major patrons of American science. Rockefeller Foundation endowments of seed money were granted to California's Scripps Institution of Oceanography, Massachusetts' Woods Hole Oceanographic Institution, and the University of Washington to strengthen their still fledgling oceanographic research and teaching facilities. The first two soon became major East and West Coast centers of marine science. This generous injection of fiscal hormones, however, did not continue. Then, with World War II, knowledge of the undersea environment suddenly became crucial to both offensive and defensive submarine warfare. When submerged, a submarine can be detected only by underwater acoustics, and then with great difficulty. Research was energetically pushed to study the propagation, reflection, and diffraction of sound through the water column and seabed, in relation both to ranging of a potentially hostile sub and effective concealment of a "friendly." After the war, despite the enhanced status of the submarine, further expansion of research was suffocated by the Navy's battleship/carrier mentality. Private endowments also withered. So the fortunes of oceanography declined again, seriously enough that two subsequent studies were generated by the Academy in 1949 and 1956 to stimulate support, now primarily sought from the federal government.

While in retrospect this academic neglect is surprising, it is consistent with the fact that only a handful of scientists and a sprinkling of engineers were concerned with marine science and technology, compared with a technical population of roughly 100,000 scientists and 500,000 engineers. Moreover, this body of marine-related technical personnel was almost static. The first advanced degree in oceanography was awarded in 1930, and by 1958 when 2,780 Ph.D.'s were awarded in science, only 13 were concerned with the sea. For one thing, oceanography was not a separate science but rather a family of classical natural sciences focused on a single environment. While resting on foundations of physics, chemistry, biology, geology, and mathematics, the study of oceanography was compartmented into such fields as physical oceanography, marine biology, and marine geology, and thus not fully compatible with traditional structures of host academic institutions. Moreover, cross-discipline relationships with the social sciences of economics, law and public affairs were nonexistent. The relatively few (3,000) scientists who called themselves oceanographers fell into three classes: those who began careers as physicists or biologists and later specialized in ocean-related research; those who initially studied in specialized fields of marine sciences; and those who began oceanography as generalists, succeeding a generation of marine naturalists. That division in professional identity was accompanied by a further split between those specializing in fisheries and in oceanographic research. Altogether, this enduring fragmentation created sociological problems within the profession that may have been a major ingredient of its retarded progress.

In the search for financial support, the profession was further divided on a basic issue of exploration philosophy: surveys versus research. One body of oceanographers contended that the sea should be surveyed in geographical packages, relating properties of the sea, its contents and seabed, to specific oceanic areas. The second body favored planning around such scientific questions as the sea-floor spreading associated with continental drift, patterns of oceanic circulation, or interaction between ocean and atmosphere.

Yet a third set of factors based on institutional loyalties and rivalries further splintered the oceanographic community. By the middle of the 1950s three major institutions had emerged—Scripps, Woods Hole, and the Lamont Geological Observatory at Columbia University. Each was headed by a strong leader in his field: respectively, Roger R. Revelle, Columbus O'Donnell Iselin, and Maurice W. Ewing. Each built a scientifically prestigious department that inadvertently hatched as a personality cult—with internal loyalties expected, and received, from staff and students alike, and with a clear disdain for those in rival camps. While competition and jealousies are not unknown in the professions— especially the quest for precedence in discovery on which reputation, academic position, or public prize may depend—these three laboratories practiced greater restraint than might be expected in cooperating with each other, whether on scientific expeditions or in presenting a united front to advance the general cause of oceanography. What was especially bizarre about this dissension was that all three institutions were funded substantially during and after World War II by the U.S. Navy. Even this common cause did not ease the strain.

All three shared one characteristic in style, a focus on research rather than graduate education. Students were expected to undergo unusually long apprenticeships, a practice that undoubtedly fostered their professional maturity while providing inexpensive technical assistance for the relatively few senior scientists. The growth of these institutions was thus conditioned by the number of projects and assistants a single senior investigator could supervise under the stress of unsolved problems queuing up for solution—a management restriction well known in academic circles. Within the scientific community, unsympathetic critics of this system observed that the field was frequently steered by individuals infatuated with yachting who cloaked their love of the sea with more socially relevant purpose.[2]

Thus, through the 1950s exploration of the sea was a feeble, delicate, loosely knit enterprise.

What forced a change was the evolution of this activity from "little" science to "big" science. That nomenclature, incidentally, was beginning to reflect the high costs of apparatus needed for research. Perhaps the best examples lay in high-energy physics, where equipment required to crack the atom and reveal the presence and life styles of component particles demanded an outlay more costly by orders of magnitude than the thrifty, prewar scientific community had ever envisioned. Radio astronomy telescopes also came high. In contrast, specialized oceanographic ships had been small, or merely poverty-induced conversions of retired naval auxiliaries. While never approaching the cost of accelerators, large ships had previously been well beyond the purse or even the aspirations of the oceanographic community. But now, to engage in deep sea research where longer ranges were required, and to provide a more stable platform and more productive working time for research on inhospitable seas, oceanographic ships would have to grow larger. And so would their cost. For the first time a unifying influence—the need for increased funding—began to foster interscientific cooperation.

About this same time, the governmental customers needing varied oceanographic data to satisfy their agency missions and public clientele were becoming increasingly dissatisfied with the level of federal allocations in the field. Foremost was the Navy's Hydrographic Office and Office of Naval Research (ONR), but other interests had developed in the Bureau of Commercial Fisheries (BCF) in the Department of the Interior, the Coast and Geodetic Survey (C&GS) in the Department of Commerce, and the Atomic Energy Commission (AEC) concerned with oceanic disposal of radioactive wastes. While each had statutory missions requiring a better understanding of the sea, oceanography was low on the priority lists within the parent agencies. Cabinet officers had never been fascinated with the sea; on the list of national or presidential priorities, it was nonexistent.

The problem was readily defined in terms of budgets. In 1958, oceanographic research and surveys received less than $30 million out of roughly $5 billion for research and development. Even worse, support for oceanography had increased since 1948 by only 50 percent, while over-all research and development expenditures had grown 500 percent.

It was thus not remarkable that lower level government officials and representatives of the scientific community pooled their mutual concerns and mutual interests in a quiet coalition to generate more funds. The initial manifestation of this alliance between knowledge producers and knowledge consumers was a 1956 letter from the Chief of Naval Research on behalf of himself and officials from the other interested agencies addressed to the President of the National Academy of Sciences, requesting that a major study be initiated of national needs for an expanded program of marine research.[3]

Created in 1863 by congressional charter for the purpose of bringing the potential of science to the service of government, initially in relation to the Civil War, the National Academy has continued as an advisory mechanism on a wide range of topics, and collaterally serves as a vehicle for honoring distinguished members of the scientific community.[4] Apart from the two previous oceanographic studies, the Academy had long undertaken similar inquiries on a range of federal deficiencies, including those which led to creation of the Geological Survey and the modern form of the Weather Bureau. In the spirit of this advisory pattern, the NAS President Detlev W. Bronk, responded affirmatively, even enthusiastically, to the government's requests. Bronk by coincidence had chaired the second NAS Oceanography Committee in 1949, and as an avocation had an abiding interest in sailing.

Mindful of the academic rivalry among the giants in oceanography, Bronk wisely decided to seek a qualified chairman from outside these three major oceanographic institutions, even outside the oceanographic community that was itself divided on matters of leadership. Harrison S. Brown, a young, distinguished, and energetic Professor of Geochemistry from the California Institute of Technology, was his choice. Iselin, Revelle, and Ewing were on the committee, as were representatives of smaller institutions—Fritz F. Koczy of the University of Miami; Gordon A. Riley of Yale University; Colin Pittendrigh, a biologist from Princeton; and Athelstan F. Spilhaus with the University of Minnesota, a meteorologist who had gained wartime recognition for his invention of the bathythermograph so essential to submarine operations. Milner B. Schaefer of the Inter-American Tropical Tuna Commission brought fisheries science expertise, but industrial representation was lacking from fishing, petroleum, or shipping interests. The most unlikely member, yet perhaps the most catalytic, was Sumner Pike, a canny businessman from Maine, former member of the Atomic Energy Commission, with family interests in Bay of Maine fisheries. Pike was nominated by a senior official in the Bureau of Commercial Fisheries which helped fund the study, and thus was initially resented by the NAS as "political." In time, he injected a sophistication about governmental affairs lacking in the other members and frequently served as neutral referee and diplomat to mediate clashes among other strong-minded participants. Richard Vetter, then a junior staffer in ONR, was selected Executive Secretary.

On November 10, 1957, when Bronk announced appointment of the NAS Committee on Oceanography, the ignition switch was turned on the scientific enterprise that generated many of the political events that followed, and until 1966 provided the most significant single impulse toward a new era of American exploration of the sea. That date also marked the beginning of the end to a period of intellectual isolation of those who wished to study the sea from those devoted to utilizing it more effectively to serve mankind's needs.

THE NEGLECT OF MARITIME SYSTEMICS

After its appointment, the National Academy of Science's Committee on Oceanography—NASCO as it became known—set about to implement its charter amidst a host of subtle problems. The oceanographic enterprise was not only small and segmented, it was largely ignored by the broader technical, industrial, and political communities. With rare exceptions—notably certain elements of the Navy—communications with users of the sea were scattered, infrequent, and tense. Oceanographers were largely ignored by practical-minded fishermen and seamen, even by the pragmatic oil industry newly gone to sea. Perhaps in frustration, oceanographers seemed to assume a cloak of proprietorship about research in the sea, and disdained involvement with commercial interests.

With few exceptions, university scientists were unwilling to share their oceanographic data with the oil industry, contending that industry would not reciprocate with information that it regarded as proprietary. Too, the older and larger oceanographic institutions did not welcome the entry of new research organs because, when funding was scarce, the small, low-calorie pie had to be sliced into still more pieces. Sharing of limited funds could seriously dilute the excellence that had been generated in building a research capability with a balanced mix of ships, shore facilities, and talented staff.

Adding to the isolation of marine sciences and cleavage within, a further split developed, once the study began, between those interested in the inshore environment and those who went to sea. The inshore scientists who worried their way through tidelands in antiquated motor boats were a breed apart from those ensconced in richly equipped, large ships on prestigious deep ocean expeditions, fraternizing with foreign scientists and addressing scientific questions of planetary dimensions.

Then there was a pecking order within the scientific community itself. High-energy physicists seemed buoyed to the top and the oceanographers ballasted near the bottom. This distinction was further inflamed when the NASCO report was released and gained publicity in 1959. Marine scientists were not merely ignored by influential leaders in the scientific community, they were quietly deprecated as not being good scientists, and as using antiquated experimental techniques compared to the sophisticated analysis and elaborate apparatus of particle physics. The point was made that very few oceanographers had earned election to the Academy.

This was the uneasy milieu for the third NAS investigation on needs and opportunities in oceanography. It is difficult to forecast what would have been the outcome in that disquieting atmosphere in the absence of another singular and ostensibly unrelated event—the Soviet launching on October 4, 1957, of the first operational spacecraft and its reinforcement a few weeks later by the orbiting and safe recovery of a live dog. Sputnik and Mutnik made world-wide headlines, immediately magnified by the near hysteria of citizen and politician alike. With one count down, science was launched from the quiet and seclusion of the laboratory into the orbit of national policy. It is ironic that the great $50-billion Soviet—United States space race budded from a 66-nation cooperative scientific venture, the International Geophysical Year. Several nations (but especially the United States and U.S.S.R.) planned together to orbit some modest instruments for geophysical surveying; after secretly accelerating their program, the Soviets had unilaterally stunned the world with a breakthrough in space technology.

Energetically propagandizing the feat, the Soviet challenge flung at the United States was more than a scientific competition, as interesting as that horserace might be. Rather, it was a naked political challenge, on the basis that visible excellence in space technology could be equated to implied excellence in military technology. In a world already split into two ideological camps, the stakes were genuinely ominous. Many observers of this achievement speculated on what other surprises were in store, what advantages the Soviets had in numbers of trained scientists and engineers, in quality of their educational enterprise, in priorities accorded science and technology in their over-all schema. Speculation overran fact, largely because so few facts were available; the closed society of the U.S.S.R. had erected such effective security barriers that very little was known in the West of their scale of research and development, much less their level of sophistication.

The reverberations of Sputnik I were felt instantly throughout the governmental establishment and soon after throughout the scientific establishment. As a step to reassure an anxious electorate through a radio and television address on November 7, President Eisenhower announced creation of the post of Special Assistant for Science and Technology (since termed the President's Science Advisor) to bring into the White House an authentic knowledge of the substance and processes of scientific research that would be of moment to the President in the decisions he faced ahead. The selection of James R. Killian, then president of the prestigious Massachusetts Institute of Technology, was intended to reassure the public that the very best of advice was being solicited. A second step on November 22 moved closer to the White House a little publicized but highly influential panel of technical experts that since 1951 had served the President through the Office of Defense Mobilization on national security affairs, especially on nuclear weapons, missile delivery, and defense systems. That 18-man presidentially appointed body was designated as the President's Science Advisory Committee (PSAC), with its chairman elected by the membership. Without exception, the President's Science Advisor has been so chosen. Staff to support the Special Assistant and PSAC, housed in the White House complex, completed the genesis of governmental apparatus to provide advice to the President on science affairs.

Following the Soviet space initiative, numerous governmental instruments were also created by legislative action: the National Aeronautics and Space Administration (NASA), an independent agency in the Executive Branch formed by reconstitution of the long-standing and widely respected National Advisory Committee on Aeronautics (NACA); the National Aeronautics and Space Council, an advisory body on space affairs positioned in the Executive Office of the President and comprising the Administrator of NASA, Secretary of Defense, Secretary of State, Chairman of the AEC, and four miscellaneous presidentially appointed members, and chaired by the President;[5] the Aeronautical and Space Sciences Committee in the Senate; the Science and Astronautics Committee in the House of Representatives; and the post of congressional science advisor through provision of a senior specialist in science and technology located in the Legislative Reference Service (LRS) of the Library of Congress.

Quite apart from these new additions to the federal apparatus, changes were occurring in existing elements. Department of Defense (DOD) and AEC budgets grew sharply. The space program convincingly justified its budgets to a willing Congress, to grow from its precursor NACA annual level of $80 million to almost $1 billion in 1961. Other agencies began to ride the crest of the wave of concern over Soviet technological prowess. For example, budgets for scientific research in the National Science Foundation that had grown with agonizing hesitation since the agency's inception in 1950, suddenly doubled in the two years after Sputnik, tripling again in the succeeding five years.[6] And support for graduate education that had long been opposed in and outside of government blossomed in the National Defense Education Act.

Science and technology as a whole were the beneficiaries of the first shocked reaction in political circles that lagging the Soviets in any technical area was a potential serious threat to our national security. One effect reaching to the oceans was like a time bomb that did not detonate for nine years. Only after space affairs had led to creation of a new federal agency was the pattern set for a similar evolution in marine affairs that earned the dubious appellation of a "wet NASA." For the first eighteen months after Sputnik, however, oceanography was left in backwater; the oceans seemed a forgotten environment as the nation and its government gazed at the stars.

On February 15, 1959, NASCO delivered its landmark report *Oceanography 1960–1970*,[7] rushed to completion when word leaked out that the Navy was soon to release its 1958 analysis of oceanographic needs in a projected *Ten Years in Oceanography* (TENOC). NASCO recommended increased funding for marine research consistent with assessed urgent needs and inherent limitations on the rate at which ships and laboratories could be built and new oceanographers trained. Justifications were enumerated in terms of defense, but also out of concern for improved productivity of fisheries and protection of the environment from radioactive waste disposal at sea.

Five general and twenty specific recommendations emerged. The first three proposals advocated a doubling over ten years of the 1958 level of federal funding for basic research and ocean-wide surveys. This expansion was projected on a base of support that NASCO estimated in 1958 to be a trifling $24.1 million annually.[8] The fourth general recommendation concerned the various federal agencies and what would be expected as their share of funding. The fifth was a pro forma exhortation that "private foundations, universities and state government should all take an active part in the recommended program of expansion."[9] Said the Committee: "Action on a scale appreciably less than that recommended will jeopardize the position of oceanography in the United States relative to the position of the science in other major nations, thereby accentuating the serious military and political dangers and placing the Nation at a disadvantage in the future use of the resources of the sea."

Within the expanded envelope NASCO democratically nodded at all claimants—basic research, applied research, and surveys: "The key to the growth of oceanography in the United States lies in basic research —research which is done for its own sake without thought of practical application. . . . The rate of progress in the applied marine sciences will be determined in the long run by the rate of progress in the basic marine sciences." To satisfy a vocal minority on NASCO plumping for systematic surveys on an "ocean-wide, ocean-deep basis," they proposed that "the increase in support of basic research should be accompanied during the next ten years by a new program of ocean-wide surveys." Third, they noted that "the United States should expand considerably its support of the applied marine sciences, particularly in the areas of military defense, marine resources and marine radioactivity." But they were deliberately vague about how large an increase was warranted, cautious about inadvertently provoking a competition between out-of-house university and in-house government laboratories that were thought of as functionally specialized between basic and applied science.

Twenty specific recommendations detailed where the additional funds should be invested, in new ships and shore facilities, manpower training, and education. The primary, uninhibited thrust was toward nourishment of the feeble research enterprise and its essential tools. While the many nonmilitary uses of the sea to which the research might be applied were perceived, they were not effectively articulated. The anticipated funding would thus go primarily to the private research institutions—the same ones represented on the committee.

Twelve back-up reports were subsequently issued by NASCO in support of primary conclusions, dealing with the importance of research to accelerate uses of the sea. But there was little mention of policy or policy issues, of the legal, economic, and political implications, or of the tapestry of capital investment, technology, organization, and management necessary to realize these benefits. It was as though scientific knowledge alone would catalyze the complex endeavors and institutional framework needed. Implications for resource development or coastal management that later arose as key concerns were posed as decorative rhetoric rather than as irresistible arguments on this nation's stake in the sea. No unifying concept of maritime systemics yet emerged to blend knowledge with use. These deficiencies, however, can be judged only by hindsight. In its time, the report packed a wallop, and 1959 was the heyday of blind faith in science.

Dean Don K. Price in his incisive assessment of the scientific establishment[10] used the oceanography program as a case study in describing professional peerage and politicians, and their linkage through "administrative gadgets." Noting that it was easier to "recognize the unity of nature than to develop unity in the oceanographic program," he characterized the NASCO gambit as a strategy "to muster the kind of support within the scientific and professional communities that can produce political action," namely, to expand research.

What followed next was an unparalleled effort by a committee of the NAS to market their report. The endeavor actually succeeded, but ironically in the legislative rather than in the executive branch that had originally sponsored the study.

THE FRAGMENTED FEDERAL BUREAUCRACY

American involvement with the sea extends back to the very origins of the country, the founding of the Colonies and their dependence upon maritime transport of supplies. Of necessity seeking foreign exchange to nurture its economic independence, our fledgling government fostered a merchant marine, a fishing and whaling fleet that pressed its goods and services on overseas customers. The sea became primarily of interest to its employers, and these were private traders and seamen, not the government. Protection on the high seas against either acts of piracy or acts of war could not, however, be met by individual defense. Neither could costs of preparing navigation charts be borne by individual traders. So requirements for a Navy emerged in the Atlantic and Great Lakes followed in 1807 by the first marine-related agency, the Coast Survey, to provide assistance to the merchant marine.

The federal role at sea continued to grow in a random succession of separate legislative assignments. The catalogue of functions and agency responsibilities shown in table 1 reveals how wide and thinly spread was the aggregate of marine activities by 1970. Not only were new missions provided for, but organizations created for their implementation were repeatedly shuffled around. For example, the Coast Survey was established in 1807, redesignated the Coast and Geodetic Survey in 1878; subjected to consolidation of previous authorizations in 1947; incorporated in the Environment Science Services Administration (ESSA) formed July 13, 1965 by Reorganization Plan 2 of 1965; and finally embedded in the National Oceanic and Atmospheric Administration (NOAA) formed October 3, 1970, by Reorganization Plan 4 of 1970.

The Bureau of Commercial Fisheries began February 9, 1871, as the

TABLE 1

Federal Marine Science Activities, 1966–70

Agency	Mission
Department of State..............	Participation in international organizations; support of international fisheries commissions; international marine policies.
Agency for International Development.	Foreign assistance and food resources for developing nations.
Department of Defense........... (Navy; Advanced Research Projects Agency; Army Corps of Engineers.)	All phases of oceanography relating to national security; naval technology; statutory civilian responsibilities: Great Lakes, river, harbor, coastal, and ocean charting and forecasting; Great Lakes, river, harbor, and coastal development, restoration, and preservation.
Department of the Interior......... (Geological Survey; Federal Water Pollution Control Administration; Bureau of Commercial Fisheries; Bureau of Sport Fisheries and Wildlife; Bureau of Mines; Bureau of Land Management; National Park Service; Bureau of Outdoor Recreation; Office of Saline Water; Office of Marine Resources.)	Management, conservation, and development of marine natural resources; lead responsibility for coastal zone management planning and research;[a] measurement and enforcement of water quality standards; acquisition, preservation, and development of coastal areas; identification and development of technology for evaluation of mineral resources; identification of sources and interrelationships for supply of fresh water.
Department of Commerce......... (Environmental Science Services Administration; Maritime Administration.)	Lead responsibility for air/sea interaction program and marine environmental prediction program;[a] tsunami and hurricane warning; charting and mapping of coastal and deep-ocean waters; research on ship design, shipbuilding, and ship operations; marine transportation and port systems.
Department of Health, Education, and Welfare. (Public Health Service; Office of Education; Food and Drug Administration.)	Human health, healthfulness of food, biomedical research, and support of education.
Department of Transportation...... (Coast Guard; Office of the Secretary.)	Safety and protection of life and property in port and at sea; delineation and prediction of ice masses; navigation aids; oceanographic and meteorological observations; transport systems analysis and planning.
Atomic Energy Commission........	Radioactivity in the marine environment; development of marine nuclear technology.
National Aeronautics and Space Administration.	Feasibility, design, and engineering of spacecraft and sensors for ocean observations.
National Science Foundation........	Basic and academic oceanography; lead responsibility for Arctic research and the International Decade of Ocean Exploration;[a] facilities support; Sea Grant colleges and programs.
Smithsonian Institution............	Identification, acquisition, classification, and ecology of marine organisms; investigations of the geophysical factors of oceanic environment.
National Council on Marine Resources and Engineering Development.	Assist the President in planning, development of policy and coordination of federal marine science activities.

[a] Lead agency designation involves federal leadership for stimulating the exchange of ideas with other interested agencies; planning in terms of the identification of goals with the advice and assistance of other agencies; and negotiating interagency support required for the achievement of goals in these areas—goals to be achieved through multiple, coherent activity of the interested agencies—funding gaps, where possible. No transfer of any agency's statutory responsibilities is involved.

Source: MSA, 4th report (1970), pp. 2–3.

independent Office of Commissioner of Fish and Fisheries to deal with declining fish catches; it was renamed the Bureau of Fisheries and placed in the newly formed Department of Commerce and Labor in 1903, and remained with Commerce when Labor was separated in 1913. In 1939 the migrant Bureau of Fisheries was moved to the Department of Interior and in 1956 was reorganized as the Bureau of Commercial Fisheries. In 1970, it was transferred to NOAA in the Department of Commerce and redesignated the National Marine Fisheries Service.

The U.S. Weather Bureau, established in the Department of Agriculture on October 1, 1890, was transferred to the Department of Commerce by Reorganization Plan 4 of 1940, consolidated into ESSA by Reorganization Plan 2 of 1965, and then into NOAA in 1970.

The Coast Guard had its beginnings in 1790 and is now a united service composed of the old Revenue Cutter Service and the Lifesaving Service, the former Lighthouse Service and Bureau of Marine Inspection and Navigation. It was consolidated in the Department of the Treasury in 1915, and became a component of the newly created Department of Transportation on April 1, 1967.

The Department of the Navy and the Office of the Secretary of the Navy were established April 30, 1798, having existed for nine years previously by legislation of August 7, 1789, that placed conduct of naval affairs under the Secretary for the Department of War. After several stages of reorganization, the Navy was incorporated in the National Military Establishment in 1947, which was later redesignated the Department of Defense. The Navy's responsibilities for systematically studying the oceans were assumed in 1842 when Matthew Maury was appointed Superintendent of the Department of Charts and Instruments. On June 21, 1866, this unit became the statutory Hydrographic Office charged with "improvement of a means for navigating safely the vessels of the Navy and of the mercantile marine." In June 1943, the Hydrographic Office was given responsibility for furnishing oceanographic information required by all armed services. In 1966 it was upgraded to join several new elements in the newly created Office of the Oceanographer of the Navy.

The complex taxonomy of federal marine science before the birth of NOAA in 1970 hints broadly that the organizational structure was never carefully designed to fulfill a coherent set of marine requirements. A better premise is that the array grew piecemeal in response to some pressure or crisis not being met by existing apparatus, accompanied by a game of bureaucratic musical chairs as agencies were expanded, modified or transferred. Once spawned, each organism began to lead a life of its own, with its individualistic life styles and instincts for survival sharpened by the competitive conditions under which the bureaucratic game must be played. Institutional power and prestige are not static qualities guaranteed by legislation but must be animated by leadership within the agency or by pressures from without, some of these being artfully energized by an incumbent bureau chief.

Whether an agency flowers into adulthood depends on its nourishment. All the authority to conduct this or that mission fades into abstraction unless adequate funding permits the reasonably prompt implementation of missions. Inevitably, the aggregate of agency funding proposals is far greater than any President can approve or the country afford.

Nothing succeeds quite like success in the budget battle. Influence follows money. In marine affairs, the large number of federal agencies involved carried little weight because of the meager funding associated with each. In 1959, when NASCO delivered its report, marine science represented by the aggregate funding in all agencies was a poor relation at the federal budget table. Compared to the $4.8 billion of federal research and development, the 24.1 million represented only one half of 1 percent. Among the array of marine participants, the Navy was and still is dominant. In 1960, the first year that marine affairs funding was extracted from the budget for special analysis, the Navy accounted for 65 percent. By 1970 its share was still 56 percent (but by FY 1973, with growth in civilian agencies, it had dropped to 37 percent).[11]

Until 1966 the full implications of government-wide marine science budgets were not influential in any decision-making locus of government; tabulations served primarily as an accounting convenience. Each agency component was developed, rationalized, defended, and approved as a separate entity. Within its own organizational sphere, it was tested against functionally unrelated competitors. The support each received depended both on the validity of the individual request and on the eloquence of its advocate in the councils of government.

Characteristically, the further down the administrative ladder a program is enunciated, the lower is the power of the protagonist. And the lower down the ladder, the more vulnerable it is to competition with alternative elements and to a longer sequence of reviews, any of which can deal a crippling or fatal blow to the proposal. Each echelon can find reasons to say "no" more easily than "yes," and one "no" can cancel any number of earlier affirmations.

The budget pressures from lower echelons, however, are aided and abetted from within and without. Any high official, up to and especially including the President, who has looked with favor on a proposition can inject his casual support, and in this game can successfully move even a weak budgetary element to higher levels with private or public endorsements. It is only when all officials in the chain of command are facing in the same direction that support is guaranteed. And the circumstance most likely to induce that positive response is neither orders from above that may be frustrated in the unwieldy process of making hard choices nor monochromatic ideology. As with iron filings, it is the magnetic power of compelling crisis or the influence of special interest groups that seems to align the otherwise random orientations.

In marine affairs, the military threats of 1812, 1860, 1916, 1941, and 1951 produced such response. Our merchant fleet as well as the Navy underwent cyclic growth and decline.[12] But these intervals of maritime affluence were short lived. After the Korean War, and despite evidence of potency of the nuclear submarine and the Polaris fleet ballistic missile, even the cold war threats proved unenduring as a stimulus to undersea research. Even within the Navy, the largest potential sponsor in the government family, lobbying for oceanography met indifference; the post of Hydrographer of the Navy was once considered the last port before retirement for less promising candidates for flag rank promotion.

Support for oceanography from special interest clientele was virtually nonexistent. The only two industries that could mount the organizational and propaganda apparatus that qualified them as lobbies were the fishing industry and the merchant marine. Both industries were sick. Neither cared much about research, at least not enough to promote it strenuously along with demands for more direct governmental assistance.

It is no wonder then that those governmental agencies whose specialized responsibilities drew on marine science were small, low in the hierarchy, unnoticed in the public press, and while quietly doing their job were unable to muster internal support. It is also interesting that federal officials having key responsibilities for management of marine affairs and troubled about budget doldrums looked to the Academy of Sciences for assistance, since in its last study of the languid state of oceanography in 1949, NAS had done nothing to reverse the neglect. Could it have been that the scientific community was the maritime bureaucracy's only reliable constituency?

Related to the meager funding and influence of marine agencies, individually and collectively, was the fragmentation among federal agencies that matched the handicaps from division in the marine science community outside of government and spawned additional obstacles to progress. Eleven agencies were involved; except for the common medium in which they conducted their business, there were few incentives or mechanisms for cooperation. Even within a single agency, differences in viewpoint were reflected in wobbling and conflicting directions. Thus eleven agencies had the potential of heading in twenty-two directions. To make matters worse, with funds tight, each agency eyed the others suspiciously as rivals in a zero-sum game. Whatever crumbs agency A might sweep up after the bigger and more powerful engines of bureaucracy had feasted at the budget meant less for agency B. And there were few penalties for lack of coordination. Conspicuously lacking was leadership, or even a podium for leadership. All that gave the fractured governmental oceanographic activity some internal order was the informal fraternization of mid-level technical staff. Recognizing that research conducted by one agency could aid understanding of scientific questions relevant to a second agency's mission, they sought to exchange data. They also exchanged such technical gossip as the performance of research contractors. This communication was regularized about 1956 under the rubric of an informal Coordinating Committee on Oceanography, established by the Office of Naval Research. But there were no formal appointments of members, no assignments of authority or responsibility for action. According to Vice Admiral John T. Hayward, the viability of the CCO was attributed to its informality and to rotation of the chairmanship among member agencies.[13]

Such informal coordinating mechanisms abound in the federal government without benefit of clergy. To a considerable extent, it is these interpersonal relationships that make the system run when so many watertight compartments impede communication, when cabinet officers feud, and when the utter complexity of interlocking programs sags under dinosaurian inertia. The reward structure in a bureaucracy in general prompts loyalty first to a bureau chief, then in some order to a cabinet officer, to the administration, and to a constituency, including Congress. Bureaucrats may admire rivals in other agencies but seldom cheer at their victories. With so much energy demanded of staff for troubleshooting a system failure or in fighting fires ignited by external crises, little is left for long-range planning, conceptual development of new programs, or assessment of past performance, much less for coordination. Any breakthrough of the system usually requires a strenuous act of leadership at the top, a feat demanding creative energies, willingness to cope with the barricades of bureaucracy, and a stern sense of inner security.

On this issue of coordination the administration was complacent, and that posture was to prove stubbornly durable. At an interdepartmental meeting convened February 24, 1959, by Secretary of the Navy Thomas S. Gates to consider NASCO's revelations about coordination, the minutes report that "a formal interagency committee on oceanography . . . [was] not required at this time," although the CCO should be enlarged and retained. Further, "it was agreed that the Department of the Navy would take the lead in requesting the establishment . . . of another committee to be formed at the next higher echelon of government . . . to meet once or twice a year to consider recommendations of the CCO." These were the origins of the 1959 Interagency Committee an Oceanography (ICO) and of a Navy initiative in the nick of time to take over leadership.

While the presence of leadership for coordination is a necessary condition, it is not sufficient. The leader must occupy a podium with sufficient authority, prestige visible to the outside community, and access to top management inside that he can control the tendencies toward anarchy beneath. It is difficult to estimate what level of officialdom constitutes the threshold of functional visibility. But for the marine science functions in government, it was evidently too low.

The large number of separate soft voices were thus not singing the same song, but even if they had been, their harmony would have been drowned in the noisier clanking of bureaucratic machinery.

CONGRESSIONAL INITIATIVES

Given this unpromising situation, what response could the NASCO report hope to fire up in the executive branch? For one thing, each bureau chief would have new ammunition for his arsenal of small-caliber weapons in seeking greater support up the line within his department and eventually in the Olympian review by the Bureau of the Budget and the President. But unless historic patterns of indigence were to be reversed, the proposals would be derided, rejected, or simply ignored. The track records of responsible incumbents in the various marine-related posts were not impressive, despite heroic attempts to animate budgets on earlier occasions.

By 1959, thanks largely to Sputnik, such proposals might fall on more receptive ears. Federal funding for research and development rose in that fiscal year to $7.1 billion, a 42 percent step up. But all the science-related bureaus and their supporters in the scientific community were amplifying their claims for increased support.

Marine sciences, however, did not share in the reaction to Sputnik. With well-developed geopolitical instincts, Harrison Brown had anticipated this difficulty and generated an alternative strategy. When he assumed the cloak of NASCO chairman, he decided that his product was not to be deposited for someone else to promote or neglect. Washington was already burdened with thick advisory reports, quietly gathering dust. Recognizing this statistically probable fate, he gained concurrence from Academy President Bronk that he would be permitted to propagandize committee findings. Instead of waiting for his executive branch clients to act or react, Brown went directly to the Congress. When the report was in its final stage of drafting, he and certain colleagues having access to key congressional offices opened a quiet and unprecedented sales campaign. Although Senator Henry M. Jackson of Washington was the first to meet with the committee, targets of systematic advocacy were the Senate Commerce Committee and the House Merchant Marine and Fisheries Committee, both of which had legislative oversight over a number of the federal agencies involved in marine sciences. As openers, House and Senate staffers John Drewry and Dan Markel together with a number of Congressmen were invited to Sumner Pike's estate in Lubec, Maine, flown up in a Navy plane; all were subsequently invited to NASCO sessions in Washington.

The response NASCO evoked was sudden and positive. In a few weeks, Congressmen were making speeches, introducing resolutions and bills, and holding hearings, all focused on uncritical implementation of the NASCO proposals. Even the authors of the reports were surprised. The first recognition accorded NASCO on the floor of either House came not from the jurisdictionally concerned committee chairmen, but from an unlikely midwestern source, Hubert H. Humphrey, then the junior senator from Minnesota. On February 17, based on a meeting with Harrison Brown, Senator Humphrey declared in a statement entitled "Importance of Studies in Oceanography," "I came to realize how pitifully inadequate was our knowledge of the seas which traditionally have been the lifelines of the Western World," and went on to deplore the national ignorance of the sea, the failure of antisubmarine science to keep abreast of international progress, and the threat of the Soviet submarine fleet. The statement included a gracious tribute for the "monumental service [rendered by] Dr. Harrison Brown and his distinguished colleagues . . . in proposing hard, concrete, specific measures by which we can regain the lost initiative in this critically important field of knowledge."[14]

The urgency of undersea defense was echoed in the House of Representatives by Representative Hastings Keith, who on March 9 warned that the "United States is losing to the Soviet Union the biggest and most important sea battle in mankind's history—the contest to unlock the ocean's secrets for use in peace and war." In a statement on June 1, Senator Warren G. Magnuson said: "Soviet Russia is winning the struggle for the oceans. . . . Soviet Russia aspires to command the oceans and has mapped a shrewdly conceived plan, using science as a weapon to win her that supremacy."

This mode of response was encouraged by NASCO and its chairman in hearings. Said Harrison Brown: "I believe that in another decade or so warfare is going to be conducted primarily in the oceans. . . . If we are going to fight a war in the oceans, it means we have to know something about the oceans."[15] The threat of projected Soviet power at sea was credible because it was cast by the Navy in terms of submarines carrying intermediate range ballistic missiles that, launched off our shores in international waters, could reach not only our coastal cities but St. Louis, Denver, Chicago, and Albuquerque. Then, too, the Soviets were parading their new nuclear icebreaker *Lenin*, the 5,900-ton oceanographic ship, *Mikhail Lomonosov*, and a fisheries submarine, *Severyanka*, each unmatched in the United States. Congressional response to strengthening oceanographic research was building up momentum; with conspicuous lapses, it would carry through the next decade.

The chairman of the Senate Commerce Committee was Warren G. Magnuson from the state of Washington who was elected to the House of Representatives in 1937, to the Senate in 1944, and had risen to Committee Chairman in 1955. Known for his liberal views and leadership in a variety of causes, he had been sponsor, with several colleagues, of the National Science Foundation Act of 1950, following Vannevar Bush's monumental proposals in "Science—the Endless Frontier." Although his committee's jurisdiction was exceedingly broad, the field of oceanography had never been high on the agenda. Nevertheless, Magnuson was a willing listener when the NASCO delegation met with him informally. The nautical language was not new; Magnuson had spent time during World War II as a Lieutenant Commander in the Navy, his constituency included a fair share of fishermen, and he had cultivated communication with the faculty in oceanography and in a separate, unique fisheries college at the University of Washington. Milner Schaefer on NASCO was a graduate of the fisheries college; Roger Revelle's father had been a law colleague of Magnuson. An appointment for NASCO representatives to brief the Senator was easily arranged.

In the House, warm hospitality was also unexpectedly encountered. Chairman of the Merchant Marine and Fisheries Committee was Herbert C. Bonner, a lawyer from North Carolina's outer banks, whose primary committee interest was the merchant marine. His constituency included fishermen and the Coast Guard families who manned the Hatteras lighthouses. As with its Senate counterpart, this committee had never placed oceanography on its agenda. However, just a few weeks prior to meeting NASCO representatives, Bonner had urged his chief counsel, John Drewry, to examine the scope of committee jurisdiction to determine how it might be strengthened. Bonner realized that among the constellation of sixteen standing committees that had been distilled out of a much larger, unmanageable flock by the Legislative Reorganization Act of 1946, his committee retained the most limited range of legislative responsibilities.[16] By a coincidence, Drewry was working on legislation for a nuclear-powered icebreaker and in that research had met a reserve naval captain, Paul S. Bauer, who volunteered to serve as oceanographic staff for the committee at no salary; Bauer assisted Drewry in wooing Bonner's interest.

Perhaps awed by the sophistication of scientific nomenclature and scientists, Bonner was nevertheless captivated by the opportunity to enlarge committee influence in the new and intriguing field of oceanography, an inclination spurred by recognition that another House committee had a competing jurisdictional claim—the Science and Astronautics Committee. Spawned in response to Sputnik, the latter was designed primarily to exercise legislative oversight of the American space program, both civilian and military. But like its title, the terms of reference of the committee extended over a far wider spectrum of scientific disciplines. Impressed by the implications of the Soviet achievement and suddenly viewing the entire panorama of science as an arena for competition with the U.S.S.R., Speaker John W. McCormack and House members handed to the new committee full jurisdiction over the whole range of science. (This included legislative oversight of the National Science Foundation, which had previously been under cognizance of the House Committee on Interstate and Foreign Commerce.)

The chairman of the Science and Astronautics Committee was a robust Louisiana bayou politician, Overton Brooks. While claiming no education in or understanding of science-related issues, Brooks thoroughly enjoyed the aura of prestige then reflected upon all who associated with science or scientists; at that time, pictures taken in company with national scientific figures for his local press had genuine political value. Bonner guessed, and circumstances later confirmed, that once the oceanographic cat was out of the bag Brooks would endeavor to add inner space to his sphere of authority over outer space.

So Bonner made his move first. The congressional stage was set in private for the February 15 release of the NASCO report. Two days later, Chairman Bonner announced creation of the Special Subcommittee on Oceanography, to be chaired by Congressman George P. Miller of California; two weeks later, on March 3, the first hearings were convened.

Miller, gentle, courteous, and constructive, was one of the few members of Congress trained as an engineer. He had been Executive Secretary to the California Division of Fish and Game and had a sincere affection for this new field. With help from Bauer and Drewry, he opened the subcommittee with a strong statement of national need. Some forty-two witnesses appeared, representing the scientific community, government agencies, and certain special interests. Repeatedly enunciated in testimony was the single theme—strengthen our oceanographic research capabilities. The effect of the hearings was largely educational and their tone sympathetic. Stirrings of interest by Congress were evident in this new and beguiling field, and an unprecedented courtship began between Congress and members of the oceanographic community. From the colloquy at hearings, a new literature on the sea was devolving. Verbose, unstructured, ephemeral, and sometimes dull, it nevertheless illuminated the issues and the personalities dealing with the relationship of the sea to man.[17]

The jurisdictional rivalry anticipated by Bonner surfaced soon after the Miller hearings, with Brooks' ploy on April 13 of introducing the very first piece of oceanographic legislation—H.R. 6298 to amend the National Science Foundation Act of 1950. Quite apart from the merits of this bill to provide explicit financial assistance for university teaching facilities in oceanography, its NSF peg was a foot in the jurisdictional door. It afforded the traditional vehicle for holding hearings and a hunting license for the committee to go well beyond the scope of the draft legislation. August 25 hearings were brief, however, reflecting the tremendous work load assumed by the committee in its annual authorization of the pubescent, unfamiliar space program. NSF strongly opposed categorical support for any field of science, and the bill died. Exactly one year later, Brooks uncorked further initiatives.

In the meanwhile, the Senate adopted a different strategy. Wanting to show support for NASCO and less threatened by the jurisdictional question because the Senate Aeronautical and Space Sciences Committee, then under Lyndon B. Johnson, had no broad powers in science, Senators Magnuson, Clair Engle, and Henry M. Jackson introduced Resolution 136 on June 22 setting forth a national policy to strengthen the entire field of oceanography. Encouraged by his administrative assistant, Gerald Grinstein, who foresaw a favorable response back home, Magnuson became the Senate's "Mr. Oceanography."

The Senate resolution began with a statement of the problem: "Whereas expanded studies of the oceans and the ocean bottoms at all depths are vital to defense against enemy submarines, to the operation of our own submarines with maximum efficiency, to the rehabilitation of our commercial fisheries and utilization of other present or potential ocean resources, to facilitating commerce and navigation and to expand our scientific knowledge," and "whereas several other nations, particularly the U.S.S.R. are presently conducting oceanic studies of unprecedented magnitude on a worldwide basis. . . ."[18] It then unabashedly plagiarized NASCO wording to increase funding for research and for research tools. The measure was immediately and unanimously passed by the Senate, a pattern that reveals senators' respect for initiatives by committee chairmen, at least on less controversial matters. Actually, resolutions carry only a "sense of the meeting" without any direction to the President, although the Tonkin Gulf Resolution highlights the opportunity given a President if he wishes to exploit the action. Apart from commending the NASCO report, Senate Resolution 136 took note of the widely dispersed legislative authority of numerous agencies to engage in oceanographic research and proposed coordination through "some method of interagency cooperation—possibly through an Oceanographic Research Board or Commission"—an issue that was to be massaged often during the following ten years. The resolution also urged increased international cooperation and exchange of data on a "carefully supervised and reciprocal basis."

On August 4 a bill to remove geographical limitations on operation of the Coast and Geodetic Survey was introduced by Senators Magnuson and Engle in a bipartisan move characteristic of much of the Senate Commerce Committee style.[19] It passed on August 19. An identical bill was introduced a week later by Bonner. These bills, drafted in the Survey, reflected the initial behind-the-scenes response by the executive branch to congressional enthusiasm—bald moves to expand agency jurisdiction. Inconsistently, at the same time fragmentation was being criticized by Congress, this step would extend the duplication in authority of individual agencies even further. In this instance, a geographic as well as political expansion would be authorized for the Survey that historically had confined its activities to mapping and charting of the inshore waters and coastline to aid safe navigation. Now it was handed a carte blanche to extend its range into the deep sea where the more muscular brand of oceanography had been practiced under the Navy. The Navy furiously attacked this intrusion into their domain, but without success. As they expected, with this new authority, the Coast and Geodetic Survey began construction of longer range vessels, a ploy to carve out a larger niche in multiagency marine activities.

Key players in the Congress continued to urge government-wide expansion of the program. Reflecting on the large covey of participating agencies, however, they harped on needs to improve executive branch coordination and to create a focal point for responsibility that would arrest buckpassing when Congress had to poll eleven agencies for a complete answer to a single question.

By this time, incidentally, science and technology were blossoming rapidly in almost every federal body, fueling latent concerns of "unnecessary duplication." That overworked phrase had long been a hallmark of congressional anxiety or at least of its rhetoric. Occasionally, duplication was confirmed by horrible examples of poor management, but the mere existence of a plurality of activities automatically elicited the specter of waste. But with science and technology, the rapid growth was almost unprecedented. What was worse, to the Congress, it was so complex as to be unintelligible and so diffuse as to be unmanageable.

One of the earliest to sound the congressional alarm of duplication in science and technology was again Senator Humphrey. As early as 1957,[20] armed with reports from Government Operations Committee staffer Walter Reynolds, he introduced S. 3126 to consolidate many research activities into a Department of Science and Technology. That proposal was to find many opponents and few adherents. Two well-known supporters were Lloyd V. Berkner, before his death Director of the Graduate Research Center of the Southwest, and Wallace Brode, a former science advisor to the Secretary of State. Berkner argued that many environmental agencies "are stepchildren of considerable nuisance value to their individual departments"; since they are not strongly related in an organic sense to functions of a single department and are at a vital disadvantage in gathering budget support, they should be consolidated. Curiously anticipating later events, he included oceanography along with meteorology, physical sciences and standards, mapping, scientific and technical information, time, geodesy and astronomy, radio and upper atmospheric research, polar activities, resources, fish and wildlife.[21] On the basis of these arguments, Humphrey reintroduced the bill in the 86th Congress.[22] While consolidation then gained little support, the coordination issue became the byword and was reflected in all oceanographic legislation.

S. 2692, entitled the Marine Sciences and Research Act of 1959, introduced September 5, 1959, by Senator Magnuson and twelve others on the Commerce Committee, contained a specific declaration of policy to strengthen oceanography so that the United States would not be excelled by another nation. Still uncritically following NASCO and TENOC recommendations,[23] the bill detailed a comprehensive ten-year program of research and surveys; authorization for construction of new ships, new facilities, and new instruments; the development of manpower; the establishment of a national oceanographic records center; and improvement of international cooperation. It also made a point of emphasizing research to improve economic and general welfare related to living marine resources. Government-wide planning and coordination were assigned to a new Division of Marine Sciences in the NSF, authorized to develop a continuing national policy and a program that included the Navy's TENOC; to recommend contracts and grants for education and research; to encourage cooperation of participating federal agencies, the NAS, and universities; to foster information exchange; and to evaluate scientific aspects of programs sponsored by the federal government.

That NSF was singled out for a leadership role is of special interest. Ordinarily this conspicuous designation might be attributed to quiet lobbying by the agency itself. Rather surprisingly, NSF had neither such ambitions nor an aggressive style. Its first and reappointed Director, Alan T. Waterman, was a keen intellect and a respected scientist but the antithesis of the stereotyped agency head. Personally modest, he gave to the Foundation a quiet, dignified posture of serving as a funding agent for academic research and education, wherein excellence rather than size of program was paramount. To him, this mode of operation seemed threatened if NSF were to lead an active political life that might have produced budgetary and power structure rewards, but only at the cost of bruising conflicts with stronger agency competitors and of vulnerability to pork-barrel pressures from congressmen bent on support for a favorite project or constituent institution. To avoid internal conflicts with the scientific community whom he respected as his constituency, Waterman strove consciously to avoid favoritism by balancing funds equally for research versus education, and within each category to fund approximately the same fraction of proposals for each scientific discipline. Thus categorical support for oceanography and anointment with a leadership role were something of an anathema to NSF. As to coordination of other agencies, NSF had been previously empowered to develop a leadership role by its basic legislation, but for any federal operating agency to control its sister was a bureaucratic fiction. NSF demurred and parenthetically, it was this unused basic authority that was later extracted by the Reorganization Plan of 1962 to form the President's Office of Science and Technology.

Magnuson's proposal thus did not originate at NSF; neither did it gain that agency's support. Only Magnuson's parental interest in NSF and the needs of underfunded oceanographers made it the target for this role.

The second session of the 86th Congress opened with a bill by Washington Congressman Thomas M. Pelly identical to S. 2692;[24] it went to the House Merchant Marine and Fisheries Committee where hearings by the special subcommittee were resumed. Soon after, Congressman Miller introduced a different bill to establish a public policy and federal coordinator of its own surveys through the new seven-agency Committee on Oceanographic Surveys.[25] S. 2692 was passed by the Senate on June 25, but it died in the House because of an early adjournment for the presidential election. Under any circumstances, its fate would have been uncertain, because all of the executive branch agencies opposed it as a bad bill—far too explicit in its authorization for ships, in its breakdown of funds by agency, and in its questionable assignment of coordinating responsibility to NSF.

At that point, the smoldering jurisdictional dispute between the House Science and Astronautics Committee and the Merchant Marine and Fisheries Committee took a serious turn. While the question was pending of which committee should dominate, the Merchant Marine and Fisheries Committee had been moving inconspicuously and only via a special subcommittee not requiring authorization by the House leadership, to avoid precipitating open hostilities. In April 1960, Overton Brooks signaled his intention to capture jurisdiction with a three-pronged attack: he convened general hearings on "Frontiers in Oceanic Research," introduced new NSF oceanographic legislation to assure referral to his committee, this time assigning responsibility for national direction to a special committee in the National Science Board associated with NSF, and released a major report analyzing oceanographic funding and coordination issues, with committee imprimatur. Charles S. Sheldon III, staff director of the Brooks committee, called on the author as first incumbent of the congressional science advisory post in the Legislative Reference Service, to assist in the jurisdictional game plan by outlining the key issues of marine science, planning the hearings, selecting witnesses, preparing questions for Brooks to address to the witnesses, and to draft the report.

In the hearings, Brooks followed the suggested strategy by placing industry into the picture for the first time, with witnesses discussing undersea pipelines, tunnels, mining, salvage, and submarine cables as well as incentives for industrial sponsorship of research and marine technology.

The report, begun in secret, was released in July 1960. One hundred and eighty pages in length, it was entitled *Ocean Sciences and National Security* in keeping with the still high level of apprehension over Soviet capabilities.[26] The term "national security," however, was employed in the broadest sense since the report expounded the potential of the oceans to supplement continental sources of food, minerals and fuel; its relationship to weather and climate; and knowledge required for safe disposal at sea of radioactive waste, at that time the major pollution concern. The report assessed the issues and added a factual inventory of U.S. capabilities that had not previously been published, together with comparisons with the Soviet capabilities that provided a more analytical perspective on the emotional question of "who's ahead." The Soviets were not. Nevertheless, separating out the Soviet submarine threat, it concluded that "the Soviet Union has accelerated its effort in oceanographic research in what may be a deliberate attempt to overtake and surpass oceanography in the United States. Scientific achievements in this field would provide ample potential for propaganda, which following its actions in outer space, the U.S.S.R. may be expected to pursue."[27]

Packaged with the report were the Committee's own conclusions and recommendations, extracted from the report's earlier arguments. The most significant concerned expansion of federal level of effort by a factor of four over the next ten years, thus upping the NASCO ante. And they emphasized sponsorship of long-range and systematic coordinated planning "to be sure that its program has unity, a sense of purpose, coordination, and vigor [to assure] that the goals are met effectively and with due regard to thrift." The Committee also complained that "the Navy, until recently, has not pursued a broad program in oceanography," and that "other civilian agencies of Government having jurisdiction in the ocean sciences have also not moved spontaneously to fulfill the potential that lies in research in the sea." In its diagnosis, the Committee suggested that the major cause for a laggard program "is the highly diffused responsibility for various sectors of research in the sea throughout a large number of Government agencies, and (at least in the past) the absence of effective coordination." The executive branch had trotted out the ICO as evidence of in-house management, but the Committee felt that that group had operated too short a time for adequate evaluation of its effectiveness and that its lack of statutory basis might inhibit its success. The Committee then said: "To remove the disparity between the current level of effort and the needs for national security, it is necessary that the Federal Government organize, manage, and coordinate the necessary program." Ignoring Brooks' own bill, the Committee asserted: "A major study is deemed necessary of Federal organization for oceanographic research. The question should be explored whether at this time, the objectives of program planning and coordination would be best accomplished by the formation of a new, separate agency, having responsibility to coordinate all elements of the program." Two sets of advisory committees were proposed, one for inside coordination, the second to obtain advice from both science and industry.[28]

These recommendations were to be realized over the succeeding eleven years, but in sequence rather than in concert.

As congressional reports go, the report received a fair amount of favorable publicity, enough to disturb the Miller subcommittee and force into the open the issue on committee jurisdiction between Brooks and Miller. Bonner had hoped to head off such an impasse in his selection of George Miller as subcommittee chairman, because Miller served also on Brooks' Space Committee as the second most senior Democrat. But to no avail. When it became clear that progress in the field would suffer if this jurisdictional battle broke out into the open, an effort was made at compromise. From my LRS post, I was called in as an oceanographic consultant to determine whether the field could be rationally split so that each committee could have a piece of the action. It did not take extensive analysis to reply: one committee could take over oceanography in the Atlantic and the other in the Pacific; or if they split cognizance horizontally, one could deal with the sea from its surface to its average depth of 12,000 feet and the other from there to, and including, the seabed. Such ridiculous suggestions made the point; Miller and Brooks decided it had to be one or the other. Miller won, and for two years he exerted courtly House leadership in marine sciences. Miller's sweet victory was to prove ironic. Upon Overton Brooks' death in 1961, Miller succeeded to chairmanship of the Space Committee, where he found that he had been so successful in locking up jurisdiction in Merchant Marine and Fisheries that he could not take it with him. Congressman John D. Dingell of Michigan became head of the Oceanographic Subcommittee.

With a presidential election imminent, the 86th Congress adjourned early. No significant legislation was passed but Congress had kept the pot simmering. With the opening of the 87th Congress, action in both Houses swiftly heated up. Following the victory in jurisdiction, the Merchant Marine and Fisheries special subcommittee was converted on

February 15, 1961, to a full standing subcommittee. In the Senate, Magnuson introduced the Marine Science and Research Act of 1961 (S. 901), identical to that in the prior Congress. While the issues of establishing national policy and generally invigorating government-wide programs floated in a dialogue of exhortation and abstraction, steps to strengthen individual agencies continued. On March 2 Senator Magnuson followed the Coast and Geodetic Survey's example of enlarged jurisdiction, introducing S. 1189 to extend Coast Guard authority to engage in oceanographic research. While thus improving utilization of its patrol ships, paradoxically another research agency was added to the already long list. Soon after, he convened extensive hearings on S. 901 and S. 1189 and reported both out favorably. Both passed unanimously in the Senate, again indicating the power of a committee chairman in gaining concurrence from colleagues who respect both the expertise and the prestige of a bill's sponsor.

The Coast Guard bill became law; S. 901, however, ran into heavy seas in the House. Among other factors, all agencies of the executive branch uniformly opposed new legislation on the grounds that ICO was a satisfactory coordinating apparatus,[29] and that the bill was defective with excessive programmatic detail. Congressman Miller endeavored to circumvent substantive objections to the excessively detailed legislative proposals by using a scalpel instead of a sword for corrective surgery on the flagging activities of the executive branch. His Oceanographic Act of 1961, H.R. 4276 introduced on February 13, 1961, would expand and develop aquatic sciences of the United States by creating a National Oceanographic Council of four cabinet officers and two department heads, one of whom would be chosen by the President as chairman. The Council was directed to develop long-range plans, to coordinate interagency programs, to establish a new data center and an instrumentation center, and to report annually to the Congress. Incongruously, it singled out the Smithsonian Institution for strengthening. This vague proposal for a new machinery encountered even stiffer opposition from the executive branch.

During this entire interval of two years, the executive branch had been relatively silent in responding to the NASCO report and to subsequent congressional calls for action. Agency heads tripped up to the Hill as witnesses and tripped down again, each adding detail to the tapestry of oceanic activities and broadening the understanding of diligent congressmen intrigued by this new arena of research. But none of these witnesses spoke for the President. None discussed policy. All were in agreement on the need to strengthen research and to improve coordination. All opposed the pending bills but suggested no alternatives. Most important, President Eisenhower did not publicly speak to the question nor did the budgets for oceanographic research undergo any spectacular increase. Meanwhile, impatient at words without deeds, the Congress fumed at not finding a magical legislative key to the problem.

CONFLICTS BETWEEN THE PRESIDENT AND CONGRESS—ROUND 1

While Capitol Hill was both enjoying and fussing over an exotic new issue, what was going on downtown?

Executive branch recognition of the emerging importance of oceanography occurred twice in the six weeks prior to release of the NASCO report. First the Navy released its *Ten Years in Oceanography* on January 1, 1959. Whether prompted by concern that the imminent NASCO report would trigger a set of embarrassing questions about Navy readiness in marine sciences, or whether by coincidence, TENOC was an in-house appraisal of Soviet capabilities, an expression of U.S. needs, and an earnest attempt at long-range planning. It temporarily enjoyed unprecedented high-level Navy interest but it then proved ineffective in kindling support from the House Armed Services Committee. Almost simultaneously, on December 27, 1958, President Eisenhower had released a report prepared by a PSAC panel chaired by Emmanuel Piore on *Strengthening American Science*.[30] The report acknowledged lack of progress in some scientific fields because of inadequate tools— citing meteorology, geophysics, radio astronomy, biophysics, linguistics, and social psychology—and then went on to say: "Oceanography is another promising field which has received inadequate attention. For the study of the oceans the United States has only a few research vessels, all inadequately equipped. A vessel specially designed and constructed for oceanographic research has not been built in this country since 1930."

Among other issues, the PSAC report dealt with the proliferation of agency involvement in science and technology. The coordination enigma was intensified because basic research conducted in a special field of science under one agency's sponsorship has potential value to others. Scientific information is neither perishable nor destroyed by use, and different clients unknown to each other may find a result independently relevant. The government's earlier experience with a postwar Research and Development Board was instructive since that agency soon grew to a cumbersome accounting machine that was an irritant both to those asked to bank information through voluminous questionnaires and to those who sought to withdraw information. In 1953 it was abandoned. Although not exactly a successor, the Interdepartmental Committee on Scientific Research and Development (ICSRD) was established in 1947 and augmented in 1954 as a communication link between agencies, but now depended on direct interpersonal communication of priority messages rather than on a massive project library with associated administrative machinery.[31]

It was thus a natural step for PSAC, mindful of its own elevation from a lower-level advisory body to a White House level, to consider the analogous elevation of on-the-shelf in-house apparatus. They proposed a subcabinet level Federal Council for Science and Technology (FCST). It would thus upgrade the lower-level ICSRD.[32] In great measure, it followed earlier recommendations of the Steelman Report of 1947.[33] Eisenhower accepted the proposal, and on March 13 Executive Order 10807 created the "Science Cabinet." Chaired by the President's Special Assistant for Science and Technology, now George B. Kistiakowsky, the FCST was composed of senior policy-level officials in each department and agency dealing with science. In some cases, these were at the level of assistant secretary. In the case of NASA, AEC, and NSF, they were the agency heads. While there had been speculation that this initiative was sparked by the threat of a congressionally imposed Department of Science and Technology being promoted by Humphrey, it is apparent that the steep rise in research funding and diffusion of programs were of deep concern to a fiscally conservative administration. Moreover, this step satisfied Eisenhower's propensity for strong staffing.

The Federal Council's mandate potentially ranged across all fields of science and technology and all federal functions. Its capacity to implement, however, was another matter, depending upon inclinations of its chairman and staff. Care and feeding were to be assigned to Robert N. Kreidler, who came to the office fresh from duty in the Marine Corps and brought great intellect, objectivity, and toughness to deal with a cranky bureaucracy. But at the time, considering the heavy attention still on military security issues, Kistiakowsky considered the Federal Council to be subordinate to PSAC in importance as well as creativity, and Kreidler's asssignments included a variety of wide-ranging issues beyond the Council. Nevertheless, with the NASCO-triggered noise generated on Capitol Hill concerning the splintered oceanographic bureaucracy, it was natural to examine the FCST's coordinating potential in marine sciences.

In May 1959 a Subcommittee on Oceanography was established under the FCST Standing Committee (later to be separated as the Interagency Committee on Oceanography), and the Assistant Secretary of the Navy for Research and Development, Dr. James H. Wakelin, Jr., was appointed chairman. Additional appointments to the committee were made of the senior operating personnel in agencies having oceanographic missions, such as Rear Admiral H. Arnold Karo, Director of the Coast and Geodetic Survey; Donald L. McKernan, Chief of the Bureau of Commercial Fisheries, and others from nine agencies.[34] Without benefit of specific mandate, the subcommittee sought immediately to collect qualitative information on various federal programs and to quantify needs, thus to respond to the NASCO proposals. Products of the subcommittee work were initially used within the administration to brief President Eisenhower and his cabinet on implications of the NASCO recommendations. According to Wakelin,[35] the cabinet primarily focused on the high-cost items—ships; debate at two cabinet sessions revolved around the pros and cons to meet growing needs by either building new ships or rehabilitating older vessels destined for retirement or the scrap yard. Marine issues of substance were yet to be identified.

For the remainder of the Eisenhower administration, no public announcement of policy or programs indicated the direction or pace of response. In 1960, the oceanographic subcommittee was given full status as the Interagency Committee on Oceanography (ICO). Wakelin was called to testify before various congressional committees reviewing the state of oceanography's health, but he was never in a position to do more than deliver an inventory to pacify an impatient Congress. Quiet, authoritative in bearing, and informed, Wakelin answered questions in plain language, never prompting those pangs of congressional inferiority that frequently attend presentations by scientists unable to avoid specialized jargon or the classroom lecture style. As the man having his hand on the throttle of the largest governmental agency sponsoring marine research, Wakelin also engendered deference from congressmen who believed he could influence the destiny of laboratories in their home districts. Nevertheless, there were expressions of concern that the base of federal funding, identified for the first time for Fiscal Year 1961, grew only about 10 percent, from $55 to $60 million in 1961. Moreover, while Congress politely listened to Wakelin offer bland assurances that the program was thriving and coordination was adequate under ICO, they were genuinely skeptical.

During this interval when enthusiasm for oceanography was beginning to churn on Capitol Hill, why was there so little action at the White House? First, it must be recalled that the mood of the Eisenhower administration was neither venturesome nor expansionist. Firing up the space program had been shared with Congress, but once it had been begun, a cautious administration was in no mood to respond to the steady goading from the Democrats in the Congress to pour more money into catching up with the Soviets. Eisenhower had begun his second term by cutting R&D, and ended up by quadrupling support. Little wonder he complained on leaving about the military-industrial complex. Thus, the White House was not looking for new marine initiatives, and the protective Bureau of the Budget found it easy to turn off the entrepreneurs in government. But then, few signals to expand oceanographic research were being transmitted up the line by cabinet officers from subordinate bureaus concerned with marine affairs. The marine constituency outside of science was quiet. So was the science pipeline in the White House. Amid the increasing swarm of bees buzzing around the President because of growing terror of nuclear weapons and anxiety about preserving a strong lead over the Soviets, Kistiakowsky's advice was sought by the President. Though not a member of the President's political entourage, he had ample access, an asset certain of his successors lacked. But he did not back oceanography. In discussions with the author, Kistiakowsky admitted that the NASCO entrepreneurship was not welcome. The report captured a great deal of press attention, which at that time was regarded as stretching the traditions of scientific ethics. Many scientists in other fields inwardly and outwardly not only winced at the marketing of the report—a practice they later sought to emulate with varying success—but they were also struck by apprehension that their more justified and underfunded projects might suffer in the competition. Members of PSAC privately contended that oceanographers were not first-rate scientists and that oceanography was therefore undeserving as compared with materials research or construction of accelerators for particle research. The chairman of PSAC found it difficult to ignore such advice.

Since Kistiakowsky concurrently served as chairman of FCST, pressure coming up that route from the Wakelin subcommittee was blunted. For one thing, Wakelin had a major responsibility in the Navy Department and only a small portion of his efforts could go to oceanography. His naval aide, Commander Merle Savington, and later Captain Steven Anastasion, spent considerable time mapping government-wide oceanography programs and fostering a new confederation of agencies, but in 1959 they were not asked to generate new programs that Wakelin could push. Moreover, Wakelin's own gentle style did not dispose him toward championing causes in-house, in light of the administration's reticence. No backdoors to the President could then be opened because the three entities created to advise the President on science affairs—the Office of the Special Assistant, PSAC, and FCST—were all headed by the same man, and he was opening none.

Not surprisingly, NASCO Chairman Brown, unable to rouse White House allies, chose the Congress as the most sympathetic market for NASCO's proposals.

In summing up the status of oceanography in the two years following the NASCO report, two small bands of oceanographic specialists were prodding diligently, one inside the government at a third or fourth echelon, and one comprising research interests outside. One had little rapport with its policy-level superiors; the other found open hostility from the scientific community at large. To complete the picture, industry was almost completely silent.

But in November 1960, after eight years of Republican rule, the electorate listened to the vibrant words of a new young leader and voted a change; on January 20, 1961, John F. Kennedy was inaugurated as President. His message was electrifying. There was no mention of pet projects or issues but a new spirit was on the land. It would not be long before it graced the study of the sea.

THE KENNEDY–JOHNSON YEARS

While snow was drifting in to paralyze traffic of a surprised city on inauguration eve, 1961, many residents of Washington were drifting out. The ritual of partisanship was being repeated. With every change in presidency, the policy-level cream of public servants is precipitously skimmed off and replaced, although the opportunity to reward the party faithful with government jobs may be more limited than folklore suggests. Nevertheless, the wholesale shuffle of top officials generates extreme confusion and a great potential for mismanagement. Civil servants who survive these calamitous events endeavor valiantly to keep the machinery running while drivers are changed with the car in motion, and during an interval when no one is steering, or everyone is trying to. It was not until 1968 that new legislation was enacted to ease this transition by providing funds for a President-elect and his incoming officials to familiarize themselves with new jobs and predecessors between election and inauguration.

The 1961 shift in management marked the beginning of a period of great effervescence and turmoil, some of it because of the change in incumbent style as well as in personnel. Of some importance was Kennedy's decision not to "sweep the rascals out" by replacing all Republican appointees with party faithful. Especially, he instructed his aides to recruit for technical positions on the basis of competence, free of strict partisan considerations but with no less attention to the matter of loyalty. A few senior civil servants were moved up to policy-level jobs, and in the process, Wakelin stayed as Assistant Secretary of the Navy, aided by interventions from Sumner Pike and Senator Jackson.

Jerome B. Wiesner replaced Kistiakowsky. An engineer turned physicist at Massachusetts Institute of Technology, who held senior responsibility of M.I.T.'s wartime radar laboratory, Wiesner had been a member of PSAC during Eisenhower's term. Unlike most of his colleagues, he actively engaged in the political campaign of 1960 along with such academic advisors to then Senator Kennedy as John Kenneth Galbraith, Walter W. Rostow, and McGeorge Bundy, from the home base of Massachusetts. Wiesner's selection was announced months before the inauguration, so with that lead time, and with Wiesner's prior familiarity with the office through his PSAC membership, he was ready to energize the system the day he took over. What gave Wiesner even more of an edge in getting his activity promptly off the ground was a recommendation to the President, which was accepted, that all of the White House staff employed in the science advisor's office under Eisenhower be continued. By their interests and personality, if not by party label, this staff was more in tune with the activist, innovative style of the new President than with the more tranquil mood of his predecessor. Very little turnover occurred in PSAC; Kistiakowsky was asked by the new President to serve out his previously unexpired PSAC term. Membership in the Federal Council composed of presidential appointees was, however, completely new.

Wiesner concentrated on sustaining the close personal rapport that he had developed during the campaign with both President Kennedy and his lieutenants. Almost daily at around 5 P.M. Kennedy would gather in the oval office with a select group that included his science advisor, and exchange banter along with deeply serious discussion of life and death issues. It would be difficult to estimate how many policies were shaped in that informal atmosphere, but it became an institution, and probably far more important than cabinet meetings, which served to ratify decisions already made.[36]

Early in February, Theodore C. Sorensen, the President's chief counsel, legislative generator, and speech writer, invited the five o'clock club to nominate legislative proposals that Kennedy could move up to Capitol Hill to implement with specifics the glowing prose of the inaugural address. Wiesner enjoined junior staff to submit proposals. Bob Kreidler, while no specialist in oceanography, was intimately familiar with the unfulfilled NASCO propositions. He had also read the congressional statements of position, including my report for Brooks that went beyond NASCO in funding proposals. In consultation with Sorensen, he learned that the President might be sympathetic to a push in marine affairs. Among other factors, Sorenson is reported to have said that with his challenging, albeit grueling, experience as PT boat commander during the war and interest in yachting, Kennedy was hospitable toward a push involving the sea. In consultation with Wendell Pigman, a junior officer in the Bureau of the Budget, Kreidler swiftly generated an incisive proposal. Roger Revelle, who was soon to be brought into the administration as science advisor to Secretary of the Interior Stewart Udall, was recruited to detail the rationale. Wiesner gave the initiative his blessing, and with no further bureaucratic ado it went to Sorensen. There it ran into trouble, not because it was unconvincing but because it did not go far enough. All that was being requested from the Congress was money, whereas Sorensen was looking for a thrust of new policy that would have Kennedy's trademark on it. Nevertheless, Kreidler and Pigman made a sale.

The vehicle for execution was a special message by the President on natural resources, delivered on February 23, 1961. To project the innovative "new starts" character of his administration, President Kennedy tackled the tough nut of bringing together "the widely scattered resource policies of the Federal Government [that] . . . have overlapped and often conflicted" in the interest of utilizing natural resources more wisely. The National Academy of Sciences and the Federal Council were requested to assist in determining the research needed for improved policy planning and for better coordination. Then the President turned to specific areas deserving action, the oceans included. After tracing a whole gamut of possible benefits, from minerals to fish for protein deficient lands, he promised concerted attention to the national oceanographic effort, and the special need for coordinated federal efforts. In closing, he stated his intention to "send to the Congress for its information and use in considering the 1962 budget, a national program for oceanography, setting forth the responsibilities and requirements of all participating government agencies."[37]

Soon after, the Kennedy words were set to music. On March 29, in special Executive Communication 734 (see Appendix 13), the President explained why the oceans were so important to the nation as to warrant a new era of support. The budgetary proposals for oceanographic research and surveys in all federal agencies lumped together were to escalate over the prior year's level of about $60 million to over $97.5 million—a quantum jump that was to symbolize the President's interest in meeting oceanographic needs and opportunities. To those who may have assumed the "add on" was the product of long, extensive, staff work and argument, this account may be a source of amusement, and perhaps a lesson.

The reception on Capitol Hill was unanimous praise. In fact, Congress appropriated $6 million more than the President requested. In the initial political honeymoon with Kennedy, no eloquence could exceed commitment of funds. However, budgetary impetus to oceanographic research did not counter the earlier and chronic apprehension that because so many agencies had their hands dipped into salt water, inefficiency and waste were inevitable. Instead of fretting publicly and cooking up new legislative proposals, the concerned members of Congress initially voiced their anxieties privately to the President and his immediate staff. Among these callers was Senator Humphrey, not pursuing oceanography specifically, but rather his broader interest that coordination should be improved in all fields of science. Wiesner gave Humphrey assurance that this administration would do better than its predecessor and sought to fire up the dormant Federal Council mechanism.

Several impediments had prevented the Federal Council from achieving the mandate of President Eisenhower's Executive Order that had created it. Kistiakowsky and his very able staff lieutenant, David Z. Beckler, had little patience with the slower processes of the bureaucratic mill. But since the Science Advisor lacked any authority to intervene between a cabinet officer and the President, his success in fostering coordination depended wholly on his ability as well as willingness to invest time to wheedle and cajole, to gain concurrence, then to monitor agreements made in the President's office to be sure they survived the countervailing influence of other pressures. A second handicap of the Federal Council under Eisenhower was its membership. Although the Executive Order stipulated that it include the senior policy-level officers having science-related duties in such departments as Commerce, Interior, Agriculture, and Treasury, the men bearing these responsibilities were usually trained as lawyers or businessmen, ill prepared to engage scientific issues. Finally, despite his rich talents, Kreidler was never able to devote any sizable amount of time to his secretariat role because of other key assignments from Kistiakowsky.

The Congress had heard little about the Council and that only from Wakelin on oceanographic matters. One reason for the silence was a policy during the Eisenhower years dictating that Kistiakowsky, as a White House aide, should operate under executive privilege and refuse congressional calls to testify. While this angered an already polarized Congress, it permitted the science advisor not only to keep a low profile but also virtually to ignore the role of Congress in affairs of government.

As a far more political animal, Wiesner adopted a different tack. While publicly retaining immunity to congressional call, he privately conferred with many congressmen on the phone and in person; Humphrey thus had ready access. Humphrey was an informed and persistent critic of weak executive coordination and moreover, he had new ammunition. Based on a study prepared by the Legislative Reference Service, he found that $12 billion for research and development was probably spread among 160,000 projects. While a number of well-meaning attempts to keep book had been initiated within such agencies as the National Institutes of Health and the Air Force, there was no central, government-wide inventory on who was doing what, where, and at what government expense. In attacking this issue, Humphrey picked up the recommendation to strengthen an ongoing biomedical project registry and to broaden it for all fields into a Science Information Exchange under the Smithsonian Institution, thus helping practitioners as well as research administrators.[38]

So Wiesner promised Humphrey that he would attack these management problems. First, he began a campaign to establish a new post in every department for an Assistant Secretary for Science and Technology who would operate at a policy level to shepherd technological activities within the agency. Collaterally, that official would be the agency's representative on FCST, equipped by training and strategic bureaucratic position to contribute to debate on science policy within the Council, to commit his agency, and to aid in subsequent implementation of Council decisions. While the only immediate success was in the Department of Commerce, other departments created posts of science advisor to the Secretary.

Wiesner also decided to spend more time himself on Council affairs and to have a full-time Executive Secretary. With Kreidler's decision to return to graduate school, Wiesner extended an invitation to the author, in July 1961, to fill that post. Because of some nagging concerns about the embryonic capabilities Congress had created for itself in science policy, I asked for time to think it over. My modest shop in the Legislative Reference Service was already generating ammunition for congressmen from both sides who were preparing, once the honeymoon was over, to assert legislative branch oversight of executive branch performance, including marine coordination. But with encouragement from legislative friends and in the belief that the White House office offered the greater challenge, I accepted. When I came aboard, Wiesner's only instruction was "make the Council work." The laconism characterized the communications I was to have with him over the next two and one-half years.

Wiesner's mind was more than the proverbial steel trap; it was a magnificent tape recorder. He had mentally filed everything he had ever learned, a resource that added to his impatience to get on with "what's new." In briefings, a few carefully selected words instantly triggered his memory, and from then on, he wanted no detail of past history, only elaboration of the issue: its implications, facts, alternatives for action— and no small talk to clutter up dialogue or slow down his pace of dealing with twenty or more separate issues a day. His acceptance of my staff work, particularly on issue identification, brought more assignments— special studies for Kennedy on manpower, long-range planning, and congressional relations.

But the Council was my first obligation. Meeting almost monthly, the Council was best able to dispose of its complex business by reacting rather than acting, based on issue papers brought to it by staff and by a reinforced committee structure. Such committees, including ICO, were created to deal with what had been termed "national programs." Their common characteristics were interdisciplinary content, broad scope involving multiagency jurisdictions, a need for strong leadership, and pronounced importance to the nation. Atmospheric sciences, water resources, and materials development, as well as oceanography, had earned this designation. Implications of priority status that attached to the term were later lost when it proved ambiguous because such programs were federal rather than national.

All committees had equal claims on the Executive Secretary's time, but soon the ICO claimed special attention because of three separate circumstances. First, Wiesner had promised Congressman Dingell, the new chairman of the Subcommittee on Oceanography, that his office would assume oceanographic leadership. Dingell was a determined activist, leading the charge to strengthen oceanography;[39] he was familiar with the record, especially of questions of coordination. When his telephoned questions to Wiesner became too antagonistic and threatening, Wiesner asked me to join him in a private meeting on the Hill, to cool off what looked like an early and unwelcome public hearing. The meeting included Drewry and Bauer from Dingell's committee staff. Wise about oceanographic management dilemmas, Bauer took for himself the role of hairshirt for the bureaucracy—readying himself for attacks by scouting for skeletons in many closets. These were often disclosed in private conversations with lower-level officials who felt that such volunteered information might pressure their bosses to pay attention to their program's poverty-stricken plight. Bauer's intuition and his "conniving society" were productive in preparing Dingell for the fray. Wiesner's answer was to instruct me in Dingell's presence to make sure the ICO worked better.

A second personal motivation to spend time with ICO was to study the sociology of its operation and perhaps to discover some basic principles of bureaucratic behavior that could be fed back into guidance of the other FCST committees. So ICO became a pilot model, then a showcase as to how coordination could and should be enhanced.

Finally, I found the ICO subject matter familiar and enticing, largely from having been exposed to the milieu of ocean exploration in 1958 while designing the deep-diving research submarine, *Aluminaut.*

Now the Federal Council offered a unique opportunity to lend momentum to a desultory enterprise. Obliged by Wiesner's instruction to shed my neutrality about marine affairs, I became an advocate—the only one within the President's science advisory maze.

That interest in marine affairs swiftly proved needed. Historical coolness to oceanography in the science advisor's office was chilled further as the Kennedy budget proposal triggered envy among scientists from other fields. Wiesner was now receiving proposals from group after group of scientist citing their neglected needs and urging nourishment at the budget trough. Some adversaries felt the name of the game was to seek an advantage by chopping off the competitor. So new attacks on oceanographic spending were mounted within PSAC even before the funds were through the appropriation process, allocated to the agencies, and subject to obligations. Wiesner, trying to meet congressional pressure, was caught in the middle. The NASCO experience, however, stimulated imitation. To many scientists, the pattern for success in gaining presidential attention was now clear: arrange for a study to be requested of the NAS; then appoint a group of distinguished experts to sort out needs and opportunities and set these forth in a readable, widely disseminated report, with collateral lobbying for adoption in federal budgets. Machinery to facilitate such projects had been fortuitously established in efforts to strengthen NAS considerations of public policy. Their Committee on Science and Public Policy (COSPUP) was appointed in 1962 with George Kistiakowsky its first chairman.

After its 1961 spurt, oceanographic funding leveled off the following year to more modest growth rates. But Congress was not satisfied and they peppered Wiesner with complaints. The steam also began to dissipate from NASCO's own initiatives. Brown was back at Cal Tech; Revelle was in the government as science advisor to Udall; others on NASCO were less dedicated to carrying the political torch or were committed to other causes. Strangely, the NASCO role of leadership was moving to the ICO and eventually to the White House Office of Science and Technology. Between 1962 and 1964, with increasing skills and energies of its effervescent executive secretary, Robert Abel, ICO flourished both as a coordinating agent and as a national promoter. To carry out Wiesner's commitment to backstop ICO, my first step was to understand the White House milieu and measures that could be appropriately taken by the FCST Secretariat to help the ICO.

While the White House appears to float tranquilly on a sweeping green lawn like a cubist iceberg, that mansion is not only the presidential residence, it is an institution as well. And again like an iceberg, most of it is hidden. It is through the activities of the anonymous staff that the President holds the reins of power, using his immediate associates, their aides, and their aides' aides to help him perform both his constitutional functions and those pragmatically evolved and now hardened into custom. He is chief executive of the bureaucracy, Commander in Chief of the Armed Services, primary representative of the nation in relations with other heads of state, leader of his political party, and ceremonial symbol of the democracy. Political scientists increasingly talk of the dilution of presidential power as our nation spawns an ever wider range of cultures and aspirations amidst the growing complexity of institutions created to satisfy them, but no one could sensibly deny that the President is the single most powerful individual in the country.

Presidential staff include those in the White House and a second set further removed from the seat of power in the Executive Office of the President. Both sets, however, partake of roles and relationships simply by functional proximity to the presidential office, and to this exposure they react with widely different behavior. While all are transformed by their appointment, some become stimulated to heights of intellect and inspiration to serve a higher cause; others become bullies; most develop their own special fields and continuously seek to influence as well as promote the presidential decision. Each coterie of White House staff has its own style.

Serving as an intelligence apparatus, the staff aids the President by collecting information on what is going on within the massive administrative establishment that he heads, within the Congress, the body politic, and in fact, elsewhere on the globe. With this sorted, filtered, and packaged data, the same staff further helps the President in making decisions by identifying issues requiring his attention, illuminating alternatives and the consequences of each, including whose toes are stepped on and at what political cost. After a decision they aid in implementation by steering it through a herd of headstrong agencies that are ostensibly controlled by the President but which actually are split in their loyalties to the congressional committees who control their fate and constituencies that they serve and yearn to harmonize with.[40] These coalitions create severe strains on White House staff seeking consistency with presidential policy and priorities, especially since these may have been hammered out of fiery controversy and have left a residue of dissatisfaction, disaffection, and disunity within the bureaucracy.

Deliberately cloaked in mystique and often mystery, the staff strenuously aspire to protect the President from error or embarrassment and to buttress the privacy that he seeks from the steady and skillful probing from interest groups hoping to influence presidential decision. The jockeying among contenders for the President's ear becomes a spectator sport for the Washington press corps but a matter of serious concern for advocates within the government or lobbies among the governed, habitually eager to identify which White House staffer is the action officer for their special area, how close he is to the disciples who sit at the President's knee, and how to gain an audience by phone or visit to add some "body English" to the package as it approaches that flash point of decision. As recent analysts have noted, staff protectiveness has led to presidential isolation from the constituency.[41]

Executive Office staff work closely with White House staff on management of the bureaucracy, especially budgets and new legislation. Unlike White House staffers who are patently partisan, only a few officials in the Executive Office are presidential appointees. Most are subject-matter specialists who owe their allegiance to the presidency rather than to the incumbent. Because their roles are more highly stuctured and regularized, these staffers provide continuity when presidents change, serving as walking encyclopedias of past events, legislative origins, conflicts and personalities, which cast shadows on future initiatives. Created in 1939 the Executive Office includes the Bureau of the Budget (extended in 1970 to a new Office of Management and Budget), and a variety of other advisory offices and councils. By law, all are equal; but in practice, the first among equals has been the BOB. With its proximity to the President and its hand on the money throttle, the Budget Bureau that was designed to assist the President on internal management affairs of government has gradually assumed a role in policy affairs as well, foreign policy and defense being two major exceptions. Their fiscal produce is propelled into view by the President's budget annually transmitted to the Congress in a budget message rhetorically lubricated to foster adoption. The vast budget document itself, with a telephone-book size appendix of details, is supplemented for mere mortals by a mercifully short summary. By this one act and one document, the President has synthesized all the multidimensional wants and needs, hopes and aspirations, of American society into a one-dimensional quantitative priority list—a total carefully modulated by resources at hand.

The pie and the size of its slices were decided upon by one man, but it would be a fallacy to believe that he started from scratch to arrive at these judgments. To a substantial extent, the President is a captive of history, bound by fixed charges of debt retirement, social security, and veterans' benefits, and commitments of past programs that continue into the new fiscal year. This leaves about 25 percent to presidential choice, but probably 10 percent reflects linear projections of the past where practical politics suggest little if any change. The full spotlight of attention then falls on the remaining 15 percent of vulnerable items, together with another set of new proposals of about equal magnitude— which is what the budget exercise is all about. The battle lines for this slice are clear and drawn early.

Because proposed budgets always exceed the available purse, the Bureau's melancholy role is to rationalize essential cuts in terms of proposal weaknesses rather than arbitrary limits imposed by the President. Instructed by the agency's submission, by separate intelligence sought by the Bureau staff from informers within the agency or from outside expertise, the Bureau is in a position to shoot down flock after flock of propositions, many of which are flying on one wing anyway because of lame premises that cannot be concealed from these sharpshooters. But many valid proposals are also cut down by enfilade fire.

"When I pay taxes, I buy civilization," said Oliver Wendell Holmes, but the budget process itself is as uncivilized as any within government. For one thing, in endeavoring to maintain a flexible and independent position in advising the President on each individual project, the Bureau tends to overlook past decisions and commitments that hang like millstones around agencies' necks, sometimes forgetting it was the President who found it expedient to back the project earlier. Often tempted to inflict its own judgments in the priority setting, even to generating programs of its own choice, the Bureau sometimes sacrifices its expertise and becomes inevitably a partner in a political decision. When a budget examiner hints what proposals would get through his gate, he ends up steering the agency that he was only supposed to examine. Agency chiefs are constantly alert to such cues, even at the expense of a pet proposal; for once the money is in hand they can bootleg their original scheme or, if not, simply bask in the glow of increased funding. When budget examiners substitute their organizational position for knowledge and assert their judgment more in playing god than on the basis of fact, the President loses the intended expert capability.

When the Bureau first encountered a higher octane of science and technology in the aftermath of Sputnik, it became clear that the three or so technically trained examiners out of a staff of 450 would be easily out-gunned by agency specialists. So when the Office of the Special Assistant for Science and Technology was created, with technically trained staff amplified by both the intellect and prestige of PSAC members, the Bureau welcomed this team as allies in the budget review, especially to balance the propaganda of agency salesmen and biases triggered by dreams of power. The agencies quickly recognized the caliber of the opposition and were obliged to respond by sharper proposals. The bureaucracy came to consider the office more foe than friend. Wiesner, for example, became characterized as a "czar" by outside critics like Philip H. Abelson[42] and insiders like James E. Webb. When preferences of PSAC entered the examining process too strenuously and the scientific community first privately, then publicly regarded the Special Assistant as "their man" in the White House,[43] the Budget Bureau became ambivalent in its attitude toward the science advisor's office. And when competition for the President's ear became intensified as the science advisor squared off in defense of favored programs against the Director of the Bureau of the Budget on equal terms before the President, the institutional tension became chronically inflamed.

Soon after my appointment to Wiesner's staff, it became apparent that the single injection of budgetary hormones to oceanography in the Kennedy add-on had not built commitment to a goal or a plan. Neither were projects being buoyed along to higher levels by impending catastrophe nor pushed from below by agitation of an effective interest group. NASCO was silenced, or at least silent. And the fancy of the citizen was only lightly tickled by the entertainment value of turquoise waters and playful porpoises. No public support was evident. With oceanography lacking horsepower to compete in the power struggle of the bureaucracy, but with Wiesner's commitment to Congress, the only strategy left was to employ the White House level Federal Council and the Interagency Committee on Oceanography as our chosen instruments to rationalize the budget. While it was the White House staff that moved marine sciences to higher ground early in 1961, the problem was how to strengthen the ICO as a more effective instrument to earn continued presidential support.

My apprenticeship first required learning about the budget rhythm —the various steps involved and their timing. The second lesson concerned the dynamics of budget justification. And the third concerned the tactics of persuasion, especially in the heat of final action when the tradeoffs and compromises are shaped to fit beneath whatever ceiling the President imposes.

ADDING WARP TO ORGANIZATIONAL WOOF—THE INTERAGENCY COMMITTEE ON OCEANOGRAPHY

All committees of the Federal Council faced an enormous challenge of coordination: to establish a coherent set of government-wide goals; to inventory government-wide plans; to identify deficiencies in the aggregate of individual programs; to rationalize the budget increases warranted to meet government-wide objectives against those set for one agency at a time; to identify agency responsibilities, considering that each budget element would have to be assigned to some existing operating component. Any government-wide program having survived these rigors must then be sold up the line from the Federal Council Secretariat, with the components in each agency simultaneously going up their normal budget ladder to their cabinet officer.

Ideally, the Bureau of the Budget would receive two compatible budget submissions—the aggregate of individual agency components via the cabinet officers, and the collective government-wide total from the President's science advisor, prepared by ICO and endorsed by the FCST.

A neat trick if it could be turned!

Early in 1961, the ICO pulse began to quicken. The budgetary infusion gave it new éclat. Wiesner gave Wakelin explicit sailing orders that structured the ICO responsibility for the first time. The annual inventory of research was steadily improved. But to achieve budget success, the ICO would have to be a program generator as well as coordinator par excellence. And it would need guidance and a firm assist from the Council Secretariat because its parent Federal Council would serve as reviewing agent, both validating internal balance and content and assessing external priorities when comparing oceanography with all the other competing programs. With Council endorsement of the total, each Council member was expected to advance his agency's component aggressively. In turn, the Council's chairman was expected to parallel the normal budget process in individual agencies by defending the Council's position to the Budget Director and, if necessary, to the President himself.[44]

This unorthodox process could not be carried out without active cooperation by the Bureau. During the first several years, such broad-gauged, experienced staff as Enoch Leroy Dillon and Hugh Loweth were given unusual dual roles as examiners for both government-wide and single-agency programs. They were critical about the unwieldy mechanism and the backdoor suction on the treasury. They were also constructive, apparently eager to observe how well the ICO process worked because similar budget disciplines might be applicable to many new families of scientifically based programs crossing agency lines.[45]

What worried Wiesner and budget officers alike was that inputs to this process that might earn a high-powered endorsement from the Federal Council were largely based on submissions to the ICO by representatives of the very agencies which would be recipients of funds requested. To counter a long history of mutual back-scratching and trading in interagency committees, the Council Secretariat recruited a panel of experts outside of government who poked and prodded with questions to detect soft spots in agency proposals and to help the ICO to strengthen (or discard) the vulnerable ones. After its submission to ICO for action, this critique was included in the package that went before the FCST.

This budget exercise became the primary task of but also the primary impetus to ICO. Members now felt they had a legitimate backdoor to an increase in budgets—a vital incentive, since collaboration was seldom rewarded at the individual agencies. Eventually, members gained some satisfaction when the ICO reports became internationally known. This intricate maze of budget process, however, should have been a clear indication of the need for consolidation of these programs into a smaller number of agencies—a proposal I made to Wiesner in 1963. But 1963 was not the time for a new oceanographic agency; that step was another seven years off.

In 1962, another development occurred regarding the President's science advisory office, that was to affect the interagency management of oceanography. By Reorganization Plan #2, President Kennedy established and Congress accepted an Office of Science and Technology (OST) in the Executive Office of the President.[46] This legislative device, incidentally, permitted the President to separate and regroup existing agencies and missions into new organizational entities better suited to changing conditions.[47] In this instance, like Eve out of Adams's rib, OST had been formed out of unused policy planning and coordination authority belonging to the National Science Foundation. This action gave statutory legitimacy to advisory machinery that had previously existed as three presidential inventions. In fact, it replaced none of them, and with the presidential appointment of the OST director, the incumbent proudly wore four rather than just three hats. OST thus housed the staff team for the FCST Secretariat and opened the way for congressional communication and requests for advice in science and technology across government that had previously been denied them when the apparatus was shielded by a White House imprimatur.

In 1962–64, ICO continued to gain skill and momentum as a co-ordinating device. Its agenda focused on ways and means of strengthening the tools of research—providing greater ship capabilities, more space in shore-based laboratories, and research funds to support an expanding enrollment of students and their faculty. While this machinery elicited funds to build research capabilities, I became increasingly concerned about resistance at the very top. The President and the Congress were asking questions about the relevance of oceanographic research to national goals, and the four-year honeymoon of Congress with science was about to end.

Late in 1963 the House of Representatives went into full cry in a sweeping investigation of all government research.[48] By good fortune, the potential of a witchhunt was defused by selection of a wise and perceptive chairman, Carl Elliott of Alabama, for a new Select Committee on Government Research. Wiesner and I were invited to testify on science policy and on coordination, respectively, the first hearings on such issues ever convened. The ICO was trotted out as a showcase.[49] Elliott's committee published eleven comprehensive reports, including one on coordination, noting ICO's progress but also questioning its limitations.[50]

Appointment of the Select Committee aroused jurisdictional instincts in the Science and Astronautics Committee, which in concentrating on the space program had neglected science. Chairman Miller promptly appointed a Subcommittee on Science, Research, and Development that would later earn its mark under Congressman Emilio Q. Daddario of Connecticut. The House Government Operations and Armed Services Committees and the Joint Committee on Atomic Energy took similar defensive measures by creating science subcommittees. Collectively, Congress was expressing a concern that the scientific establishment, supported by public funds largely on faith, was without adequate political control. The direct access of science to the ear of the chief executive was a symbol of political status, yet Congress had a growing skepticism as to what this arrangement provided for the health of the nation as well as the health of science.

To assist Wiesner in meeting these questions, I recommended that OST respond substantively by publishing an annual OST report. While not fully concurring, he endorsed the preparation of an annual report of the FCST that we were able to release late in 1963.

As to oceanography, the FCST Secretariat now urged ICO to turn its attention from the build-up of oceanographic research resources to the more basic question: Why do we need the information? The summary of the 1960 congressional report had stated, "A major question which through pending legislation is on the agenda of the 86th Congress concerns *how* this Nation can best mobilize for a concerted, long-range program of research and exploration of and in the sea." Now, I was proposing a major executive branch inquiry as to "*why?*" This question unexpectedly opened up a buzz saw of anxiety in the outside oceanographic community and a spinning of wheels in agency circles. As many observers had pointed out, the scientific community had steadfastly argued in accepting research funds that the generation of new knowledge should not be bound by concerns for its applicability.[51] Much research actually does resist labeling, and whether it is basic or applied depends upon the eye of the beholder. (Sponsors of research when justifying their budgets may list as "applied" activity the same projects that a university would extol as "basic" in citing the academic excellence of its institutions.) Semantics aside, the actual dichotomy broke into open conflict between the academic oceanographers and their government sponsors. Each had a point. Sponsors could very well impose exorbitant demands on the research team for quick and dirty results, thus squeezing out the undirected but often more potent, long-range fundamental questions. On the other hand, spurred only by curiosity the researcher could go in hot pursuit of a bevy of new questions without ever satisfying the sponsor's scientific needs. Additionally, government and academia were divided by mutual disrespect as well as mutual misunderstanding. Many academics on NASCO regarded science as pure and appplications impure, and treated with disdain those with less visible credentials of academic accomplishment. Government personnel responded to the tensions with a smug knowledge that they controlled the funds and that few scientists had the capacity to negotiate bureaucratic obstacle courses. What was overlooked, of course, was that different kinds of intelligence were at work. Each group reluctantly realized it needed the other, and with clenched teeth the ICO and NASCO sublimated their hostilities.

Faced with that mood, how was the FCST to introduce the notion regarding the practical value of oceanographic research? The opportunity came unexpectedly. After the ICO met congressional complaints about fragmentation by submitting its Federal Council-endorsed, government-wide budget proposals, it encountered a new question. Congress asked how it could evaluate the merits of a government-wide plan, no matter how attractive its internal logic, unless it could be tested against long-range goals and a long-range implementation program. Lacking such a test, the annual review could be innocently heading into unprofitable lower priority areas, leaving more important objectives unattended. Moreover, single-year commitments of funds often had built-in future commitments that advocates would be unlikely to volunteer unless required to show their full hand. And finally, with the rationale of support still echoing the NASCO and Kennedy theme of building up a research capability, and against the backdrop of congressional concern that led to the Elliott Select Committee, Congress was determined to know more about what the research was for.

Congress was asking for a long-range plan, and until then, none had existed. Not that the question hadn't been raised. For years the Bureau of the Budget, with ample sour experience to sustain their ever-present suspicion of the agencies, had required that annual submissions be supplemented with future projections to illuminate hidden commitments. But with rare exception these projections were linear extrapolations of past activity, seldom if ever substantiated with a plan, much less with long-range goals.

Concurring privately with the congressional logic, becoming every day more disturbed by the absence of agency plans, and discerning ICO's inability to articulate something that was more than the sum of the parts, the Council Secretariat persuaded Wiesner to undertake two simultaneous exercises. The first was a generalized inquiry as to planning methodology for science and technology; the second was development of a long-range plan for marine sciences.

To undertake this first enterprise, a Long-Range Planning Committee was established in the FCST, composed of the most senior planners of technical programs in each agency. The NSF provided technical support in formulating a questionnaire to elicit planning information from agencies, then in packaging individual submissions in an effort to make government-wide sense. PSAC was skeptical; so was Wiesner. They were right, but for the wrong reasons. To the scientists, long-range planning was an anathema in raising another specter of unwarranted governmental meddling. But what we discovered was that the government was then incapable of interfering even if it wanted to. Only two major agencies produced anything resembling a plan for their programs; the others never could muster the foresight to look ahead to future requirements and resources needed for their achievement. The confusion of the agencies was intensified by the NSF's lack of practicality in fashioning a questionnaire that sought statistical answers to questions not yet carefully propounded, and expecting of those answers an impossible precision. After 18 months, the committee was abandoned. The exercise, however, had two durable products. Out of the projections came the first panoramic view of what was going on for the year 1962. The focus was not only on "subjects" of science, as had been the NSF custom in its annual cataloguing of federally supported research and development, it was also an inventory of activities categorized by "objects" of science. Secondly, the dollar figures were accompanied by a second dimension, specialized manpower, which at that time threatened to be in shorter supply than funds.

The second enterprise was formulation of long-range government-wide goals and plans in oceanography, for the first time inquiring deeply as to objectives for research and the resources needed for their achievement. The ICO was asked to draft such a report. Noble was the effort, but the twelfth draft still looked like patchwork. The seams between individual agency pieces were highly visible and the jurisdictional disputes unresolved. But more serious than that, as the purposes of the research were discussed, the paths chosen years before appeared to be blindly followed, with only faint refreshment as to purpose.

Desperate to lay a planning report before Congress while the FY 1964 budget was in the mill, the FCST Secretary called in a consultant, Douglas L. Brooks, who, as president of Travelers Research Center, had published a provocative booklet: "An Outsider Looks at Oceanography," to join in redrafting the report. The product of collaboration, "Oceanography, the Ten Years Ahead . . . A Long Range National Oceanographic Plan 1963–72," was the beginning of a renewed dedication of the nation to use as well as to study the sea.[52] The report claimed to be "neither a rigid blue print to be followed slavishly, nor a single master document. Rather, it is a restatement of national objectives that depend on oceanography, an assignment of relative priorities expressed in terms of levels of activity associated with these different goals, a projection of the growth necessary to achieve these goals, expressed in terms of required research resources—funds, manpower and facilities." The report went on to preach a gospel of national goals in oceanography: "To comprehend the world ocean, its boundaries, its properties, and its processes, and to exploit this comprehension in the public interest, in enhancement of our security, our culture, international posture, and our economic growth."[53]

The study argued that the level of funding by 1972 should grow to $350 million annually. Innocently, it unleashed the fire of the Budget Bureau. They did not object to its logic; in fact, it may have been even too compelling. They wanted to bury it because, in their view, it committed the President to future expenditures, and they considered one of their tasks was to leave open all the presidential options, including the unwitting creation of chaos from turning the water on and off. But in 1963 the FCST-ICO long-range plan was cleared. Unprecedented in its statement of mission-related goals for the government as a whole, it was transmitted to Congress by President Kennedy through Wiesner, and was followed by the budgetary proposals for FY 1964. The first budget breakdown to identify funds by national goal, FY 1965, was also the first Federal Council annual report on marine sciences to be transmitted to Congress by the President,[54] a pronounced escalation of Federal Council visibility.

In April of that same year, a frightful event brought the oceans into the headlines. The submarine *Thresher* was lost off Cape Cod during a trial dive after overhaul, and the stark tragedy in loss of crew when intense pressure of the sea crushed the hull was compounded by the frustration of months of fruitless search for the wreckage. This was a clear command to accelerate study, not only of the sea, but also of technology for its mastery. Repercussions of that disaster were still ringing eight years later.

CONFLICTS BETWEEN CONGRESS AND THE PRESIDENT—ROUND 2

From early 1961, Congress had kept an eye on evolutions in the executive branch, but their quest for a stronger marine science program was never fully satisfied by the ICO mechanism with its limited authority. By late 1964 that apprehension erupted publicly. First, Congress doubted that the efforts at interagency coordination were overcoming the old bugbear of duplication. Here the Chief Executive was held guilty until proved innocent, and plurality of sponsorship made such proof patently impossible. Second, the Soviet blanket of secrecy over their research caused great anxiety that their excellence in space technology might be matched in all disciplines. But primarily, and earnestly, Congress sought leadership in marine sciences, and that elusive quality was missing.

In the space program, the imprimatur of power had been bestowed by statute on the administrator of NASA. Skillful, articulate James Webb, appointed to that difficult post by President Kennedy, had been active in government as Under Secretary of State, then as director of the Bureau of the Budget under President Truman. As an officer in a petroleum firm he had come to appreciate the substance as well as the role of engineering in industrial management. He was bright, aggressive, and he comprehended the bureaucratic geography. Oceanography had no counterpart.

The major contact Congress had with the oceanographic program was through testimony of James Wakelin in his role as ICO Chairman. Each year, hearings were convened to review the annual reports prepared under FCST auspices. These were in effect an accounting of all agency transactions and should have given Congress assurance that every player in the game was working with the other players to keep duplication to a minimum. Wakelin was a cooperative witness, but he was also frank in saying that he exercised no real authority over the member agencies in ICO. Nevertheless, the fact that a government-wide package had earned so much attention all the way up the ladder through FCST, OST, the BOB, and the White House was rather exceptional. ICO was beginning to gain national recognition with its reports. Bob Abel was devoting enormous energies to giving speeches on the glorious future of oceanography. Some members of Congress were convinced that all was under control; but on the matter of ICO's political horsepower to deal with the centrifugal forces of interagency life, most remained doubtful.

By coincidence, Congress had recently dealt with the emulsion of science and politics, but in a broader context that led to the creation of OST. In unprecedented hearings on policy-making machinery, and with perceptive analysis by staff, Senator Henry Jackson in 1960 had drawn a careful bead on certain weaknesses at the very pinnacle of the bureaucracy in need of shoring up to meet the complexities and pace of issues that in the main were external and being generated by the cold war. With regard to science advice, stronger machinery was urged to meet four objectives: to intensify and strengthen forward planning for science; to improve coordination and quality of agency programs; to increase the number and depth of full-time counselors; and to ease strains in executive-legislative relationships in science.[55]

For science as a whole, political authority of the President was considered to need help of new science apparatus. On June 9, 1962, OST was born. Congress then attempted to overcome its frustration in centrally positioning government-wide leadership in oceanography by employing the OST vehicle with a reverse twist—to endow existing technical competences (the FCST/ICO) with political clout. On July 18, Congressman Dingell zeroed in on the fledgling organization as the logical centrally located, potentially powerful, and organizationally neutral body to carry the statutory responsibility for coordination. H.R. 12601, the Oceanographic Act of 1962, was introduced to develop and maintain a coordinated, comprehensive, and long-range research program. It authorized a national program in oceanography through a newly established statutory post of assistant director of OST (for marine science). The bill called for an annual report on government-wide programs and budgets and, to aid OST, set up a statutory advisory committee of university and industrial scientists.

On August 14, OST released its first report, setting forth the FCST/ICO catalogue of oceanographic activities projected for FY 1963, claiming that it represented "not only a consolidation of mission-oriented agency plans, but also a cooperative blending of independent efforts toward achievement of common goals." The program further reported an increase in presidential requests for funding, up a substantial 25 percent from the prior year; then it went on with a note of pride to defuse any budget cutter provoked by such an increase by assurances that "efficient use of resources, by investing them in important programs and by avoiding waste and duplication is a primary ICO aim." Of the

$126 million requested, the Congress later appropriated $124, not a bad batting average. This support, incidentally, could not be attributed to the persuasiveness of the OST report, because the August release was too late to influence appropriations significantly.

The report was a little late in mollifying Dingell. For on August 14, the House Merchant Marine and Fisheries Committee reported the Dingell bill out favorably and on August 20 the House passed the measure. In a bid for Senate abandonment of a completely different bill, the House engaged in some operational sleight-of-hand. It put aside H.R. 12601 and invented its own version of S. 901 (discussed previously), superseding Senate language by the H.R. 12601 text. The new bill was passed on September 27 by both Houses prior to adjournment of the 87th Congress.

Then it was pocket-vetoed by President Kennedy.

This action was bitterly disappointing to Congress but not totally unexpected. While no public statement clarified the basis for rejection, two major objections could subsequently be pieced together from other fragments of testimony. First, to give OST supervision over other agencies violated the principle that no agency should be interposed in line of authority between department heads and the President. Second, the bill provided for a special staff position in OST for one field of science, and this could lead to a proliferation of such positions in OST for many special fields. There was a third reason, however; such an organizational ploy would have given oceanography a program priority, and the administration was not so disposed. Within the executive office, I was the only supporter of the legislation, in the genuine belief that some centralized coordinating authority was necessary and also in some apprehension that a veto would tarnish the favorable image being established that the White House cared about oceanography. In fact, the veto got very little play in the press except for protests from the oceanographic trade organs.

Rebuffed, stubborn members of the House opened fire on January 9, 1963, two days after the new 88th Congress convened, by reintroducing the vetoed measure. Dingell, however, angry and frustrated from his defeat, chose to work in another sector and the Oceanography Subcommittee got a new chairman: Alton A. Lennon of Wilmington, North Carolina. Lennon was a former judge, with quiet style and a preference for negotiation and compromise in contrast to his hard-driving, mercurial predecessor. Early in his new role he called Wiesner to come over for a chat about how to put Humpty Dumpty back together again. Wiesner took me along, and it was my lot to defend the Kennedy veto that I had privately opposed. In that session, language was forged that would meet the original congressional objectives and at the same time meet White House objections, primarily by giving to the President, with

OST assistance (rather than to OST directly), all responsibilities assigned in the earlier bill to OST.

On June 12, Lennon introduced the revised bill after taking the precaution of extracting from Wiesner a written commitment of support. Action in the House was not long in coming; H.R. 6997 was reported out favorably by the Merchant Marine and Fisheries Committee on July 31 and passed by the House on August 5. There it wilted and died. And Congressman Bob Wilson's innovative proposal to consolidate rather than coordinate the federal herd through a National Oceanographic Agency went unnoticed.[56]

No action was taken by the Senate. Magnuson had been busy with other Commerce Committee business and moreover, traditional jealousies between the two Houses had been inflamed. Although the Senate had passed the revised S. 901 that Kennedy later vetoed, they never completely forgave the House for junking its original language without a trace. Thus in the 88th Congress no one in the Senate had introduced the counterpart of H.R. 6997. The scientific constituency was silent; all the punch of NASCO had evaporated; and in a succession of new chairmen none had ever been able to emulate Brown. By then, too, quarrels among scientists over the Mohole project left dead bodies over the scientific landscape,[57] and because leaks by dissidents from another NAS committee were partly responsible for the Mohole debacle, NAS leadership prohibited direct contact with congressional committees and staff.

The only hope in the Senate seemed to be Magnuson's staff specialist on oceanography, Dan Markel. A political reporter who had attracted Magnuson's attention in Seattle, Markel joined the Senator's personal staff in 1958 and was assigned oversight of oceanography in 1960 when it became hot. Perceptive about the workings of politics, of oceanographic scientists, and of his boss, Markel was nevertheless reluctant and not in the best position to start the ball rolling. Frankly needling him from my post in OST over Senate apathy, I was convinced that some legislative adrenalin was essential to reenergize the marine field to a better administered, more durable activity than the executive branch was then able to offer within constraints of structure and leadership.

A number of events had contributed to the aura of pessimism. A President staunchly interested in the seas had been assassinated, and his successor's priorities, personal interests, and associations did not share a salt-water content. Wiesner, who backed the ICO, had resigned, shaken by the tragedy and by exhaustion from almost three years of intense involvement. His successor, Donald F. Hornig, brought a different point of view to the office. Among other factors, Hornig was given a private lecture by PSAC to the effect that his predecessor had drifted away from concerns of the scientific community into the bureaucracy. The needs of science should be examined and supported more vigorously; but in this context oceanography was not high on the list. PSAC was still reluctant to commit itself to the field, and Hornig, as a result of his personal wartime assignments at a major oceanographic institution, admitted to a disappointment in performance in this area.

Then, too, the budget climate became increasingly resistant to growth in research and development, which had been climbing at a phenomenal rate of 30 percent annually. In the executive as well as the legislative branch, it was clear that stronger justification was needed in terms of direct benefits to the nation, not simply benefits to science. This attitude, incidentally, further reflected the coolness of the new President, Lyndon B. Johnson, toward science. No longer was the presidency receptive to arguments that a knowledge-based society needed automatic buttressing of that base.

Further growth of funding would require a far sharper analysis of the importance of the oceans to the national interest than had been marshaled in the FCST long-range plan. The FCST Secretariat endeavored to signal this change in climate to ICO as well as outside interests. By then congressional support for appropriations was faltering. Of the $156 million requested for FY 1964, only $124 was appropriated, essentially level with the funding for prior years. Among other factors, the ICO report had reached the Congress too late to add ammunition to budget defense by the participating agencies, and neither House nor Senate legislative committees concerned with marine affairs had interceded in the cloakroom with their economy-minded colleagues on appropriations committees.

For FY 1965, the President requested only $138 million, not even as much as had been requested in 1964, but the Congress came through with $135 million. By that time, the ICO was unable to rationalize further growth. Despite its striking record as a body for information exchange and program coordination, it could not seem to raise its sights to consider either policy needs or plausible ways to implement the long-range goals set out the year before. By spring 1964, Wakelin's briefing to the Federal Council was unconvincing; and in contrast to support rendered in three earlier years, they agreed to endorse only $2 million out of his proposed $50 million growth. But Wakelin was distracted with other problems on TFX and trimmed-down Navy budgets, and, disappointed at lack of White House rapport, he was on the verge of leaving government.

In August 1964, I had accepted a long-standing bid by the Legislative Reference Service (LRS) to return and assemble the stronger science advisory apparatus that Congress felt it needed to avoid being continuously outmanned and outgunned by executive branch experts. Charles V. Kidd, the new FCST Secretary whom I had strongly recommended, had intimate knowledge of governmental processes that well qualified him for the Council functions, but he had no interest in marine affairs. Hornig then recruited oceanographic scientists to supply that need. But by early 1965 marine advocacy lost steam within OST, partly because the technical staff watching over the field did not fully engage the bureaucratic machine, and partly because the new chairman of the FCST, its secretary, and the ICO chairman had other priorities. Unprotected by OST, ICO went into sharp decline.

In a full cycle, it appeared that further advancement of the enterprise now lay in Congress, and its pilot would have to be Senator Magnuson. The Lennon bill, H.R. 6997, that had passed the House, then languished in the Senate. Markel sought another route, that, apart from intrinsic merits, might be attractive to his boss because it differed from the House approach and conceivably might earn support from the President because it paralleled the Space Council that Johnson had personally activated.

On July 9, 1964, Magnuson introduced S. 2990, the Act of 1964 to establish a cabinet-level National Oceanographic Council for policy planning and coordination, chaired by the Vice President. This would eventually become the National Council on Marine Resources and Engineering Development. Differing in many respects from Miller's 1961 proposal, the legislation reflected growing skill in combining a statement of national purpose with the machinery to implement it.

Now both Houses had proposed springboards for action: central coordination in the Executive Office of the President. With the life of the 88th Congress ebbing fast in a presidential election year, neither measure passed.

Dean Don Price's diagnosis of this era, necessarily based on available public evidence, has identified a set of three strategies cleverly designed by a coalition between scientists and Congress both to inflate the oceanographic effort and to tighten the congressional halter on the President.[58] The sequence moved from the 1960 Magnuson proposal to give NSF lead authority, to the Miller proposal for a Council, to the Dingell assignment of leadership to OST. While admitting that the strategies were not "the product of any very systematic advance planning or of any explicit constitutional theory," Price considered that the proposals attempted to satisfy special interest programs and to apply congressional leverage on presidential priorities. If the ICO were made a statutory power, he foresaw "truly terrifying" consequences in terms of a mix-up of operating agencies, independent advisory bodies, presidential powers, congressional ambitions and the scientific estate. From the inside, however, that ostensibly conspiratorial process was far more random, accidental, and opportunistic.

The 89th Congress opened with a louder beating of oceanographic drums. On January 4, 1965, Congressman Bob Wilson lost no time in reintroducing his consolidation proposal.[59] Lennon reintroduced his bill on January 11 to give the President coordinational leadership with OST assistance, followed by identical bills by several committee colleagues.[60] Hearings were reconvened with good attendance and participation, notably by Republicans perhaps eager to make this a partisan issue over failures of a Democratic incumbent. Lennon, however, soothed any itch for partisan advantage by consulting his senior minority member Charles A. Mosher from Oberlin, Ohio, on his entire strategy, thus insuring bipartisan support. Mosher had been a newspaper publisher, the Republican's Vietnam dove. Wise in science affairs from years as Daddario's partner, perceptive in political process and constructive in viewpoint, Mosher joined Lennon in a productive union to push the Executive.

Magnuson reintroduced S. 944, an expanded version of his Council bill, on February 1, 1965. Senator E. L. "Bob" Bartlett of Alaska introduced a measure to stimulate marine exploration and development of resources of the continental shelf with new emphasis on marine technology,[61] which he credited to a speech I gave on concepts of ocean engineering. His program would be planned and conducted by a new federal agency with an AEC-commission-type management. Congressman Thomas L. Ashley proposed a marriage of the Lennon and the Magnuson bills by locating the Council in OST.[62] Before long, identical bills popped up in both Houses. Simultaneously with the concentration of bills on the federal management issue, other measures were introduced: to exempt oceanographic research vessels from inspection laws intended for passenger ships; to survey fishery resources of the United States; to initiate a federal-state cooperative program for conservation and enhancement of anadromous fisheries; to prohibit fishing by foreign vessels in U.S. territorial waters; and to conduct a study of legal problems of management, use, and control of the national resources of the ocean and ocean beds.

This latter responsibility Lennon proposed be assigned to the Coast Guard rather than, as suggested initially by Congressman Richard T. Hanna, to NSF.[63] His rationale was that the bill would thus be referred to the House Merchant Marine and Fisheries Committee rather than to Science and Astronautics.

What signaled a new cycle of congressional intent was the convening of the Senate hearings early in the session—February 19—after more than two years of silence. The major focus of attention was S. 944, to establish a council. It was now clear that Magnuson was serious. Congress watchers had come to recognize the Magnuson style—to hover in the background as an uncertain issue blossomed until he could determine the merits of the proposed legislation and the alignment and strengths of allies and opponents. Then action. Moreover, to conserve his own energies that could readily be dissipated on a wide range of issues within his committee's purview, he would ordinarily agree only to hearings in which he felt some final target of import could be carried to a successful resolution.

With three days of hearings, Magnuson had collected sufficient testimony to make his case. Carefully selected advocates recited the unfilled promise of the sea, and government witnesses stumbled sufficiently on questions of coordination to verify the unevenness and uncertainty of top-level authority. Hornig's testimony was particularly damning when he was obliged to admit that the Federal Council had recently chosen not to endorse feeble ICO proposals.[64] The curtain was inadvertently lifted on the growing confusion and lack of direction in marine affairs that had been claimed by ICO.

But Magnuson did not bring his bill out of committee. He was awaiting House action to see which direction they might move, planning to consider provisions that might gain their support for his proposed legislation.

That signal came on June 15 when Congressman Paul G. Rogers, one of the more interested and articulate members of the House Subcommittee on Oceanography, introduced H.R. 9064 to establish a presidentially appointed National Commission on Oceanography. The gist of that proposal had been floated just a few days earlier at a national meeting in Washington of the newly formed Marine Technology Society (MTS) by its incoming president, James Wakelin. Wakelin had left government just a few months before. Once outside and still wanting to gain support for oceanography, especially for industrial interests who were angry that increases promised by the Navy in the aftermath of *Thresher* were not materializing, Wakelin tried a back door. Mobilizing support from a dozen industrial executives under the banner of the National Security Industrial Association, he drafted a petition to the President urging a stronger marine effort and offering NSIA services to undertake a major study on requirements in marine affairs that could lay the groundwork for next steps. The petition went unanswered. After failing to gain an audience either with Hornig, who thought Wakelin's move a personal criticism of OST, or with White House staff, Wakelin received a curt *pro forma* note of "thanks, but no thanks," from a junior, unknown staffer.

Rebuffed, Wakelin caucused with several colleagues in and out of the Marine Technology Society as to next steps, refusing, incidentally, to lend his support to the Magnuson proposal because he felt it would have rough sailing in Executive Office seas. His alternative was to advocate further study under a presidentially appointed Commission that might give birth to a stronger marine technology program and perhaps to some new oceanographic agency. Almost for the first time, the pressure outside of government was mounting. Strangely, leaders of the charge through MTS were two stalwarts who had seen their previous efforts to advance oceanography fail when they were insiders—Wakelin was one. The other was Rear Admiral (Ret.) Edward C. Stephan, who had headed the Deep Submergence Systems Review Group, appointed to recommend measures to meet the technological deficiency illuminated in the *Thresher* search, and whose proposals in 1965 for a sharply expanded naval technology for submarine search, rescue and salvage were brutally sliced in the Navy.

Hornig, meanwhile, had sensed the coolness of congressional response to his defense of executive branch custody of oceanography, and undertook his own study of the problem through PSAC. Gordon J. Mac-Donald was appointed chairman of a panel to examine program and organizational needs with the help of a distinguished group of scientists. Staff assistance was provided by OST through Henry W. Menard, on leave from Scripps, and by the Navy. Their product was intended to head off creation of Magnuson's proposal for the Council. Arriving in July 1966, it was too late.[65] While PSAC was at work, however, the Senate Council bill and House Commission bill were on collision course, so the long standoff between Congress and the President was reinforced by a conflict between the two legislative houses.

MARINE SCIENCES MANDATES OF 1966

The Senate bill to create a Council reported out of committee was substantially different from the one going in. First, its scope was expanded beyond the largely scientific orientation of its precursor. The title itself accentuated a broader scope, having been changed from the "National Oceanographic Act" to the "Marine Resources and Engineering Development Act," and the evolution was further reflected in its declaration of policy and purpose. Apart from its more encompassing theme, the fundamental planks were assembled from legislative lumber of earlier aborted proposals. The statement of national purpose extracted language from the statement of policy of the space act, then from proposals of the Senate, later refined by the House bill that Kennedy vetoed. These were embellished subsequently with bits and pieces from minor bills that also failed. Some of this broadened concept that emphasized the engineering and technology content—a feature that was injected into the title itself—came from other legislation such as the Bartlett bill, some from LRS staff studies. Provisions for machinery were taken from the Space Council, OST legislation, and precursors in House bills.

What these various steps reveal most is a shift in justification from a knee-jerk response to a Soviet threat to a broadened rationale for civilian uses of the sea. Emphasis in subject matter changed from oceanography, marine sciences, or aquatic science, to marine resources, technology, and the marine environment.

And while the scope was broadened, the emphasis on policy was escalated. The President of the United States was made bandmaster. The Vice President was charged with writing the music. And existing agencies of government were expected to play on their old instruments, but in harmony, a new tune—hopefully *fortissimo*.

The Senate passed the bill on August 5 and sent it to the House for action. In the meanwhile, several members had introduced versions of the Senate Council bill in the House,[66] and with a favorable report out by the House Committee on September 16, 1965, of their version of S. 944, expectations were high of early enactment. But the House and Senate versions disagreed on fundamental principles, and the growl of an ugly impasse began to emerge from behind closed committee doors. The conflict arose from the predominant emphasis put on the Council by the Senate and on the Commission by the House. The Senate felt that the issue had been studied to death, that the time had come for action placing the responsibility for a coherent, coordinated, aggressive program on the one man charged with responsibility over all the operating departments—the President. And to be assured of action, the Senate wanted to create strong apparatus to advise and assist the President in carrying out the provisions of the act assigned to him. In the Senate's view, the Commission would serve as advisory body to the Council, injecting interests, viewpoints, and expertise from the entire marine community—industry, academia, and government.

The House, on the other hand, was skeptical of the Council—partly because its Space Council model had such a checkered record of performance. In fact, the Space Council provided for in the 1958 Space Act to be chaired by the President remained unused until a space-oriented Vice President, Lyndon B. Johnson, urged his boss to activate it and alter its provision so as to name the Vice President as chairman. The House also took note of the quiet lobby against the Council bill mounted by the Executive Office forces—the Bureau of the Budget and OST. According to these presidential advisors, there was a high likelihood of presidential veto, and the House, still recalling the setback of 1963, did not want to face a second crushing failure they thought would be incited by strong emphasis on the Council. They thus built into their proposal a parity in function between Council and Commission that was potentially to plague operation by posing a case of apparent duplication; in fact, the House never expected to see the Council energized. Both sides stubbornly refused to compromise.

During this time, lobbying for the Commission became even more furious. Officers of the MTS began to circulate on Capitol Hill carrying rumors of oceanography's doom if the Council bill were enacted. They were reinforced by a new entrant—the Navy. Operating agencies have a long congenital dislike for the Executive Office of the President, inflamed annually during the Bureau of the Budget review. But there was also a history of kibitzing from OST that left scars. Thus none of the agencies was wildly enthusiastic about a Council, notwithstanding its limited powers as advisory to the President; but most strenuous in its concerns was the Navy. A coalition pushing for the Commission of Wakelin, Stephan (then released from his assignment with DSSRG), and the naval aide to Wakelin's successor, Captain Edward Snyder, was joined informally by Hornig and by Harold Seidman from the Bureau. Many outsiders who learned of this impasse thought it best to stay out of what they considered to be a congressional dogfight.

So the Senate stand on the Council was a lonely stronghold. During this time, I was asked through LRS to examine alternative after alternative for both the House and Senate committees in relation to the problems that choked oceanography's progress. Each analysis came to the same conclusion: executive indifference would not be corrected by more study nor by maintenance of the status quo. The Congress felt that the prestige and influence of OST were declining; the performance of ICO that had been trotted out in 1963 as a showcase of coordination and which led the Elliott Select Committee to call attention to its role as a model proved ephemeral. Markel held his ground; so did the committee counsel, Gerald Grinstein; and so did Senator Magnuson.

Conferees from both House and Senate met April 28, but recessed when agreement failed. On May 24 they finally found the route to compromise. The trigger that released the conferees was an agreement by Magnuson that the Council should be clearly temporary, or at least interim—set to expire, in fact, 120 days after the Commission delivered its final report. Apparently the Bureau of the Budget, impressed with the stubborn Senate position on the Council, communicated to House conferees that they would not press for veto if the Council were temporary and thus probably ineffectual. The House conferees felt satisfied. The final bill was reported out by both Houses and sent to the President on June 2. Signature or veto? No one knew. But quietly, on June 17, the President signed P.L. 89–454 into law.

As I learned subsequently, what caused LBJ to reject initially unanimous recommendations by his own family to veto had little to do with budget and staff recommendations.

In the smoke of congressional battle over the Marine Sciences Act, a new issue was introduced with the Sea Grant proposal by Senator Claiborne Pell.[67] S. 2439 would authorize establishment and operation of the Sea Grant colleges and programs, initiating and supporting programs of education, training, and research in the marine sciences and a program of advisory services relating to activities in the marine sciences. The nucleus of this concept was first proposed by Athelstan Spilhaus in a speech at the University of Rhode Island, using an analogy to the land-grant concept that had a long and widely applauded history. One goal was to facilitate the use of government-controlled offshore lands by granting income from exploitation of these submerged resources to educational institutions for the type of research, education, and extension services that had characterized agricultural development. The pace of congressional interest accelerated; so did its breadth. Members of Congress of both parties, from California, North Carolina, Washington, Alaska, Florida, Massachusetts, Ohio, Pennsylvania, Texas, Virginia, Maine, Hawaii, Rhode Island, New York, Connecticut, and Maryland introduced or supported bills and otherwise were drawn to this legislative magnet. For the first time, more than a handful saw that their constituency had a stake in the Sea Grant outcome.

For a time Sea Grant was obscured. Bills identical to S. 2439 were introduced in the House, but all engendered difficulty. Since the original legislation would create a new program of federal grants, it required some federal office for administration. Because of the breadth of subject matter involved, none of the familiar operating agencies—Interior, Commerce, Navy—seemed an appropriate home. The National Science Foundation was the most neutral in its ecumenical approach to support of science and was pegged as the custodian in the Pell bill, which explicitly amended the National Science Foundation Act of 1950. Forgotten in the rush of legislative enthusiasm was the fact that both Senate and House bills to expand NSF authority would be handled by committees other than those overseeing marine affairs. In the Senate they would be handled by the Labor and Public Welfare Committee, with whom Commerce had worked out an agreement for joint jurisdiction; in the House, it was ironically by the Science and Astronautics Committee, the old rival of Merchant Marine and Fisheries. And sure enough, that is where the original Sea Grant bills went, much to the chagrin of their backers and allied enthusiasts.

Not only would the bills be at the mercy of rivals to marine committees, but if passed would add to their prestige and power in overseeing implementation of the Sea Grant program, including the fringe benefit of a geographically distributed pork barrel not open to other marine affairs legislation then in the mill. As an initial ploy to overcome that penalty, Representative Paul Rogers, on May 23, 1966, went to the extreme of attempting to amend the Merchant Marine Act of 1936, in order that his committee could assert jurisdiction over a lost prize. Passage of the Marine Sciences Act provided an unexpected but welcome solution.

At the July 13 commissioning of the ESSA research ship *Oceanographer*, President Johnson surprised the marine community by his first public act or statement of support of marine affairs—a landmark address emphasizing freedom of the seas that was to become the springboard for highly significant policy development some years later. He also cleared away uncertainties as to the fate of the Council, which had multiplied after rumors of veto and presidential silence at the signing of the bill. In his speech, he called on the Council to get going immediately, even to bring in initial recommendations by January. With news of Council activation, the oceanography committees discovered the necessary vehicle to retain jurisdiction. On July 26, Rogers introduced H.R. 16559 to amend the Marine Resources and Engineering Development Act giving administration of Sea Grant to NSF as originally intended but providing for policy override from the Council. This step had a second advantage. Congress had been worried about whether their Sea Grant creation would flourish in the rarified atmosphere of NSF. For one thing, NSF was oriented toward basic, rather than applied, research; it had an aversion to categorical programming, preferring instead a bland support for all without favoritism; and it lacked the flash and fire that Congress hoped would attend its creation. Congress found no great interest in the NSF at that time for the Sea Grant legislation, and all parties felt its residence there might be temporary, pending the outcome of the Commission study that everyone secretly hoped would beget a new operating agency. The Budget Bureau joined in drafting with Pell's staff member, Fitzhugh Green, a strategy to elicit action in the House and assure freedom from jurisdictional competitors.

On July 27, H.R. 16559 was reported out favorably and passed by the House on September 13. The next day the Senate acted, substituting language of S. 2439 as amended that differed from the House version. Again an impasse. This time, the arguments melted faster and both Houses agreed on a compromise that was transmitted October 5 to the President. It was signed October 15 (P.L. 89–688).

By then, the bill was vastly different from its original version. It had the objective of aiding development and exploitation of marine resources by federal support of (1) education of skilled manpower, including scientists, engineers, and technicians; (2) appropriate research to develop techniques, facilities, and equipment for resource exploitation; and (3) advisory services to disseminate findings. Differences between House and Senate largely centered on level of authorization, the House demanding sharp increases that the Senate thought unrealistic, and on the role of the Council in policy guidance. It is interesting to speculate what the House would have done with Sea Grant if the Council they opposed had not been created. Difficult as was passage of the legislation, its implementation was to prove equally viscous—regarding both program content and level of funding.

On October 15, President Johnson signed a second, far more powerful, measure that was to have unexpectedly significant ramifications for marine affairs years later. That time bomb was the creation of the Department of Transportation (P.L. 89–670), to which the Coast Guard was transferred from its parent of one hundred years, the Department of the Treasury.

Between 1960 and 1966 the increase in research funding and the enactment of new legislation provided two complementary sets of tools. The funding for ships, instruments, and research grants to train manpower had greatly strengthened the intellectual resource base for study of the sea. Twenty new ships were operating. The number of scientists doubled; the number of students tripled.

The new marine affairs legislation afforded a mandate to associate the seas with our national interests. What proved remarkable was that this dual accomplishment stemmed largely from initiative by the Congress.[68] In this enterprise, the leadership in both Senate and House committees had carefully cultivated bipartisan support.

The ammunition Congress fired had been generated initially by a small group of scientists; executive-branch troops and a lively concern among a few Executive Office generals kept the feeble enterprise from stagnation. While hopeful, none could foresee where the trajectory of action was to carry a new, still embryonic, endeavor.

With more powerful tools and determination, the nation was now poised to act.

CHAPTER 3

Implementing the Legislative Mandate

THE DECISION TO IMPLEMENT

In a new nation recoiling from remote domination by an autocratic monarch, founding fathers took the precaution of delimiting governmental action by the Constitution in such a way that new functions could be exercised only by overt legislation. To be sure, activist presidents periodically stretch that authority until they are reined in by a jealous Congress or a watchful Supreme Court. No such danger existed with marine affairs. Without advice, much less initiative, from the executive branch, and after six years of ferment, on June 2, 1966, twin but lonely epicenters in Senate and House had brought forth an unprecedented oceanic mandate. Terse and unadorned, it required no legal brokerage for interpretation.

Said the Marine Resources and Engineering Development Act of 1966:

SEC. 2. (a) It is hereby declared to be the policy of the United States to develop, encourage, and maintain a coordinated, comprehensive, and long-range national program in marine science for the benefit of mankind to assist in protection of health and property, enhancement of commerce, transportation, and national security, rehabilitation of our commercial fisheries, and increased utilization of these and other resources.

(b) The marine science activities of the United States should be conducted so as to contribute to the following objectives:

(1) The accelerated development of the resources of the marine environment.

(2) The expansion of human knowledge of the marine environment.

(3) The encouragement of private investment enterprise in exploration, technological development, marine commerce, and economic utilization of the resources of the marine environment.

(4) The preservation of the role of the United States as a leader in marine science and resource development.

(5) The advancement of education and training in marine science.

(6) The development and improvement of the capabilities, performance, use, and efficiency of vehicles, equipment, and instruments for use in exploration, research, surveys, the recovery of resources, and the transmission of energy in the marine environment.

(7) The effective utilization of the scientific and engineering resources of the Nation, with close cooperation among all interested agencies, public and private, in order to avoid unnecessary duplication of effort, facilities, and equipment, or waste.

(8) The cooperation by the United States with other nations and groups of nations and international organizations in marine science activities when such cooperation is in the national interest.

The full text of P.L. 89–454 is given in Appendix 7.

The declaration of policy, Section 2, stated "what" should be done. The assignment of leadership to the President in Sections 4 and 7 asserted unambiguously "who" should carry out these directions. The authorization in Sections 3 and 6 of an interim National Council on Marine Resources and Engineering Development chaired by the Vice President to advise and assist the President, and in Section 5, of a public advisory Commission on Marine Science, Engineering and Resources to recommend long-term organizational structure answered the question "how." The boundaries of cognizance were given in the definition of marine sciences of Section 8, and the scope of implementing instruments was qualified by fiscal authorization of Section 9. This was the recipe to correct the vacuum in leadership, the evils of duplication, and the limpid priority for ocean affairs.

When this triumph of long-range vision, skilled draftsmanship, and potent statecraft was set before President Johnson for signature, it was met with ominous silence. Was the President showing disdain for oceanography or a pique that had been signaled by rumors of veto? Would he actually veto? If not, would his grudging approval be followed by reluctant implementation?[1] The silence was all the more foreboding in view of the history of the Space Council on which the projected National Council on Marine Resources and Engineering Development had been partially modeled. Created by the same legislation that established NASA in 1958 (P.L. 85–568), the National Aeronautics and Space Council had been pushed by Congress over strong objections of President Eisenhower and his staff. But once this bill was enacted, Eisenhower yielded to recommendations of the Budget Bureau and the White House science advisor to leave the Space Council unactivated. On January 14, 1960, Eisenhower urged its abolition. Not until 1961 was the

Space Council energized, when Lyndon B. Johnson as Vice President urged his boss to amend the law by transferring Council chairmanship from a very busy President to a very unbusy Vice President. Considering LBJ's identification in the Senate as "Mr. Space," such initiative was not surprising. This action comprised, however, the first major legislative assignment to the Vice President since the Constitution. For the Vice President, the Space Council also promised fringe benefits. The office of Vice President had never been funded adequately, either by a frugal Congress or by uninterested presidents, for its incumbent to carry out more than minimal ceremonial duties, and the Space Council offered a convenient stable for a reinforced personal staff. Thus, even assuming that Johnson signed the Marine Sciences bill, three possible fates might await the Marine Council it would create: (1) complete neglect (too obvious a reaction even if there were no intent to implement); (2) token activation, with expectations that it would provide a staff arm to Humphrey as the Space Council had done for LBJ; or (3) full implementation in spirit as well as substance. Considering that the massive opposition to the Marine Council from Johnson's staff paralleled Eisenhower's experience with the Space Council, alternative (3) was the least likely. Yet it proved to be Johnson's choice.

Senator Warren Magnuson and staff aide Dan Markel had guessed correctly that the Johnson who favored activation of the Space Council would not block its marine counterpart. Moreover, as a former Senator, Johnson would think twice about undermining the prestige that rides on a committee chairman's initiative, a matter that Magnuson brought quietly to Johnson's attention as the bill worked its way toward passage. Especially with new maritime commerce legislation imminent and historically attended by a public brawl, Johnson would need a powerful congressional ally. Next, Johnson and Magunson were good personal friends: the President had been best man at the Senator's wedding. On that point, Magnuson confirmed to me a supposedly apocryphal story that during a White House reception, while the unsigned bill lay on his desk, President Johnson was confronted with an innocent query from Jermaine Magnuson as to whether the President would scuttle a measure that her husband had worked so hard to formulate. To which the surprised President is supposed to have said: "Honey, for you I'll sign it."

Members of the House who pushed for the Marine Commission out of anxiety that the Council would never make it should have had more faith.

Late on the evening of June 17, the bill became law. There was no public signing ceremony embroidered with fanfare, picture taking, and souvenir pens to invited witnesses; no statement for the press; no executive order providing a clue as to implementation. The oceanographic community was exuberant but still apprehensive about the President's future intentions. On the assumption that the President could not ignore the community of interests pressing for this policy, again significantly including the Commerce Committee Chairman, the Washington grapevine began to buzz with speculation on the next important steps: presidential appointments of the Council's Executive Secretary and of the fifteen-man Commission. Because of the administration's known objection to the Council and rumored support of the Commission, most public interest was directed toward promoting candidates to the White House for the Commission. Contrary to these public expectations, the priority behind the scenes was in activation of the Council. The office of Science and Technology (OST) reversed itself and urged implementation, worried that the congressional heat it received on a debilitated Interagency Committee on Oceanography would continue. And the Budget Bureau saw the Council as a delaying ploy to hold off a possible new agency they feared worse. According to Humphrey his role at this stage was minor, aware of the action only from his vantage point as President of the Senate. Considering the always delicate relationship between a President and his Vice President, any move by Humphrey to request Council activation might have been interpreted as a self-serving grab for even a tiny morsel of independent political power. A Marine Council was, however, so intriguing that Humphrey's staff began delicately probing the White House staff as to presidential pleasure. This new venture, only the second administrative responsibility ever accorded the Vice President, offered an opportunity to assemble a fresh piece of governmental apparatus that could reflect Humphrey's ambitions to contribute to affairs of state and bear his style and imprimatur. (Quite different had been the case of the Space Council. When Humphrey succeeded to its chairmanship in 1964 it had already been so well endowed with personal staff appointed by Vice President Johnson that Humphrey had been more its captive rather than its leader.) Moreover, Humphrey had an abiding conviction that science and technology were positive forces for human progress. The Vice President, however, needed more than desire, for no matter what vehicle Congress designed, only the President could give its driver the ignition keys.

That there was to be a Council was not revealed to Humphrey until July 13, the day the public learned of the President's decision at the commissioning of the ESSA research ship, *Oceanographer*. Not only was there to be a Council, but it was to move swiftly. In the President's words, typical of the administration's early activist style,[2] "The Council [is] to provide me with its initial recommendations not later than January [15], 1967, so that appropriate legislative proposals can be made early to the next Congress."[3] Through the same speech, Johnson released the report, *Effective Use of the Sea*, prepared by his Science Advisory Committee in a belated effort to stem congressional displeasure at OST's limitations in mounting the desired programs.

So here was an initial assignment to sort out more than one hundred PSAC proposals, to select priorities, and to develop whatever rationale was necessary to earn presidential support in time to secure a place in the President's January budget. This requirement for an activist policy-planning body to lend impetus to a demanding new venture depended upon a presidentially appointed Executive Secretary and staff of specialists in marine affairs who were also wise in the ways of government. Efforts to fill this position began in earnest.

Two of Humphrey's staff, John Stewart and Alfred Stern, became Humphrey's spotters, soliciting names of candidates, then interviewing. Aware of the dangers facing a wayfarer in the old bureaucratic woods, they alerted him to a statement by the Naval aide to James Wakelin that ICO had expanded personnel to about fifteen to fulfill the Council staff function, and stood ready to perform. Humphrey, having learned from the takeover tactics of the Defense Department, rejected the offer.

In the course of headhunting, my name had come to Humphrey's attention and he was interested to find that my experience in ocean engineering and science policy included dealings with both the executive and the legislative branches. A further poll of his Senate and House colleagues uncovered bipartisan support, and Humphrey's staff began a courtship. The thought of working under such dynamic leadership was intriguing. I had met Humphrey only briefly, when advising him from my anonymous perch in the Legislative Reference Service (LRS), and had been impressed by his vitality, his willingness to do battle even in unpopular liberal causes, and by his copious intellect, which could pick up facts like a vacuum cleaner and structure them to support a position. Nevertheless I begged off on the basis that the appointment was temporary and that the new science advisory post that Congress had worked so hard to create still needed care. To counter these concerns, it was promptly arranged that the LRS would grant an 18-month leave of absence. And when I determined that the administration was really serious about a "gung ho" approach and that I would be left free to select Council staff for competence, not patronage, I accepted the offer.

In preparing the critical recommendation to President Johnson, Humphrey was gratified to learn I was known to the unpredictable President, again through LRS.[4] My apointment was approved overnight and was announced by the White House on August 13.[5] On August 17, the Council met for the first time and swore in its Executive Secretary. Later, its chairman held a news conference announcing its activation, explaining its role, and making it clear that in the haphazard agglomeration of federal agencies, the Council was a "new boy on the block."

After stressing that the President attached "the greatest importance to the subject matter of this Act," Humphrey announced the impending January 1967 deadline for the Council's initial recommendations and said emphasis would be given "those activities and programs which promote international understanding and cooperation." Private industry, the academic community, and the state oceanographic commissions were all called on to join in the over-all goals.[6]

The oceanographic press serving an expectant but frustrated marine industry gushed with enthusiasm as though Atlantis had been discovered. The *New York Times* took solemn note of the happening,[7] while the major journal serving the scientific community, *Science*, astringently observed:

The council, a temporary body unless made permanent by some later act of Congress, would advise the President on the planning and coordination of the overall national oceanographic effort—an effort which many people in and outside of Congress believe would be larger if it were given more attention at the highest echelons of government. . . . At the moment, the long-range outlook on the make-up and management of the national oceanography program is still hazy.[8]

GENERATING A CONCEPT

Successive waves of refinement in earlier legislative proposals had the same beneficent effect that aging produces in whiskey. The original immature and unnecessarily detailed language was converted into a seasoned mandate. But the resulting framework of legislation was just a beginning and left extensive detail to be filled in; a range of legitimate interpretations was open to the executive branch. Not surprisingly, after Humphrey received the Council ignition key examination of the Act disclosed only a rough blueprint for the car, plus reams of standard administrative regulations. The garage held no chassis, wheels, or engine; much less did it provide gas for the tank—not even funds to pay staff salaries. Two steps had to be taken immediately. The Council vehicle had to be assembled, but even more urgent were start-up decisions on where the car might go and how.

The design concept for building policies and strategies and for setting Council style during the Johnson administration resulted from two preliminary hour-long conferences between the Chairman and the Executive Secretary. These conversations focused squarely on the problems of our nation and of society that intrinsically had no maritime content, but whose solutions could be facilitated by utilizing the oceans more effectively. We discussed the potential of the sea to mitigate world hunger and provide fuel and minerals; the increasing pollution and abuse of the coastal zone; the decay of the waterfront; the hazards of turning the seabed into yet another arena for nuclear armament; the festering conflicts among nations competing for common-property resources; and we pooled our knowledge of impediments, strife, and false starts. To Humphrey, the zest of what could otherwise be dull machinery for coordination lay in the challenge of the opening passage of the Act: "to develop, encourage and maintain a coordinated, comprehensive and long-range national program in marine science for the benefit of mankind"—a rhetoric, incidentally, borrowed from the precursor Space Act. His first decisions placed emphasis on identifying these potential benefits, an exercise in harmony with Humphrey's idealism and interests in meeting the dilemmas of modern society.

A second decision Humphrey made at the outset was that the Council should assume an activist style. The Council had two components—one was a nine-man policy board composed of the heads of agencies that comprised the separate operating arms to carry out the government's marine-related missions; the second was the staff secretariat. Not only was the Secretariat to serve the President by recommending priorities to the Council members; staff was to follow through, by threading the approved program over, under, around, and through the maze of bureaucratic impediments that seem invented to test the energies of an advocate if not the merits of a program itself. By such strategy the agencies would be supported in sharpening the cutting edge of the nation's marine interests, to strengthen the management of a roughly $400 million enterprise that constituted the base of an on-going effort in our constituent agencies, and most important, to advance to new goals.

For marine affairs, the President was the locus of decision-making. But because of competition with so many other issues, the challenge was something like a game played on amusement park merry-go-rounds, where passengers snatched at rings as they whizzed by a rack. The rings were of wood, except for one beautiful, shiny, and sought-after prize—the brass ring! In Washington, a similar game is played by government officials and special interests alike. In straining for visibility and funds for favorite programs, the brass ring is public support by the President. A paragraph, a sentence, or even a phrase in a key presidential speech brings joy and prestige to the advocate and therapy to his ego. Best of all is mention in the State of the Union Address, but such an Olympian satisfaction is seldom more than a dream of jostling contenders. So the coveted prize of presidential recognition was now to be sought for the previously neglected field of marine affairs. This meant tuning Council antenna onto policy goals and moods of the Johnson administration to which the sea could genuinely contribute. We were aware that we faced problems: a coolness to funding for science as a whole, inflation that was generating shivers through the economy, the issues of Vietnam and civil unrest capturing press and citizen attention.

In this regard, the Council tried to keep marine affairs in reasonable perspective: the sea simply was not at the top of the nation's wish list, but we were determined it not remain at the bottom. We also recognized that P.L. 89–454 had to be implemented in harmony with a whole battery of nonmaritime national objectives. But within this complex array of goals, programs, and participants, we tacitly adopted five unifying themes to guide development of policy.

First, science and technology were key ingredients for effective use of the sea. Second, the federal government would have to continue to provide national leadership to finance oceanic research and exploration, just as for science and technology generally. Third, as with land resources, competitive private enterprise would undertake marine resource exploitation but in a public/private partnership. Fourth, the geographically differentiated problems of managing the inshore marine environment would require state and local solutions to land and water management, but with federal guidelines. Fifth, in view of the stated national policy to study and use the sea for the benefit of all men, and considering the inherent international character of scientific exploration and deep ocean resources, a multinational approach was essential.

After dealing with strategic considerations, the Chairman and Executive Secretary briefly touched on tactics to orchestrate the bureaucracy, to synthesize the disparate competences that could be integrated in a government-wide capability. We viewed the Council as more than an umpire at bureaucratic games. The Council's key function was to assist the President in carrying out his statutory responsibilities of leadership, not to serve as an operating agency. In helping the President steer marine affairs, the challenge was to bring together a wide range of diffuse public purposes, specialized bodies of knowledge, institutions, and federal agencies, and to transform them from a fragmented, unsteady, and loosely knit caboodle into a broadly based, coherent, system sparked by a sense of urgency. In keeping with legislative instructions to assist the President, the Council would have to:

—identify unmet national needs and technological opportunities to which federal marine science programs could be directed, especially gaps in programs that cross agency lines;

—recommend priorities and identify impediments to progress and strategies for their circumvention;

—develop policies by which the objectives and programs of one agency would not inadvertently conflict with equally valid but independent activities of another;

—coordinate, through a committee structure, programs that were of concern to many agencies, and in those cases where missions of several agencies might overlap, recommend that one agency assume a lead responsibility for government-wide action;

—insure that the appropriate resources of the federal government were brought to bear on mutually agreed upon goals;

—evaluate programs so as to eliminate marginal activities;

—develop background, legal, economic, and technological studies identifying alternative policies and criteria for choice; and

—coordinate federal programs of international cooperation (such as the International Decade of Ocean Exploration).[9]

The Council was also to carry out policy-planning functions for the Sea Grant Program that was soon assigned by P.L. 89–688.

This was an ambitious charter, to keep eleven agencies from sailing off in different directions, but both of us were persuaded that the Congress had been correct about the need for integration and policy leadership. Well aware of the limited success of interagency mechanisms for coordination much less for government-wide inspiration, we were confident that this vice-presidentially geared instrument could do it. The unifying role of the Council was conceived in an engineering analogue to the strength of a beam loaded at its center and supported at its ends: if comprising two flat laminations, one on top of the other, the beam would carry a certain load before failure. Simply by gluing the laminations together, the load could be remarkably increased. The Council as a kind of adhesive could provide an effect far greater than the sum of the parts.

We were not sanguine, however, about the stresses that could develop. In their ardor to enforce territorial prerogatives and in their fierce resistance to external guidance, agencies could be expected to play a rough game of judo, possibly leading to destruction of the Council itself. Humphrey's initial warning to me was to be recalled often: "Don't let the pirates board our ship." Ominously, only two weeks after the Council was formed the Navy announced its reorganization to upgrade the chief hydrographer's post to "Oceanographer of the Navy." This was a constructive and much needed step, but it had been perhaps initiated for the wrong reasons—to reassert the identity of the Navy as the major performer. Apparently, elephants can overreact to mice.

Humphrey had one other word of advice: to remember that the Council was a congressional creation, and to keep Congress appropriately informed. By no means was he a strict constructionist regarding separation of powers of the two branches.

The philosophical compatibility established at these first meetings cemented a harmonious relationship between a part-time Chairman and a full-time Executive Secretary. Humphrey seemed to feel comfortable in delegating Council functions, confident that I would be loyal to his principles and concepts. In turn, I felt that at those frequent decision points where his overcrowded schedule barred consultation, my perception of the issue and resultant decision were likely to be compatible with his instincts and would gain post facto endorsement. An alliance with his staff was quickly established that facilitated direct and hospitable access to the Chairman, and rapport with his aides for press relations, travel logistics, scheduling, and speeches. Later, when the Council Secretariat endeavored to construct a comparable relationship with staff under a new administration, it was to meet with far less success.

The core concept we developed was reflected in Humphrey's statement of purpose read before the Council at its inauguration August 17, 1966, and released to the press that afternoon. In the statement Humphrey touched on a number of goals, including the nourishment of scientific capabilities and development of ocean technology through a partnership of federal, state, university, and industrial interests. He hinted at the possibility of a new agency expected from the Commission study to follow the demise of the Council after its allotted 22 months. He then mentioned the urgent deadline for the Council's first report, noting that the administration "is taking the challenge of ocean exploration seriously." With a heartfelt request that each of the many involved agencies carry its share in "the great responsibility to plan ahead and plan together," Humphrey concluded that we draw "bright new opportunities from the oceans in the spirit of exploration that has characterized this Nation's entire history."

The membership of the Council was established by legislation and was ready to be convened.[10]

With Humphrey's symbolic admonition of "damn the torpedoes, full speed ahead," the next step to follow preparation of a travel itinerary for the Marine Sciences Council was manufacture of the staff vehicle for implementation—the Secretariat.

WEIGHING ANCHOR

When a new agency is created, its presidentially appointed steward has the irksome housekeeping chore of putting together the machinery authorized by Congress, while at the same time clientele expect it to be operating as though it had been in business for a decade. So while generating Council destinations, milestones, and travel routes on the basis of early conversations with Humphrey, I was simultaneously faced with the onerous task of putting together a new, albeit minuscule, independent agency. Humphrey cared little for these details and delegated the entire responsibility. First priority was to seek enough interim funds to cover staff salaries and expenses for telephone services and wastebaskets while we got started. This was done by borrowing $50,-000 from other agencies. Direct appropriations would then be requested as part of supplemental requests that the President fired up to Congress each fall to cover unexpected emergency demands. As noted later, this fiscal plan backfired and became a painful initiation into the cabal of the congressional appropriations process.

Finding office space was the second priority. We then faced the need to establish all of the administrative paraphernalia of a much larger agency: security clearances, personnel files, payroll, purchasing, accounting.[11]

Of all our tasks, the most important was to recruit personnel. Criteria for selection of presidential advisory staff had been tested during my indoctrination by Wiesner: integrity, intelligence, initiative, sophistication, objectivity, self-confidence, and ability to work under stress with a high tolerance for frustration. But because of the short projected life of the agency and the arduous task of dealing with the variegated red tape of the bureaucracy, recruitment was formidable. Few candidates combined professional skills with breadth of policy experience who could be readily budged from their current posts. On the plus side, the Council offered prestige, the excitement of being close to where decisions were made, as well as high salaries within governmental scales, and exceptional support services. Moreover, it promised a "can-do" environment sparked by vice-presidential gusto. From universities and industry, other government agencies and state government, we assembled a remarkable cast of twenty-five, with expertise in economics, law, engineering, foreign affairs, as well as oceanography. They made the Council work. Enoch L. Dillon, recruited from his position as defense research and oceanography examiner in the Budget Bureau, became the Council's first staff officer, Florence Broussard and Charles S. Barden, personal assistants. (Upon my departure in 1970, Dillon was appointed acting Executive Secretary until Council retirement fifteen months later.) As successive individuals arrived, they learned to swim by being abruptly plunged into hot water.

In the course of this riotous growth, the system proved to be far more rational, pliable, and resilient than alleged. Indeed, the conventional wisdom that damned the Civil Service Commission as the source of red tape proved untrue. Usually the complications and impediments originated with some agency official who did not choose to exert initiative to reduce delay, or to risk possible failure when the question did not safely fit familiar precedents.

On two matters, we strictly followed the rules: security clearances and fiscal accounting. The Secretariat must have reviewed five hundred files, some containing several hundred pages of information collected by the FBI or civil service investigators. The investigative procedures, if somewhat naïve, were thorough and fair; the decision on clearance was indeed left to the judgment of the agency head doing the hiring.

Problems in fiscal accounting proved more troublesome. Of the many razor edges on which agency heads walk, the most agonizing is that of meticulously staying within appropriations. Even in this age of computers, keeping books on agency obligations is difficult. As midnight June 30 approaches each year, a sliver of unexpended funds must be retained. There is the option of playing it safe simply by forgoing obligations. Since any unexpended balance must usually be returned in July to the Treasury leaving worthy projects unfunded, our accounting discipline attempted to leave only a small cushion in the budget. During the first year, with the additional precaution of a separate audit in June, we celebrated July 1 with about the planned margin. That joy was short-lived. In August, an unrecorded obligation for several times that safety allowance reared up and triggered almost the greatest alarm of bureaucratic life. General Services Administration lawyers, frantically consulted in the face of this emergency, discovered a life preserver in obligating funds of the succeeding year that were planned for another project. But the experience was a sharp lesson, in terms of how far the system could be pushed before the simple limitations of human fallibility would signal "tilt."

As Council operations came into gear it seemed best to amplify the capabilities of the relatively small staff through use of outside consultants. These were appointed from industry or academia on the basis of personal qualifications, not of institutional affiliations nor partisan politics. By testing agency or Council ideas against this panel, we could simultaneously isolate weaknesses in concept and errors in fact, and discover whose toes outside of government might be inadvertently trampled.

Further picking brains, we instituted contract studies under authority providentially written into the legislation. The Secretariat identified key undeveloped areas of potential import to marine affairs, confirmed that studies were not under way within the establishment, then sought the best talent we could find. Perhaps naïvely, we kept hands off contractor findings, for, in fact, several ran counter to our own views. But all were released publicly.[12] This device also served to advertise more widely the horses on which the government might be placing future bets and thus legitimately signal industry and universities where to build up a marine affairs expertise. About 40 percent of our budget was devoted to renting talent outside the agency.

The triad of Secretariat components that included in-house staff and outside consultants was completed with a committee infrastructure to draw on talent within the member agencies. This coterie reviewed most of the policy and program proposals to come before the Council in plenary session, especially to comment on "how" to do things rather than "what" to do, and to identify problems of implementation; the Secretariat might serve as naval architect, but the agencies were constructors. Committees were also an authoritative source of information as to varying points of view among different clientele the agencies felt they had been created to serve. By this machinery also there was a fluid exchange of information on current plans, on future plans, and on gaps, a safety valve to ventilate and resolve internal discrepancies. Initially, the Council sought to utilize the ICO and revitalize its panel structure. When that broke down because of inevitable conflicts between competing power centers, the Vice President appointed four Council committees: Marine Research, Education and Facilities; Ocean Exploration and Environmental Services; Food from the Sea; and Multiple Use of the Coastal Zone. Additionally, at the request of the Vice President, the Secretary of State established a fifth committee on International Policy in the Marine Environment to serve the mutual interests of the Council and the Department of State. Each was chaired by a senior officer of an agency having a major interest in the subject area, so as to give him leverage in developing cooperation from sister agencies. The responsibilities of these committees, appointed on August 16, 1967, are outlined in table 2.

With its administrative machinery functioning, the Council took off at full speed. John Daley summed up the situation neatly before a Navy League audience early in 1967. The takeoff of the Council, he said, was comparable to a stagecoach driver trying to contain a team of wild horses over an unpaved roadbed with unmarked curves and mudholes, and with sharpshooting bandits at every pass. Indeed, the result was a tremendous bumping and jolting for all its passengers.

THE MARINE AFFAIRS TRANSITION

Following Humphrey's concurrence on strategy, policy planning began to examine how the sea relates to human affairs, to identify priority issues and to formulate policy alternatives, program, and budgetary responses. This scope considerably stretched the frame of reference adopted by NASCO and ICO, which dealt primarily with research activities possessing salt water as a common denominator. But by law, the Council was to deal with "ends" as well as "means": regarding mankind's shared concerns about a more peaceful world community, famine, ill health and disability, a better quality of life itself, and regarding economic growth. The Council philosophy, adopted in its early plenary sessions and revealed in the Council's first annual report, was succinctly and felicitously expressed in Humphrey's response to Alton Lennon's request for comment at hearings on the Council's first anniversary.

Wrote Humphrey:

August 16, 1967

Let me briefly review some of these challenges that face our Nation and the world today:

TABLE 2

COMMITTEES OF THE MARINE SCIENCES COUNCIL

Committee	Responsibility	Chairman[b]
Committee on Marine Research, Education, and Facilities.	Basic oceanographic research and engineering, and related vessels, equipment and facilities for Federal activity in marine research; manpower and specialized education in the marine sciences, including the Sea Grant Program.	Robert A. Frosch, Assistant Secretary of the Navy for Research and Development.
Committee on Ocean Exploration and Environmental Services.	Exploration, description, and prediction of the marine environment and the ships, buoys, satellites and other facilities required for these purposes; mapping, charting, and data management activities.	Robert M. White, Administrator, Environmental Science Services Administration.
Committee on Food from the Sea.	The broad, long-range program, adopted by the Council and the President in early 1967, to exploit the oceans in every reasonable way as a source of food, including the Fish Protein Concentrate Demonstration Program.	Rutherford M. Poats, Deputy Administrator, Agency for International Development.
Committee on Multiple Use of the Coastal Zone.	Environmental planning, conservation, and development; water pollution, erosion control, and shore development activities; channel and harbor development and redevelopment; marine ecology; and recreational development of marine areas.	Stanley A. Cain, Assistant Secretary of the Interior for Fish and Wildlife and Parks.
Committee on International Policy in the Marine Environment.[a]	U.S. foreign policy pertaining to the marine environment; international activities and initiatives pertaining to the marine environment, including cooperation by the United States with other nations and participation in international organizations and meetings.	Foy D. Kohler, Deputy Under Secretary of State.

[a] Established by the Secretary of State, at the request of the Vice President, to serve the mutual interests of the Council and the Department of State.
[b] Initial appointments by Vice President Humphrey.
Source: MSA, 3d report (1969), p. 193.

There are one and one-half billion hungry people in the world. The full food potential of the seas, seriously neglected in the past, must be realized to combat famine and despair. Technologies now at hand can be directed toward increasing the world's fishing catch and enriching the diets of the underfed.

Seventy-five percent of our population lives along our coasts and Great Lakes. Nine of our fifteen largest metropolitan areas are on the oceans and Great Lakes and three are on ocean tributaries. Twenty million children live in these metropolitan areas within sight of potential water recreation areas but are often denied their use. Only three percent of our ocean and Great Lakes coastline has been set aside for public use or conservation.

More than 90 percent by value of our intercontinental commerce travels by ship. Although there have been rapid changes in the character of ocean cargoes and technologies of cargo handling, the average age of our port structure is 45 years and the average age of our merchant ships is 19 years.

The continuing threats to world peace require our Navy to maintain a high level of readiness and versatility through a sea based deterrent and undersea warfare capability. Middle East conflicts following closure of the Gulf of Aqaba vividly emphasize the urgent need for a strengthened code of international law of the sea.

Thirty million Americans swim in the oceans, eleven million are salt-water sport fishermen, and eight million engage in recreational boating in out coastal States, yet industrial wastes being dumped into ocean tributaries will increase seven-fold by the year 2000 unless there are drastic changes in waste handling.

Ocean-generated storms cause millions of dollars of damage annually along our coasts, but marine weather warning services are available to less than one-third of our coastal areas.

During the past year I have discussed these challenges with scientists, engineers, business leaders, and local, State, and Federal officials here in Washington, at oceanographic installations in nine coastal states, and in the capitals of six countries of Western Europe and one in Asia.

The problems of the sea are complex, and they involve every type of concern and institution that exists on the landward side of the shoreline. Thus, we must solicit the varied ideas, the advice, and the participation of universities, industry, and all elements of government, just as we have found this mixture an essential ingredient for the vitality and progress of our Nation on shore.

Pure logic and practical economics dictate this program. However, not to be forgotten is man's compelling desire to explore and to understand the world around him. The spirit which has carried us to rugged mountain peaks, remote polar icecaps, and distant reaches of outerspace now propels us to the ocean deeps. This spirit is fortified with a confidence developed by past contributions of science that we will not only conquer the ocean deeps but will use them in satisfying the needs of our society.

In concluding, may I say how much I welcome this continuing interest by the Congress in what is both an enormously complex set of issues and an untapped set of opportunities to study and utilize the sea to serve man. This is a program that has support by both Executive and Legislative branches of Government, free of partisan controversy, and I look forward to our working further with the Congress in serving our mutual interests.

Sincerely,

HUBERT H. HUMPHREY[13]

For a Vice President to submit testimony as an officer of the executive branch before either House of the Congress was virtually unprecedented, but his acquiescence to this request helped immensely to melt the House's initial aloofness toward the Council because it had been a creature of the Senate.

Applying knowledge of the oceans to public purposes impelled a deepening blend of science and society. We were deliberately forcing a transition from what previously had been termed "oceanography" (or "oceanology," to use an archaic but more accurate term that was ex-humed and promoted by Senator Claiborne Pell) to a completely different universe of activities that we termed "marine science affairs."

The transition was cultivated in two quite separate directions. First, a traditional research activity that had been narrowly oriented toward describing phenomena of the sea was now broadened to embrace a constellation of activities that anticipated engineering application and legal, economic, and political considerations. Secondly, we endeavored to upgrade technical program planning to a higher level of policy planning, with a correspondingly elevated hierarchy of participants. This level of concern extended to the President of the United States. Internationally, it would come to include the Secretary General of the United Nations and the heads of foreign governments.

This transition was also intended to stretch the involvement of private interests. Activities had formerly been of greatest moment to oceanographers and fisheries scientists who composed the knowledge-producing core of the enterprises. Now it was hoped to engage an entirely new cast of knowledge-consumers: engineers, economists, lawyers, public administrators, foreign affairs specialists, bankers, businessmen, industrialists, and some old-fashioned explorers.

The earliest vehicle for exhibiting this transition was the first annual report of the Council, March 1967. It was entitled "Marine Science Affairs," to portray the body of oceanic activities that linked science and society. It also rejected any euphemisms as to intent with a subtitle: "A Year of Transition." This broadened scope was further delineated in terms of funding. The "National Oceanographic Program" that had been previously coordinated by ICO and reported annually by the Federal Council had been contained within categories of research, surveys, engineering, instrumentation, ship construction, services, and facilities. Additional compilations of government-wide activities embraced by the Marine Sciences Act now included classified naval research; ship and vehicle research; technological developments in fishing, marine minerals, and energy resources; and seashore land use and recreation. A "map" was included, showing for the first time where the money went, now in terms of major purpose.

For purposes of comparison, a base was chosen of Fiscal Year 1967, the year immediately preceding the Council's creation. Then, total federal funding of these activities was $409.1 million; of that amount, $228.5 million corresponded to what had previously been defined as the "National Oceanographic Program," a distinction carefully stated to head off criticism that future growth was the result of artful redefinition rather than the genuine expansion that marine advocates sought.

The total proposed by the President for FY 1968, as shown in table 3, was $462.3 million, up 13 percent. The distribution by percentage for each major purpose is shown in figure 12.

This aggregate of activities was found easiest to rationalize by distinguishing three stages, as suggested in figure 13. The first two stages constitute a general purpose capability of academic research, manpower training and education, general purpose ships, environmental data acquisition, facilities, and instruments that provide a reservoir of information and techniques. The third stage, functionally dependent on the first two, includes applied research, mission-oriented development, and other marine technology programs directed to explicit public needs such as national defense and resource exploitation. Funding for each

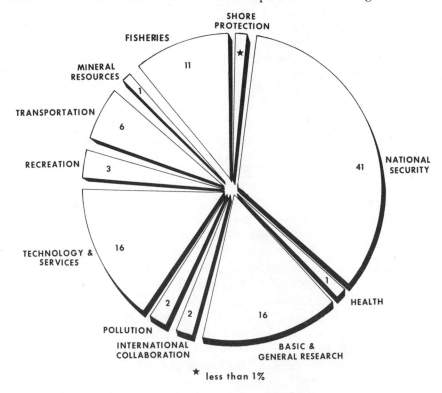

Fig. 12. The marine science and technology dollar. Source: *MSA*, 1st report (1967), p. 28

TABLE 3

FEDERAL MARINE SCIENCE PROGRAMS BY MAJOR PURPOSE

	Fiscal year 1968 program estimate (in millions)
National security	$191.6
Fisheries development and sea food technology	49.2
Transportation	27.8
Recreation	13.6
Pollution abatement and control	9.5
International cooperation and collaboration	7.4
Minerals, chemicals, water, and energy resources	5.8
Health	4.2
Shore and harbor stabilization and protection	1.7
Multipurpose activities:	
Oceanographic research	73.2
Mapping, charting, and geodesy	39.1
Ocean environmental observation and prediction services	21.1
General purpose engineering	10.5
Education	5.5
Data centers	2.1
Total	462.3

Source: *MSA*, 1st report (1967), p. 26.

category of development delineated by the vertical bars is justified separately, depending on the relative priority of different goals, anticipated payoff, and on the opportunities to realize the sought for benefits.

The breakdown by purpose revealed the feeble support for civilian goals. Past programs had been predominantly focused on national security objectives, with well over 50 percent of all federal funding for marine sciences budgeted by the Department of Defense. Beginning in Fiscal Year 1968 the Council's influence was deliberately exerted to utilize marine sciences in meeting industrial, economic, and social goals as well as military. This civilian growth was not to be at the expense of military effort, however, because each activity was judged on its own merits. But the Council's patent challenge was to gain presidential support for more funds, and to do so required persuasive targets.

NEW INITIATIVES: PRESIDENTIAL CHOICE OF MARITIME GOALS

If putting more fuel in the oceanography tank was the name of the game, what purposes would justify that addition? The fragmented multiagency program that the Council inherited had two characteristics: by virtue of ambiguous missions, every agency was doing a little bit of everything. And all were projecting future growth that in many cases was "more of the same." Critically needed were government-wide priori-

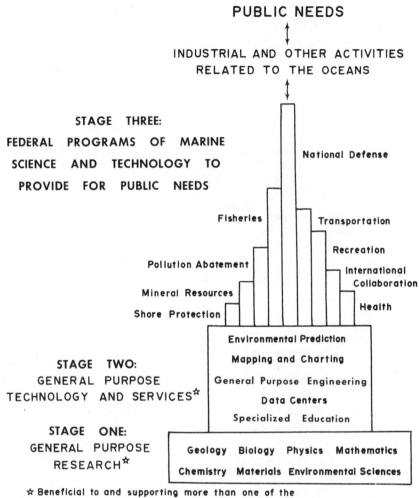

Fig. 13. Functional relationships of marine science and technology. Source: *MSA*, 1st report (1967), p. 29

ties that would fill gaps or lack of balance and would justify growth in selected areas. There were two management choices. First, we could staple together existing programs of the participating agencies and lend support to expansion of all. We could even refurbish meritorious rejects from prior unsuccessful budget sorties. Or we could consider the full range of needs and try to formulate programs for these targets, using a rifle rather than a shotgun. The Secretariat chose the latter approach, to reconnoiter for explicit "new initiatives," rather than asking for increased funds across the board. After analysis of current efforts, the Secretariat concentrated on selecting priority areas, developing implementation programs, clarifying agency responsibilities, and formulating means for interagency collaboration. These selected enterprises had four characteristics: The objectives contributed to such broad national goals as pollution abatement and the war on hunger; the priorities represented a consensus of the senior government officials responsible for marine science affairs, most of whom reported directly to the President as individuals (as well as collectively through the Council); benefits were clear and means for implementation were immediately available; multiagency collaboration was to produce multiplier effects from small investments by deploying existing capabilities more effectively.

Where missions overlapped, a "lead agency" was designated to be responsible for government-wide planning, guiding, and coordinating, and to assure fiscal support through its own funding when gaps were detected in dovetailing components. The Council assumed responsibility for assuring effective implementation and for assistance in budget defense.

Prompt action was essential in the fall of 1966 because budget planning for Fiscal Year 1968 was well under way. There were ample proposals, in fact an embarassment of riches: one hundred items from PSAC and some new ones from ICO and government agencies. The Council Secretariat made a selection among these offerings and generated some itself. Nine priority projects worked through the agencies and the Council were presented to and approved by the President. Of the $53 million increase in funding recommended for FY 1968, approximately $41 million represented program areas selected by the Council for emphasis.

The priorities for marine science development began to take shape, as they were discussed in successive years by the President and by the Council through its vice-presidential spokesman. The earliest pronouncement came from President Johnson in his initial message to the Congress, March 1967, where he singled out eight explicit projects: to counteract worldwide protein deficiency, to implement the Sea Grant program, to improve data management, near-shore weather prediction, environmental protection of estuaries (with emphasis on Chesapeake Bay) and resource exploitation, and to strengthen deep ocean and Arctic technologies. A ninth goal he cited, a desire for international cooperation, was more a benign principle than an explicit policy, but the theme emerged in his peroration as a "determination to work with all nations to develop the seas for the benefit of mankind."[14]

In its accompanying report to Congress, the Council set forth its concept of the same priorities. International cooperation was first on that list, particularly to reduce the "resources gap between rich and poor nations," followed by elaboration of the eight that the President had announced. The Agency for International Development was given **lead** responsibility for the Food From the Sea project; the Council was to contract for a study of producers and consumers of oceanographic data; the Corps of Engineers was to spearhead the Chesapeake Bay study, using a projected scale model of the Bay's hydrodynamic behavior. The loss of *Thresher* and an unarmed H-bomb off Spain were cited as motivations for expanding technologies for search and recovery at great depth.

By the time of the second Council report, the Chairman was mentioning in his letter of transmittal the four major planks of the underlying philosophy. Topping his list was the desire to "strengthen our economy" by identifying marine resources, strengthening marine technology and enlarging U.S. maritime enterprises. Then followed the matters of enhancing the quality of urban living by coastal awareness, of strengthening world security, and of fostering the education of specialists.[15]

In his second report, the President again accepted the recommendation of new initiatives, but shifted the order, placing education and research in marine sciences at the head. His second goal stressed development of fish protein concentrate. An entire section was devoted to a promise that this nation would seek to strengthen international law to reaffirm traditional freedom of the seas, encourage mutual restraint to avoid military conflict, and attempt to insure "that ocean resources are harvested in an equitable manner and in a way that will assure their continued abundance." The Secretary of State had been asked, he said, to "explore with other nations their interest in joining together in long-term ocean exploration." Added to the prior year's list were development of ocean buoys and spacecraft oceanography and the prototype estuarine project was expanded by general attention to the coastal zone, Great Lakes and ports, and to oil pollution.[16]

In the third annual report, January 1969, President Johnson sustained most earlier initiatives and added the International Decade of Ocean Exploration, separately announced in March 1968. But now he gave top place to coastal management,[17] and this priority also appeared in the March 1970 message of his successor, President Nixon.[18] For that one year, Nixon sustained continuity in the "initiatives" approach, additionally supporting Arctic research, restoration of the Great Lakes, and strengthened the data collection, with special emphasis on funding for the Decade.

Coincidentally with this selection, alignment, and realignment of priorities, the basic philosophy of marine science development was also gaining definition as a national policy. In the first report of the Council, President Johnson had compared the exploration and development of the seas to the opening up of the American West.[19] A year later he again spoke of the "expanding frontier" and of the seaward quest for fuels, minerals, and food, as well as for recreation.[20] Humphrey's letter of transmittal accompanying the second Council report stressed how "the power of science, transformed . . . to a technology, may serve our Nation's diverse interests."[21] And in his third transmittal, the Council's Chairman dealt optimistically with the work to be accomplished in focusing "the resources of eleven Federal agencies toward common objectives, through improved internal management; and [to] reduce institutional impediments that retard effective development of marine resources."[22]

Over a period of nearly five years roughly 65 policy issues considered by the Council (some repetitive) were reflected in presidential statements, in new initiatives set forth in annual reports, and in budgetary emphasis. One exception was the merchant marine, however, where the Council seldom chose to displace the firm jurisdiction exercised by White House and Budget Bureau staff.

The Council had become a maritime presence in the White House and in the agencies in gaining a higher degree of sustained awareness of the oceans, perhaps the most important result of establishing the Council at the highest possible level in the Washington firmament. Magnuson understood this advantage to marine affairs. Close proximity to the President clearly fostered agency cooperation; none wanted to risk presidential displeasure. While that locale alone did not guarantee easy access, it did afford greater insight as to the presidential processes and timing and the opportunity to follow an issue through this maze to the important finale of action. Even an issue that has become a candidate for presidential support must be shepherded assiduously right up to mention in the State of the Union address, for example, or in other important messages to the Congress, where the President asserts his priorities. Drafts go through successive steps of revision to engender felicitous expression appropriate to presidential dignity and also to give the text the precise meaning (or fuzziness) demanded. The message must nod appropriately to outside interest groups whose support is hoped for; tread on few toes; recognize the subtleties of congressional interest and jurisdiction; substantively be mature, but contribute to political fortunes. At key moments the presence of the Council in Executive Office territory facilitated Secretariat access to key aides such as Joseph A. Califano, Jr., Walt Whitman Rostow, and (under Nixon) to Peter M. Flanigan, John C. Whitaker, Henry A. Kissinger, Arthur F. Burns, Robert Ellsworth, and Lee A. DuBridge.

A similar opportunity to defend marine initiatives was presented during the budget season. Usually the Chairman forwarded proposals directly to the President. As the moment of truth approached—the sign-off by the President of his recommendations to the Congress—more and more sacrifices had to be offered up to bring proposed expenditures into closer alignment with projected revenues. The Budget Director would repeatedly test a late total with the President and come away again with instructions to cut more. The Director then had the unhappy chore of passing such cuts on to the agency heads, listening sympathetically to their agonies but stating firmly that their appeal, "reclama" as it was called, had to be made directly to the President. This would make them think twice. As members of the President's political family, their fortunes were tied to his, and a reasonably balanced budget was in their interest. With favorite projects threatened, they had to decide whether to yield or to face presidential wrath and possibly to lose ground in next year's competition for presidential support.

Seldom were marine affairs at the top of a cabinet officer's list, and when the Secretariat learned of threats to such items in those last tumultuous days of budgetary striving, I would urge a cabinet officer to reclama, or appeal directly to the budget director. But there was always the backstop of Humphrey to coax the President, a role that could not have been effected from any other echelon in the bureaucracy. Our role as advocate stemmed from the basic legislation that had assigned the responsibility for advancing marine affairs explicitly and unambiguously to the President, with the Council to advise and assist. In the final clash of competing interests and wills we saw a balanced defense of marine affairs threatened unless some voice, albeit weak, were heard. Interestingly, while the Budget Bureau criticized this intrusion into what they considered a tidy process of objective budget decisions, they approved publication of the Council's role of advocacy in one of the annual reports:

The Marine Sciences Act reflected Congressional concern over the need for high-level advocacy for the field. Notwithstanding the popularity that the field enjoys and the heightened popular interest in the scientific mysteries of the oceans, there had been no central advocate within the Federal family for examining and responding to the question of "What portion of the Nation's energy and wealth should be devoted to ocean endeavors?" Thus, in addition to its primary responsibility of assisting the President in carrying out the provisions of the Marine Sciences Act, the Council has served as an advocate—acting through its Chairman, its members, and its Executive Secretary, who in communications with members of Congress, speeches, public articles, communications to oceanographic institutions, and talks with foreign leaders have urged wider understanding of the relationship of the oceans to people and their concerns.[23]

That role of advocate, incidentally, was continued for a while in the Nixon administration, not to market new Council proposals but rather to advance a few carefully selected from the previous administration or from unfinished business of the Commission recommendations, delivered by its Chairman, Julius A. Stratton, in January 1969. (As described in chapter 8, the Commission proposal, with support by the Council Secretariat, ultimately brought forth a new National Oceanic and Atmospheric Administration, but that evolution spelled doom for the future of the Council itself.)

Because of disparity between the nation's needs and the size of its purse, the Council was expected to act not only as advocate but also as a responsible member of the President's family in streamlining costs and improving management. Although in a conflict between turning money on and turning it off, the Council Secretariat resolved to contribute to the decision-making process by developing creative but realistic options for the President and facilitating implementation so as to deliver intended results by reducing management ambiguities through use of lead agency concepts and government-wide utilization of ships.

The structure, organizational position, agenda, and style of the Secretariat as an unusual form of systems manager and alternative to crisis management are analyzed in chapter 9, but whatever the qualities of the Secretariat to accomplish any of its tasks, the performance of the Council depended on its Chairman.

THE VICE PRESIDENT AS CHIEF OCEANOGRAPHER AND ASSISTANT PRESIDENT

Humphrey needed no education for the role he was to play as the new leader of an old cause. In essence, he was to be the nation's chief oceanographer. Until the presidential campaign of 1968 absorbed his energies, he was to serve that cause with sensitive leadership and zest.

The drive for marine visibility opened in September 1966 with his visit to the Scripps Institution of Oceanography while he was in California, followed by a speech in Seattle. Later he was to host a seminar in Houston, to visit labs in Galveston and labs and *Aluminaut* in Miami, to dedicate a Navy undersea range in the Bahamas, and to take an overnight cruise on the magnificent new Woods Hole oceanographic ship, *Atlantis II*. While his speeches were largely part of a crusade to amplify interest in the oceans, he took the occasion of a visit to the University of Rhode Island laboratory to open the campaign for coastal management and at Bowdoin College in Maine to underscore his support of international cooperation. He wrote signed articles for a number of journals. Of his dedication to the job, there could be little doubt, but few of his ocean speeches gained national headlines; no small wonder his staff sometimes urged that he invest that precious time elsewhere.

When in Washington, D.C., Humphrey felt it wise to be far less conspicuous and never to garner a headline that might have gone to his boss, and he reminded me to do the same. But increasingly, speaking requests made to him were bounced my way. Accepting only about one in three in the conviction I could do more at home to help the President and the Council Chairman, I nevertheless crisscrossed the country with about 100,000 miles per year of such travel.

While the public appearances stimulated Humphrey, his strongest personal contribution was in private—guiding the one-half billion dollar program of marine science and technology. As a first step, Humphrey wanted to fulfill Johnson's initial request for immediate recommendations. But at Council startup time, budget decisions in individual agencies were almost crystallized. To generate new proposals, to gain Cabinet-level consensus before the Council on government-wide priorities, then to invade the budgetary process in every participating agency and in the Bureau of the Budget to gain concurrence by December was feasible only because of vice-presidential interest, prestige, and, when required, advocacy.

In almost every case, the agency bearing major responsibility for implementation and assigned a leadership role by the Council was pleased that the Council could aid as well as thwart bureaucratic ambitions. One notable exception was the NSF. By September 1966 it was clear that the Sea Grant Act was picking up steam on the Hill, likely to be enacted that session, and deserving support via the presidential initiative route. Several times, its energetic sponsors—Senator Claiborne Pell of Rhode Island and Congressman Paul Rogers of Florida—feared it might be neglected because a categorical program especially of applied rather than basic research was contrary to NSF's funding style. Humphrey gave Pell assurances that the Council would do its best to inject the necessary hormones. When the bill was signed October 15, Pell promptly collected on that promise and Humphrey made the shocking discovery that the Foundation had no serious plans to implement the act once it became law. The Bureau of the Budget also was opposed to providing immediate budgetary support, almost cynically overlooking any implied commitment by the President to carry out a program he had just signed into law.

On December 2, I noted in a memorandum to Humphrey that Sea Grant was in jeopardy, the only one of the nine Council initiatives the Budget Bureau was then reluctant to support. By December 16, when budgets are almost always put to bed, Humphrey communicated directly with Bureau Director Charles Schultze in an effort to override the examiners' bearish outlook. Schultze told Humphrey that salvage of this opportunity would require the crash development of an acceptable program. Under the cloak of the Council's policy role, Humphrey invited NSF to submit its plans for review and Council endorsement, but when these hastily drawn plans were undraped at the Secretariat before a hardheaded group of consultants, they were found simply another pocket of funding for academic research. So back to the drawing board. The only shortcut within the deadline was for Council staff and consultants to draft a program that, if acceptable to NSF, could be submitted to the Council as their own. Unable to muster an alternative and aware of congressional displeasure, NSF acquiesced. To meet the lag in the budget cycle, the President was asked to request Congress to re-allocate $1 million from current NSF appropriations for FY 1967 to initiate the program immediately, with $4 million for FY 1968. The President concurred, and the Sea Grant concept published in the Council's first annual report became the official guide for institutions aspiring to qualify for the Sea Grant funds. Soon after, under pressure from interested members of Congress, NSF Director Leland Haworth established a separate Sea Grant Office and offered Robert Abel the position of head. Regretting the loss of his long experience in federal oceanography, we yet urged him to accept in the belief that Sea Grant would flourish in NSF only under someone with his vitality and commitment to marine sciences. As it was, Humphrey's continued interest in Sea Grant was required twice again to dislodge funds from a reluctant Budget Bureau.

In moving initiatives through the Council, up to Johnson, and over to Congress, the budget was the barometer of success. In 1966, when the Council was spawned, support for marine affairs had been creeping upwards but with agonizing uncertainty. A striking change took place after presidential support was gained via the Council's focus on new areas of priority. The continuing increases in funding for marine affairs were the more remarkable in view of the decline of research and development budgets in general that had begun in 1963. In the case of marine affairs, as shown in tables 4, 5, 6, and 7, there was also a conspicuous growth in civilian components, especially in categories of coastal zone development, transportation, and international cooperation.

Ultimately, funds for new programs had to be folded into the budgets

TABLE 4

FEDERAL FUNDING FOR MARINE SCIENCE AFFAIRS,
COMPARED WITH TOTAL BUDGETS

Fiscal Year	Total Budget Outlay (millions)	R/D & Plant Obligations (millions)	R/D Obligations as % of Total Outlay	Marine Affairs Obligations (millions)	Marine Affairs as % R/D
1966	$134,654[a]	$16,179[a]	12.0	333.4	2.1
1967	158,352[a]	17,149[a]	10.8	438.0	2.5
1968	178,862[a]	16,525[a]	9.3	431.8	2.6
1969	184,556[a]	16,306[a]	8.9	463.5	2.8
1970	211,425[b]	16,392[a]	7.8	513.3	3.1
1971	236,610[b]	15,733[b]	6.7	518.5	3.3
1972	246,257[b]	17,108[b]	7.0	594	3.5
1973	270,898[b]	18,608[b]	6.9	658	3.5

[a] Source: Federal Funds for Research, Development and Other Scientific Activities, Washington: NSF, 1970.
[b] Source: President's Budget for FY 1973.

TABLE 5

FEDERAL MARINE SCIENCE AND TECHNOLOGY PROGRAM BY MAJOR PURPOSE

(In millions of dollars)

	Estimated FY 1967[c]	Estimated FY 1968[c]	Estimated FY 1969[c]	Estimated FY 1970[c]	Estimated FY 1971[c]	Estimated FY 1972[d]	President's Budget FY 1973[d]
1. International Cooperation and Collaboration	7.1	9.6	8.4	10.0	8.8	9	10
2. National Security	161.8	119.9	127.2	127.0	101.5	95	97
3. Fishery Development and Seafood Technology	38.1	40.1	45.3	49.8	50.8	49	62
4. Transportation	11.9	11.1	16.7	23.5	36.1	64	70
5. Development of the Coastal Zone[a]	21.4	27.6	32.1	43.5	45.8	65	84
6. Health	6.6	5.3	6.0	5.4	5.9	6	5
7. Non-Living Resources[b]	7.2	7.3	8.0	10.5	10.6	16	20
8. Oceanographic Research[b]	61.5	78.1	78.4	78.4	104.4	124	126
9. Education	4.0	7.0	6.7	8.2	7.8	7	9
10. Environmental Observation and Prediction	24.4	28.8	33.7	39.8	41.1	48	52
11. Ocean Exploration, Mapping, Charting and Geodesy	77.4	75.7	79.7	89.9	73.3	89	101
12. General Purpose Ocean Engineering	14.8	19.2	19.1	24.7	29.4	29	30
13. National Data Centers	1.8	2.1	2.2	2.6	3.0	3	4
Total	438.0	431.8	463.5	513.3	518.5	605	667

Note: Many programs of the Departments of Defense, Commerce, Interior, and Transportation, and other agencies closely related to marine science, are not included. Programs supported by ARPA on advanced surface platforms and by Interior's Office of Water Resources Research on marine research are included for the first time in Column 3 and subsequently. Details may not add to totals because of rounding.

[a] Includes Shore Development, Pollution Management, Recreation.

[b] Research beneficial to more than one of the headings above.

[c] Source: Annual Reports of Marine Sciences Council, 1967–71.

[d] Source: Preliminary draft of Marine Sciences Report of the President, April 1972.

TABLE 6

FEDERAL FUNDING FOR MARINE SCIENCE AFFAIRS,
SHOWING REDUCTIONS IN THE APPROPRIATIONS PROCESS

Fiscal Year	President's Request	Estimate (1 yr. later)	Estimate (2 yrs. later)	Total Increase in President's Request (over prior yr. estimate)
1967		409.1	438.0[a]	21.0[b]
1968	462.3	447.7	431.8	53.2
1969	516.2	471.5	463.4	68.5
1970	528.0	514.5	513.3	56.5
1971	533.1	518.5	517.4	18.6
1972	609.1	605		90.6
1973	667			62

[a] Includes shore development, pollution management, recreation, not included in scope of prior estimates for the same year.

[b] Based on comparisons extracted from successive reports of the ICO on the National Oceanographic Program.

Source: Marine Science Affairs; annual reports of the President to the Congress on Marine Resources and Engineering Development, 1967–71.

TABLE 7

COMPARISON OF DEFENSE AND CIVILIAN COMPONENTS IN
MARINE SCIENCE BUDGETS

Fiscal Year	President's Request	Department of Defense	Percentage of Total	Civilian Agencies	Increase in Civilian Agencies over prior year
1967[a]					−4.1
1968	462.3	258.7	55	203.6	30.3
1969	516.2	297.7	58	218.5	27.7
1970	528.0	297.9	56	230.1	20.9
1971	533.1	239.7	45	293.4	42.7
1972	609.1	245.8	40	363.3	69.7
1973	667	252	38	415	51.7

[a] Prior to FY 1968, Presidential budget requests did not include a breakout for marine science affairs, as required by P.L. 89–454.

Source: Marine Science Affairs; annual reports of the President to the Congress on Marine Resources and Engineering Development, 1967–71.

of operating agencies, and the Council came to recognize that a sizable fraction of agency time and energy is devoted to raising funds. Proposals must run a tortuous obstacle course of review and approval, first inside the agency and then outside it.

All the lower level proposals are tailored by the departmental budget officer, then trimmed by a cabinet officer. Winners in these semifinals confront a skeptical Budget Bureau where the submission is further squeezed, only to be cut again when the President makes his final decisions on the basis of expected receipts and last minute realignment of priorities. Ultimate benediction resides in the "President's budget," which becomes the Holy Grail of bureaucratic fervor and the symbol of accomplishment. Each January the net product is convoyed across Washington to enter the labyrinthine and mysterious ways of congressional action, there to be further dissected, analyzed, assayed, and possibly dismembered.

Appropriated funds are almost always less than the President has proposed. In the executive branch efforts are unbelievably renewed to shave further; the Budget Bureau releases appropriations in quarterly allotments, a trickle down process that affords them additional opportunity to withhold amounts so long that all cannot be obligated before the fiscal year expires. Under severe budget pressures during the year, the President may cut appropriated funds by a fixed percentage; or the Congress in a fiscal crisis may impose similar restrictions by making presidential cuts of last year's appropriations a condition of next year's. Finally, Congress may have acted on the appropriation long after the new fiscal year begins, during which time agency financing has been permitted at the prior year's level on a "continuing resolution" so as not to completely paralyze governmental functions. Because originally proposed increases often cannot be committed after such delays, the appropriations may be trimmed in proportion to the time lost—a maneuver destined to frustrate operating officials that have suffered both the uncertainties of delay and anguish of getting a program accepted. Finally, Presidents have been known to impound appropriated funds to release in a cornucopia of plenty in an election year. No small wonder agencies had so little time, energy and inclination to try anything new.

The tensions developed in such an atmosphere baffle most attempts to get anything done, so the Council and its Chairman tended to give priority attention to programs that crossed agency lines, because their multiagency nature often meant support from none. A second role of the Council was to help individual agencies defend their participation in multiagency programs, when their assigned role failed to earn sufficient priority at the Cabinet level, but warranted support within the government-wide marine mission that the Council had developed for the President. On several occasions, the Secretariat defended a budget in the White House or on Capitol Hill.

In view of the absence of crisis and the impotence or lassitude of the special interests, this growth was an anomaly that could be partially explained by the introduction of a new factor into the government's marine family—they now had a chief oceanographer as advocate.

The Vice President was concomitantly serving marine affairs as an "assistant President," and it was in this management role that Humphrey made his most telling contributions to marine affairs.

The most critical decision elements of Council operations demanding the Vice President's attention were the plenary sessions. Each was a carefully planned production, with a scenario, a script, and a stellar cast. The central plot was to transact business over such counterplots as a constant testing by recalcitrant players of the Chairman's determination and leadership, and a pulling and hauling by agencies as competitive as Gimbel's and Macy's. The Executive Secretary's role was to ferret out the issues deserving action at a high policy level, to fashion the appropriate agenda, and to prepare advance papers with recommendations for action. In preparing these we meticulously tried to list all known facts and background, for to have done otherwise would have eroded the integrity and effectiveness of the office. We ran an honest game but definitely not a neutral one, since we studiously avoided the low impact compromise that characterized many interagency bodies. We sought decisions. The Secretary's horror of spinning wheels occasionally upset the Chairman, who once complained that too many moving targets had been set up on the rifle range at one time.

Working papers were manufactured in a variety of shops—by Council committees with the aid of their staffs, by lead agencies, by Council staff or consultants. Preparation often took months, especially if the issue were controversial enough to warrant interagency consultation at lower levels. As deadlines relentlessly approached, the draft paper was often found incomplete, temporizing, or lacking a government-wide perspective. The product was delivered by hand to the principals about ten days before a meeting, but this was no guarantee of adequate preparation before a session. Every principal had a kaleidoscope of tasks, and usually the marine affairs homework was shunted down to senior civil servants, to gain both their expertise and their advice on how best to defend agency interest, thence back up through the bureaucracy. A goal of ten days for preparation seemed unrealistic in view of this procedure, yet the hierarchy was spurred by the urgent need to brief their cabinet officer or policy-level representative so he would not make a foolish commitment at the Council meeting. Such constant goading roused some snapping back by the bureaucratic dinosaur.[24]

Quite correctly, Council members assumed that staff recommendations for action openly reflected my own views. Humphrey's position, however, was guarded. Even more opaque were the views of agencies hoping to gain support for some surprise alternative. The fact that during a meeting all agencies had to lay their different cards on the table for a public vote was the way a pluralistic society should find expression. The Council Secretariat assumed there would be differences and it sought to reconcile these as essential to public process. But it was a painful discipline. Secret backdoors were closed because higher-ups among presidential staff who might have helped recalcitrant agency heads were now present as observers and even if members did not commit themselves to Council proposals, they were at least expected to set forth their reservations and not to undercut the Council privately at a later date. In fact, that occasionally happened.

Humphrey's style was not simply to preside but to participate in the meetings and even to steer the dialogue. Thus before each session he expected from the Executive Secretary a thorough briefing on the issues, their ramifications, and the desirable outcome. He was provided all the papers distributed to Council members, drafts of statements he could use in opening discussions, and a confidential aide memoir explaining each issue, the alternatives, differences among agencies and reasons why (if they could be flushed out by Council staff). The briefing books were always delivered the night before a session, but there were times he entered the Council chambers with the book virtually unopened, and some disaster seemed inevitable. That none ever occurred was probably due to the Chairman's remarkable memory. At the meeting, Humphrey was quick, intuitive, and surefooted where the diverting rhetoric of participants might have tripped a less able chairman. Attendance by Council members or their alternates at a high policy level was excellent, partly the result of personal letters of invitation from Humphrey, partly because of the sense of collegiate effort that elicited willing participation. Observers were also added to the list of statutory members from operating or Executive Office agencies having interest. Names are listed in note 10 to this chapter.

At the session itself, the dialogue was spontaneous. After Humphrey opened with his own paraphrasing of prepared material, he asked for my précis of the issue, then discussion. Participants having a direct interest in the issue usually volunteered comment; if they did not, Humphrey called on them in sequence in efforts to prevent later obstructionism. He was always fair and considerate. But if there was visible or suspected fence straddling, his humorous probing soon brought out the various points for resolution. Among other benefits, this type of debate raised a cabinet member's perspective to presidential-level sights, temporarily subordinating normal parochial concerns.

As a Council session neared a point of decision, Humphrey would phrase the decision and ask for dissenters to make their views known. This immobilized those who had reservations primarily on the basis of narrow self-interest rather than logic, and the die was cast.

Occasionally the issue paper proved defective and would be withdrawn to the kitchen for further baking; occasionally candor elicited a deadlock, in which event action was deferred until the next meeting. In that interval the Secretariat was expected to develop a basis for agreement. Seldom were there confrontations between Council members and Chairman or Secretary: Both Humphrey and I knew that too many of these would erode the Council's effectiveness. But meeting after meeting we were in the role of forcing a decision where circumstances and the President's interests warranted pressure.

Our most unpleasant moments came when a Council member's position was contrary to what a subordinate had asserted or been persuaded to adopt at a preparatory session in Council committees. Although a subordinate's principal was not committed at these lower-level negotiations, it was assumed that he had been consulted and tacitly had approved. The uneasy suspicion emerged and was frequently confirmed that the earlier position had not been overturned; the subordinate had simply chosen to defer candid disclosure at lower-level sessions when he felt he could not hold his own, in the hope that his Council principal would be more stubborn. That pattern occurred often enough that Council staff sought more effective tactics for the early detection of opposition.

When a Council member actually made a commitment, we could count on his integrity in not going behind the Vice President's back to the President with a plea to overthrow a privately disliked decision. Budget Bureau and Office of Science and Technology representatives, however, traditionally upheld their privilege to remain silent at the meeting. My task was to caucus with them privately to learn if they planned to differ with the Council in the privacy of the President's office, then to seek our own avenue of appeal, either Humphrey or I to White House staff, or Humphrey to the President. Only under Nixon did I communicate directly with the President in a confrontation with an Executive Office counterpart.

Humphrey and I knew the fate that Council decisions could suffer if unrecorded, wherein invisible opposition might quietly slip out of commitment, so we agreed to record minutes. However, to encourage candor and spontaneity, a summary from the verbatim stenographic report was prepared, and the original was never disclosed. Instead, the précis was circulated for review of all participants and adopted at the next meeting. As a further protection to participants, it was never released publicly. When Congress once called for the minutes, I exercised presidential privilege by turning aside the request as tactfully as possible, explaining the potential damage of disclosure to Council operations, and the possibility that minutes would not in the future be taken at all. These minutes formed memoranda of understanding that, so far as I know, were generally honored.[25]

For the Secretariat, an issue was usually only begun in plenary sessions, with follow through for the President a collateral responsibility. Herein lies one little-publicized weakness of government, the wobbling implementation by the lower echelons of the bureaucracy of agreements made by agency heads. Often communications from higher authority fail completely, sometimes because of distortion in transmittal. Or lower levels may resent a decision affecting their operation on which they may not have been directly consulted. Or a Council member may make a commitment, even on internal advice, that cannot subsequently be honored because funds or manpower are inadequate—sometimes as a result of presidential cuts after the commitment is made.

In such cases of bogging down, the Council staff assisted in the presidential responsibility by monitoring implementation, sometimes by debriefings for lower-level agency staff eager to fulfill commitments but floundering from poor communication. Following each session I wrote confirmatory letters to affected parties to rivet commitments, but no procedure was foolproof.

In this succession of Council meetings, a steady stream of recommendations to the President emerged, usually relayed by Humphrey. Virtually all were accepted, although several of the budget proposals were trimmed down, and several policy proposals to Johnson late in his administration were deferred for action by his successor. Subsequently, most of these were independently reviewed by President Nixon and adopted. Apart from special messages to Congress, the annual report transmitted by the President to the Congress was the primary vehicle for public announcement.

As is explained later in this chapter, the role of Vice President as chief oceanographer and assistant president subsided early in the Nixon administration, and by April 1971 its statutory delineation expired. The President's responsibilities under P.L. 89–454 remained, however, including that of submitting an annual report to Congress.

PUBLICIZING DECISIONS

Most of these internal decisions, plans, arguments, successes, and failures of the Council were shielded from public view until the distilled results were released in the President's annual report. Its preparation was a nightmare. No other Council exercise demanded so much in the way of staff creativity, animal energies, skill at negotiation, and talent in communication.

By P.L. 89–454 the report was required to contain: a description of government-wide programs with funding and activities delineated by purposes as well as by agency; an evaluation of agency programs in terms of statutory objectives; justification for new federal initiatives and the marine science program as a whole; and recommendations for legislation. In addition we included an abstract of individual agency accomplishments, including a progress report on implementation of earlier Council initiatives and a discussion of unresolved issues, especially those deserving further study by nonfederal as well as federal interests.

Aware of the turgid prose of the typical government document, our ambition was to produce a literate—and readable—compendium that would portray the broad context in which marine-related decisions were made, including background facts, long-range goals, priorities, and statistical tables helpful to those who wanted to make their independent assessment of action.

The annual report was thus somewhat unusual: its chapters were organized according to major uses of the sea rather than by agency cognizance; it was prepared by independent presidential staff instead of by the agencies themselves (which had often resulted merely in a stapling together of sectorial and occasionally self-serving statements); it was analytical in identifying issues and unsolved problems.

Each report detailed Council decisions made during the year, not only as information but as a mechanism to reinforce and clarify issues and internal assignments of jurisdiction that had been fuzzy, in doubt, or subject to conflict. Drafting by the staff was based on material from agencies, and edited by the Executive Secretary, who determined scope, theme, content, and language. Copies were then circulated to our consultants to be checked for technical accuracy and balance, and also to some sixteen agencies for comment and clearance. This latter process often became a test of wills. Almost all agencies replied with the admonition, "The report is too long, but you didn't say enough about us." Then they would submit additional self-serving details that they insisted we use, many of which collided with those from competing agencies. Our ground rules were to correct all errors of fact but to utilize prose only if it contributed to the overall quality of a presidential document. Only once did an agency go over my head to complain to Humphrey.

The Budget Bureau had its own turn at criticism. They would pool the remarks of a dozen or so subject specialists who were examiners for all the agencies emcompassed by the Council, then demand their adoption. Negotiation was strenuous, because their representative felt his prestige was at stake if he yielded on recommended items. In the main, the Bureau wanted to reduce the commitment of the President, either in terms of funds or programs or by blurring policy directives. In their view, that was his best protective stance. Mine, on the other hand, was to state what had been proposed by the Council and agreed upon, in terms of the President's responsibility under the Act.

So the Bureau and the Council staff walked down the same road, not hand in hand but jabbing each other in the ribs, each believing sincerely he was serving the President the best. In retrospect, there is an element of pathos because the two were as disparate in muscle as David and Goliath. Bruises from that contest were accepted by Council staff as part of bureaucratic life. A second type of Bureau intervention was harder to accept when their representative adopted a stance of superior knowledge and disparaged Council staff as not knowing their subject matter. Considering the limited technical training and policy experience of many examiners, and their isolation from challenges to defend their own positions, it seemed they assumed that proximity to the throne conferred a mantle of infallibility.

While the report was ostensibly that of the President informing the Congress, it enjoyed a far more catholic audience. This included the federal staff who could learn what policies were formulated vertically to which their individual missions should respond, and also what activities were being conducted horizontally in sister agencies to which theirs should articulate. Then there were the outsiders: scientists, industrial executives, and technicians, officers of state government, and interested parties abroad who needed to know what was going on in Washington and why, so that they could consider their own program targets, funding sources, and opportunities for collaboration. Finally, we saw an opportunity to give the average citizen a deeper appreciation of science, not as a "gee whiz" spectacle, but as an important step in dealing with complex issues in a technological age.

The popular and trade press carried announcements of the report's release, usually with complimentary remarks. In fact, Council relations with the press were uniformly harmonious, with our attempting to reciprocate their fairness by candor at press conferences.

Occasionally, we goofed. My most embarrassing moment was when I carried copies of the first year's report on the European trip with Humphrey and met with the senior officials to interpret the new legislation, especially to allay apprehension that we might be launching out into space-scale programs requiring some response that they could ill afford. A few hours after presentation of the report in Bonn, our embassy officials called me aside to say that the West German Ambassador in Washington had just filed a protest because a listing of foreign research ships in the appendix had referred to the East Germans by their selected designation of "German Democratic Republic," which the West Germans at that time insisted was the "Soviet Zone of Germany." I was shaken at having potentially contributed to an international incident just when we were seeking a marine sciences rapprochement, but relieved to learn this was a common mistake and forgiven readily.

A STRATEGY FOR THE OCEANS

It is not surprising that both insiders and outsiders drew analogies between the fledgling oceanographic and the robust space enterprise, although one was burgeoning under competition with the Soviets while the other was just leaving the nest. Both dealt with unexplored, uninhabited, and hostile environments; both were charged with high-voltage rhetoric about their promise for humanity. Some of us sought, however, to educate even our enthusiasts about the major inherent differences in the programs. The space effort was then entirely funded federally, whereas fiscal support for marine resource development was expected from private as well as public sources. In addition, space development had been genuinely motivated by concern that the Soviets could exploit their space superiority at least to political advantage and possibly to military advantage as well. No such crash program could be justified in marine affairs from either public knowledge of Soviet military capabilities or intelligence reports. Finally, the marine activities directly involved far more social institutions than was the case in space.

The space analogy seemed a poor basis for beating drums. Nevertheless, we were aware of the growing world-wide influence of the Soviet Union as a maritime power. Japanese shipbuilding was also coming on rapidly, and we had dropped to sixth place in fishing. Paradoxically, the land-oriented Soviets with a minuscule seacoast had gone to sea while the maritime United States continued its perplexing indifference. There was thus a nagging question as to a fundamental relationship of the oceans to U.S. interests that both encompassed and went beyond an aggregate of separate rationales that sparked the various initiatives.

What I felt needed was a richer, more powerful concept: an updated Mahan doctrine that would reflect contemporary technology, external world affairs, and national life styles that Admiral Mahan defined in the nineteenth century as tenets of seapower.

The quest began for a "strategy of the oceans."

I first sought consultation with Admiral Samuel Eliot Morison, Admiral Arleigh Burke, Jerome Wiesner, Carl Kayson, and others, but resultant suggestions for unifying the disparate activities at sea proved to be elusive. No set of policies was readily deduced from basic principles to displace those already derived inductively.

Seeking to buttress the array of policy initiatives by a central theme was something like the futile quest for a philosopher's stone. Nevertheless, we were convinced that a systematic rather than atomistic approach to the role of the sea in world affairs deserved a continuing search. Robert Kay, the most strategically minded of our Council staff, was put in charge of a major exercise dubbed "Sigma M" to extract the broadest long-range considerations on which an American strategy of the oceans could be built. The project uncovered more potent arguments for strengthening maritime enterprises, but without any new unifying themes.

In the concluding chapter of the first annual report entitled "Looking

Ahead," we stated one philosophical underpinning for what eventually became a strategy but more in terms of process than substance—the role of longer range planning:

> Looking ahead, forward planning on a Government-wide basis will complement our shorter term initiatives, especially to encourage an influx of fresh ideas and provide policy officials with a greater awareness of the impact of new options afforded by the marine environment.
> Long-range perspectives will then reflect (a) future commitments for funds that arise from current decisions; (b) the time interval necessary for training of manpower and development of scientific resources; (c) the long time framework associated with interactions between various economic, cultural, and political institutions, and (d) the longer term impact of man's actions on his environment, so as to anticipate and moderate detrimental effects.[26]

By the second annual report, the quest for doctrine had shifted to a quest for key questions—those that focused on fundamental national goals. Until that time the Commission had still been uncertain as to its own targets, and those posed by the Council were potential guidance. The last chapter, entitled "The Nation and the Sea: Questions for the Future," stated:

> While the course for the next 18 months—until the end of FY 1969—is reasonably clear . . . we must now examine the more fundamental question: "What portion of the Nation's energy and wealth should be devoted to ocean endeavors, and how?" Precise quantitative answers may elude us, and our best estimate today will be slightly altered tomorrow. Nevertheless . . . our task is to inquire about the role of the oceans in contributing to the world of the future.
> To establish the future direction and pace of a truly national—and not just Federal—marine science program, we must examine and re-examine the same key questions about both ends and means that were posed in the earlier chapters as a basis for contemporary decisions. . . .
> The Marine Sciences Council is thus endeavoring . . . to evolve a forward-looking and durable backdrop of policy—a possible strategy for use of the oceans.[27]

In the last chapter of the third report, entitled "Looking Ahead," the Council reported that a further step toward a strategy lay in better integration of ocean-related institutions:

> While the responsibility for leadership must largely rest on the Federal Government, our ocean activities embrace the States, industry, the academic world and other nations and involve complex interrelationships and interdependencies among these participants. . . .
> These activities and institutions may be thought of as being in two classes: the first are knowledge-generating; the second are knowledge-consuming. Effective articulation between the two is obviously required to match needs and opportunities. Policy planning has focused on imperfect articulation between participants, to identify and where possible dissolve inadvertent impediments to the smooth application of marine discoveries.

Within the Federal Government, interaction between knowledge producers and knowledge consumers has operated most effectively in the case of the Navy [which] historically has been this country's major sponsor of marine research, both basic and applied. Our progress in civilian fields has largely stemmed from this naval research base. . . .

In the case of civilian uses of the sea, however . . . the number and variety of participants, their frequent competition of interests, the need to introduce non-technical considerations of law, finance, and foreign relations, and the setting of priorities in a more public forum have been accompanied by new communication problems between science, technology, and marine affairs.[28]

That third annual report released January 17, 1969, exactly three days before inauguration, concluded by calling attention to the Stratton Commission report, which appeared almost simultaneously. Their document, "Our Nation and the Sea," painted a sweeping panorama of individual opportunities for this nation to realize its stake in the oceans and cited attendant problems. Mature, comprehensive, and courageous in many of its stands, the report contained no prescriptions for a coherent strategy, but rather put its eggs in the basket of federal reorganization.

The fourth and fifth Council reports with presidential statements were released under the Nixon administration in March 1970 and April 1971. In both, there was a conspicuous absence of a last chapter; there was no "Looking Ahead." That silence from the Nixon administration was broken in June 1970 with adoption of NOAA. But its over-all concept for the oceans was unstated, in harmony with the administration's determination to keep a low profile, generally.

BUREAUCRATIC JUDO

The Council Secretariat invested much effort in soliciting agency understanding of its role and their cooperation, realizing that success depended critically on diplomacy and on harmonious personal relationships as much as its power and position. We never forgot it was the agencies that had the funds and the structure to carry out whatever policies and programs the Council generated. Nevertheless, the toughest, bloodiest, most enervating elements of Council life centered around the invisible struggle within the bureaucratic family. No federal agency had backed Council creation. To them the Council loomed as another layer of Executive Office red tape. Even when, in response to President Johnson's request for initial program recommendations, the Council had demonstrated it could loosen up purse strings, gain presidential attention, and overcome congressional budget cuts, many agencies still regarded the upstart with jaundiced disfavor. Up the lines of communication came nautical admonitions of "Don't rock the boat," and "You're making too many waves." Exhausting as it proved, the Council staff actually was setting tides in motion through the bureaucracy. The pace of Council actions was both a necessary stimulus to action and a defense. We knew we were making the bureaucracy uncomfortable, and we felt our best protection against the tactics invented to slow us down to a traditional speed was to make it difficult to hit a moving target.

The Secretariat was in a struggle with almost every member agency at one time or another, but warfare with the Navy was the most exasperating. For years, the Navy had been the "big boy" in the oceans— running ships, sponsoring most of the research, influencing Congress, and capturing the headlines. Suddenly, competition appeared in the form of a mouse that could potentially roar. Since the Council was never conceived of as an operating agency, it could never be a budget competitor, but it could get public attention and it could capture the ear of the President. Up to this time the Navy had been on the bridge of the entire government's marine program via its chairmanship and staffing of the FCST Interagency Committee on Oceanography. It had tried to influence the PSAC study on oceanography. It had tried to name the Council's Executive Secretary and, failing that, to staff it. (In 1967 it tried to influence the Commission by again volunteering ICO staff.) Now, a naval officer involved with ICO and without knowledge of higher civilian authority as to tactics began to undercut the Council with direct and contradictory contacts with Congress, hearty press relations, and publication of Council-initiated reports with their continued ICO identity. Hornig, advised of the problem, seemed in no hurry to control one of his FCST committees and lose his only instrument of salt-water cognizance. Only after a year of jousting did Assistant Secretary of the Navy Robert Frosch and Hornig finally agree to the transfer of the ICO from the Federal Council to become the Marine Council's first committee. When cooperation still was withheld, Humphrey intervened, persuading Frosch to mothball the ICO; Humphrey appointed a new committee with Frosch its chairman, but now one of four under Council cognizance. The ICO, with its hidden control by a naval officer, died hard.

Harassment of the Council by some of the Navy's young stallions mirrored an attempt to freeze this intruder out of the Navy's activities on grounds that the Council would dictate the Navy's program. When the Navy had trouble with their own appropriations, they refused Council offers of congressional intervention, yet spread the rumor we had contributed to their problem by not supporting their budget in the Council! Later, these officers complained of Council favoritism, cutting the Navy's budget to enhance the civilian Environmental Science Services Administration. Not only was this not done—it couldn't be done. The marine science pie of whatever size becomes an aggregate of separately rationalized components, rather than a fixed entity that is later subdivided.

On at least three occasions, policy echelons in the Navy rejected offers of Council assistance, especially to expand deep-ocean technology. Again, when preparing FY 1971 budgets, pressure from Nixon or Laird had resulted in strenuous dehydration of all non-Vietnam defense programs, and the Navy's hydrographic and charting services became a conspicuous target. Ships were to be laid up and staff laid off. Council intervention was rejected by the Oceanographer of the Navy.[29] Whatever satisfaction the Council felt in advancing capabilities in marine science affairs was tempered by not having been able to break through the Navy's defensive attitude or obtain its support.

Another conflict lay in the repeated wrestling match on Sea Grant between the Council and NSF. The first encounter resulted from NSF's slow implementation after passage of the Act, the second from delays in considering applications from universities. The Council's March 1967 report stated that "criteria by which proposals for such grants would be judged . . . [were] stated only in general terms" and that elaboration was "soon to be announced by the Foundation."[30] The appetite on many campuses was whetted, but by summer no announcement had been released. Would-be applicants were communicating their dissatisfaction to our office, highly muted out of concern for offending a potential sponsor. And Congress was needling. Calls to Abel revealed his draft stuck on the NSF Director's desk, awaiting personal review and editing. Humphrey wanted action, and we successfully blasted the plan out by scheduling a presentation by NSF at the August 31 Council meeting. That winter congressional growls became steadily louder because a year after the legislation had been enacted, NSF had not made a single institutional grant. At the August Council meeting, Leland Haworth had stated that "tremendous interest has been shown in the program in all kinds of institutions . . . throughout the country," and that "even though NSF has thus far discouraged formal proposals, a half dozen formal and about twenty informal proposals [have been] received," and that "by mid-fall the first grant should be made." One cause of delay was the poor quality of university submissions for interdisciplinary, institution-wide projects. Nevertheless, the Council Secretariat found itself caught between NSF resistance to nudging and a staccato of congressional inquiries as to progress, and also aware of its requirement to assist the President.

As the deadline approached for release of the second annual report, sixteen months after enactment of Sea Grant legislation, a telephoned plea to Philip Handler, Chairman of the National Science Board brought prompt action. On February 21, 1968, ten days prior to report publication, the NSF announced awards to the University of Rhode Island, Oregon State University, and the University of Washington.

The next hassle occurred in the summer of 1969. The Budget Bureau was becoming apprehensive about the possible ultimate size of Sea Grant, concerned about how many institutions would be involved and with their support levels. Uncertain as to marine affairs commitment under the Nixon administration, the Secretariat requested NSF to update the 1967 policy statements with details as to program goals to entice Bureau support. In an emotional explosion, the NSF representative at a Council's preparatory committee attacked the Council with a threat that revisions would be sought in Sea Grant legislation so that the Foundation could free itself from interference. But there was no program proposal until the Secretariat stubbornly persisted. Three times the Council successfully rescued the Sea Grant program from an impending Budget Bureau slash that was threatened because plans were unconvincing.[31]

Many arguments are honest differences of opinion, but in the bureaucratic structure a substantial fraction of conflict arises from rivalry based solely on instincts to protect sectoral views. Occasionally, the issue became one of life or death for the Council itself. The main threat came from the Bureau of the Budget. Initially, the Council's statutory life span of only twenty-two months was an asset, a legitimate excuse for deadlines to our members. But when it became clear that Council expiration would leave a hiatus unless Commission proposals were acted upon within the last four of the twenty-two months of Council life provided by P.L. 89–454, an unlikely prospect, our suggestions in the spring of 1967 for extension were abruptly snuffed out by the Bureau.

But delay in Commission appointment, and their discovery that eighteen months was too short a time for them to respond, then inadvertently stretched Council life to June 30, 1969.[32] As the 1969 death knell approached, and among all the uncertainties of a new administration, the Bureau acted vigorously to head off further extension. Nixon's prior commitment and the favorable reputation of the Council proved too strong, however, and a scant two months before the ax was to drop a second brief extension was granted, to June 30, 1970. Even this small triumph was jeopardized when the Bureau proposed to cut budgets by half, requiring major reductions in staff that inevitably would shake the morale of those left. Fortunately, at this early stage of his involvement Vice President Agnew was interested in his Council responsibilities and joined me in intervention. All too soon, the June 1970 deadline brought another threat to the Council's life. Surprisingly, the Bureau this time reversed its longstanding opposition and approved extension and even a reasonable budget, presumably because of a fervent hope that this action might head off some of the pressure for NOAA, which the Budget Bureau opposed even more energetically. But in these episodes rivaling the old Saturday afternoon *Perils of Pauline*, the heroine failed to survive. By 1971 Agnew had abandoned his predecessor's active role of support for the Council. A House Appropriations Committee grown skeptical of Council influence cut the budget so far that even partial restoration in conference with the Senate made its continuation improbable. Within the administration, the Council was being privately opposed by the Secretary of Commerce and Acting NOAA Administrator Robert M. White, who did not want a kibitzer in the Executive Office (shades of Navy experience). Lacking congressional support from Lennon and Mosher because they had no indication of Vice Presidential interest and could not understand why a successor to the Executive Secretary was never appointed, the administration gave notice on January 5, 1971 that the Council would be terminated April 30.

Most of these collisions represented conflicts between agencies where representatives did not speak as individuals. Occasionally, the issue became personal. One example of such a contest arose with Vice Admiral Hyman Rickover over the Soviet's offer to buy *Star III*, a miniature research submarine designed and built by Electric Boat Division, the nation's leading shipyard constructing nuclear subs. *Star III* was a two man craft, capable of submergence to 2000 feet or so, typically employed in underwater archeological research. In an effort to break into the commercial market abroad, Electric Boat placed the craft on display at a trade fair in Frankfurt in 1965, incidentally sponsored by the Department of Commerce. There was one offer to buy—from the Soviet Union. With the knowledge that the Soviets were trailing the United States in the technology of research submarines, Electric Boat consulted governmental channels responsible for clearing sales to Eastern European countries of materiel that might be an inadvertent aid to a competitor's armament. Since there was no precedent for this kind of craft, it needed special study. The matter came quickly to the attention of the Council because of the interest of several agencies—Navy, State, Commerce—and because it had high-level policy implications. Coincidentally strong in submarine expertise, the Council staff examined *Star III*'s design to determine if any classified technology was employed in its construction. The result was negative; the welding technology that was later to be at issue was published years before in the Soviet Union, and the hull itself fabricated in a commercial shop in Houston. Independent studies within the Navy came to an identical conclusion.

By that time, however, Vice Admiral Hyman Rickover, who operated outside of channels with his nuclear submarine responsibilities, was interested. Rickover unequivocally said "no" and let his wishes be known to the Navy Department, which reversed its position. By then he discovered that the Secretariat had been developing a recommendation to the President and was encouraging the Deputy Undersecretary of State, Foy D. Kohler, not to bar the sale precipitously. The transaction offered hope for a small beginning to that international comity in marine affairs that we had been talking about so earnestly in policy papers. We considered the Soviets the other major participant in the oceans, and here was an opportunity to show good faith. The only concern I had was with possible use of *Star III* by the Soviets for snooping at our underwater acoustical surveillance gear, and that possibility seemed remote. The boat had such limited range that it would always be near a conspicuous mother ship, and if the Soviets dismantled it to copy, they would be using a technology at least five years behind what the United States was using in its later, higher performance research submarines. So I recommended sale, but with one proviso. Curious about what the Soviets would do with the boat, I thought we should maintain a warranty provision that all servicing and provision of spare parts should be through United States channels, thus at least keeping track of where the boat was used—Black Sea, Pacific, or wherever.

When Rick caught wind of my recommendation he telephoned. "Wenk," he said, "are you a patriotic American?" My response was, "I don't know of any reason why my loyalty is being called into question." His rejoinder was, "It will be, if you persist in your recommendation to sell to the Soviets." I presented all the arguments on why no element of national security was involved, recalling my own extensive experience as the Navy's specialist in this very field, even the fact that I had been in charge of the first deep dive of *Nautilus*, the Navy's pioneering nuclear sub. Nevertheless, the next day in closed session with the Joint Committee on Atomic Energy, Rickover changed the scheduled subject to spend an hour lecturing on the hazard to our security if the sale were made. Senator John O. Pastore called later to ask if the White House had studied the problem. I assured them it had and that I was prepared to defend the position with facts. The next week I left (with Humphrey for his fifteen-day whirlwind tour of Europe to explain Johnson's Vietnam policy) to open discussions with European officials on cooperation in the sea. On returning, I found a *fait accompli*. Rickover had prevailed over the Assistant Secretary of the Navy Frosch who personally saw no security threat; Kohler had rendered a negative decision, unwilling to fight with Rickover before the Congress on this issue. And since Electric Boat would rather lose a trivial sale than anger Rickover, the case was closed.

While these experiences confirmed that bureaucratic family life could be painful, as a government official I was aware of another and more public arena for strife—on the other side of Washington. The Council was a child of Congress and they were continually concerned for its care and feeding. There was, however, a second and even more trenchant explanation for their interest. Because it cannot deal with overlapping committee jurisdictions, Congress has difficulty with overlapping interagency programs. The only way government-wide activities can be reconciled is under the President, so Congress is obliged to rely on a politically ticklish discipline of leadership in the executive branch. In marine affairs, the Council was the chosen instrument. Moreover, the congressional committees involved appreciated rapport with an element of the President's office that had all too frequently been solidly isolated by executive privilege.

From our side of Washington, the President always expects his team to extract support, and if possible, praise, from the Congress for his initiatives. Rapport with Congress was essential.

RAPPORT WITH CONGRESS

Thus, under Humphrey's tutelage, the Secretariat deliberately sought cordial but correct relations with Congress. Indirectly, some 30 subcommittees in both Houses were involved. As a step toward knitting that relationship, in his first marine affairs message President Johnson paid Congress tribute for "responding to the challenge of the oceans by enacting the Marine Sciences, Sea Grant, and FPC Pilot Plant Acts."[33] That literary nod did not sit well with the Bureau of the Budget, who seemed to dislike legislative politics. While the Bureau staff has a poorly concealed distrust of politics and politicians as natural enemies of the tidy processes of efficiency the Bureau espouses, they were especially apprehensive that the congressional contacts Humphrey and I had might dilute loyalty to the President. What seemed to nettle the Bureau most was the fact that the Vice President, as President of the Senate, had a completely legitimate pipeline to Congress. With Humphrey, this was no ceremonial responsibility. He had loved his club and continued to build respect among his colleagues by use of his legislative office, literally and figuratively. As presiding officer in the Senate, the Vice President of the United States has a formal office in the Capitol just off the Senate chamber, far more impressive than the suite in the Executive Office Building and useful when vice-presidential prestige might foster some negotiation.

This proximity to Senators was indeed convenient for discussing marine affairs. Congressional complaints that a program was flagging could be made to its statutory chieftain. In turn, Humphrey was able to drop hints to members of the Senate Appropriations Committee when some element of marine affairs was fractured in the House and needed to be restored to health.[34]

Formal engagement with Congress took place through their hearings. In three years as Executive Secretary I was to appear as an administration witness twenty-two times, to interpret and support the President's government-wide programs or positions on new initiatives, to defend Council budget requests or legislation, and to speak on behalf of other agencies' programs.[35]

My baptism on October 10, 1966, however, was nothing short of traumatic. I was defending a $1.3 million supplemental appropriation request by the President on behalf of the Council and the Commission before the House Interior Appropriations Subcommittee that was to be our banker.[36] The chairman was Winfield K. Denton, and members included Julia Butler Hansen from coastal Washington, an outspoken supporter of fisheries development and an expected friend. I thought I had done my homework. But the hearing had just opened when the first needle was injected. Inquired Chairman Denton, "If you were to make a recommendation regarding the oceanographic activity of the Defense Department, and the military experts were opposed to such a recommendation, what would happen?" Admitting that disagreement could occur with any agency, I observed that since arguments could arise from different bases of fact, the first step was to assure that both sides had equal access to information. If a genuine difference of opinion remained we would attempt resolution between the Vice President and the department head. Failing that, the issue would go to the President.

After that challenge to the Council power structure had been weathered, the next blow was shattering. Congressman John O. Marsh, Jr., first drew me out on how I was getting started with $50,000 borrowed from the Navy and the National Science Foundation. Then he dropped a bomb, implying that I had been operating illegally because of "the continuing resolution of the Congress on the 30th of June that no funds should be used to initiate or resume a project that was not already initiated or underway in FY 1966," a resolution frequently adopted to keep the government machinery running after June 30 because of delay in congressional action on appropriations, but admitting of "no new starts." Pleading ignorance of their caveat, I reported having consulted counsels of both the General Services Administration and the Budget Bureau on the legality of interchange of funds among agencies, and cited the relevant section of the U.S. Code as authority. Apart from legal considerations, I tried to point out my desire to get on with the job, an alacrity I thought Congress would applaud after six years of executive malaise. Moreover, according to P.L. 89–454, we were likely to be in business only twenty-two months and the President had wanted immediate recommendations. Marsh returned to their blanket injunction of "no new starts." Next I tried the tack that the Marine Sciences Bill became law June 17, which predated the June 30 barricade, and involved coordination of activities that were in business years before. Still no sale.

Then came the revelation of what was really bugging the Committee. The fact was that the President often cuts appropriations after congressional action, embarrassing the Appropriations Committee as though it had not done its job. So on principle, they did not want him starting, much less stopping, any activities without their Committee's permission.

What other committees wanted in marine affairs was irrelevant. Said Mrs. Hansen:

We wouldn't be quite so perturbed if programs already going on, and which were most important, weren't already curtailed. They were appropriated, funded by the Appropriations Committee, and then the Bureau of the Budget . . . curtails these activities. . . . Then in this instance they permit the beginning of a new activity in a time that has been stipulated . . . that no new activity shall take place. I think this is one of the very maddening instances of the executive branch not recognizing the legislative branch and the authority vested in the Congress.[37]

I was caught in a crossfire that had nothing to do with marine affairs. And I discovered again how nonpartisan members can be when, outgunned by the executive, they close ranks. Moreover, there was no special courtesy accorded Executive Office components. On the contrary, between the lines I could sense their resentment of any meddling by presidential staff in the comfortable relationships between senior legislators and agency professionals who dominate programs regardless of who is in the White House.

The House cut the Council budget by $500,000 on grounds that our contract studies were unnecessary, slapped my wrists further for getting started without their permission—stating that although financing "was 'within the framework of legality,' the Committee is perturbed with this procedure and wishes to go on record that it does not look with favor on this 'left-handed' or 'side door' funding of new activities"—and embarrassed the Council by prohibiting return of the borrowed money to Navy and NSF.

Damned if I had not got started promptly and damned if I had.

Despite my appeals to Congressmen Lennon and Mosher for help, our hope for restoration lay in the Senate. Humphrey addressed a personal letter to Senator Pastore, who was to chair the supplemental appropriations hearings supporting the full budget. I wrote to Senator Carl Hayden, along with a respectful letter to subcommittee chairman Denton reiterating the legal validity of my original stand, now backed by further research: Navy and NSF funds used were "no-year" appropriations carried over from earlier years; the functions were not new, and the Act was signed into law before the date of the continuing resolution. Nevertheless, Senator Stephen Young remarked that the "Appropriations Committee members on both sides don't like to have a department of government embark on a new program without their getting the amount in advance"—a cue that perhaps I should have consulted these Appropriation members before doing anything. Then Pastore set the stage for the Senate mood in noting, "Even if there was an error, it was done in good faith." The Senate restored the full amount, and the compromise was unexpectedly more than halfway: $1.1 million.

Apart from the hearing episodes, I had made a special point of regularly informing Lennon and Mosher of our progress and problems, mindful of their scars from battle with the Senate over the Council and their skepticism that the Council could attract high-level administration attention. Once the bill was signed, I tried to suppress the fact that the Council was the Senate's child. House apprehension apparently turned to admiration as they read the first annual report and they even felt a surge of pride in having been one of its parents.

Apparently that effort paid off when our progress in fostering harmonious relations with the Congress became known on August 17, 1967, when the first anniversary of the Council was celebrated by a hearing. Lennon opened, saying:

. . . Some of you will recall that there was a difference of opinion with respect to creating by this act a so-called National Council. There was some apprehension on the part of the administration that it would be repetitious if the individuals designated by these appointed to the Council at the Cabinet level would be the same individuals who have comparable or similar positions under the White House program for marine sciences.

I want to make this public statement here that through the inspiring leadership and, I think, dedication and complete interest under the Chairman of that Council, Vice President Humphrey, we have seen a new order and a new day . . . and I know that each meeting of the Council has been well attended by the members of the Council or someone at a responsible level. . . .

They have done, in my judgment and in the judgment of those on the subcommittee, Mr. Mosher primarily . . . a very fine and excellent service to this great field.[38]

The hearing itself was not entirely a love feast, however. Congressman Richard Hanna expressed concern on the seabed legal regime issue, discussed in chapter 6. Strongly opposing the February 15, 1967, resolution of Senator Frank Church that was suddenly recalled as Malta Ambassador to the UN Arvid Pardo sprang his surprise proposal at the UN for internationalization of the deep sea bed, Hanna admonished the government not to give away its birthright. My reply was, "I can say positively that the administration has taken no position at all on this proposal." Congressman Ed Reinecke from California, reflecting an even stronger conservative bent, pushed harder on the seabed issue, then changed direction and flatly charged that "the Council has just decided to take over the Commission." Apparently he had been briefed by some of our detractors, for at that time at least three Commissioners and their staff seemed persuaded that there was a competition between Council and Commission that would produce a winner and a loser. Reinecke insisted that the Commission was "our real breath of hope over administration threat of veto." Congressman Thomas Pelly corrected that view, and Mosher so staunchly defended Council activities that Reinecke retreated, but not happily.

Having served as a congressional advisor, I well understood and respected their role in questioning our performance. For my part, I hoped that my deportment in energetically and candidly defending the administration's activities would be regarded favorably by my bosses. There was no doubt that such appearances before Congress have a healthy, sobering influence, reminders that power stems from the people.

For a period, the Congress felt satisfied, and the initiative to advance marine affairs that resided in the Congress from 1959 to 1961 and 1963 to 1966 seemed now to be in the appropriate hands of the implementing agent. Lennon and Mosher became strong Council supporters, and it was their initiative in writing President-Elect Nixon that kept me in office at the transition. But in 1970, as discussed in chapter 8, congressional apprehension was to erupt anew.

RECRUITING A CONSTITUENCY FROM THE 51ST STATE

Notwithstanding a new trajectory of marine activity beginning in 1966, many external events were affecting the program. The first of these concerned our country's domestic affairs. Amidst growing social unrest, magnified by the drain of resources for Vietnam, the nation discovered it had a long agenda of unfinished business. Against these needs no explicit ocean crisis existed, such as in 1957 when Soviet initiatives spurred our own space program, to warrant priority attention.

The second problem was ocean science's failure to relate its capabilities to these turbulent times. In fact, science as a whole had difficulty asserting its relevance. The cleavage between the scientific community and the government opened as early as 1963, with storm signals that Congress was no longer willing to support research in unquestioning faith that it was good for the country. And they were alienated by the scientific community's rebuttal.

In the case of marine exploration and development, we had from the outset emphasized the economic and social benefits to be anticipated. Nevertheless, the net momentum and constitution of a marine-science program depended on two concurrent forces: the "pressure" of science-generated basic knowledge clamoring for application and the "suction" of unsatisfied social needs its programs could meet. But with the tide running against science, the first of these two voices was muted in the noise level. The second was no stronger.

Among the fragmented ocean-related interests outside of government, there was neither inclination nor aptitude to function as a legitimate constituency. The President had stressed the public-private partnership involved in effective use of the sea in each annual presidental message. In his first, Johnson said: ". . . appropriations to support marine science activities . . . will enhance the capabilities of local government, universities, and private industry to join in this vital enterprise."[39]

Then in 1968, the President said: "While we strive to improve Government programs, we must also recognize the importance of private investment, industrial innovation, and academic talent. We must strengthen cooperation between the public and private sectors."[40]

Again in 1969: "We engaged the ideas, encouraged the participation and focused the investments of our Federal Government, states, industry and universities on more effective and intelligent use of the marine environment."[41]

In 1970 President Nixon echoed the same theme: "The Federal government will continue to provide leadership in the nation's marine science program. But it is also important that private industry, state and local governments, academic, scientific and other institutions increase their own involvement. . . ."[42]

Beyond this rhetoric, the Council endeavored to mobilize a closer partnership with outside interests. But what we recognized as essential in the competition for federal attention, priority, and funds was effective advocacy. In our pluralistic society, the achievement of public policy goals normally stems from the combination of a politically effective lobby outside the government and a strong organization within. In the case of marine science affairs, both of these were lacking. One could have hoped for coalitions among the different marine interests: scientists, aerospace, fishing, offshore oil, minerals, and shipping industries, and conservationists interested in preservation of the coastal zone. We even hoped the highly atomized contributors to marine affairs would have learned the truism about strength in unity, but as that fragmentation proved incurable, we sought to awaken each interest separately to its stake in the ocean.

With the scientific community, we worked through the Committee on Oceanography of the National Academy of Science. That voice of marine science in Washington that shouted during the late fifties and roared during the Kennedy administration was no longer booming. NASCO was also licking its wounds after scathing criticism by the press and James A. Crutchfield[43] of its 1964 report[44] on economic return to be expected from the sea, and retreated into a shell of more familiar technical consideration. It was further enjoined by NAS leadership from its earlier posture of advocacy because of the Mohole incident.

Meanwhile, a related interest group was born in the newly created National Academy of Engineering. In the wake of the Marine Sciences Act that gave unprecedented prominence to ocean engineering, the NAE in 1967 appointed its first governmental advisory Committee on Ocean Engineering. They immediately embarked on their own study of engineering needs and opportunities in marine affairs, but the Stratton Commission's report had the effect of upstaging their analysis. Efforts to interest them in policy issues were somewhat blunted by their inherent professional orientation toward technique and an ideological shyness about the role of government. Their comprehensive report, discussed in chapter 10, was released in 1972.

Finally, there was COLD, the Council of Oceanographic Laboratory Directors, leaders of the twelve largest oceanographic laboratories. Several times I met with them, mainly to receive complaints that increases in federal research support were inadequate, a condition that by 1970 was all too true. Not interested in policy issues nor viable as a group, by 1970 their cohesion was so tenuous that at one informal conversation no one could recall who had last been elected chairman.

Contacts with industry also produced mixed results. By far the largest private sponsor of ocean exploration and source of heavy capital investments was offshore oil. That industry had long developed its own contacts in the federal government, especially in the Department of Interior, and was not eager to see familiar pipelines of communication disturbed. Early in Council life, they had been apprehensive as to whether the government would extend its interests beyond some arbitrary public-versus-private boundary, especially in offshore geophysical surveying. Quietly, they kept an eye on Council initiatives. With the appointment to the President's Marine Commission of three individuals having direct understanding of the industry, and with Council utilization of petroleum consultants, that anxiety was somewhat mitigated. But they were hesitant about more direct discussion with an agency having complete independence from special interests, and in 1968 the temperature dropped further when the National Petroleum Council (NPC) recommended limits of offshore national sovereignty that collided with policy trends of the State Department that NPC believed to be forged in the Secretariat.

The fishing industry was fragmented. West Coast fisheries disagreed with East Coast; salmon coastal differed with tuna deep sea interests; shrimp competed with finfish. Endeavoring to gain their involvement, even in their own interest, proved futile.

Maritime shipping saw little that the marine sciences could do for their industry. The maritime unions, however, did. As early as 1964, organized labor tried to unionize oceanographic ships through the National Labor Relations Board, an action fought bitterly and with uneven success by the oceanographic institutions who wanted both to retain a more direct say in hiring crew and to keep down costs by having crew double as technicians. The union leadership was imaginative enough to recognize the potential: more marine activities would spawn more ships and thus more jobs. But any coalition between maritime unions and industry toward a common purpose or effective use of the sea was undermined by the continuous warfare being waged on issues of ship automation.

Efforts at rapport with port and harbor interests were only moderately productive, and later evaporated entirely in their candid opposition to Council initiatives for study of regional ports and coastal management under state government. They did not welcome disruption of the status quo of their alliances, and felt threatened by prospects of national planning.

The aerospace, high technology, and ocean services industries were keenly interested in diversification and took some bold fliers into the marine market. A few succeeded, and some key technological innovation resulted. In research submersibles, for example, numerous companies followed J. Louis Reynolds' private financing of *Aluminaut*, in expectations of growing markets. Many were lured by the sales pitch of enthusiasts; other deluded themselves. Most lost money. In frequent contacts with this sector, the Secretariat became aware of that industry's capability to lobby for individual contracts, but its ineptness in supporting a field.

The third year of Council life was nevertheless focused on improved communication with nonfederal institutions. The annual report was subtitled, "A Year of Broadened Participation." In January 1968 representatives of twelve leading mining firms were invited to present views on the appropriate federal role in encouraging development of marine minerals. Also in January more than forty senior representatives of government, industry, and universities exchanged views on the International Decade of Ocean Exploration. A March conference at the University of Washington, partially supported by the Council, assembled more than 250 representatives of all parts of the seafood industry. In April the directors of nonfederal marine research laboratories met with the Vice President and senior government officials to discuss programs such as the International Decade of Ocean Exploration, Sea Grant, and federal support of academic research. In October, representatives of key organizations were invited to comment on national marine policies to the Council and agency officials: National Petroleum Council, American Mining Congress, National Fisheries Institute, American Merchant Marine Institute, Aerospace Industries Association of America, Marine Technology Society, National Security Industrial Association, and National Federation of Independent Business.

The trade press breathlessly seized every opportunity to postulate a pot of gold at the end of the marine rainbow. Many of them knew of its uncertainty but had resolved to excite prospects in the hope of generating a self-fulfilling prophecy. Notwithstanding that zeal, such newsletters as *Ocean Science News* and *Oceanology*, the journals *Undersea*

Technology and *Oceanology International* functioned as aggressive needlers to all participants, government included, or as brokers of important but unpublicized intelligence on marine activities. All had to struggle courageously through the drought of advertising. (By 1972, only two had survived.)

By late 1968 a glimmer of hope for stronger outside involvement appeared in another untapped constituency—the coastal states. In November, a symposium in Williamsburg, Virginia, was organized by the Council's Committee on Multiple Use of the Coastal Zone to provide a forum for more than sixty representatives of state and local governments, as well as individual specialists, to discuss governmental mechanisms for enhanced use of the Coastal Zone. The potential coalition of the twenty-two states fronting the Atlantic Ocean, the Gulf of Mexico, and the Pacific Ocean, plus the eight states bordering the Great Lakes had been long in the making. The states had dealt with their own coastal resources, but always in a traditional mode regarding fisheries, offshore oil, and recreation. Only a few had seen the prospects of enhancing economic development via marine affairs or of growing conflicts in coastal functions. As early as 1964, Governor Edmund Brown of California had been prevailed upon by alert staffers in his and Speaker Jesse Unruh's office to convene a conference based on that state's interest in the new spirit of oceanography, fanned by recognition that aerospace funding could not support future growth of the state's economy. Soon after, Brown created the Governor's Advisory Commission on Ocean Resources that was continued, albeit modified, by his successor. While echoing some of the poetry and prophecy that characterized promotional activity in the national scene, California got down to brass tacks rather swiftly in examining both the potential for economic development and the emerging conflicts in use of coastal resources. Other states followed, particularly Florida, whose emphasis was almost exclusively on wooing marine related industry gearing up in response to the Navy's encouragement in the aftermath of *Thresher*. Enactment of the marine sciences legislation spurred state initiatives further, and before long signals were radiating from governors in Maine, New Hampshire, Massachusetts, New York, Delaware, Maryland, Mississippi, Texas, California, Washington, Hawaii, and Alaska. This appetite was further whetted on the East Coast by prospects of siting a regional laboratory proposed by ESSA that could serve as a magnet for industry. The relevance of Sea Grant to state interests were stimulated by government enthusiasts such as Harris Stewart and Robert Abel.

Council efforts, however, were to direct state attention to the coastal zone problem rather than to the illusive hopes of swift economic return. Humphrey first spoke to this issue at Kingston, Rhode Island, in the summer of 1967, and with the near-catastrophe of the *Torrey Canyon* fresh on the public mind, people began to listen. By 1968, numerous state mechanisms that dealt with recreation, commercial fishing, nonliving resources, maritime commerce, waste disposal, and conservation were examining the shorelines anew.

In November 1968 the time seemed ripe to catalyze this interest into a coalition of coastal states, and a Miami conference convened by Governor Claude Kirk provided the occasion. After the Council's support had been informally sought and assured, he proposed that a coastal state organization be formed. I spoke in direct endorsement and read the influential congratulatory telegram from Humphrey. The response on the floor was encouraging, but the effort subsequently fizzled when individual states were reluctant to provide the money to establish a modest central staff. The objection was the old one of sovereignty—they were concerned over domination by Florida. In the meantime the National Governors' Conference had been encouraged to consider the issue, and it took root finally in 1970 with the creation of a Coastal States Association. The lure of possible federal grants-in-aid proved irresistible enough to overcome interstate rivalries, and leadership from Virginia's pioneering coastal research Institute of Marine Sciences and its director, William J. Hargis, Jr., helped it get moving. But its marginal influence on coastal management legislation in the 91st Congress proved it was still an infant.

The possibilities remained of pooling outside support in such organizations as the National Security Industrial Association, Marine Technology Society and the National Oceanography Association, of which NSIA was by far the most effective in providing advice and assistance. As early as 1964, a Committee on Ocean Sciences and Technology had drafted a "National Ocean Program," but with its own visions of sugarplums five times the ongoing level of support. It was especially valuable as a clearinghouse for experts, but its charter was limited. Conceived in 1944 by James Forrestal, it was aimed at a peacetime coalition between industry and the Armed Forces to foster military preparedness. MTS had been founded in 1963 when it was recognized that oceanographic interests crossed so many disciplines that a central forum was needed for technical papers. Through officers like James Wakelin, Edward Stephan, and Paul Fye, it saw the need for a coalition of professional groups to promote ocean interests but resisted entrance into the arena of public policy. Initially they argued that their tax-exempt status under Internal Revenue rules prohibited their engagement in lobbying, but this caution became a convenient shield from controversial issues that could potentially tear apart their membership.

The NOA, on the other hand, was formed unequivocally as a lobby.

Fired by the initiative of a public relations firm that saw this field as the source of a prospective clientele, NOA developed a charter broad enough to encompass the hoped-for coalition, and membership was solicited from all sectors of the marine industry. Council staff met with their leadership to offer encouragement, but its potential screamed to a dismal halt in early 1967. Almost immediately after the Pardo proposal on an international seabed regime, elaborated upon in chapter 6, and learning of keen interest in the issue within the administration, NOA staff on their own initiative and apparently without instructions from NOA's officers fanned out on Capitol Hill to lobby vigorously—but to lobby by charging the United States was giving away what they regarded as a national birthright. One of the Neanderthal political concepts that sprang forth was that the Pardo proposal should be countered by planting the U.S. flag on the seabed (a unilateral action shortly after undertaken by Governor Kirk on *Aluminaut*). By sensitizing the latent views of several conservative Congressmen, NOA succeeded in triggering a paroxysm of floor statements that shook the State Department and would probably have paralyzed their further initiative if the Council had not goaded them to continue. But NOA's ideological coloration and the high handed action by its staff snuffed out rapport with the Council until they recruited new staff in 1968. Despite tacit support from the oil industry through its president, Thomas A. Barrow of Humble, NOA scarcely stirred a ripple in plumping for ocean priorities. Its consolidation with the public-education oriented American Society for Oceanography in 1970 was unproductive. By 1971, the marine affairs momentum of the previous four years took such a severe downward turn that NOA's industrial sponsors faltered in their support, notwithstanding its president being chief executive of one of the nation's largest petroleum companies. ASO left NOA to join MTS. Quietly, NOA went out of business.

By 1972, the constituency of a 51st state remained unmobilized.

THE JOHNSON–HUMPHREY YEARS

Translated into political terms, the question of how important the oceans are to the nation can be directly measured by how important the President believes they are. Of all the officers of government, the President has by far the greatest power to define the nation's major political goals, to synthesize divergent interests into a public interest, and to develop strategies and tactics to accomplish his programs. He is manager of the bureaucracy, obliged to enforce a coherent unity to the fractured internal machinery that endeavor to respond to signals from clientele they serve, and to resolve disputes; and he seeks new authority when existing powers of the Executive are inadequate to fulfill the agreed upon goals. In marine affairs, Section 4(a) of the Marine Sciences Act assigned these responsibilities explicitly to the President, whereas previously they had been implicit.

Nonetheless, whatever policy it may enclose, a law remains amorphous clay until sculptured by the Chief Executive according to his assessment of priorities and his style. On priorities, no enthusiast for the oceans even in wildest flights of fancy would ever have pegged his darling at the top of the list, in terms either of national priorities or of science priorities. But on President Johnson's list, the question was not how high ocean affairs might be, but whether they were on his list at all. The biographers of Lyndon Johnson agree that he was heavily influenced by his involvement in public policy during the Roosevelt administration. His sentiment was strongly populist; his highest concern was for fundamental needs and wants for the underprivileged, underfed, underhoused, undereducated, those often denied equitable access to natural and human resources. Whatever contributions the ocean could make to these problems seemed remote to a product of the Texas hill-country.

Thus, we searched deliberately for opportunities to appeal to Johnson in areas that he recognized were of pressing importance. Johnson had said himself, "Next to the pursuit of peace, the greatest challenge to the human family is the race between food supply and population increase. That race is now being lost." The pointer spun to our Food from the Sea initiative to attack protein malnutrition. In the area of urban decay, we argued in the first annual report that "urban development does not end at the water's edge. As municipalities look increasingly seaward for new inhabitable areas, or perhaps offshore airports, correlated planning of uban and ocean activities becomes essential." What was on our mind was the pattern painfully repeated in almost every major American metropolis that the worst part of the city begins geographically at the water's edge. We searched for marine solutions in coastal management and pollution abatement. Potential programs in tune with Johnson's interests were also examined in relation to underemployment and economic distress. Aspects to be considered were the problems of American fishermen; deterioration of the coastal economics; lost lives and property damage from coastal storms; long-range weather forecasting that could be improved to reduce the misery from unexpected and prolonged drought—a matter well understood in the heart of Texas. International cooperation in space had been one of Johnson's proudest achievements in his influence on drafting of the Space Act, and it had a marine analogue. And, of course, the Sea Grant program was in harmony with Lyndon Johnson's deep-seated interest in education.

One avenue we deliberately suppressed was bald support of science for science's sake. That sector had found a warm reception in the intellectual ferment of the Kennedy administration, both because the need to strengthen our oceanographic research capabilities was so conspicuous, and because science was selling. But the broad rationale that science was good for you because its fruits of knowledge would eventually benefit mankind was tattered and torn. In September 1965 Lyndon Johnson shook the sanctum sanctorum of health sciences by asking questions about the validity of such a large health sciences budget. This was a reasonable question from a pragmatic leader, but it was widely flaunted in the scientific community as evidence of Johnson's hostility to science.[45] It was increasingly clear that Johnson was disenchanted with the science advocates and lukewarm to his own science advisor. Several circumstances opened that breach. First, Johnson's agonies over Vietnam brought about a sharp intolerance of dissent, and the university community was polarized on this issue. Aware of that break, Johnson tried to bring into the White House staff some leading intellectuals who could be thought of as representing the academic point of view at high councils of government, and thus to soften the edge of that dissent. No doubt he looked to his science advisors with their ample credentials and prestige in academia to reinforce that function, but such expectations were unrealistic. Don Hornig became less and less at home in the White House. Moreover, on the theme of, "What have you done for me lately?" Hornig was under pressure to demonstrate the contribution science could make to victory in Vietnam through defoliation chemicals and remote acoustical surveillance. In that antagonistic atmosphere, it was clear that emphasis in our marine program had to be placed clearly on social, economic, and political goals of the nation as interpreted by LBJ; support for science, per se, was not on that list.

The oceanographic community was disturbed by lack of explicit science support in the Council's first recital of initiatives, and we were never able to explain the climate in the White House that led to our decision. In drafting Johnson's message and in the text, we went out of our way to reflect the importance of the scientific base, mentioning the NAS and PSAC several times. But it took three years for them fully to comprehend why the program was generated and explained in categories of public need rather than by fields of science and in terms of "ends" rather than of "means." NASCO's 1969 report, *An Oceanic Quest*, revealed a new, enlightened stance.

By January 1967, five months after the Council was activated, it had sown seeds that Johnson blessed, then fertilized with funding in Food from the Sea, coastal management, Sea Grant, data management, expanded ocean prediction observation and mineral surveys, deep-ocean technology, and Arctic research. Despite heavy opposition from the Bureau of the Budget, Food from the Sea was mentioned in the Foreign Aid message; legislation for estuarine and coastal management appeared in his special message on the environment. Port development and safety at sea, including increased attention to engineering standards for off-shore oil development, never made the President's personal "Hit Parade" in terms of his major messages, but they popped up again as areas of priority emphasis recommended by the Council itself and integrated into the President's budget and program for FY 1969. As will be outlined later, all survived and gained public support except one—the proposal to develop a national port system. By the middle of 1967 we were able to begin a campaign within the White House for support in another area of key priority on which we expected to find resonance with the President, the role of the oceans in contributing to world order. The impact of Johnson's July 1966 declaration of "no colonialism on the seabed" was pervasive, and we were endeavoring to build on the concept. But we were to discover that theme was already threatened by unsympathetic officials within the bureaucracy.

Two unrelated events opened the way to what proved a milestone of presidential action. In August, Ambassador Pardo from Malta made his proposal to turn seabed resources over to UN jurisdiction, with benefits to be distributed to the developing nations. At that 22nd UN session, fifty-eight nations spoke in the debate, some advocating that title to the seabed be vested in the UN, others calling for a moratorium on unilateral exploitation of seabed resources. Most agreed on the urgency for a freeze on claims of national sovereignty, but some even opposed UN consideration. Fortunately, we had not been caught completely unpre-pared, but just how far our delegation at the UN would be authorized to go was being heatedly argued in private in Washington. The opportunity for presidential interest was clear.

The second circumstance emanated from a peculiarity of Johnsonian style, his predilection to develop and then announce surprise initiatives in his State of the Union or some other message. The opportunity arose in 1967 when Joseph Califano, the President's number one Special As-sistant and ubiquitous manager of the bureaucracy, had added to his other backbreaking chores the nomination of candidate topics from se-lected officials in the administration. After selecting possible winners, he sought implementation plans, cost, and estimates of political value. All in secret. My invitation to participate arrived with the stirring "Eyes Only" admonition on the outside envelope—a term reserved for only the most sensitive political or military papers that are not intended to be read or staffed out by anyone other than the recipient. "Eyes Only" ac-tivities were designed to gain priority attention, and they succeeded.

By August 25 I had generated 22 potential initiatives for the Coun-cil to propose to the President. At Califano's request, these were weeded to eight; five were pushed by the Secretariat energetically; three were adopted. Of these, the International Decade of Ocean Exploration became our greatest triumph by its identification in the 1968 State of the Union message.

President Johnson employed the authority of P.L. 89–454 to name a Commission on Marine Sciences, Engineering and Resources, announcing its appointment in January 1967 under chairmanship of the distinguished Julius A. Stratton. Their report proposing consolidation of numerous marine bureaus into a National Oceanic and Atmospheric Agency was delivered only days before Johnson left office and so awaited the response of Richard M. Nixon. Chapter 8 recounts the fortunes of that Commission invention.

Whatever opportunities might have opened up in the last year of the Johnson administration were aborted by his withdrawal from the presidential contest, by Humphrey's absence from Council affairs during his own campaign, by the lame-duck environment after elections that included Johnson's policy to defer any initiatives so as not to tie his successor's hands, and by heated objections to a new agency, proposed by the Commission.

The durability of earlier initiatives and the high batting average at implementation preserved momentum of the program almost until inauguration. Defeated in the campaign, Humphrey nevertheless returned to fulfill his vice-presidential responsibilities, chaired the fourteenth Council meeting on January 10, 1969 that produced the agenda of unfinished business for Nixon, and there received the plaudits and expression of affection and admiration from the Council through a gift and citation.[46] During the two and one-half years of Council existence, it had met fourteen times. Attendance by high-level officials had been excellent—better than 90 percent by the Council member or his policy-level alternate. Humphrey chaired all sessions until his campaign, when Glenn T. Seaborg and Thomas O. Paine substituted.

On January 14, upon release of the Council's third report, Humphrey met the press, displaying a phenomenal ability of recall and interpretation that further advanced the recognition of the importance of the sea. Most important, the objectives of the Marine Sciences Act were beginning to be realized, in direction, momentum, expansion. The ledger of Johnsonian accomplishments included many plus signs in this area: presidential support of a foreign policy that would preserve freedom of the seas and initiate steps toward development of a legal regime for the deep-ocean floor and seabed arms control; launching an International Decade of Ocean Exploration; development of fish protein concentrate to meet world-wide malnutrition; illumination of coastal zone problems with encouragement of states to engage the issue by planning; support of regional or intergovernmental approaches such as with the Chesapeake Bay and the Great Lakes; instituting measures to insure safety of life and property, including countermeasures to meet pollution emergencies; acceleration of technologies for deep-ocean operations, satellite and buoy observation networks, precise navigation, and for data management; a start toward formulation of policies and plans for the Arctic; and general strengthening of the base of science and technology, including utilization of Sea Grant. Apart from specific measures, there was new federal leadership and a heightened response to maritime opportunities.

On the minus side, little progress was made in rehabilitation of the fishing industry, in financing the International Decade and civilian marine technology. More generally, with Vietnam draining off funds, marine investments that had grown in the civilian sector by 60 percent in less than three years could not keep pace with the new array of goals.

Marine affairs in 1969 were nevertheless maturing, moving at a pace and along a trajectory unknown in 1966. With a new captain about to take the bridge, there was new uncertainty as to how he would chart his course.

THE NIXON–AGNEW YEARS

The inauguration of a new president is a major happening. Supporters celebrate. Losers convalesce from the bruises of battle. And bureaucrats develop a protective stance with the lowest possible profile, to await the sweeping of a new broom, the giddy flutterings of neophyte presidential appointees, and sailing directions from a new boss. The machinery of government clanks on with routine concerns; but at the highest levels the most incredible confusion prevails. Newcomers first locate the closest men's room, then the xerox room, then take stock of their symbols of prestige to make sure that room size, plushness of rug, and visible accouterments (the carafe having replaced the cuspidor) assert appropriate ranking in the power structure. Then comes establishment of lines of communication—up, down, and laterally. A layer of mutual suspicion develops along the boundary separating appointee from civil servant. The latter tests the knowledge and courage of the man he must arm to represent the agency; the appointee, aware of his limitations, assesses political loyalties and the proverbial sluggishness of his public-servant subordinates. Despite efforts to ease the transition of political roulette, to preserve the continuity of statutory functions of government, and not to leave a void vulnerable to exploitation by foreign or domestic interests, the event is a strenuous and muddled affair.

Considerable speculation revolved around whether marine science affairs would maintain the same stature they had enjoyed under the Johnson administration. As a candidate, Richard M. Nixon had issued a forceful statement in Miami on October 30, 1968, strongly acclaiming the role of the oceans in national affairs and asserting his support for its further development.[47] To be sure, the statement was prepared for local consumption during a campaign ringing with promises from both candidates, and support for oceanography appeared on both party platforms in parallel statements that had been well prepared by ghost writers who obviously knew their subject. But campaign rhetoric is an art form to be taken with a grain of salt by speaker and listener alike—a kind of political science fiction more to be enjoyed than believed. Thus the implications of this commitment were highly uncertain.

A second factor in the guessing game was the Stratton Commission report, newly printed and unstained by any actions of the prior administration that would have stamped it partisan. Since the report's major recommendation for consolidating several agency functions into a new one had been deliberately buried by Johnson and his White House aides for reasons to be elaborated later, there was even hope that it would find favor under Nixon, who was expected to draw contrasts with his predecessor. Nevertheless, the Stratton Commissioners had been appointed by Democrat Johnson, and the recommendations would cost money. The incoming administration was very unlikely to be looking for new obligations.

A third factor that could be potentially instrumental in sustaining momentum from the Johnson-Humphrey era was the presence of outside interests. As noted before, the marine affairs constituency was weak, inarticulate, fragmented, and leaderless. But there were three sets of advocates who had some access to the new President. One set comprised partisan supporters such as John H. Perry, Jr., the Florida publisher who had been a member of the Commission; Chalmers G. Kirkbride, Vice President of the Sun Oil Company and a leader of ocean-related interests in the National Security Industrial Association; James Wakelin; and others. A second set of advocates with access came from Republican members of Congress. By now, this group of supporters included such influential members as Rogers Morton, Chairman of the Republican National Committee; Bob Wilson, Chairman of the Republican Policy Committee; John Anderson from the Appropriations Committee; as well as Charles Mosher. The remaining set of advocates who might gain the presidential ear included certain interested professionals from science, engineering, and technology.

Members of each set probably had a hand in establishing the fourth —and perhaps decisive factor—of Council continuity. This was the Nixon decision that the Council's Executive Secretary under Johnson should retain his post under the new administration. The decision was made known in an exchange of letters after the election between Mosher, who took the initiative to urge this continuity, and the President-Elect. The President-Elect also signaled this intent in a letter to the Secretariat. Despite these assurances, however, tenure of a carry-over was a highly delicate affair, vulnerable to political winds blowing from unpredictable directions. But the Council Secretariat was retained; and by comparison with other fields, marine science was thus blessed by continuity in its director and professional staff. Bonds with the new Vice President, nexus with White House staff, ligaments to new Council members, and communications with their immediate staff had yet to be established.

The first step was to convince the new administration that they could expect loyalty to the President and zealous implementation of the Marine Sciences mandate. Second, we hoped to establish our credibility as an alert, competent, and objective staff arm. Third, we hoped the President's instructions to his Vice President, Spiro T. Agnew, would ignite the same interest in fulfilling his role as Council Chairman that we had enjoyed with Humphrey. Finally, we secretly hoped that the rich agenda of unfinished business shelved at the end of the Johnson administration would capture the imagination of the new President.

During the succeeding year, we were to find greatest success achieved in the fourth objective, the least in the third.

After an initial interval of mutual inspection, communication became fairly relaxed and closer than in the prior administration with such senior White House staff as Robert Ellsworth (later our Ambassador to NATO), Arthur Burns, Peter Flanigan, John Whitaker, Science Advisor Lee A. DuBridge, and their assistants. Initial contacts were also sought with Agnew's staff, especially Stanley C. Blair, who had been Governor Agnew's Secretary of State, a man of great integrity and management skill. Most were new to this arena; all were in a state of suspense, inwardly nervous about their own role in a power structure but outwardly maintaining a diffidence that was to be the new trademark of White House style. Staff briefings on marine affairs had been started with vice-presidential aides during the transition, with my liaison ricocheted successively to three different aides and finally lodged in the hands of Jerome Wolff, who had been Agnew's Commissioner of Roads in Maryland, and was now his "science advisor."

Within a week I gained an audience with Agnew, who endorsed the President's decision for me to continue, a necessary step if we were to develop a close working relationship. Soon after, he was briefed on the scope of subject matter within his Council's purview and discussed pending business. I recommended that he seek formal instruction from the President to activate the Council, then grasp the reins promptly by an early meeting that would familiarize new Council members with their responsibilities. Discussions of substantive concepts underpinning marine affairs and problems facing the new administration were brushed aside. Agnew lacked any prior interest in this area and had no special inclinations to engage at those hectic moments of transition in complex issues of setting Council philosophy.

In fact, the administration chose at the outset to proceed without any strong philosophical themes that could guide the orientation of marine affairs. Whereas the Johnson program gave one the impression of being pragmatic, populist, and spontaneous, the style adopted by the Nixon forces seemed one of anxious caution. Cutting Johnson's budget was the main preoccupation. Among some staff, there was implicit planning for 1972. Thus there was no doctrinal framework to ignite or to dampen renewed action on issues in marine affairs, often euphemistically phrased in terms of the Commission's proposals rather than Johnson's. Receptivity to proposals for action was, however, enhanced by several other events. The first was an exercise by Arthur Burns, then in an unprecedented post as the President's Counselor, to identify key items of new business and to match each with a determination of who among the incoming cabinet officers would be responsible, and with what deadline. The Stratton report was fortunately on his list. By oversight, however, its analysis was assigned to Secretary of Interior Walter J. Hickel instead of to the Vice President as Council Chairman. By chance, I caught this slip before any public action was taken that later could have proved embarrassing to Agnew.

Meanwhile, the Council was quickly put to work. I received letters from the President and the Budget Director requesting participation in a government-wide analysis of the Johnson budget, to review and ferret out soft spots for possible cuts and thus to leave room for add-ons should the new administration wish to put its own brand on the government promptly. The latter intent soon disappeared into a studied effort to halt new initiatives on any but the most urgent national issues. The Council staff were to recommend cuts of $25.9 million, partially offset by proposals for increases of $8.4 million to salvage good programs cut by the prior administration. But by April the President had agreed to a $4 billion cut across the board, and all marine programs were in jeopardy. Under greatest attack were the demonstration plant for fish protein concentrate unsupported by Hickel, the leasing of underutilized, privately funded research submersibles sacrificed by Defense Secretary Melvin Laird, and the national buoy program offered up by Transportation Secretary John Volpe, none of which the Council Secretariat had proposed for cost cutting earlier. Efforts to sustain these programs were only partially successful. The buoy program I was able to rescue immediately; Congress later rescued the fish protein plant; the submersible program was lost. The cut in submersible leasing that Johnson and Clark Clifford had approved came as a surprise, incidentally, because even that small plum would have been welcomed by industry, which was agonizing over lack of marine expansion. To the Nixon staff at that time cuts in the budget were more important than pleasing one industrial constituency.

In turn, I propelled the necessary paperwork up the line to press for a budget extending the Council beyond June 30 and to set forth possible initiatives, as had been the style of the preceding administration. These were strongly influenced by the Stratton report, which was gaining public view and had sparked congressional interest to such a degree that more than sixty bills related to marine affairs had appeared in the opening-week hopper of the 91st Congress.

At that stage, Agnew seemed to enjoy the prospects of Council life. The President had already given him several parallel assignments based on a similar premise that he would function as an assistant President, dealing with issues that cross agency lines. Because of his origins in a coastal state, I had hoped this field would prove attractive. I had been a little shaken by his absence from his own Governor's Conference on the Chesapeake Bay in August, but since that had occurred during the campaign, his delegation of responsibility to Rogers C. B. Morton (in whose district it was convened) was understandable.

On January 28, an event occurred that was to prime the public interest in coastal pollution and lend point to our encouragement of the new administration's attention to marine affairs: a gas well five and one-half miles out in the Santa Barbara Channel blew out with a violence that spewed oil over the beaches, entrapped birds, and opened a major issue on safety of off-shore drilling. Aware of its political potency, I alerted Ellsworth to what could be a serious federal problem. If badly handled, the oil leakage could do lasting damage to both natural and political environments and could ensnare the White House. When the seriousness of the accident became apparent, Ellsworth asked for advice on action and was referred to the national pollution contingency plan that had been completed the previous November, and which assigned responsibilities for containment and cleanup countermeasures to the Interior and the Coast Guard. He asked that the Secretariat represent the White House to make sure that appropriate action was taken. We found the Coast Guard had been notified by the Union Oil Company but that, in the early learning stages of new appointees, Secretary Hickel and his staff had not realized their key responsibilities. They were so alerted. Council staff monitored the activities for several weeks, and kept Ellsworth informed as to the adequacy of agency response. So the White House staff began to look to the Council Secretariat as an ally in protecting the President.

The Council was activated by President Nixon, as he had promised prior to inauguration, but by verbal request via White House staff rather than by a letter to Agnew as I had proposed, to follow the pattern of Johnson to Humphrey. Soon after, on February 23, the Vice President received instructions to review the Stratton report, thus correcting the earlier erroneous assignment to Hickel. However, it set an impossible

March 10 deadline, which was extended after my protest. Indeed someone on the White House staff wanted action, but this style was soon to
change, especially after Burns was promoted to Board Chairman of the
Federal Reserve System and his staff dissolved. The President's full energies were focused on Vietnam and foreign policy, and sleeping domestic
issues were to be let alone.

To implement the President's request, Council members were individually requested to assess the Stratton recommendations. Their
views were then to be distilled into a set of recommendations for the
Vice President to transmit to the President, by a committee consisting of
DuBridge, Ellsworth, Burns, Robert P. Mayo, the new Budget Director, with myself as chairman. The response from Council members
embroiled in taking over departmental reigns was, as expected, based
on the same staff advice that had been given to their predecessors. After
four drafts the recommendations for Agnew's consideration were completed. Discussions were aggravated because of a contest between the
Bureau of the Budget and me on the NOAA recommendation—whether
to bury it altogether as BOB hoped, or to make a decision. The compromise that at least kept it alive was to recommend its study in the context of a broader review of government reorganization that was given
birth with the April 5 appointment of a Council on Government Organization, chaired by Roy L. Ash.

Agnew's March 23 and March 27 letters to the President on the
Commission report, detailed in note 32 to chapter 8, were to elicit a
May 19 response from President Nixon for the Council to examine all
recommendations other than the one—the major one—on reorganization.
Even that action was almost blocked by the Budget Director. After approving the draft letter which Agnew sent to Nixon, he interposed a
personal note to the President expressing reservations. Among other
developments that irked the Bureau was our proposal for Council extension. Strongly backed by Agnew, this matter assumed a new urgency
because Congressman Lennon had already taken such an initiative.

The House passed H.R. 8794 on April 21, and its Senate counterpart S. 1925 was introduced the next day. By letting the House take
initiative and credit, the Senate thus headed off repetition of the contest
that had erupted in 1965–66. Passed May 14 and signed June 9, the
bill resuscitated the Council a scant three weeks before its scheduled
demise. Staff drew a breath of relief.

With steps completed to keep the machinery intact, I next turned
to specific issues that had been deferred by Johnson: the seabed jurisdiction, updating the oil spill contingency plan on the basis of the Santa
Barbara experience, the future commitment of funds to the Decade,
development of policy on coastal management, and the seabed disarmament issue. Staff work for these issues had been completed by the Coun-

cil Secretariat, but now in a new administration had to be worked through the agencies a second time, through the Cabinet, through the Budget Bureau, through the Council, and thence to the President.

On February 26 the Council held its first meeting under Agnew in the White House (where Agnew initially had offices). The Council had not enjoyed that status symbol under Humphrey when meeting in the Executive Office Building next door, but that blush of prestige was to be short-lived. After an extensive set of briefings, the Council accepted recommendations to proceed on four "carry over" issues, while developing commentary on the Stratton report and awaiting further presidential instructions. The disarmament issue was extracted for cognizance of the National Security Council and Henry Kissinger. The legal regime issue was attacked for the Council by the Committee on International Policy in the Marine Environment, jointly serving the Secretary of State and the Council Chairman, that was continued under chairmanship of Deputy Undersecretary U. Alexis Johnson. When the previous administration dealt with this issue, controversy had arisen between Defense and State on the one hand, favoring a narrow seabed jurisdiction, and Interior and Commerce on the other hand, favoring a wide jurisdiction as proposed by the National Petroleum Council. Exactly the same agency alignment took shape under the new committee on February 28, but with explosive vigor. Johnson soon gave up hope for resolution at his committee level. With the UN Seabed Committee due to meet March 8 the vacuum of a U.S. position seemed ominous. Full resolution by the President the next year is recounted in chapter 6.

As for the other pending issues, Hickel was requested to convene his interagency committee provided for in the contingency plan, with assistance from a DuBridge panel; Russell Train, then Undersecretary of Interior, was asked to chair an interagency study on coastal management; and Robert White, ESSA administrator, was asked to prepare a paper on the Decade, with assistance from the Council committee he had chaired in the previous administration.

Across the board in marine affairs, the new administration was getting its ducks in a row and was in no position yet to reflect publicly where it was going or how fast. Nevertheless, immediately after the election Agnew accepted an invitation to address the American Management Association on January 30. Such early exposure seemed hazardous, but his commitment cut off the options of silence. My draft for Agnew was a carefully worded résumé of the issues, especially of coastal management, that would have committed the administration to nothing but would have revealed the speaker as an interested and knowledgeable Council Chairman, readily stepping into his predecessor's shoes. His staff writers were still insecure in their new role and unwilling to have active participation by an outsider. Claiming to know his style, they vastly altered the text. In short, it stated that under the fiscal constraints looming ahead no promises could be made for growth in marine affairs or in response to the Stratton report.

That pessimistic tone was just what an audience of frustrated industry people did not want to hear, but it reflected an insistence of the new Vice President on blunt candor, self-consciously in contrast to the Johnson administration in not promising anything that could not be instantly delivered. The trade-press reaction was not good, and it became downright hostile five months later when he addressed an MTS meeting in Miami. Attempts to have his speech reflect at least a personal commitment to continued leadership in marine affairs failed, again because of a rewrite by his staff. Its gloomy tone was unnecessary because the Nixon administration had not yet made budgetary decisions that would arrest growth in this field. Now MTS was dismayed. In the meantime, Agnew began to enjoy a smidgen of publicity in this area—pictures with aquanauts from the Tektite project. But he decided against any more public exposure until the administration acted.

Although the Nixon administration had put the Commission report on its agenda promptly, the impatient Congress took steps to warm up marine activities further. On February 20, Lennon announced hearings. I was invited to testify as administration spokesman, but was able to negotiate delay until June. The Senate passed Concurrent Resolution 72 in support of the Decade, and an identical measure was introduced in the House. By the middle of March congressional committees had requested the Council to comment on twenty-two bills, ranging from federal reorganization of marine activities to marine sanctuaries, pollution from oil spills, submersible safety, boating safety, use of Outer Continental Shelf revenues for marine research and legal studies of the seabed.

Another external pressure on the administration was the imminent meeting of the Eighteen Nation Disarmament Committee of the UN, to which the U.S. had the opportunity of introducing proposals for seabed arms control. Henry Kissinger was staffing out the issue in preparation for a March 6 preparatory meeting of the National Security Council. Kissinger extended an invitation to the Secretariat to participate. The issues and conflicts between the Arms Control and Disarmament Agency and the Joint Chiefs of Staff are discussed in chapter 6.

The second Council meeting in the new administration, scheduled for May 23, had a vital agenda, between the carry-over issues and hot news four days before that the President had responded to the March letter on "next steps." But the session proved to be a fiasco. Preparation had followed past practice: position papers were drafted and circulated in advance, an elaborate briefing book was prepared for the Chairman, and our staff was deployed in efforts to determine where the bureaucratic trouble spots might occur. In a very short time alarms were streaming in and all sounding one note—vast apprehension within the agencies as to threats to their priorities if the Council pushed for new initiatives, especially the Decade, because the agencies feared that funding would "come out of their hide." The convulsions that had shaken the establishment in gathering victims for the sacrifice of $4 billion in an over-all cut of Johnson's budget were fresh on every one's minds. So was recognition that the administration was still on a shakedown cruise. My reassurances telegraphed to agency staffs that the Council could protect marine elements if the administration bought the new programs were met with hearty skepticism.

The bad news gathered only a day before the Council meeting made it clear that the meeting would be tense, with much depending on the Chairman's leadership. As the session approached, time to brief him evaporated and my concern grew. Humphrey had let his own lack of discipline over his schedule repeatedly squeeze out the one hour preparatory session before Council meetings that I had scheduled, but his familiarity with the issues and rapport with members proved strong compensation for not completing all his homework. Agnew had no such advantage. He went into the May meeting with only a five-minute private briefing by his own apprehensive staff aide, whose concern led to some advice to expect trouble but not on how to handle it.

At the meeting, White's briefing on the Decade was criticized as especially lacking in relevance to federal missions. Then Hickel and several other members, who had been briefed by their own agency staffs to make no commitment unless the administration promised an add-on, gave the Chairman little encouragement. With agency support rapidly shrinking, Agnew unexpectedly walked away from the Decade, saying that the administration should not mislead the public in supporting a project it could not fund. The Vice President's message was delivered with passion: "This administration is going to reverse its predecessor's style of promising without delivering and thus demonstrate its absolute honesty." He would not be party to any duplicity. No one could quarrel with that axiom of virtue. But where did we go from here? He then left the question of Decade funding up to the agencies, rather than have the Council accept responsibility to determine the merits of the proposal and to lay it on the President's desk as an option to be accepted or rejected along with other alternatives. His unwillingness to sway Cabinet officers was instantly apparent, and everyone sensed his growing panic, even a desire to stop the meeting.

The discussion ended with the Secretariat receiving an assignment to extend planning of the Decade but with a far stronger rationale than

White had presented. The question of seeking funds would be re-examined later.

Agnew was bitterly disappointed at the lack of support from Cabinet officers. His subsequent instructions to his staff were to have the Executive Secretary prepare a scenario for all members at future Council meetings so that the outcome would be known; otherwise, he would not preside. I registered the impracticability of that demand. Trying to write script for any Cabinet officer with a guarantee that he would follow it was beyond my capabilities and, moreover, seemed feasible for only the most uncontroversial issues.

The Council did not meet again in 1969, but its theater of action was transferred to a lower-level Committee for Policy Review (CPR) that by coincidence was created at the May session. After a letter poll of Council members, Agnew established this new body at an assistant secretarial level to replace the committee infrastructure that had been created under Humphrey.[48] The action was taken partly to give the organization a new look, partly to improve the inputs to the Council because some of the older committees had run out of steam while other topics went untended because of lack of an appropriate committee to assign them to. Agnew appointed the Council's Executive Secretary as CPR Chairman. Although I had originally proposed creation of CPR to work up all the proposals coming to the Council and gain the benefit of interagency education, and where possible obtain consensus, I was now gloomy about its prospects. For without an active Council Chairman, CPR could become a junior-league council, like its analogous FCST. Some agency staff hoped for that evolution. Many times they complained about the problems arising because membership on the Council was too senior; Cabinet officials understood too little about the issues they were obliged to act on; vertical communication in the hierarchy was poor, both up the line and in transmittal of Council papers down the line for consultation. Finally, since Council members were permitted to bring only one aide to meetings, the people responsible for implementation were often deprived of knowledge of the decisions and especially of innuendoes of commitment. Since assistant secretaries or their equivalents in most departments were line officers, they were thought to be at the right layer, high enough to reflect policy and at the same time low enough to be technically posted.

For all these advantages, the problem of a junior council was that members at this stratum were not really able to commit their agencies, nor would their collective advocacy bear the same weight as a consensus of Cabinet aristocracy. Moreover, there was generally a tendency for Federal Council members to send alternates junior to their own rank. No matter how high up these might be on the civil service ladder, they would prove even more timid about commitment and less prestigious in rank. At least when a Cabinet officer could not attend, his alternate would usually be an assistant secretary.[49]

The new CPR was thus asked to consider the Stratton proposals that the President had referred to the Council May 19. At its first meeting, June 19, it wrestled with and adopted a formal statement of its terms of reference and decided to form task groups to study a number of projects: Stratton proposals for coastal management (Under Secretary of Interior Russell E. Train); the concept of National Laboratories (Al Berman, Director of the Naval Research Laboratory); needs for a national capability in marine technology (Assistant Secretary of Commerce Myron Tribus); elaboration of the Decade concept (Council Executive Secretary); and requirements for programs on man-in-the-sea (John Billingham, a senior research physician from NASA's Ames Laboratory with similar expertise on man-in-space). The CPR meetings were conducted in exactly the same pattern as those of the Council, but with a freewheeling candor that became so exciting agency staff endeavored to attend because it was the best show in town.

On September 9, 1969, a new firecracker was handed the Secretariat, with a very short fuse. White House staff on behalf of the President requested identification of priority areas from the Stratton report that deserved immediate emphasis. With the administration relaxing the caution that had marked its first eight months, there was a hint that whatever was adopted now would be announced promptly and not delayed until the January budget transmittal. Apparently the congressional pressure for a new marine agency had hit a White House nerve. To consider whatever proposals evolved, an ad hoc committee was appointed, with White House staffer Will Kriegsman as chairman, to consider Council staff inputs and draft recommendations to the President.

The Secretariat nominated seven areas of which five were later adopted. The two that failed were proposals to assist the domestic fishing industry and a program of leasing out research submersibles that were being seriously underutilized while scientists could not satisfy research objectives for want of funds. Both failed because of an adamant position by the Bureau of the Budget. In their view, the fishing industry was beyond repair and research from submersibles was of low priority. For the five that made it, and with an assist from Lee DuBridge, I extracted a commitment of $25 million over and above the base program, which itself might grow or shrink depending upon its merits and the budgetary discipline that the President would impose on the entire government. The end result for FY 1971 was an increase of about $19 million over FY 1970. Details of implementation were fortunately available from the CPR task groups, already appointed. These prospects were then run through CPR to gain advice from agencies, then by mail through Council members to gain government-wide endorsement. The product was given to Vice President Agnew to transmit to the President whose concurrence was already assured.

On October 19, six weeks after the President's request, the exercise was completed with Agnew's announcement of a five-point program. It covered coastal zone management, establishment of coastal laboratories, a pilot technological study of Great Lakes restoration, the International Decade, and Arctic environmental research. (See Appendix 18.) In many respects, the five-point program was a breakthrough, in reinforcing priorities for marine affairs, in the selection of key areas for injecting funds and in the commitment of the additional funds themselves. The Decade, for example, was finally energized. Joy returned to the oceanographic community; and the Executive Secretary of the Council, believing that the unfinished business to foster a renewal of maritime interest had been transacted, submitted his resignation to President Nixon.

By April 1970, when the Council's fourth annual report was released, some of these commitments in the five-point program were weakened, especially on coastal management and research. By 1971, most had been placed on the back burner. But by then, the Nixon administration had faced up to some hard choices and had made the landmark decision to consolidate many marine functions in the new National Oceanic and Atmospheric Administration. Paradoxically, the new agency was assigned none of the administration's priority programs, and there was a slowdown in funding in FY 1971. In FY 1972, funding resumed the rising trajectory for civilian marine affairs started in the Johnson administration, although the Navy's marine science and technology fortunes relatively declined. There were major accomplishments on seabed arms control and on a draft treaty for the seabed legal regime—two momentous steps that capped initiatives begun years before.

Except for the foreign policy implications, the oceans generally dwindled in 1971 in the list of presidential priorities. The Council had met only once, in pro forma session, in its last eighteen months. Its chairman did not speak on marine affairs after June 1969. With retirement of the Council in April 1971, four years and eight months after its birth, there was a presidential gesture of gratitude "for a job well done [in] the development of marine science policy, coordination of Federal programs, and the effecting of an orderly transition during the reorganization. . . ."[50] Except for announcement of the decisions mentioned earlier, Nixon's only reference to marine sciences was made during a hasty trip to Savannah on October 8, 1970, to lend support to the Republican candidate in the governor's race, when he unveiled the architect's drawing of a new building for the Ocean Institute of the Atlantic. Once marine affairs had been regularized with creation of a nominal operating arm through NOAA, it was dropping from presidential notice.

From this discussion of the governmental taxonomy and processes of policy planning, some image emerges as to how two presidents, assisted by the Council, carried out their continuing legislative mandate under P.L. 89–454. That account is continued in relation to the birth and rearing of a major new civilian instrument—NOAA—in chapter 8. But while that symphony orchestra of federal agencies is important, it is the music of policies and programs themselves that reveals the evolution of marine affairs. Four of those compositions representing case studies of key issues follow in chapters 4 through 7.

CHAPTER 4

Managing the Coastal Environment

THE NEGLECTED BAND WHERE PEOPLE MEET THE SEA

It is rather curious that until 1966 the scientists and congressmen who fired volleys of rhetoric on more effective use of the sea scattered buckshot at the entire global ocean as a target but never rifled in on the regions of special importance near our coasts. The politicians were preoccupied with Soviet undersea threats to national security. The scientists conferred their highest prestige on research in the deep ocean. The narrow coastal band was silently neglected. Yet these 17,000 miles of ecologically unique estuaries, headlands, beaches, and lagoons constitute the primary intersection of the oceans with human affairs. Along the shorelines of thirty coastal and Great Lake states is a preponderance of our industrial investment: manufacturing, refining, power generation, shipbuilding, offshore oil and gas development, and fisheries. Three quarters of our present population concentrate on 10 percent of the land at the coastal margin, which promises to nurture all the megalopolises projected to the year 2000. In contrast to these mundane values, the natural beauty of the shoreline offers recreational and aesthetic values for a harried urban people, with irreplaceable joys for future generations as well. Paradoxically, this heightened utilization provokes abuse and depletion of the resource itself. Wetlands are bulkheaded and filled, beaches disrupted, channels dredged, ports expanded, and the exquisite littoral wilderness destroyed by urban sprawl. From the cities, the chemical wastes of an affluent society drain into coastal waters in such profusion that they can no longer be safely accommodated by tidal pumping.

Indifference to the shoreline was subtly epitomized in the 1968 report by the President's Council on Recreation and Natural Beauty.[1] Entitled *From Sea to Shining Sea*, that persuasive argument for environmental conservancy focused largely on the physiographically varied land mass between our coasts and ironically devoted but 11 of its 300 pages to the marine borders suggested by the title.

Although hints of a coastal dilemma were portrayed by naturalists much earlier, and highlighted in 1964 by Roger Revelle at the California Governor's Conference,[2] policy level attention was then focused largely on port development, erosion control, and water quality. The first political initiative drawing attention to neglect of our shoreline regions as a whole and their intimate relationship with human activity was taken on behalf of the Marine Sciences Council by Vice President Humphrey on July 27, 1967. Addressing the Marine Frontiers Conference at the University of Rhode Island, he said:

When we think of our marine environment, we tend to visualize the Pacific Ocean, the Atlantic Ocean, the Gulf of Mexico or the Indian Ocean, and the Caribbean. Though we speak in these vast terms, the most useful and important portion of that environment—both actual and potential—is the cities with their harbors and estuaries, the beaches and boating facilities, our centers of shipbuilding and shipping, the waters to the edge of the Continental Shelf, the Great Lakes, and our adjacent shoreline lands. . . . We must . . . solve the related problems of conflicting uses of our shoreline and the pollution of our estuaries and streams.[3]

The conflicts referred to by Humphrey were largely impelled by marketplace economics. In the absence of planning, coastal resources tend to become pre-empted on a "first come, first served" basis that spawns a bruising anarchy, for in the competition for scarce shoreline, one use may impede or completely block a second. Solutions are hampered because the coastal margin is a thicket of vested interests, complex public and private ownerships, amidst a tangled skein of legal jurisdictions. Users are often absentee landlords or transients. Nowhere is there a surrogate for the public interest.

This issue was put near the top of the Council's priority list from its origins in August, 1966. Subsequently, Council self-education was followed by proposals that drew responses from both President Johnson and President Nixon. The Council labored hard to define the issue, to diagnose and to prescribe remedies. But we realized the need to go beyond policy analysis; it was essential to generate enough public awareness to excite political action. As it turned out, headlines reporting massive oil spills and damaged beaches solidified a growing ethic of environmental stewardship. By 1970 the need for a tougher policy to manage coastal resources as a public trust became widely recognized, as were corrective measures. But as the contests between interested parties on local as well as federal levels grew, resistance to controls grew even faster.

As the Council Secretariat fortified its intuition with facts about patterns and trends of use, it began to examine the littoral itself and the historical roots of its development. Natural shorelines were found to vary widely—from placid, coral-enclosed lagoons of tropical Hawaii to wave-scarred cliffs of California's Big Sur, from ice-scoured margins of Arctic Alaska to wind-molded dunes of sandy Cape Cod, Hatteras, or Padre Island. Sculptured by waves and weather, the shoreline uncloaks the geology of its landward parent. But much more than inanimate rock and clay at its boundaries, coastal waters are vibrantly alive, perhaps the most biologically fertile, varied, and potentially productive of all our ecosystems.

An estuary is wings whispering over marsh-fringed waters . . . muskrats scurrying through reed, sedge and grass; shorebirds feeding busily on a tidal flat; trout noisily foraging in eelgrass . . . sea oats bending in the ocean's breath on a lonely dune . . . succulent oysters, toothsome shrimp, tender flounder. . . .[4]

Serving as nursery grounds for most coastal fisheries, these tidal waters are fed by fresh-water streams and rivers, and twice a day renewed as fresh nutrients are flooded into the area by tides. Here shellfish flourish throughout their full life cycle, and anadromous species such as salmon transit the zone to and from their spawning grounds. These same wetlands function as sanctuaries for waterfowl and shorebirds during migration or for secluded residence. The entire mix of marine-related life lives in harmonious ecological balance.

The coastal wilderness was initially probed by settlers with but a single objective, to locate safe anchorages for ease of access to the interior. For explorer and colonist, the coastline was a gangway for settlement. The estuaries were merely gateways to the hinterland beyond, which was to be won from nature, wild beasts, and savages. Pioneer diaries rang with the theme that the unbroken and trackless wilderness had been "conquered" and transformed into fruitful farms and flourishing cities.[5] Taming of the primeval, barbaric wilderness by Bible-toting frontiersmen epitomized progress; there were surely rewards in heaven for such toil. Except in tidal Maryland and Virginia, where tobacco plantations fronted on comely Chesapeake waters that provided local transportation, the coastal areas were largely ignored.

With independence, a new American nation entered upon a course of nationalism. Settlements along the East Coast engaged in flourishing maritime activities—commerce, fishing, whaling, clipper-ship building, and the obscene triangular trade of cotton, rum, and slaves; all earned foreign exchange and economic as well as political independence. Further reflecting nationalism in the 1840s, the doctrine of "manifest destiny" proclaimed that Americans were to occupy, civilize, and populate the vast continent between Atlantic and Pacific. The effect was to accelerate the surge of intellectual and physical energy away from the coast where it first clustered. Behind the ocean buffer, the country flourished, free of the conflicts and aggressions that plagued the growth of Old World cultures and lulled into the first of repeated cycles of maritime indifference.

The permanence of eastern ports was anchored by the needs of the landward settlements, however, for on their docks bulk cargo from the ocean was broken down for transshipment inland. Similarly, the Gulf ports became gateways to the Mississippi and the Missouri-Ohio river systems, and points of transshipment for cotton. The West Coast ports were classic springboards for exploitation of interior resources of minerals, timber, and fur, with consumer goods coming in and raw materials flowing out.

Coincidentally, eastern rivers provided a line of waterfalls near the coast for reliable cheap power that spawned manufacturing centers. The Industrial Revolution, exported to the United States in phase with its territorial expansion, accelerated their growth. Yet in each case these cities occupied but a small fraction of the coastline. While man was decimating buffalo and plucking out virgin timber, the natural beauty of the coastal margin remained largely undisturbed. In the drive to develop the interior wilderness, much of the shoreline was inadvertently left in its nascent state. That accident of history proved a salvation, until unwitting initiatives of the last decade began to gnaw away the natural legacy of the shore.

Relentless Economics of Coastal Modification

In the twentieth century, it was again the gateway characteristic of coastal cities that triggered a new era of growth, urban concentration, conflict in use, and shoreline modification. But now instead of funneling goods inland, railheads handled exports of bulk coal and grain. On the Great Lakes, ships moved iron ore to smelters closer to the coalfields. During World War II, the coastal ports disgorged mountains of supplies to feed and fuel overseas armies in both Atlantic and Pacific theaters. After the war a new wave of heavy industries demanded great quantities of chemicals, ore, and fuel, bulk goods that could be transported so economically by ships. Capital investment of industry logically sought the coastal area; labor to serve these industries prospered there, creating more local markets. By 1970, 70 percent of the nation's population lived within one hour's automobile drive of an estuary, ocean or Great Lakes. Inexorable urban growth brought pressing demands for fresh water, satisfied by streams and rivers as they raced toward the sea. With that fortuitous physiographic setting, the energetic supply-

and-demand mechanisms of the American free market system relentlessly and efficiently modified the coastal landscape.

This utilization of the coastal zone intensified in bewildering variety. Some uses were fully dependent on proximity to water: port terminals; industries such as shipbuilding, commercial fishing, and naval bases; extraction of salt, bromine, and magnesium from sea water; sport fishing; sanctuaries for waterfowl or nursery grounds for fish. To these can be added new ventures: aquaculture and offshore oil and gas development. A second class of uses was preferentially dependent on proximity to coastal waters: oil refineries; sand and gravel mining; housing; cooling for power production, especially of nuclear plants; mining of heavy minerals; waste disposal; recreation.

While the economic importance of the coastal margin is dramatized by the enormous investment of tax-producing industry, residential development imposes by far the greatest demand for shoreline mileage. The pressure of demand seems irresistible and waterfront property everywhere is priced far above comparable plots inland. As soon as the net of roads, water, sewer, and powerlines is superposed on the natural area, the economic values skyrocket. Recovered wetlands sold by Maryland's Ocean City to real estate interests for $100 per acre brought $500 per front foot when developed. In the virtual absence of planned and controlled development, the ecological inventory is depleted as market values increase. Ironically, the roads built through coastal marshes to increase access to frontal beaches destroy the very resource that justifies these expenditures.[6] The unnoticed trends of utilization are described in tables 8–11.

This inventory of shoreline use dramatically reveals that of the 60,000 miles of irregular shoreline in the United States, only one third is suitable for recreation, and of this third less than one twentieth is

TABLE 8

REGIONAL SHORELINE ALLOCATION

Shoreline Location	Detailed Shoreline (statute miles)	Recreation Shoreline (statute miles)	Public Recreation Shoreline (statute miles)
Atlantic Ocean	28,377	9,961	336
Gulf of Mexico	17,437	4,319	121
Pacific Ocean	7,863	3,175	296
Great Lakes	5,480	4,269	456
U.S. total	59,157	21,724	1,209

Source: Commission on Marine Science, Engineering and Resources Panel Reports, *Science and Environment* (Washington: GPO, 1969), vol. 1, p. III-17.

TABLE 9

ESTIMATED MILEAGE, BY STATE, OF THE U.S. RECREATION SHORELINE,
BY TYPE, OWNERSHIP, AND DEVELOPMENT STATUS[a]

STATE	TOTAL (miles)	TYPE			OWNERSHIP			
		Beach (miles)	Bluff (miles)	Marsh (miles)	Public		Privately owned (miles)	Development status
					Recreation areas (miles)	Restricted areas (miles)		
Alabama	204	115	. . .	89	3	1	200	Low
California	1,272	283	883	106	149	100	1,023	Moderate
Connecticut	162	72	61	29	9	. . .	153	High
Delaware	97	41	. . .	56	9	9	79	Moderate
Florida	2,655	1,078	406	1,171	161	122	2,372	Low-mod
Georgia	385	92	. . .	293	5	. . .	380	Moderate
Illinois	45	13	32	. . .	24	4	17	High
Indiana	33	33	. . .	. . .	3	. . .	30	High
Louisiana	1,076	257	. . .	819	2	. . .	1,074	Low
Maine	2,612	23	2,520	69	34	. . .	2,573	Low
Maryland	1,368	40	912	416	3	113	1,252	Low
Massachusetts	649	240	288	121	12	6	631	High
Michigan	2,469	292	1,959	218	357	. . .	2,112	Low
Minnesota	264	22	175	67	19	. . .	245	Low
Mississippi	203	134	. . .	69	. . .	25	178	High
New Hampshire	25	7	9	9	3	. . .	22	Very high
New Jersey	366	101	33	232	18	15	333	Very high
New York	1,071	231	590	250	47	. . .	1,024	Moderate
North Carolina	1,326	285	260	781	139	42	1,145	Low
Ohio	275	20	195	60	9	5	261	High
Oregon	332	133	181	18	101	. . .	231	Moderate
Pennsylvania	57	9	44	4	19	. . .	38	Moderate
Rhode Island	183	39	145	4	8	10	170	High
South Carolina	522	162	. . .	360	9	10	503	Moderate
Texas	1,081	301	421	359	5	18	1,053	Very low
Virginia	692	160	118	414	2	26	664	Low
Washington	1,571	121	1,294	156	46	27	1,498	Moderate
Wisconsin	724	46	634	44	13	48	663	Moderate
Total	21,724	4,350	11,160	6,214	1,209	581	19,934	. . .

[a] Excluding Alaska and Hawaii.
Source: Commission on Marine Science, Engineering and Resources Panel Reports, *Science and Environment*, vol. 1, p. III-18. Data are for 1967.

publicly owned. Near urban areas, only an infinitesimal fraction is reserved for residential, cultural, or recreational use. In Seattle, a typical example, the waterfront is essentially denied to citizen and visitor alike; the majestic views of Puget Sound and its Olympic Range are primarily merchandised from the Space Needle. Virtually all coastal American cities at one time possessed attractive residential districts near or on the waterfront. But almost as though a social disease had crept out of fouled harbor waters and deteriorating warehouses, those neighborhoods were the first to decay. Landlords deferred maintenance; then came vacancies and violence. What should have been the best part of the city, as exemplified in Stockholm, Copenhagen, and other North European ports, became the worst part. In another manifestation of endemic neglect of the shoreline, urban rehabilitation that should have started at the water's edge skipped over it. From 1824 to 1966, the federal government invested $22 billion in coastal and Great Lake dredging and other harbor

TABLE 10

SEATTLE HARBOR WATERFRONT, LAND
USE INVENTORY (1966)

(thousands of square feet—Net Area)		
Use	Area	Percent
Residential....................	18	...
Commercial....................	9,321	19.8
Industry.......................	10,711	22.9
Transportation[a]...............	13,814	29.5
Government and Institutional.......	4,624	9.9
Cultural and Recreational..........	58	...
Undeveloped and Misc.[b]...........	8,402	17.9
Total.....................	46,948	100

[a] One fourth of this figure is for auto parking lots.

[b] One third of this is reserved for facilities already under construction or planned (1967).

Source: *Shoreline Utilization in the Greater Seattle Area*, study by Management & Economics Research Inc., January 1968, for the Marine Sciences Council.

TABLE 11

BALTIMORE REGIONAL PORT SHORELINE
LAND USE INVENTORY[a]

Use	Miles	Percent
Residential...................	147	55
Industry......................	42	16
Government...................	13	5
Recreational..................	24	9
Unused.......................	40	15
Total..................	266	100

[a] The Baltimore Regional Port Shoreline is defined as the western coastline from the Chesapeake Bay Bridge to the Aberdeen Proving Grounds, a distance of 266 miles of water front.

Source: *Chesapeake Bay Case Study*, report by Trident Engineering Associates, Sept. 28, 1967, for the Marine Sciences Council.

construction to facilitate economic development.[7] How much the government spent to rehabilitate the waterfront society is too negligible to have been tabulated.

The second important fact is that most of the estuarine areas of the United States have been subject to a creeping modification by man's activities. The severity is revealed in table 12, which classifies eleven biogeographically differentiated regions. Researchers found that 23 per-

TABLE 12

CULTURAL MODIFICATION OF ESTUARIES

BIOGEOGRAPHIC ZONE	DEGREE OF MODIFICATION OF ESTUARIES[a]		
	Slight	Moderate	Severe
North Atlantic	44	48	8
Middle Atlantic	5	68	27
Chesapeake Bay	44	50	6
South Atlantic	36	60	4
Biscayne and Florida Bay	50	50	0
Gulf of Mexico	15	51	34
S.W. Pacific	19	19	62
N.W. Pacific	13	50	37
Alaska	80	20	0
Hawaii	54	15	31
Great Lakes	35	46	19
United States	27	50	23

[a] All estuaries and subesturaries were individually rated for each zone. The percentage refers to the proportion of these individual areas that were rated as indicated.

Source: Field evaluation carried out by Fish and Wildlife Service's personnel during course of Estuary Protection Act study; *National Estuary Study*, 1:25.

cent of the estuaries have already been severely modified; 50 percent moderately, and 27 percent only slightly. Only Alaska and Biscayne and Florida bays have escaped. A more typical example is San Francisco Bay, where 700 square miles have been reduced to 400 by diking and filling of tide and marshlands. Overall, the present loss rate of wetlands of 5 to 6 percent per decade has been considered by some to be insignificant. The rate of depletion is increasing, however, and if this trend continues, all wetlands will have vanished in a century.

The economic gain has its costs, even in the short run. Dredging unsettles bottom sediments, removes bottom-dwelling life, blankets fish nests, masks out light required by aquatic plants, and smothers ecologically sensitive bottom organisms; dam construction alters estuarine salinity and replenishment of beach sand; jetty and groin construction upset beach ecology; hurricane barriers disturb circulation of bay water, also affecting marine life. Some shoreline modification, of course, results from natural forces. Surges of tides, currents, and hurricane-induced storms may erode valuable shoreland, and the detritus swept away may clog navigation channels, suffocate marine life, and affect tidal circulation. Man is not *always* the culprit. In fact, he may in such cases be the hero, as coastal engineering acts to stabilize the shore.

Headlines are not made by this slow impoverishment of the littoral, but by man's more conspicuous form of destruction, pollution. In an affluent society, most consumer goods and by-products of their production are not consumed but eventually converted to garbage, sewage, industrial chemicals, fertilizer runoff, and solids that until recently we disposed of with the expedient principle of "out of sight, out of mind." The ubiquitous coastal waters accommodated our need to hide the ever-growing effluents.

THE CONVENIENT ESTUARINE SEWER

Said the Marine Sciences Council in its fourth annual report:

Contamination of the ocean has begun. Chemical wastes from factories, heat from power plants, domestic waste and sewage from cities and towns, insecticides and fertilizers from land runoff, atmospheric fallout of gasoline vapors, low level radioactive wastes from reactors, laboratories and hospitals are all flowing into the ocean. The sheer bulk of the material disposed of and the presence of new types of nondegradable waste products are now beginning to affect the ocean at an increasing rate.[8]

With the exception of materials airborne from the continent and deposited at sea through rainfall (not an unsubstantial quantity of hydrocarbons, aerosols and lead), pollutants enter the sea via the coastal waters. The nature and world-wide volume of certain of these materials have now been estimated.[9]

Municipal sewage from American cities, some eight billion gallons per day, carries into receiving waters a flood of inorganic compounds, pathogens, suspended solids, and decomposing organic matter that depletes oxygen needed by marine life. Some sewage is untreated; some receives primary, secondary, or in a few cases tertiary treatment. In specific areas the problem is compounded by raw sewage ejected into harbors from commercial, naval, and fishing vessels and pleasure craft. Some 200,000 industrial plants add their load of wastes.[10] Some of this material is dumped into fresh water and transported downstream; some goes directly into the coastal waters. With this load, some bays and estuaries soon resemble a chemical broth. Still other materials are barged across the coastal zone and dumped at sea, some in designated disposal sites even beyond national jurisdiction, most in open coastal waters less than one hundred feet deep. The Council on Environmental Quality has declared that while the amount thus dumped is small relative to the volume of other pollutants and is currently not a serious problem, "In the future the impact of ocean dumping will increase significantly relative to other sources."[11] Moreover, the solid wastes routinely barged to sea often contain poisons that leak quietly into the marine environment for years later.[12]

Industrial wastes, increasing at three times the rate of population growth, include: dredge spoils when water depth is increased for navigation; the sludge remaining after municipal treatment of sewage; refuse or garbage; mining and manufacturing refuse; discarded automobiles; construction and demolition debris; radioactive waste from processing of irradiated fuel elements, nuclear plant operations, medical isotopes, and obsolete radioactive equipment; explosives and chemical warfare agents.

Another very substantial source of pollution is accidental oil spills. Some 14,000 of these are recorded each year in navigable waters of the United States, one third of them from ships in operational routine.[13] This is not surprising when 60 percent of the volume of all maritime cargo on the seas today is oil. Although tankers no longer purge their empty tanks en route to being refilled, it is estimated that during the handling of oil in ports, even under the best of conditions, 88,000 metric tons per year may be accidentally spilled. Then there are major calamities from ship casualties and from runaway offshore oil wells. In the last ten years, over 500 tanker collisions have been recorded world-wide,[14] and the Santa Barbara incident dramatized the hazards of sub-merged drilling.

Thus the tender ecology of the seashore is being invaded by a host of pollutants that directly threaten marine life through toxicity, oxygen depletion, biostimulation, and habitat changes. Some effects are sublethal but reduce vitality, growth, or reproduction, or interfere with sensory functions. There is evidence that concentrations of pollutants have produced local dead seas: some parts of Lake Erie, the Houston ship canal, San Pedro Bay, the upper Delaware River, off the Hudson estuary. With massive oil spills, cleanup tactics employing chemical dispersants for their cosmetic effect may menace sea life as seriously as the oil itself and the persistence or possible irreversibility of such effects is not known. In addition, there is agricultural runoff: animal and poultry ex-creta from feed lots and hatcheries, fertilizer, insecticides, herbicides, and fungicides of great variety.

Some natural species appear tolerant of such chemicals, but DDT apparently threatens extinction of certain bird populations because birds that eat fish tainted with DDT hatch eggs so fragile as not to survive normal abrasion of nesting. Ducks, ospreys, hawks, eagles, and espe-cially the peregrine falcon and the brown pelican are believed to have suffered. DDT is also suspected of inhibiting spawning in salmon, per-haps because it hinders their detection of water temperature suitable for reproduction.[15] PCB, a chlorinated hydrocarbon used in a variety of industrial products, also seems to be entering the marine environment.

When not directly affected, marine species may nevertheless serve as hazardous transmission agents in the human food chain. Coho salmon in Lake Michigan absorbed DDT in their fatty tissue in such quantities as to make them unsafe. Shellfish absorb heavy metals such as mercury, sometimes in concentrations thousands of times more intense than the dilute saline solutions that they pump to extract food. While thriving themselves, these crustaceans become poisonous to humans, partly because the originally discharged stable compounds of mercury are converted in the marine environment to more physiologically potent methyl compounds. Occasionally shellfish harbor such pathogenic organisms as hepatitis from municipal sewage. From Penobscot Bay, Maine, to Chesapeake Bay, lobster and clam beds have been withdrawn from production; one-fifth of the nation's ten million acres of shellfish beds are closed due to contamination. Beaches have been closed in New York; Atlantic City was ordered to install new long outfalls because of back-bay pollution. Apart from the hazards to marine life and public health, the economic losses from limitations in shellfish harvesting or recreation are not negligible.

The most recent form of pollution introduced into coastal waters is waste heat from nuclear power plants. Now representing only 1 percent of the nation's power generation at central stations, nearly 100 nuclear plants projected for the next ten years will increase this source to one-third the total. Although costs are only roughly competitive with fossil fuels, the ability to produce power without the usual pollution of the atmosphere of CO_2, more noxious SO_2 and other substances, makes nuclear sources unusually attractive. The disposal of waste heat in nearby estuaries, however, could locally affect marine life. In some few instances, the effects could be beneficial in promoting rapid growth of fish, but it has been recently recognized that the heat load of nuclear plants by 1980 may require one-fifth the total fresh-water runoff of the United States, with increased problems of limiting temperature effects on marine life.

Most important, all waste disposal impedes other uses of the coastal margin; in fact, the range of utilization options is sharply curtailed by pollution. As discussed later, this problem is being attacked by prohibition at the source, but no one has estimated when the deterioration of coastal waters will be reversed, much less their purity restored.

Conflicts Among Uses and Jurisdictions

In light of the natural and the altered ecology of the coastal margin, the Council Secretariat turned to the incompatibility of human applications. It found that industrial development is inimical to nature conservancy, waste disposal to housing, sport fishing to commercial fishing. Filling of marshlands and dredging of channels may deplete fishery stocks; offshore drilling rigs increase hazards to navigation and risk oil pollution on beaches or of shellfish beds; unplanned expansion of private ownership restricts public access to beaches; oil refineries inhibit residential development; physical destruction and pollution destroy a marine region's usefulness for all kinds of employment. Coastal engineering, while reducing shoreline erosion or storm damage in one area, may inadvertently alter shoreline circulation and inhibit natural replenishment of sand in another.

This pattern of interactions is clear in figure 14. In the early stages of a shoreline's development, scattered individual actions are relatively innocuous. But as pressure of numbers grows, each user influences both the ecology and his neighbor. In time, the resource base may be irreversibly dissipated: the coastal band is literally consumed.

Four specific areas of conflict are involved regarding: the space itself; use of resources; present dedicated use versus future necessary use of the same area; and uses at some distance away affected by uses that modify the coastal environment.

Single-purpose uses can be rationalized one at a time, but rarely have attempts been made to harmonize compatible multiple uses with options kept open for the future. This checkerboard of use is complicated by a tapestry of public and private ownership, further confused when private development is deliberately encouraged of waters and submerged land that are themselves public domain. The conflicts are as much between different private users as between public and private.

Even public cognizance is hopelessly fragmented by political subdivisions involving a mix of municipal, county, facility district, port authority, state, multistate, and federal jurisdictions, each having specific, limited, functionally specialized responsibilities. Some of the coastal zone even lies beyond national sovereignty. One observer estimated that public regulation of Seattle harbor involved 15 entities. Yet fish, wildlife, water, ships, people, and pollution cross these boundaries freely. Users are mainly transients. Only a fraction reside there.

Because of the maze of functions and legal jurisdictions impinging on international and federal-state relations and involving ambiguities in agency missions, the federal government has never been able to agree on an explicit definition of the coastal zone. Despite Marine Council promotion of that term, all five annual reports ducked this sensitive definitional question. The Stratton Commission viewed the coastal zone as limited to: "(1) seaward, the territorial sea of the United States, and (2) landward, the tidal waters in the landward side of the low water mark along the coasts, the Great Lakes, port and harbor facilities, marine recreational areas, and industrial and commercial sites dependent upon the seas or the Great Lakes."[16] They added the caveat that "each coastal state, however, should be authorized to define the landward extent of its coastal zone for itself."

In an address before the American Society of Civil Engineers in New Orleans, February 3, 1969, the Executive Secretary chose an operationally viable if cartographically diffuse definition:

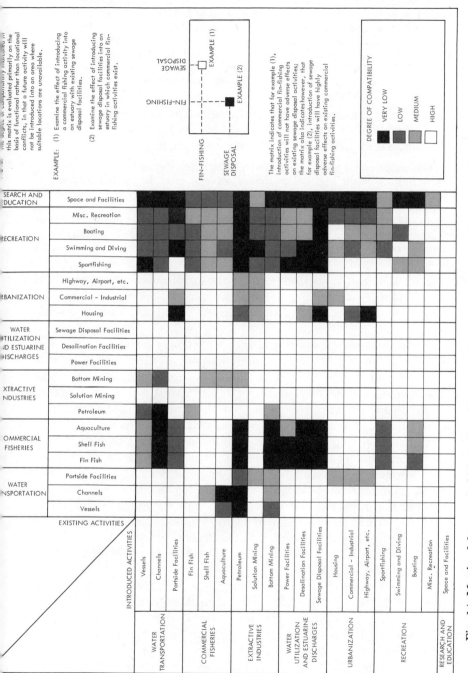

Fig. 14. Matrix of degree to which an introduced activity is compatible with an existing estuarine use. Source: Wilsey & Ham and Department of Interior, Bureau of Sport Fisheries and Wildlife

Conceptually, the coastal zone is a band of water and land surrounding the continent in which the sea exerts a measurable influence on uses of the land and on its ecology; and the land exerts a measurable influence on the sea and on its ecology. It extends offshore at least to the outer edge of the continental shelf and onshore at least to the upper reaches of lunar tide and shore areas. Although the characteristics and problems of the Great Lakes may differ considerably from those of a seacoast, they may be appropriately included in our definition.[17]

In existing legislation, the administration and Congress continued to be vague.[18] And the problem continued in new legislation. In 1969, for example, a draft bill (S. 2802) went further, stating:

The term coastal zone means lands, bays, estuaries, and waters within the territorial sea or the seaward boundary, whichever is the farther offshore, of the various coastal States and States bordering the Great Lakes and extending inland to the landward extent of maritime influences. . . . The term landward extent of maritime influences means such amount of land running back from the high water mark which in contemplation of human and natural ecology may be considered to come under the direct and immediate influence of the adjacent sea or lake.[19]

S. 582 of the 92nd Congress set the landward limit arbitrarily at seven miles, a definition removed in a later version, S. 3507. As discussed later, the location of the landward boundary is a key political issue.

As to the seaward boundary, the coastal zone involves five marine regions within and beyond a nation's sovereignty that require elaboration: internal waters, the territorial sea, the contiguous zone, the high seas beyond, and the continental shelf. The rivers, lakes, and canals on the landward side of the low-water line along the coast and waters landward of closing lines across bays nominally constitute the coastal nation's "internal waters."[20] On the seaward side of that baseline, lies "a belt of sea adjacent to its coast," described as the "territorial sea." Under the International Convention on the Territorial Sea and the Contiguous Zone, each nation's sovereignty embraces that territory. The width of this belt is neither specified nor agreed upon internationally, but claims vary from three to two hundred nautical miles. Rather wistfully, the United States has maintained that no claim to territorial sea breadths over three nautical miles is sanctioned.[21] International implications are discussed in chapter 6.

The International Convention defined the high seas as including "all parts of the sea that are not included in the territorial sea or in the internal waters," declaring that they "are open to all Nations" coastal and noncoastal, and sustaining a concept of freedom of the seas that no nation "may validly purport to subject any part of them to its sovereignty." To complicate delineation of these marine regions further, the convention provided that within the high seas is a contiguous zone "which may not extend beyond twelve miles from the baseline from which the breadth of the territorial sea is measured." Each coastal nation may exercise within this contiguous zone the control necessary to prevent infringement of its customs, fiscal, immigration, or sanitary regulations and may punish any such infringement committed within its territory or territorial sea. The United States, following other nations, has passed laws and regulations prohibiting foreign vessels from fishing in its twelve-mile "exclusive fisheries zone" without permission,[22] but some coastal nations have unilaterally redefined the contiguous zone to claim exclusive access to living resources of the sea up to two hundred miles.

For the seabed, limits of national sovereignty do not coincide with those for the waters above. The continental shelf is defined by a separate International Convention as "the seabed and subsoil of the submarine areas adjacent to the coast but outside the area of the territorial sea, to a depth of 200 meters, or, beyond that limit, to where the depth of superjacent waters admits of the exploitation of the natural resources of said areas,"[23] a definition subject to such technologically induced ambiguities as to incite a variety of passionate interpretations. The Convention recognizes the sovereign rights of the coastal nation to explore the shelf and exploit its resources; no one else may do so without express consent. The resources, incidentally, include mineral and nonliving, and also living resources that at "the harvestable stage either are immobile on or under the seabed or are unable to move except in constant physical contact with the seabed or subsoil."

Within the sovereign limits of the United States, a division of authority over these offshore areas exists between federal and state governments, but most of the governmental functions performed in the coastal zone fall comfortably within the ambit of the powers constitutionally reserved to the states by the Tenth Amendment. Title to submerged lands below mean high water was, by English law, vested in the king. Following American independence, this title was assumed by the states as successor to the king. At times title was granted to private individuals or delegated to local political subdivisions, but in some cases this is being recaptured. Under the Constitution, the national government was granted jurisdiction over interstate and foreign commerce, including navigation, and national defense and federal property. In recent years the defense and commerce clauses have been broadly interpreted. By virtue of the Submerged Lands Act of 1953, the coastal states, except Texas and Florida, own the living and nonliving resources of the seabed and subsoil of the sea out to three nautical miles. Texas owns resources out to nine nautical miles from its coast, as does Florida on the Gulf Coast.[24]

In general, the states decide how seabed minerals are to be exploited, how marine fisheries are to be utilized or conserved, how land and waters may be altered to benefit their citizens, and which uses should receive preference in the inevitable resultant tradeoffs. They also bear responsibility for pollution control, though this is now shared with the federal government. The states have principal zoning jurisdictions for land use within the coastal band; but as to water use, the 1899 River and Harbor Act giving the Corps of Engineers authority to issue permits regarding impediments to navigation also provided permit authority over any dumping of pollutants. Recently, the Corps has exercised this long dormant provision (and paradoxically in the process has been accused by conservationists of licensing pollution). Thus, the states hold in trust the submerged lands of coastal and navigable waters, except some ceded to federal or private interests, but the federal government has primary control of the water above. Intertidal land is usually considered part of the submerged land held by states, although in some cases it is the property of riparian owners.

In most states, authority over utilization of marine resources and zoning of land and water use has been delegated to local units of government, so that jurisdiction resides at the county, municipal, or even tax district levels. That power has been utilized effectively in some places but not in others. Often local jurisdictions have adhered to a laissez-faire policy of squatters' right or understandably have yielded to the application for industrial or commercial use that offered highest short-term economic return to the area, because they rely on property and real estate taxes for revenue. Until alternative sources of revenue are provided, they are likely to respond to such seduction.

There are also regional entities created by interstate compact that have some powers over coastal affairs; the Delaware River Basin Commission and the Susquehanna River Basin Compact, for example, are vested with authority for regulation, development, and operation as well as planning. These units are explicit acknowledgment that ecologically homogeneous regions do not begin or end at state boundaries.

Until recently, few states or local areas were awakened to their responsibilities for coastal management. Prior to 1969, only Massachusetts, Florida, and New Hampshire had started even to control drainage and filling of wetlands, and especially to require permits for alterations to private coastal property. From 1969 to 1971, Connecticut, North Carolina, New Jersey, Maryland, Georgia, Delaware, and Rhode Island enacted wetlands legislation. The citizen is only now aware of the debate and the broader issues at stake, as, for example, in the state of Washington, where in 1970 a voluntary Washington Environmental Council garnered 143,000 signatures on a petition for comprehensive shoreline protection.

If the states had this authority over coastal affairs, why have they been so slow to act? First, perception of the problem was slow in coming except in crisis areas such as San Francisco Bay. By 1970, however, the Committee on Natural Resources and Environmental Management of the National Governors' Conference polled its members to enumerate concerns and assign priorities. Next to the question of water resources, "coastal zone management" topped the list. If this were weighted to reflect the fact that only thirty of the fifty states have coastlines, the coastal issue might have been first. That same committee then asked: "What are the factors responsible for inadequate resource management?" Lack of funds led in responses, followed by lack of motivation and concern by the public as a whole; lack of technology and research; lack of proper plans, goals, objectives, and priorities.

This list does not adequately reveal major shortcomings in present state authority over coastal zone activities.[25] First, in most state governments there is no single focus for guiding the rational development of the coastal zone. The various functions of industrial development, natural resources conservation, port and harbor development, water pollution control, community planning, tourism promotion, and highways are aggregated in a bewildering variety of combinations. Internally, they reflect the same conflicts in goals projected externally by their discordant clientele. Second, although state resource agencies in general recognize the need for reserving and preserving common coastal resources for future generations, they are handicapped by shortage of funds for land acquisition (where purchase of the fee or some lesser interest is indicated); by insufficient knowledge to permit accurate predictions of the effects of alternative land uses; by complex and confusing estuarine land laws. Third, the states are hampered by the geographical limits of their jurisdiction. Marine fish are largely migratory, moving freely through the waters of various states and between the territorial waters and the contiguous zone or high seas. State jurisdiction over minerals beneath the territorial sea merges (along some yet-to-be-determined line) with federal jurisdiction on the outer continental shelf. Ambiguous and overlapping jurisdiction between local, state, and federal government creates problems because political fragmentation makes no sense in an environmental continuum.

To this list could be added certain political facts of life that legal control over land use is often ineffectual. Broad planning, much less zoning of land and water permits for explicit projects, will never be effective unless states are prepared to enforce decisions. And with money and people flowing freely across state lines, competition for both is fierce among the individual states. Controls that throttle any economic growth have been unpopular.

In this confusion, we have neglected one essential function of planning. Thoreau once admonished, "What is the use of a house if you haven't got a tolerable planet to put it on?" Who in the political process represents the public interest for rational coastal management?

The states are keenly aware of these problems, but initiatives to develop a collective position through a Coastal States Organization were frustrated until one was created by the National Governors' Conference in 1970. Meanwhile, citizen groups were focusing interest on coastal preservation and supporting political leaders who advocated rational management. But all seemed waiting for federal leadership in generating a national policy to help the states solve their dilemmas. Energizing the multitude of separate federal jurisdictions and aligning them in a common drive, circa 1966, was one of the major concerns of the Council and its Secretariat.

FEDERAL INVOLVEMENT

Federal legislation aimed at coastal activities goes back to 1899, and evolved as haphazardly as other marine statutes.[26] No single federal body had jurisdiction coinciding with boundaries of the coastal zone, since responsibilities of individual agencies devolved from varied functions performed there: water-quality research, planning and enforcement; waste management; control of beach erosion; improvement of ports, harbors, and waterways; conservation; provision of nautical charts and geophysical surveys; sea and storm prediction; acquisition and development of recreational lands and waters; commercial shipping and enforcement of maritime safety; shellfish culture sanitation; assistance for commercial and sport fishing and offshore mineral extraction; leasing of public lands and regulation for offshore safety; and finally development of scientific understanding of the coastal ecology. This fragmentation in missions rather naturally reflected the mosaic of different user groups, and in consonance with their interests, conflicts arose among the agencies as well. Prior to the creation of NOAA, these tasks were spread among twenty federal agencies represented on the Council and two that were not:

Department of the Interior: Office of Water Resources Research; Bureau of Commercial Fisheries; Bureau of Sport Fisheries and Wildlife; National Park Service; Bureau of Mines; Geological Survey; Bureau of Land Management; Bureau of Outdoor Recreation; Federal Water Pollution Control Administration.

Department of Defense: Department of the Navy; Department of the Army.

Department of Commerce: Environmental Science Services Administration; Maritime Administration.

Department of Health, Education and Welfare: National Institutes of Health; Food and Drug Administration.

Department of Transportation: Coast Guard.

Atomic Energy Commission.

National Science Foundation.

National Aeronautics and Space Administration.
Smithsonian Institution.
Department of Agriculture.
Department of Housing and Urban Development.[27]

From this catalog, it is obvious that many federal agencies had long been dealing with the coastal periphery, but the puzzle of coastal management was concerned mainly with navigation, shoreline recreation, or pollution control. Questions of fresh-water quality earned federal attention in 1948 and drew rigorous standards in 1965, but the issue of ocean quality lay dormant except where oil and radioactive waste were involved. The relatively unnoticed spills of oil were governed by some legislation, but the offshore drilling and transport in jumbo tankers that were growing in importance were, until the late 1960s, not visualized as a threat.

Genetic hazards from the accidental return of radioactive substances from the sea to man or to other biological organisms had long been recognized, and precautions had been taken to stay within the estimated capacity of the seas by limiting disposal to low-level waste such as that generated in university and industrial laboratories, hospitals, and research institutions licensed by AEC to use relatively small quantities of radioactive material.[28] As the industrial and military uses of nuclear processes increased, the Atomic Energy Commission in 1960 brought ocean disposal of even low-level waste almost to a halt, and the remaining disposal activities were voluntarily restrained. Risks to the marine environment persisted, however, from accidental introduction of radioactive substances from nuclear power plants, nuclear-propelled ships, defective artificial satellites involving nuclear elements, or the loss at sea of unarmed nuclear weapons. While the hazard of contamination, much less explosion, was remote, plans to cope with these contingencies continued.

The insidious pollution from municipal sewage and industrial waste was so dimly perceived in 1959 that it went virtually unnoticed in the National Academy of Sciences catalytic report. Official recognition appeared for the first time in the 1963 long-range plan of the Federal Council on Science and Technology. The 1966 PSAC report on *Effective Use of the Sea* also emphasized the water pollution question in its projection of needs. As late as Fiscal Year 1967, the consolidated federal oceanographic program prepared by the Interagency Committee on Oceanography showed funding levels of only $5.3 million per year, primarily for the Public Health Service in its operation of shellfish sanitation research centers, which studied botulism and the effects of pesticides on marine life.

Meanwhile, Congress uncovered and acted on the problem of coastal pollution. The Clean Water Restoration Act of 1966 (P.L. 89–753)

focused primarily on grants to states, municipalities, and intergovernmental bodies for demonstrating advanced methods of waste treatment
and purification and discharge control to attack the pollution problem at
its source. It did, however, call for a comprehensive three-year study of
the effects of pollution on the nation's estuaries. While the legislative
intent was water quality oriented, the scope of the study was stretched
to include an analysis of the social and economic importance of estuaries
and of the effects of pollution upon use; a discussion of developmental
and ecological trends; and recommendations for a comprehensive national program for the preservation, study, use, and development of
estuaries, along with responsibilities that should be assumed by federal,
state, and local governments or by private interests.[29]

An attack on the separate issue of coastal conservation was opened in
the 89th Congress by Congressman John D. Dingell, but his proposals
in H.R. 13447 were rejected in October 1966. Dingell bounced back
with H.R. 25 on January 10, 1967, which, with various diluting amendments, he fought through Committee jurisdiction battles to enactment
of P.L. 90–454 August 3, 1968.[30] The thrust of that legislation was to
extend studies of P.L. 89–753 to inventory the estuaries in relation to
conservation. Debate on this issue also uncorked action on the earlier
study which, although authorized in 1966, was not funded until April
3, 1968.

The objectives of P.L. 90–454 also signaled the contest that was
soon to envelop the nation on environmental quality versus economic
growth in referring to: "the need to protect, conserve, and restore these
estuaries in a manner that adequately and reasonably maintains a balance
between the national need for such protection in the interest of conserving the natural resources and natural beauty of the Nation and the need
to develop these estuaries to further the growth and development of the
Nation." The role of the states was underscored.

Conservation by land acquisition of undeveloped ocean shoreline with
highest value for public recreation dates back to 1934 when a National
Park survey identified selected beaches and urged state or federal acquisition. It met with feeble response. During the 1950s the undeveloped
stretches were again inventoried and in 1962 the Outdoor Recreation
Resources Review Commission called on state and federal agencies to
save magnificent stretches of unspoiled coastline. "The need is critical.
Opportunities to place these areas under public ownership are fading
each year as other uses encroach."[31] Subsequently, the National Park
Service postulated additions to its system; the National Wildlife Refuge
System was expanded under the accelerated Wetlands Acquisition Pro-
gram; the Bureau of Outdoor Recreation included marine areas in its
National Outdoor Recreation Plan.

To reduce coastal devastation from storms and to inhibit beach erosion, the Army Corps of Engineers had long been engaged in protective construction, backed by research on wave action, shore processes, and estuary dynamics.

So pollution abatement and coastal erosion were being attacked; conservation and recreation land acquisition were being promoted. But all of these spasmodic curatives were piecemeal. The more intransigent issues from conflicts in use had been neglected. As Daniel P. Moynihan has pointed out:

To be sure, government for the most part has been a collection of programs . . . authorized or required by statute . . . In terms of the social system, programs represent "inputs": so many miles of highway; so many square miles more or less of national forest . . . altogether too frequently it will be found that the actual results of a program are not all that was hoped for or promised. . . . The "successful" manipulation of national processes in one part of the ecology may lead to altogether unacceptable distortions in another part of the system. . . . We are moving [thus] from program to policy-oriented government. . . . Thus policy is primarily concerned with the "outputs" of a given system. . . . The key term here, of course, is system.[32]

In coastal affairs, one or more federal agencies cared about each element, but the separate agency programs simply did not add up to a system. The Marine Sciences Council was the first federal agency having a broad enough mandate and inclination to endeavor to deal with the entire Gestalt, to forge an integrated policy-level approach as a counterbalance to the staccato of fragmented programs responding to individual interests, and to articulate a concept of comprehensive stewardship. According to John Lear, who studied these origins in some detail, it was the Council's initiative in coastal management policy that nourished the broader gauge land-use policy proposed by Senator Henry M. Jackson in the 91st Congress and promoted in revised form by President Nixon.[33]

MARINE COUNCIL INITIATIVES AND DEBATES

Very early in the life of the Marine Sciences Council, the coastal problem was isolated as involving far deeper issues of public policy than rudimentary approaches toward pollution abatement and acquisition of recreational lands. However, we lacked even basic information and analysis that would help us ask the right questions. The evolution of the coastal management issue through the Council was punctuated by key staff papers and dialogue that revealed both the greening of the concept and the procedural impediments. In August 1966, the Interagency Committee on Oceanography was requested to assist the Council by defining the issues, but their well-intentioned dragnet on agency views lacked any policy or systems orientation and reflected internal conflicts that mirrored the external. With help from the Council staff, still in its infancy, and from such consultants as Abel Wolman, Professor Emeritus of the Johns Hopkins University, the Secretariat learned how complex and stubborn were the coastal problems. Yet it was crucial to include a wedge for coastal concern in the first collection of initiatives being prepared for President Johnson in response to his July 13 request. As we grasped at pending, albeit limited, opportunities, one was spotted. The Army Corps of Engineers had for some years been advocating construction of a three-dimensional hydrodynamic model of the Chesapeake Bay at a scale of 1:1,000 to study tidal mixing of fresh and salt water. After waiting its turn at the congressional pork barrel, the model had just then been authorized. Council staff and consultants recommended the Chesapeake Bay estuary as a pilot area for interagency, intergovernmental, and interdisciplinary study, weaving an enriched program around the original plan to answer a number of related questions, especially those of the capacity of the Potomac estuary and subestuaries to absorb pollutants, the ecology of the Bay, and its sensitivity to man's intervention. Insights from these studies might be extrapolated to other bays and estuaries in the United States and the Great Lakes.

On October 14, 1966, Vice President Humphrey submitted a progress report to the President outlining "problem areas where redirection or increased emphasis in marine science affairs would immediately contribute to broad national interests and to the Great Society concept." The first three priority areas were: (1) exploitation of food from the sea to meet world-wide starvation, especially protein deficiency; (2) abatement of pollution at estuaries and seacoasts to conserve economic, recreational, and aesthetic values; and (3) improved near-shore weather prediction, especially to protect petroleum operations, sport boating, and other marine interests.

The Council session on October 27, 1966, devoted to eight propositions opened with a staff paper largely in terms of earlier emphasis on coastal pollution: "United States estuaries are becoming increasingly polluted with serious degradation of shellfish, resources and recreation." But even at that point of origin, conflicts in use were isolated as fundamental.

A requirement thus exists for a comprehensive coordinated study of estuaries and bays, which will consider the relation of pollution, fisheries, recreation, mineral resources, navigation and beach and shore protection. Since the Army Corps of Engineers already has planning funds for a lab and model of the Chesapeake Bay, it would appear expedient to [proceed].[34]

Humphrey then imposed a gentle nudge, stating: "I have personally reviewed these staff proposals and believe all eight are of sufficient policy-level importance to warrant the time and attention of the membership of the Council." Realistic about eternal budgetary constraints and not wanting to put undue pressure on his boss, Humphrey continued:

. . .This will be a tough budget year . . . [but] we cannot stand still. If there are new programs or policies that are essential to our national interests, it is our responsibility to identify them, to determine costs and benefits, and to have the courage to defend such items even if it means dislodging some older programs of lower priority . . . that are continuing because of past momentum.[35]

That expression of budget agony and exhortation for responsible management was to be repeated often.

The Council approved the Chesapeake Bay proposal and the Secretariat was assigned over-all coordinating responsibilities, even though the Corps would be lead agency for its scale-model development. Following through in his November 16, 1966, memo, Humphrey wrote to the President that "The Council has isolated key issues and identified FY 1968 priorities that should begin to give coherent direction and purposeful momentum to what had been isolated approaches by 23 separate agencies."

The eight proposals were outlined, including the Bay study. Humphrey closed by saying:

Since the Council members concurred on the importance of these items, each will assume responsibility to provide the manpower, to seek funds and to assert leadership to accomplish and implement their program element. I shall be following up on these items to see that the members don't back off under the budget squeeze. . . . I recommend you mention these new directions in using the oceans for the benefit of man and these steps toward better Federal management in the State of the Union and a special message to Congress.

The President approved all initiatives, but our shot for the State of the Union message was not to be realized until January 1968. In his March 1967 marine affairs message to the Congress, President Johnson first reflected the chain reaction of the Council advice by proposing an attack on marine pollution with: "The continuing pollution and erosion of our seashores, bays, estuaries, and Great Lakes must be arrested and reversed to safeguard the health of our people and to protect resources of the sea. . . ."[36] He also approved broadening the Chesapeake Bay model to a full-fledged study of an estuary, but that model was destined to fail as a pilot example. Design efforts were slow to crystallize; funding estimates proved far too low; and the methodology became controversial. Interagency cooperation was still guarded.

Although this initial approach was to be delayed, the Council's attack on broad coastal policy was successfully launched. After establishing a tentative definition of the issue, the Council was faced with three problems: the need for policy planning on what to do; the need for interagency review and coordination on how; and tactically, the question of steering competitive instincts of federal agencies to a unified approach. The first need was met by adding Boyd Ladd to the staff. Ladd, a resource economist with broad governmental experience, on leave from a private research contractor, followed the pattern of amplifying Council capabilities by arranging for policy research contracts on: (1) legal problems arising out of the management, use, development, recovery, and control of the marine resources along the U.S. coasts;[37] (2) legal problems of the Great Lakes (truncated by the untimely death of the chief investigator at the University of Wisconsin); (3) the potential of coastal aquaculture;[38] (4) multiple uses of waters and coasts of the Chesapeake Bay;[39] (5) competing demands for land and water use in the greater Seattle harbor;[40] and (6) shoreline use of the Great Lakes.[41]

To engage in-house talent, we encouraged ICO to establish an inter-agency panel that in 1967 was converted to a Council Committee on Multiple Uses of the Coastal Zone (CMUCZ). It was chaired by Stanley A. Cain, an ecologist from the University of Michigan then serving as Assistant Secretary of Interior for Fish and Wildlife and Parks, with assistance from James McBroom, a senior civil servant who did much of the ramrodding of what proved to be a powerful and active arm of the Council.

The CMUCZ had fifteen members and four observers.[42] To avoid a conflict of interest with his host department when attempting to serve objectively as committee chairman, Cain asked that a second representative of Interior from the water resources area be added as Interior's committee member. Their debates became as uninhibited—and occasionally as acrimonious—as any between separate agencies. In plenary sessions October 16 and November 15, 1967, the Council with Secretariat and CMUCZ assistance probed the deeper implications of the problems and adopted the basic tenets of a policy concept: the coastal zone was to be treated as a system, with special attention to the role of the states. And as announced in the 1968 Council report, progress was made in dealing with the issue on a government-wide basis:

> An improved study and planning capability for dealing with problems of the Coastal Zone on a multi-disciplinary, multi-agency basis will be developed. . . . Information and plans will be exchanged on: environmental data needed for policy decisions concerning the Coastal Zone; establishment of ecological baselines along the coasts; requirements for monitoring Coastal Zone phenomena; human and ecological factors in establishing quality standards; scientific and engineering requirements to abate coastal degradation.[43]

Piercing these watertight compartments may have been one of the Council's primary achievements.

In his précis of the Council's second annual report, Humphrey picked up the central theme of coastal management:

> During the past year, the agencies of our government, separately and collectively . . . have focused ideas, facilities and manpower with a clearer

sense of direction and priorities—to enhance the quality of urban living by arresting degradation and erosion of the shoreline, fostering urban waterfront development, and expanding water recreation opportunities. . . .[44]

In transmitting that report to Congress, the President supported "increased research and planning to improve our coastal zone and to promote development of the Great Lakes and our ports and harbors."[45] The embryonic concept of rational management of the coastal zone was maturing to a sophisticated notion of planning "for the optimal use and conservation of the coastal environment and its resources, to control pollution and erosion, and to choose among multiple and often conflicting uses through wise cooperative management" with other governmental jurisdictions.[46] Along with the words went budgetary music. Funding for estuary and Great Lakes research was increased from $33.7 million for FY 1967 to $37.2 million for FY 1968 and $42.4 million for FY 1969.

This heightened awareness of shoreline conflicts was interlaced with an equally vital perception that key legal responsibilities had been left by the Constitution to state jurisdictions. This diffusion was of special merit since both ecology and citizen preferences vary from one locality to another. But administrative fragmentation required unifying concepts. And states then lacked machinery of sufficient planning competence and regulatory authority to manage coastal areas and to harmonize competing uses. Council staff contended that as a last resort the states urgently needed federal support by a policy and a grant-in-aid program to foster creation of new machinery and to buttress the political courage of state officials wanting to do the right thing. By midsummer 1967 a proposal was generated, but there was such stern opposition to this grant-in-aid route by the Bureau of the Budget, on grounds that states lacked the needed capabilities and that the money would be wasted, that the Secretariat felt obliged to find other allies in the skirmish with the Bureau.

The dragnet of initiatives cast by presidential counsellor Joseph Califano provided an opportunity to offer the coastal management concept through another door. On July 19, 1967, I recommended to Califano that the President "initiate a grant-in-aid program for State and Regional groups to study programs of conflicting uses of the shoreline" and also to "initiate a program of harbor redevelopment—to aid modernization and relocation of port facilities to serve cargo trans-shipment better, to rehabilitate aesthetics of waterfront and even provide water-recreational facilities within walking distance of poverty areas. Engineering of artificial islands could be advanced" to provide sites for offshore nuclear power plants or unloading of supersize vessels. Califano bucked the idea to the Bureau of the Budget, where the idea was again stalled by a Bureau specialist. Nevertheless, there was a tacit admission that new statutory authorities might be needed to facilitate planning for the coastal zone, and a start was made to consider regional instead of state entities.

On October 18 I wrote Califano that these ideas had evolved to a proposal for new legislation "modeled on River Basin Commissions,[47] to establish regional authorities to deal with local estuarine problems, on a Federal grant-in-aid basis. This would provide explicit mechanism for Federal/State collaboration and head off each state going in a separate direction."

On October 26, the Budget Bureau shot this down. In forwarding staff comments to the Bureau Director, the division chief felt that steps should be taken to "help dampen any enthusiasm that might have been generated by Dr. Wenk's proposals,"[48] although in fairness, they provided copies to the Council's Secretary. Largely their objection revolved over the cost of the entire set of proposals. As to preserving quality of the coastal zone, the Bureau objected to creation of another federal grant-in-aid program and a new set of regional commissions before other alternatives had been explored, leaning to the addition of estuarine problems to the scope of existing regional mechanisms.

My November 27 rebuttal to Califano was also unsuccessful, but while frustrated in a tactical engagement, the Council Secretariat was still dedicated to improved coastal management. We were permitted by the White House to float a trial balloon:

However, additional administrative actions or a Commission similar in make-up to the River Basin Commissions may be needed to carry out such measures along the coast between estuaries, and extending out to include related activities on the Continental Shelf.

In order to maximize the benefits from the Coastal Zone, additional actions need to be examined with the two-fold purpose of developing:
—increased public awareness of the nature and importance of the Coastal Zone; and
—an integrated program of science, technology, and public administration to develop more profound understanding of the coastal regions, to sharpen our awareness of biological interdependence, and to plan and implement measures and mechanisms for carrying out a policy which will open new opportunities on the one hand, and preserve that which is best in the environment on the other.

It is now also apparent that while the Federal involvement in the uses of the Coastal Zone has been growing, an institutional framework for management of specific segments of the Coastal Zone as regional entities may be needed if we are to realize the maximum social benefits from this vital natural resource.[49]

At that stage national policy was driven by a compelling logic dealing with ecological units that crossed state lines, and also favored the utilization of existing instruments. By mid-1968, however, the Council Secretariat, now buttressed with CMUCZ consensus, concluded that the original concept of state-managed coastal authorities provided a more viable means of political mediation, notwithstanding a residue of inter-state problems.

The next step by the Council Secretariat was to entice Cain's CMUCZ to "draft legislation, executive orders and Departmental regulations . . . that will provide for due consideration to all potential users of the coastal zone when determining whether Federal action should be taken affecting the coastal zone." The Committee was skittish on how far it wanted to go, partly because of internal conflicts between Cain and others in Interior generated during the preparation of the *National Estuary Study* by the Federal Water Pollution Control Administration (FWPCA). For what had begun as a "comprehensive study of the effects of pollution" metamorphosed into a much broader inquiry including recommendations for a "comprehensive national program for preservation, study, use and development of estuaries of the Nation, and the respective responsibilities which should be assumed by Federal, State and local governments and by public and private interests."

On August 2, 1968, the Council in plenary session approved a call for government-wide action by December 1, explicitly requesting and receiving concurrence from Lieutenant General William F. Cassidy, Chief of the Corps of Engineers, the most powerful of the actors on the coastal stage. But while the Council as a whole was led to water, they could not bring themselves to drink; adopting guidelines for action on November 26, the Council gingerly avoided adopting policy. But the Council under the Johnson-Humphrey administration had one more crack at Olympus at its last and perhaps most effective session, January 10, 1969. Fresh from a heartbreaking defeat in his race for President, Humphrey was determined to complete unfinished business and leave the marine house in order. The session had a full agenda, including a review of the major recommendations of the Stratton Commission that by law was to submit its report "to the President, via the Council, and to the Congress."

Attending the session were Secretary of State Dean Rusk; Secretaries of Interior and Transportation Udall and Boyd; Chairman of AEC Seaborg, and Haworth, Director of NSF; Undersecretary Bartlett of Commerce, Assistant Secretary Frosch of Navy, and Surgeon General William H. Stewart for HEW. Also present were Charles E. Bohlen, Deputy Undersecretary of State, Ambassador Donald F. McKernan and Herman Pollack of State, Chief of Engineers Cassidy, Assistant Secretaries of Transportation and Interior Lehan and Pautzke, Willis H. Shapley, Sidney Galler and Irwin Hedges of NASA, Smithsonian, and AID, and a sprinkling of other officers. No one attended for the Budget Bureau.

The Council first turned its attention to the controversial proposal for NOAA, with explosive debate along bureaucratic lines, as summarized in chapter 8. Then followed a ringing declaration of policy on seabed sovereignty put forward by Rusk. Next, the Council focused on coastal management. Distributed in advance of the meeting were draft copies of the Commission report, its analysis by the Council's Ad Hoc Committee to review Commission proposals, an independent report by CMUCZ after its two years of study, a comparison of Commission and CMUCZ proposals, and a brief statement for Council action. The comparison revealed that "these two proposals, separately developed, arrived at the same basic conclusion: the State governments should be encouraged by the Federal government to take a stronger hand in controlling the use of the coastal zone." That action was felt needed because "irreversible changes are taking place in the coastal zone, largely at the initiative of the private sector, without plans or adequate expression of the public interest."

Even up to the time of the Council meeting, and notwithstanding the consensus that had been evoked in the CMUCZ report, staff were apprehensive about action. My aide memoire to Humphrey stated:

We expect serious problems with: (1) the Corps of Engineers, which is nervous about recommendations that go beyond planning to include management; and (2) the Water Resources Council, which ignored the coastal zone until our Council began to study its problems and now wants to take over the entire program (they have only planning authority now). A possible objection from Interior may suggest deferment until an FWPCA study is completed next year.

Sure enough, when Udall left early, his Undersecretary David S. Black asked that it be recognized that the Secretary of the Interior "was given statutory responsibility under the Clean Water Restoration Act of 1966 to make comprehensive studies of the estuary zone and recommend a management system for estuaries." Clarence Pautzke, Cain's successor in both the Interior and Council Committee posts, courageously defended the Council move on grounds that there was no conflict between Interior's studies and Council action. The Vice President gave dissenters their final opportunity to speak, but any earlier tendency toward rebellion evaporated, and the proposed policy guidelines were formally adopted.

Although Council members were guarding their program components, they admitted that the national welfare deserved something better. A spirit of interagency cooperation permeated the outgoing Johnson administration and in its waning moments the Council adopted a broad framework of operational considerations that had been generated and recommended by interagency task forces at the working level, to:

1. Strengthen planning and decision-making processes that rely on the expertise and knowledge of the coastal States, including the

Great Lakes States, and are responsive to national as well as local needs;

2. Expand the structure for consideration of multi-State issues in Coastal Zone planning and management;

3. Broaden Federal policies and programs—without supplanting State autonomy and responsibility—to encourage the States to develop and carry out programs of planned use of the Coastal Zone in the national interest; and

4. Develop new mechanisms for acquiring, analyzing, and distributing management information and environmental data concerning the Coastal Zone.[50]

With these recommendations were twenty-four detailed suggestions on how they might be implemented.[51]

Because of rapid degradation of all the Great Lakes except Lake Superior, the Council also endeavored to develop a more coherent approach among the regional and international agencies engaged in planning and resource management: the Great Lakes Basin Commission, the Great Lakes Compact Commission, the Upper Great Lakes Regional Commission, the Great Lakes Fishery Commission, and the International Joint Commission. Only the latter two appeared to recognize Canadian interests in decisions affecting the lakes, leading to Council recommendations that both be strengthened.

Late in 1968 steps had also been taken to test implications with those parties beyond the federal government that would be affected—state and local governance and such interest groups as real estate, industry, fisheries, oil, and conservation. Rather than poll response sector by sector, the Council through CMUCZ conducted a three-day symposium at Williamsburg, Virginia, where the tentative proposals were exposed to a small coterie of selected participants. In an atmosphere of candor, comments and criticisms were freely offered. Expected opposition from certain interests developed, and the alignment of contending parties was exposed.[52] Although this exercise had been valuable to the Council in formulating its concepts and learning who felt their toes squashed, it also illuminated the long, circuitous road ahead to gain public support. The entire Council diagnosis was laid out in November 1968 testimony on beach erosion control.[53]

During this interval when the portent of coastal management emerged, the Secretariat had been unaware of how thin was the Council's jurisdictional charge to ignite action on the coastal zone. The functional terms of reference in the legislative mandate of P.L. 89–454 were silent in regard to pollution and conservation. On the contrary, emphasis was on "accelerated development of the resources of the marine environment." Even in its delineation of geographical scope, mentioning the oceans, Great Lakes, and continental shelf, such littoral designations as bays and estuaries were omitted. The coastal zone had not yet been discovered. More significantly, no mention was made of jurisdiction over what transpired on the landward side of the high-water mark, whatever the ecological or functional interrelationships of contiguous land and water. In retrospect, the Council acted with more guts than authority. Tacit jurisdictional blessings were received, first from the President by his endorsement of the 1966 initiatives, then in the spring, 1967, when Congress approved appropriations for Secretariat proposals to study coastal issues.[54]

By early 1968, the Council was overtly challenged. The sex appeal of coastal management became evident through the entire federal labyrinth, and several agencies became privately concerned that the field might explode in importance and leave them behind. The territorial challenge came from unexpected quarters—the Water Resources Council (WRC) that had been created in 1965.[55] Chaired by the Secretary of the Interior, it dealt with the management problems of quantity and quality of fresh-, not salt-water. Contending that estuaries and coastal waters are fed by fresh water and suffer pollution if the feeding tributaries are fouled, staff of the WRC asked for and received invitations to sit as Marine Council observers, where they interposed quiet objections to our coastal initiatives. In February 1968, their campaign to take over this area went into high gear when they reckoned that if coastal zone planning grants were furnished states, the federal agency selected to interface with all the states would be equipped with the formula for erecting a power structure. They saw the Marine Council conspicuously eligible for that role. Working through their network of communications with water officials in coastal states who were apprehensive that new authority over coastal activities might be conferred on a competitive state agency, the WRC staff director tried to derail the Marine Council. He failed, partly because he was dislodged with the change in administration and a successor was not immediately appointed.

Nervousness about Marine Council jurisdiction popped up again in the President's Council on Recreation and Natural Beauty. Their report acknowledged the Marine Council's creation in the section on "Shorelines and Islands," and then coldly stated: "The Council has established a Committee on Multiple Uses of the Shoreline, but is not authorized to exercise continuing coordination of an overall coastal shoreline program."[56] In contrast, the Council's involvement and initiative were acknowledged without caveat in the National Estuary Study released in January 1970.

By the end of the Johnson administration, coastal priorities gained increasing status within the rubric of marine science affairs. The President now listed in his January 17, 1969, message as first among his administration goals the realization of the "promise for enhancement of men's lives" through actions to "enhance the many uses of our seashore and coastal waters by directing national attention to the need for skillful management of this coastal zone."[57] There was also further recognition that the state level of government was the politically appropriate site with legal jurisdiction for detailed coastal management. The Council said its member agencies were "taking steps to encourage the states to strengthen and expand the necessary institutional framework for carrying out planned use of the coastal zone in the national interest."[58]

There was clear consensus in the bureaucracy on what to do but not on who should do it. The bugaboo of jurisdiction was to rear its head with a stubborn refusal of the involved agencies to agree on a lead agency for coastal management because cooperation appeared to provide some competitive advantage to a sister agency. For fear of bloody debate, the Council also ducked the key question in this politically fertile field. And at the waning stages of his administration, Johnson instructed the bureaucracy to defer new legislative proposals that could tie his successor's hands.

A legacy in the web of coastal affairs had been spun, however, in generating first principles for a public trust.

PRINCIPLES FOR A PUBLIC TRUST

Roots of policy for coastal management were adapted from public purposes previously executed landward. Long-standing concerns for water quality were extended to estuarine waters, and for nature conservancy to include wetlands and similar shoreline regions. But the coastal band differed from its terrestrial counterpart in one major characteristic; it was also a public trust. In the three years after initial perception of this fact, swift progress was made as to both diagnosis and remedy of what was amounting to a tragedy of the commons from resource degradation by unlimited entry.

The "tragedy of the commons" was the phrase used to describe the fate of medieval pasture lands held in common, where unrestricted entry by all herdsmen caused overgrazing and eventual ruin. The same fate accompanies any overexploited limited resource and the coastal zone and its waters were no exception. Only recently had it been acknowledged that property withdrawn from the public trust was being converted to private development, usually at a handsome profit. Suburban sprawl posed the greatest threat of all.[59] Pressure to expand airports was being felt in these undeveloped areas because they were the only large expanses left near coastal cities. The battle over the Everglades jetport near Miami was just the first round.

Given the ecologically sensitive environment of finite size and the profusion of coastal interests that were in irreconcilable conflict, the primary issue was not how to preserve coastal resources in pristine purity, nor how to exploit them for economically optimum return, but how to harmonize multiple coastal demands, public and private, and still obtain the greatest long-term social and economic benefits. The Council Secretariat distilled a set of seven basic principles to guide policy development and later draft legislation:

1. A national policy is required to balance protection and development of coastal resources for this and succeeding generations.

2. Every foot of coastline should eventually be subject to a comprehensive management plan for land and water use.

3. The plan should be prepared at the state level of government subject to review and approval by the governor.

4. The state should provide and exercise the necessary regulatory authority, zoning, and land acquisition powers to implement its plan.

5. Provision should be made for public notice and hearing in development and modifications of such plans.

6. Provision should be made for conducting and utilizing relevant ecology and policy research, including establishment of estuarine sanctuaries, which will provide a factual basis for estimating the impact of man's intervention in the natural environment.

7. Provision should be made for multijurisdictional cooperation, with special emphasis on regional planning for ecological areas that cross state lines.[60]

Apart from the technical mechanics of coastal management, we had introduced the concept of citizen participation. But what sort of program was needed to convert these principles to action?

First, to meet the fragmentation of responsibilities among state agencies, some one state agency must be made responsible for administering use of the coastal zone. This agency could not be completely independent and comprehensive, for it would then duplicate functions being carried out by the other state agencies. It could not be understaffed or too restricted in authority, however, for it would then be overly reliant on the existing line agencies, many of whom represented specialized points of view and vested interests. Obviously, some middle ground must be found.

Second, to protect the general public interest, there must be means of regulating individual uses of the coastal zone. The exact mechanism is not nearly as important as its effectiveness; conceivably, as many different devices could be employed as local conditions dictated. Zoning of land and water might be used. Permits for specific projects might be issued. Master plans for coastal development might be devised and adherence fostered through a permit or zoning system. Areas of significant public value for recreation or of cultural and aesthetic importance might have to be purchased by the local, state, or federal government.

Third, there must be some means of enforcing decisions regarding use of the coastal zone. Permits are pointless if work goes on without them;

zoning is ineffective if violations occur with impunity; even public ownership breaks down if trespassers are rampant. This enforcement is not
easy; surveillance itself is a monumental undertaking; enforcement
proceedings are complex and new statutory authority has to be tested in
courts. In many cases, state/federal boundaries have yet to be determined on the land beneath the water, and private/public boundaries
represent a maze that will take years to unravel.

Fourth, more environmental data, clear ecological facts, and rational
analyses are needed upon which wise policy decisions can be made; these
require major improvements in research and monitoring of coastal
ecology. If management decisions on coastline use are not to be swayed
only by political winds, rational scientific and engineering analyses
based on comprehension of the environment would be needed to generate
options and facilitate choices among alternative regulatory actions, public and private uses.

And finally, there is the matter of money. Many states and local governments finance coastal zone activities through a variety of licenses,
taxes, fees, and general appropriations. They are assisted in these endeavors by federal funds from programs of the Economic Development
Administration, Department of Housing and Urban Development,
Bureau of Outdoor Recreation, Fish and Wildlife Service, Water Resources Council, Corps of Engineers, Soil Conservation Service, and
others. But more funds would be required.

By 1969, the nation had four major but unresolved coastal management issues on its immediate decision agenda:

1. Establishment at the federal level of national policy on coastal
affairs and adequate funding to aid the states with its implementation.

2. Assignment of key responsibility for coastal management to a
federal agency.

3. Strengthening of coastal research.

4. Action at state level to dovetail with federal policy by creating the
necessary machinery for planning, management, and public participation.

THE STAGING FOR RENEWED INITIATIVES

On January 9, 1969, the Marine Sciences Commission transmitted
its comprehensive report on *Our Nation and the Sea* to the President and
to the Congress. The coastal zone was a clear target for recommendations. Said the Commission:

1. Management of the coastal zone should continue to be vested in
the states; but backup federal legislation should encourage creation of
state coastal zone authorities empowered to carry out national objectives
by planning and regulation of land and water use, and by acquisition
and development of land in the coastal zone for public purposes.

2. University-affiliated coastal zone laboratories should be designated and supported, similar in function to agricultural research stations and extension services.

3. Representative coastal and estuarine sites should be set aside as natural preserve bases for studies to assess the effects of man's activities on the environment.[61]

4. Monitoring and research on coastal pollution should be intensified.

5. A national research project should be initiated to undertake Great Lakes restoration.[62]

The Commission also recommended that these functions be lodged in a new independent National Oceanic and Atmospheric Agency, which might take over certain marine science activities from the Departments of Transportation, Interior, and Commerce, and of the National Science Foundation. In addition, the Commission urged that: state procedures be developed for leasing of submerged lands for such new uses as aquaculture, an industry quite new to this country but having significant potential; the authority of the U.S. Corps of Engineers be expanded to deny permits for coastal engineering projects if recreational opportunities were jeopardized or other uses degraded; AEC be authorized to consider environmental effects of projects under its licensing authority.

Their report, submitted in the final days of the Johnson administration, was understandably lost in the noise level of changing presidents. But not forever.

On inauguration day, the Nixon administration inherited a full plate of explicit oceanic proposals from both the outgoing administration and the Commission. But with so many issues queuing up for attention, would coastal management qualify? There were few advocates from outside the federal government, and the Nixon administration had made no commitment to the issue during the campaign. Within, well-entrenched bureaucratic positions that had been combed to unity by the Humphrey Council were likely to resume their functionally disjointed programs and rivalry with the transition in administrations. All were in their tornado cellars.

The major advocacy for a coherent approach fell to the Council's Secretary. Using the Stratton Commission report as a cover, and knowing from earlier experiences where the booby traps lay, it did not take long to reignite action on maritime initiatives that had been idling from the prior administration. Coastal management was clearly of high priority, but when Vice President Agnew remained indifferent to suggestions that as former governor of a coastal state he could comfortably assume the cloak of activist Council Chairman, hopes for rapid advancement of this issue dimmed. Nevertheless, this was one of the key issues adopted for study in March by the special White House task group, and developed by the Council's ad hoc successor to CMUCZ under Russell Train.

By September, the Nixon administration had arrived at the same position on coastal policy that the Johnson administration had put in escrow. On October 19, Agnew announced the new administration's first thrust in marine affairs, described in chapter 3. In the five-point program, three were concerned with coastal affairs: a policy and grant-in-aid program of state management, intensified coastal research, and a Great Lakes cleanup demonstration project. Of special significance, the Department of the Interior was given lead agency status on two. In responding swiftly to their first assignment they and Council staff drafted the administration's version of a coastal bill, H.R. 14845 introduced on behalf of the administration November 18, 1969.[63] Interior's alacrity was no doubt triggered by the promise of statutory jurisdiction in coastal affairs. Then followed an intensive public relations campaign by Interior to develop an "image" in the growing competition as host agency for NOAA. Despite its territorial initiative, Interior failed to carry out its second assignment and coastal research was neglected. Henceforth, for almost one year, the administration went silent on coastal affairs. When action was resumed, it was in the form of a presidential proposal for comprehensive land use planning that subsumed coastal management,[64] and a proposal by the new Council on Environmental Quality to deal comprehensively with one isolated piece of the problem, ocean dumping.[65] Explicitly, the administration abandoned its interest in coastal management.

In early 1969, Congress was yet to be heard from on coastal matters, but history showed signs of repeating itself as it began to make loud noises about Nixon's inaction on the Stratton report. The noises were not symmetrical. The House, far more than the Senate, wanted action on NOAA, but it was the Senate that moved first, and it moved on coastal legislation. In late August 1969, Dan Markel of Magnuson's staff asked the Council Secretariat for assistance in drafting a bill to implement the Stratton coastal management recommendations, but with a major exception. Program leadership was to be vested in the Marine Council instead of in the projected but highly uncertain NOAA. David A. Adams, a former member of the Stratton Commission, a Council consultant, and now on the Council staff as Ladd's replacement, was instructed to assist, at the same time we were working on an identical measure for the administration. Although we had warned the administration that Congress was restive, Magnuson won the race. With cosponsor Senator Philip A. Hart he introduced S. 2802 on August 8, 1969, two months ahead of the administration.

In the fall of 1969, Senator Magnuson, who had previously kept oceanography as his own preserve through actions before his full Commerce Committee, unexpectedly created an informal subcommittee on oceanography chaired by the junior Senator from South Carolina, Ernest F. Hollings. Hollings, whose vitality and determination were to be reckoned with subsequently in marine affairs strove to advance S. 2802. In the second session of the 92d Congress, however, his emphasis turned to plumping for creation of NOAA, and until that victory, coastal legislation took a back seat. About the time it could have been revived in late 1970, the SST debate heated up, and since that pre-empted the full attention of Washington State senators through March 1971, coastal legislation was again deferred.

Nevertheless, S. 2802, the first Coastal Zone Management Act of 1968, was a succinct précis of all the objectives enunciated by the Marine Sciences Council and Stratton Commission alike. Its declaration of policy stated:

The Congress finds and declares that the coastal zone of the United States is rich in a variety of natural, commercial, industrial, recreational and aesthetic resources of immediate and potential value to the present and future development of our Nation; that unplanned or poorly planned development of these resources has destroyed or has the potential of destroying, the basic natural environment of such areas and has restricted the most efficient and beneficial utilization of such areas; that it is the policy of the Congress to preserve, protect, develop, and where possible to restore, the resources of the Nation's coastal zone for this and succeeding generations through comprehensive and coordinated long-range planning and management designed to produce the maximum benefit for society from such coastal areas.

In defining the geographic limits of the coastal zone, the bill took a bold step in defining the ambiguous and controversial landward extent of maritime influence as: "such amount of land running back from the high-water mark which in contemplation of human and natural ecology may be considered to come under the direct and immediate influence of the adjacent sea or lake."

In short, the bill called for formulation of a "master plan for the coastal zone" that would contain "a statement of desired goals and standards to help shape and direct future development—to promote the balanced development of natural, commercial, industrial, recreational and aesthetic resources and to accommodate a wide variety of beneficial uses." Such plans would be prepared by the states, through a "coastal zone authority—designated by the Governor," although any coastal state could designate an interstate agency of which it was a member. Public hearings would be required in formulating plans "to obtain all points of view." Subject to approval, the Marine Sciences Council was authorized to make grants up to 50 percent of the costs for both developing and implementing such plans, out of a "Marine Resources Fund" created by the initial $75 million of revenues received under the Outer Continental

Shelf Lands Act. One condition was that the coastal authority would have power "to draw up land use and zoning regulations—to acquire lands . . . through condemnation . . . to develop land and facilities . . . to borrow money and issue bonds for the purpose of land acquisition—or development and restoration projects. . . ." Thus, the state authority was to have powers to implement and enforce the plan.

Estuarine sanctuaries separately proposed by Senator Joseph D. Tydings in S. 3460 were authorized as a step toward recognition of the need for coastal research, but no other provision was made.

In testifying May 4, 1970, on S. 3180, S. 2802, and S. 3460, Secretary of the Interior Walter J. Hickel lent eloquent support to the bill on grounds of the "conservation issue." He placed special emphasis on local land use practices "too often shaped by the tax structure of local government which depend on the property tax for their revenue base." Continued Hickel: "Considering the small revenues available to local governments in the face of their many vital needs, their problem is very understandable. . . . This creates a vicious circle, resulting in unwise use of land, by encouraging local governments to zone, just to increase revenues."

Of course he supported the administration's proposal for federal jurisdiction in his Department of the Interior, rather than the Council. To nail down Interior's contention as host department for the pending NOAA, he called attention to his order creating a new Office of Marine Affairs responsible for overseeing Interior's lackluster concern for coastal research.

Many objections were raised to the administration's bill, S. 3183, because the meager sums allocated would enfeeble policy implementation.

Subsequent to hearing by Senator Hollings' subcommittee, the bill was modified on August 8, 1970, by a stronger preamble, especially to recognize that "the key to more effective use of the coastal and estuarine zone is the introduction of a management system permitting conscious and informed choices among alternative uses," and that "the absence of a national policy and planning mechanism for the coastal and estuarine zone resources has contributed to impairment of the Nation's environmental quality."

With the administration by then supporting NOAA, Hollings rewrote the legislation to transfer the lead agency responsibility from the Council to NOAA. Grants-in-aid were increased to two thirds. The designation of "coastal authority" was eliminated in favor of an agency designated by each governor. No special source of funds was designated, simply an authorization.

Stimulated by the administration's October 1969 five-point program, and worried by the Senate initiative, the House Subcommittee on Oceanography wanted to assert its interest in coastal affairs, but refrained from introducing legislation to avoid a committee jurisdiction fight that it might lose as it had lost in March over oil-spill cleanup authority. Instead, Lennon convened a two-day symposium on coastal issues in the winter of 1969, inviting guests from all the coastal states and from the Stratton Commission and the Council, including some of the same objectors and vested interests who had spoken at Williamsburg in late 1968.[66] But the House Subcommittee on Oceanography did not make its move on coastal legislation until 1971, generally following in the Senate's wake.[67]

None of the legislative proposals set before Congress provided for the estuarine research necessary for intelligent coastal planning. As the Marine Sciences Council stated:

We still lack much of the knowledge needed to provide the understanding required to assess and predict the effects of the man-induced and natural modifications of the marine environment . . . establishing baselines or standards from which we can detect and measure environmental changes . . . what pollutants and in what quantities are entering the ocean; how much pollution the marine environment can absorb without substantially harming other uses; how marine pollutants circulate and disperse, degrade and convert . . . what effect man's physical modifications of the coastline have on water dynamics, marine life and sedimentation . . . how pollutants enter the life cycle of marine organisms and what effect they have on them; and about how to treat ocean pollutants. An adequate pollution monitoring system could . . . provide the scientific basis for assessing and predicting man-made changes, identifying and controlling pollutant buildup, managing waste disposal and safeguarding the physical and biological quality of the oceans.[68]

A research capability built on existing federal, state, and private capabilities was anticipated by both the Stratton Commission and the Council. With the October 19, 1969, request by the Vice President to the Department of the Interior ignored in their Fiscal Year 1971 budget requests, staff of the NSF Sea Grant Office endeavored to establish a new role for themselves as sponsor of coastal research laboratories. Although a lame duck as Council Secretary, I opposed that action, fearful that if the needed coastal research were exclusively a Sea Grant franchise, universities would have their goals dictated from Washington, yet would be unable to give the quick response needed in coastal management. In the meantime, Senator Claiborne Pell rushed in S. 3118 to give the Sea Grant Office responsibility for creating a system of estuarine laboratories, and a new fight erupted within the administration as Sea Grant officials endeavored to build a strong political constituency in state houses as well as on college campuses.

No coastal management legislation emerged from the 91st Congress, but both Senate and House resumed initiatives in the 92nd. Senate bill

S. 582 introduced by Senator Hollings was reported out favorably with amendments on September 30, 1971;[69] but the comprehensive land-use planning bill being pushed by Interior Committee Chairman Senator Jackson created an embarrassment. Functionally, it subsumed shore-line management, but in its provocative scope, it had little chance at passage. Despite a long history of cooperation between the two Washington State senators, their respective roles as chairmen of two power-ful committees led to a conflict on this legislative plum because of the assignment by S. 582 of responsibility to the Secretary of Commerce who by then had been given cognizance of NOAA, whereas the rival bill on land-use planning recognized competences in Departments of Interior and Housing and Urban Development. Until the spring of 1972, the impasse remained. Determined to follow through on an initiative dating back to 1969, Magnuson's Commerce Committee staff under Michael Pertschuk resolved the jurisdictional differences, reported out a clean bill S. 3507, and enticed unanimous Senate action 68 to 0 on April 25, 1972. Interagency coordination, incidentally, was provided for in a mini-council, chaired by the Vice President. As a mark of admiration, the measure was amended on the floor to be entitled the Magnuson Act. Almost the next day, the House Merchant Marine and Fisheries Committee reported the companion measure, H.R. 14146, out favorably. Compromise was almost certain. Still there was no flicker of interest from the administration, although it would clearly have to face the hard choice of acceding to assignment of coastal management to the Department of Commerce or sticking with its 1969 selection of Interior as lead agency, or deferring coastal management until the more comprehensive and controversial land use measures, H.R. 4569, S. 632, and S. 992 were acted upon.

On the matter of agency jurisdiction over coastal management, out-side interests had testified at hearings as follows: the National Wildlife Federation favored Interior; the National League of Cities, the U.S. Conference of Mayors, and the League of California Cities favored HUD. The American Association of Port Authorities, the Mississippi Valley Association, the National Rivers and Harbors Congress, and the Gulf Intracoastal Canal Association, all of which communed with the Corps of Engineers, opposed any legislation.

As the government labored sluggishly toward adoption of a national policy, four other developments were having an influence on the health of the coastal zone. Some states were acting individually to protect their littoral inheritance; for example, Delaware through Governor Russell Peterson in the face of strong opposition from his industrial constit-uents.[70] Increasing amounts of oil were spilled and legislative counter-measures were being invented. Ocean dumping attracted regulation. And a new ethic of environmental conservancy was being reflected through an unexpected legislatively generated discipline: the requirement for assessment of environmental impact.

The first oil spill episode to capture public attention was the *Torrey Canyon* wreck off southwest England in March, 1967. Then the *Ocean Eagle* broke up in Puerto Rico, the *Witwater* spilled oil off Panama, the *Yukon* ran aground in Cook Inlet. Other spills occurred off coasts throughout the world. The government's reaction—formulation of contingency plans so as to have a standby capability to contain and clean up spills—was announced by Johnson on June 7, 1968, in response to recommendations by the Council in concert with the Secretaries of Interior and Transportation. Details were completed by the fall.

The problem, however, persisted. According to President Nixon in his May 20, 1970, message to Congress on oil pollution:

Seaborne oil transport has multiplied tenfold and presently constitutes 60 percent of the world's ocean commerce. This increase in shipping has increased the oil pollution hazard. Within the past ten years, there have been over 550 tanker collisions, four-fifths of which have involved ships entering or leaving ports.[71]

The U.S. Coast Guard reported for 1970 that they investigated 2,345 ship casualties involving vessels over 300 tons. While only a small fraction of these were tankers, bunker oil spilled by bulk cargo carriers was an equal hazard to local marine ecology. In 1970, 26 million barrels per day of oil were transported by sea, a volume expected to double by 1980 to 53 million barrels. The number of tankers will not double because they are growing larger, but the number of casualties may well increase because the supertankers are far less maneuverable when facing possible collision. While containment and cleanup measures had been instituted in 1968 and sharpened in the aftermath of the Santa Barbara incident, little attention was given to prevention. Amidst these hazards, again as emphasized by President Nixon in a February 8, 1971, message, the Coast Guard lacked statutory authority to control vessel traffic in inland waters and U.S. territorial seas and to oversee handling of hazardous cargo; harbors lacked surveillance radar; ships lacked means of communication to others in the traffic pattern. Few steps had been taken even to increase tanker safety through such design requirements as limitation of compartment size or total size, or the use of double-bottom hulls.

Just this kind of issue led to a round of objections to the Alaskan pipeline, because the Department of the Interior's analysis of its environmental impact neglected evaluation of the maritime extension of the pipeline from Valdez to such tanker terminals as Cherry Point in Puget Sound.[72] Requirements imposed by the National Environmental Policy Act of 1969 were a major innovation in governmental policy, for they were designed to assure advance consideration of environmental hazards.

BIRTH OF A RADICAL SOLUTION—
A NEW ENVIRONMENTAL POLICY

While the coastal management theme was maturing, two unrelated concerns were surging to a legislative crescendo, one on environmental policy and the other an assessment of the total intended and unintended consequences of technology. After a splintered evolution of air, water, and solid waste management, a ground swell had developed for a more universal and radical ethic of environmental protection.

The National Environmental Policy Act of 1969 (P.L. 91–190) was wrought, which incidentally included the offspring of the second independent activity, that of technology assessment. The Environmental Act was the political expression of heightened national awareness that our planet cannot indefinitely absorb the insults of man-induced change. A product of initiatives by Senators Muskie, Jackson, McGovern, Nelson, and Congressman Dingell, it recorded a national policy to "encourage productive and enjoyable harmony between man and his environment; to promote efforts which will prevent or eliminate damage to the environment and biosphere and stimulate the health and welfare of man; to enrich the understanding of the ecological system and natural resources important to the Nation. . . ."

Section 102b called for implementation to "fulfill the responsibilities of each generation as a trustee of the environment for succeeding generations . . . attain the widest range of beneficial uses of the environment without degradation, risk to health or safety, or other undesirable and unintended consequences . . . maintain, wherever possible, an environment which supports diversity and variety of individual choice; achieve a balance between population and resource use. . . ."

Some of these principles had originated in the multiple-use concept of water resource development, others in the more recent awareness of conflicts in the coastal zone. But this policy took the big leap forward, combining both concerns in calling for stewardship of the environment for future generations.

What was so radical in this step? For almost two hundred years, the United States had been pursuing policies (or lack of policies) rooted in the laissez-faire belief that rugged entrepreneurship and a potent technology were the primary routes to improvement of the human condition. Even when conservation was dramatically made a matter of public responsibility under President Theodore Roosevelt, the response was based on the economic philosophy of managing resources to obtain the widest use. Although George Perkins Marsh in his classic work, *Man and Nature*, as far back as 1864 had preached environmental stewardship (Stewart Udall called that work the "beginning of land wisdom in this country"[73]) it was not until January 1, 1970, that this notion took hold. As some observers noted:

There is a definite ethical basis to a national policy for the environment. This centers around the responsibility of government as the agent of the people to manage the environment in the role of steward or protective custodian for posterity. It requires the abandonment of government's role as umpire among conflicting and competing resource interests and the adoption of the total environment as a focus for public policy.[74]

Much more than management is required to determine what is optimal in stewardship: it also requires knowledge of the values sought by people in their environment.

The Environment Act (NEPA) went further in establishing apparatus to advise the President on environmental affairs. A three-man Council on Environmental Quality and staff were settled into the Executive Office of the President over initial objections by the Office of Science and Technology (OST) and the Bureau of the Budget that were strongly reminiscent of their objections to the Marine Sciences Council.

NEPA also had teeth. It set forth an administrative procedure in Section 102(2)(c) that required a detailed statement as to environmental impact to be included in any proposals for legislation and in any "major Federal actions significantly affecting the quality of the human environment." Such a statement was to outline "any adverse environmental effects which cannot be avoided should the proposal be implemented, alternatives to the proposed action, the relationship between local short-term uses . . . and enhancement of long-term productivity, any irreversible and irretrievable commitment of resources which would be involved in the proposed action."[75] Such a discipline of analysis and public dissemination of findings was to be catalytic.

This concept of looking before you leap had matured in a parallel initiative on technology assessment by the House Subcommittee on Science, Research and Development under Congressman Emilio Q. Daddario. The notion of technology assessment recognized that technology focused on an attractive social or economic goal often produces unwanted side effects that are difficult to correct economically or politically once they arise. Noise, pollution, urban blight, traffic congestion, invasion of privacy furnished potent examples. The notion also embraced the obverse—that important social needs go unmet because of inadequate transfer of technology in the civilian sector. While adverse effects, which the economists deem "external costs," had always been recognized as possible, three contemporary circumstances seem to reinforce the possibilities. First, technology had become so ubiquitous that effects may be inadvertently propagated beyond the immediate theater: locally injected pollutants travel world-wide. Second, the technological arena seemed so saturated with initiatives that all tend to become economically as well as functionally interdependent. And third, they affected institutional interests well beyond the clientele seeking the innovation.

Each action toward a new technological initiative seemed to step on someone's toes.

Technology assessment could thus be an early warning system to examine the interaction between man's tools and society, identify unwanted effects or missed opportunities, and consider pros and cons of alternatives. The process thus had a future orientation, stressing technological forecasting, planning, priorities for resource allocation, and generation of broad policy rather than piecemeal programs. But all had to be cast in a perspective of what people want, and who is affected by a new technology. Daddario grasped this issue and injected it into the legislative process. By coincidence, the author was involved in its birth.[76]

Daddario also toyed with a national policy for environmental management, not based on stewardship, but on continued "use of the environment for the benefit of all mankind" and "maximized productivity— systematic management of applied science and technology to achieve best usage."[77] The political scientist Wandesforde-Smith criticized this approach as one that "threatened to multiply rather than solve problems created in large measure by the unbridled application of science and technology to the natural world."[78] Later, Daddario was to move with the tide of environmental conservancy by way of proposals of a national policy and procedures for technology assessment of adverse effects.[79] So while his sortie into environmental affairs was blunted, seeds he planted of technology assessment blossomed in NEPA. Sadly, Daddario was to leave the Congress in 1970 to enter the Connecticut race for governor, and lose. But his efforts since 1964 as the primary exponent in Congress of science policy were a lasting contribution.

In the absence of a national policy for coastal management, NEPA proved a significant influence, for some of the major confrontations arising from its implementation erupted at the shoreline. Two environmental hazards were involved, oil and heated power-plant effluent. The Trans-Alaska pipeline system to tap Prudhoe Bay oil in the Arctic ran into a blizzard of opposition because the potential harm to the tundra was initially overlooked by the petroleum industry promoters and Department of Interior analysts who were empowered to approve the project. Court appeals by conservation groups brought the project to a screeching halt pending further study under section 102(2)(c). In January 1971, the author speaking as a private citizen challenged the adequacy of environmental analysis because the marine extension of the pipeline from its Valdez terminal to West Coast ports including Puget Sound was completely overlooked.[80] My position, however, was directed to prevention of calamity without paralysis, by a multitude of controls that would permit tanker operation with reduced risk to the environment.

Subsequent legislative initiatives developed to further protect the inshore environment from oil.[81] In response to the massive offshore oil spills in 1969 which coated beaches, damaged property, and destroyed marine life, the Congress passed P.L. 91–224. In addition to holding petroleum carriers liable up to $14 million in cleanup costs for oil spills, the legislation provided a revolving fund of $35 million to enable the U.S. Coast Guard to clean up oil spills regardless of source, subject to recovery of costs from responsible parties. The law required the President within 60 days of enactment to develop and publish a National Contingency Plan for oil spill incidents.

As to power-plant effluents, a serious question had arisen over the potential impact on estuarine waters of waste heat from large nuclear electrical generating stations. The AEC claimed that licensing requirements had been met, but in an historic decision on a Calvert Cliffs, Maryland, site, the courts stung AEC with pointed criticism of their attempt to dodge responsibility under NEPA. Some fifty-one plants under design or construction were enjoined to reconfirm consideration of impact.[82]

The National Environmental Policy Act had yet other effects impinging on the coastal zone. The Army Corps of Engineers, which had cognizance over dredging, filling, and structural changes in navigable waters, began to consider environmental aspects of permit application,[83] in conjunction with reviews by Department of Interior agencies now required by the Fish and Wildlife Coordination Act, as well as by interagency agreement. The Council on Environmental Quality (CEQ) that NEPA created took brief note of coastal management problems in its first annual report of August 1970.[84] CEQ then seized upon one discrete issue from the waning Marine Sciences Council with an October 1970, report to the President on ocean dumping. After reviewing the enormous volume of 48 million tons per year of waste dumped at 257 disposal sites, CEQ concluded that serious hazards were directed at marine biota and human health alike. They recommended strict regulation, with virtually a ban on most dumping. The report gained widespread attention and immediate legislative action.[85] However, CEQ soon adopted a low-visibility profile as advisor to the President, and the public watchdog on the environment was to be the Environmental Protection Agency created in 1970 by the President's Reorganization Plan #3. A new independent agency involved in marine affairs, it was showing signs of being aggressive in discharging its assignment in pollution control.

NEPA proved a strong element in coastal preservation. But by the end of 1971, the main issues that go beyond the environment were still unresolved. What should be the national policy? Who in the federal constellation was to be pegged unequivocally with coastal management responsibility? Was it to be Interior as tagged in 1969? A new NOAA as the Congress preferred? Or other agencies contending for position because of claimed historical prerogatives in land-use planning? Once more, Congress, the President and the varied inside and outside interests seemed on collision course.

But even more sobering, a quiet attack was being mounted on coastal management back at the state and local level in tandem with a general backlash to environmental stewardship. It was being said that steps toward environmental quality were costing jobs and profits and so should be compromised. Vested interests were pecking away as states like California failed in their attempts to meet the challenge of wise utilization in the public interests. Like beach sand sucked away by a relentless sea, the opportunity to assert a conscience of stewardship of the coastal margin for future generations was, in 1971, quietly vanishing.

CHAPTER 5

Proposition: To Unravel the Mysteries of the Sea

GENESIS OF THE INTERNATIONAL DECADE OF OCEAN EXPLORATION

Understand . . . predict . . . control. With this necessary progression to gain the Biblical "dominion over the sea"—its moods, its mysteries, and its resources—it is clear that man must first devote sizable energies to its comprehension. This requires study of a vast part of the earth's surface: to ferret out the motions and contents of the water, dissolved minerals, and gases; to probe the energy injected by the sun and re-exchanged with the atmosphere or dissipated in tides and waves; to reconnoiter the texture of the seabed and to inventory its occulted mineral contents; and, finally, to capture and classify the myriad creatures that inhabit the sea. Apart from the sheer burden of description and taxonomic classification, it is a monumental task to trace the subtle interrelationships in this kaleidoscope of marine geography. Operationally significant is, "How?"

The Marine Sciences Act recognized the cardinal role of scientific knowledge and of its generation by basic research. It also had declared that the program should contribute to cooperation "with other nations and groups of nations and international organizations . . . when such cooperation is in the national interest." These two themes of basic research and international cooperation had to be coupled if resources of the sea that were largely common property, and about which very little was known, were to be extracted in accord with that mandate "for the benefit of mankind." Moreover, President Johnson was aware that as other nations wakened to an interest in the oceans, the same conflicts and rivalries that marked the swashbuckling attack on land frontiers would certainly follow. To stave off such a collision, Johnson signaled his position on July 13, 1966, in a speech commissioning the *Oceanographer*, when he said: "Truly great accomplishments in oceanography will require the cooperation of all the maritime nations of the world. . . . I [am] . . . calling for such cooperation, requesting it, and urging it. . . . The sea, in the words of Longfellow, "divides and yet unites mankind.' " (The full text is given in Appendix 15.)

With the helm set by both the President and Congress toward international cooperation, the Marine Sciences Council sought a multinational rather than a unilateral approach as a *sine qua non* to scientific study of the oceans. That orientation was sharply conditioned by awareness of the domestic scientific climate. While both branches of government after the 1957 Soviet space initiative viewed science as an instrument of national power and prestige and ungrudgingly pumped funds into research and development, support had leveled off. The Johnson administration had grown noticeably reluctant to fund science at prior high rates of growth. Yet, advancement of marine affairs depended upon expanding the scientific base of information. It also depended on the marine scientists, for without their support and their enterprise, the legislative objectives could be jeopardized. Interested in helping the President meet the international clauses of the mandate and beginning to develop tenets for a new phase of global exploration, the Secretariat was faced with a dilemma that never evaporated—how to harmonize scientific and political goals and their corresponding institutions.

Given the international perspective, one solution lay in merging that scientific quest with two other national goals: a stable and lasting peace toward which international cooperation could provide a much needed boost, and assistance to the less developed countries by accelerated development of untapped marine resources. At that time, latent concerns for the global environment had not yet bloomed. The confluence of marine science and domestic policy thus had a clear counterpart in its interaction with foreign policy. By early spring of 1967 the Council Secretariat began to weld abstract scientific and policy goals into an explicit concept destined to interest and involve the entire domestic enterprise, to elicit policy support from both a Democratic and a Republican president, and to wash up on the shores of all coastal nations. When crystallized one year later, it was to be a proposal by President Johnson for an International Decade of Ocean Exploration.

The Council had already stated premises for such an enterprise: that expertise in the marine sciences was shared by many nations, and that collaboration was essential if knowledge of the vast marine environment were to increase within a meaningful period. In the global search for marine resources, new technology was becoming available that would make widespread exploitation possible. Toward that end, major capital investments would be required, and international laws would have to be clarified for the protection of the rights of the developers, subject, however, to the need to preserve the traditional freedoms of the seas.[1]

The Decade was thus envisioned as a period of intensified collaborative planning among nations and the expansion of exploration capabilities by individual nations, followed by execution of systematic and integrated, national and international, programs of oceanic research and resource exploration. Designed to contribute to the economic and scientific development of all participants, it was at that stage oriented as much toward delineation of marine resources as toward science. Despite this stretching of scope beyond traditional oceanographic research, existing patterns of effective scientific cooperation would provide a basis for future arrangements. Various nations would contribute their particular expertise and capabilities, assume a share of responsibility for the program, develop their manpower and facilities, and disseminate findings to others.

Most important, the Decade was not to be merely a continuation of past efforts, but had several unique aspects. The proposal anticipated a sustained, long-term exploration of the sea, planned and coordinated on a global basis, in contrast to the sporadic efforts of the past, which were developed project by project and comprised a loose collection of national efforts. It envisaged more deliberate coordination of the many interested international organizations, such as the Intergovernmental Oceanographic Commission, Food and Agriculture Organization, and the World Meteorological Organization, to gain the benefit of specialized competences and capabilities. And it looked toward more systematic collection of data and prompt availability, with adoption of internationally agreed-upon standards for data collection and compatibility of processing techniques. Finally, participation in ocean exploration by a large number of countries would be encouraged, especially those having a maritime geography but which might have previously lacked interest, trained manpower, or capabilities to explore the oceans, even near their own shores. In this way, developing nations could share the capabilities of the more developed countries, acquire contemporary technology for their own use, and increase opportunities to identify contiguous marine resources. The Council Secretariat began to delineate potential benefits:

The knowledge which will evolve during the Decade will assist nations individually to plan ocean related investments and collectively to develop arrangements for managing ocean resources, to establish baselines as a step toward preserving the quality of the oceanic environment, and to improve forecasting of ocean and weather conditions. . . .

Unused fishery resources and fuel mineral deposits exist off the coasts of a number of developing countries. Many are dependent upon maritime transportation to link coastal communities and provide the basis for foreign trade. The Expanded Program will give special emphasis to broadening the opportunities for developing nations to participate in the use of the oceans and its resources through encouraging them, for example, to map selected areas of the Continental Shelves, survey coastal fishery resources, and obtain training and experience in marine sciences and engineering.

. . . .

The oceans contain large unused fishery resources and fisheries offer an opportunity to assist in closing the protein gap with many latent fisheries lying within easy access of nations plagued by serious protein deficiencies. The pooling of knowledge about these resources by interested nations during the Decade could contribute significantly to development and management of world fisheries resources.

More accurate, timely, and long-range forecasts of climate and weather conditions . . . will benefit expanding commercial and recreational marine activities; reduce the destruction of life and property in the coastal zone and at sea; and enhance industry, agriculture, water management, and other land activities dependent on a better understanding of the interactions between the oceans and the atmosphere.

. . . .

Such patterns of accelerated ocean investigation will result in advancement of science on a broad front.[2]

Whatever the elegance of this rationale for international collaboration, both scientific and policy communities would have to be persuaded of its merits. Support had to be obtained in the United States from the President, participating federal agencies, Congress, and the scientific community; and, world-wide, from other nations, organs of the UN, and other international bodies dealing with the sea. As we undraped the roles, historical involvement, customs, life styles, and sensitivities of each institutional group, we discovered that the enterprise was to be fraught with unbelievable difficulty—conflicts between the State Department and the White House, between the Budget Bureau and Council, between the U.S. government and the scientific community, between different international organizations jockeying for position, between the United States and other nations within the UN. The conflict most to be expected, however, did not occur—that between the United States and the U.S.S.R. But if we had foreseen the minefields ahead we might well have abandoned the venture.

By early 1967, a broad strategy was evolved in the Council Secretariat that was to lay the groundwork for the Decade: (1) To characterize scientific, political and economic goals of the United States that would be enhanced by multinational exploration; (2) to estimate scientific capabilities needed to achieve these goals; (3) to examine U.S. mechanisms of expedition planning and deployment that would blend in- and out-of-house scientific interests and personnel; (4) to identify problems in gaining overseas cooperation (including problems of internal communication within other nations), the state of their capabilities, and the extent to which their interest would be confirmed by funding; (5) to consider needed and available international apparatus.

Our first steps to guide further development were to examine the kinetics of recent international expeditions, the climate of the American scientific community and, with the help of a fortuitous presidential assignment to Humphrey, the receptivity of the international community.

International communication among scientists was as important to advancement of science as was domestic. Invisible colleges freely crossed national boundaries, and during especially frigid intervals of the cold war, contacts between U.S. and Soviet scientists had formed the main thread of communication. The health of science had long been based on freedom of the investigator to select topics of inquiry and modes of collaboration. Governmental priorities reflected in funds available for research inevitably season such choices, but many scientists, feeling their international stature and the cordiality of their essential foreign contacts would be tainted if they were suspected of representing their government in international negotiations, had after World War II studiously avoided backing federal scientific programs that were clearly intended to contribute to foreign policy objectives. They often distrusted political promises, swiftly made but often slowly honored, and were wary of project control by government bureaucrats. Respectful of these scientists' views, the U.S. government considered the fruits of nongovernmental communication of such potential value to American interests and to world order as to sponsor international scientific projects without political strings attached. In that eagerness, the United States frequently picked up a disproportionately large part of the international bill and ended up supporting a composite of what scientists would have done anyway, except that government purses opened more generously at the prospects of international cooperation.[3] The resulting international comity was believed worth the investment.

The strenuous and gratifying experience of the International Geophysical Year (IGY) was still fresh in the minds of both the scientific community and the government. Eight thousand scientists from sixty-six nations had participated. Official support had been accorded this tour de force by President Eisenhower in 1954 with the United States' share of funding so large by comparison with the total NSF budget—$20 million in appropriations compared to $37 million—that it was treated as a separate "line" item.[4] A small fraction devoted to oceanography thus had become the first rung in the ladder of budgetary growth for marine sciences even before the salient 1961 Kennedy add-on.

By 1959, that injection of IGY funds into domestic scientific efforts had begun to fade, followed by withdrawal symptoms of apprehension among the scientists. Marine data continued to be analyzed by participants and deposited in IGY World Data Centers in Moscow and Washington, D.C., but the level of effort essential to progress was threatened. Three countermeasures were undertaken to prime the international research pump: as a follow-on to information switching and administrative planning of IGY, a Special (later Scientific) Committee on Oceanic Research (SCOR) was created under the International Council of Scientific Unions (ICSU), as a secretariat for international planning in marine exploration. Secondly, American and Soviet oceanographers (who had managed to establish low-visibility rapport despite the cold war) collaborated on plans for a comprehensive new oceanic body that would transcend all of its geographically limited or functionally specialized progenitors. It would be called WOO, the World Oceanographic Organization.[5] When that ambitious concept was rebuffed by our State Department on grounds that it disrupted the status quo of hard-won fishing conventions and other bilateral agreements, Roger Revelle, its American parent, and Vladimir G. Kort, the Soviet advocate, worked through their respective governments to construct a small-scale replica of WOO—the Intergovernmental Oceanographic Commission (IOC) in UNESCO. Warren S. Wooster, a Scripps colleague of Revelle, was chosen as its first executive secretary.

The third post-IGY project was an International Indian Ocean Expedition (IIOE), conceived by Revelle in 1959 and negotiated on an informal basis with marine scientists from a number of nations. Responsibility for coordinating U.S. participation and assuring fiscal support was assigned by Eisenhower in 1960 to NSF and renewed by Kennedy in his oceanographic message of March 29, 1961. Although the Indian Ocean covers 14 percent of the earth's surface, it was the least known of all the major oceans, partly because it lies in the often neglected southern hemisphere. Both scientific and economic benefits were advertised as objectives. Participating with the United States were Australia, France, Germany, India, Indonesia, Japan, Pakistan, Portugal, South Africa, Thailand, and the United Kingdom. International scientific planning was undertaken through SCOR as the planning agent for IOC. Beginning in Fiscal Year 1962 and continuing through 1966, funding totaled close to twenty million, greater than all the marine components of the IGY. The increased scope and complexity led both to greater achievement and to more pronounced frustration. Discord between individual scientists and specialized guilds was common, some complaining of coercion to participate, some believing that biological or air-sea interaction components were unduly subordinated, and some finding that available ships were inadequate for missions undertaken. On the other hand, NSF staff director, John Lyman, privately complained that the scientists in wanting a free hand never permitted NSF to fulfill its coordinating function. Under IOC, a number of other expeditions have since been mounted with more limited objectives and more thorough planning.[6]

With this undercurrent of discontent still audible and with the new, albeit uncrystallized, thrust of multinational exploration in prospect, the Council sponsored a study of the IIOE by Robert O. Snyder, who had previously headed the SCOR Secretariat. That study revealed disquieting thinness of planning and coordination by the participants; the scattered objectives, projects, scientists, ships, and languages had produced a mélange best characterized as enthusiastic chaos. The expedition was a patchwork of ad hoc arrangements between individual scientists, not a deficiency per se, since this undirected style was the only basis on which the scientific community would have undertaken the enterprise. Snyder contended that there was no prima facie evidence that the quality of scientific endeavor suffered, but felt that the aggregate of cliques did not appear to add up to a coherent whole; gaps in research remained; data were slow to be exchanged and often regarded by the scientist as proprietary; and few new techniques in expedition planning emerged.

In considering the prospective Decade as both intergovernmental and interscientific, it was clear to Council staff that a major part of the technical competence to be engaged would come from the academic scientists. We came to recognize that even though IOC was intergovernmental, representatives of member states were largely distinguished scientists; in some cases, that individual was almost the only marine scientist in the country. In the early and mid-1960s, although the U.S. delegation was led by a senior government official, position papers and prompting came almost entirely from American oceanographers, with only limited amendment by governmental interests through the Federal Council's ICO and the State Department. The scientific core thus held a legitimately powerful position not only domestically, but world-wide.

At home, the Council Secretariat had from its creation consulted with the leaders in the oceanographic community, through its advisory panels and NASCO, for advice on the government's research policies. In June 1967, we met with the council of laboratory directors (COLD)[7] and encountered strong apprehension over slow increases in funding for their academic laboratories and about the goal-oriented style of the Council that might squeeze out support for basic research. Moreover, promotion of the marine sciences in the 1960s had produced sharp increases in students and degree-granting institutions,[8] a mixed blessing because new entrants were claiming slices of a limited pie. Other meetings with scientists disclosed candid anxiety over the invasion of politics and politicians into the ocean domain, notwithstanding Humphrey's efforts to bring oceanography higher prestige and visibility. The institutional gulf separating science and government did not bode well for governmental-scientific collaboration on the prospective Decade.

An unexpected opportunity to explore the question of foreign participation first hand materialized in March 1967, when President Johnson designated Vice President Humphrey as his emissary to West European allies to explain U.S. policies on the growing involvement in Vietnam and to try to gain their understanding, if not their support. The trip was to cover the United Kingdom, France, Italy, West Germany, Netherlands, and Belgium. Humphrey realized that such other issues might arise as an alleged "technology gap," wherein Western European nations asserted that the United States was capturing their commercial markets with computers and other high-technology products against which, with their diminutive research enterprises, they could not compete. A second issue was the "brain drain" of their top scientists. Humphrey asked me to go along as his science advisor.

We both also saw this as a convenient opportunity to explain the new U.S. emphasis on marine affairs reflected in the legislative mandate, in Johnson's July speech on the seabed, in U.S. initiatives at the UN in the winter of 1966, in governmental actions by creation of the Marine Council, and in presidential requests for action. Conversely, we could learn what policies, plans and capabilities other nations might be envisioning for marine affairs, and could test the opportunities for cooperation.

With only a week's advance notice, our embassies in the various capitals arranged interviews with senior officials having marine interests and with the most distinguished oceanographers in each country. This was a magnificent job because marine interests in other governments were as fragmented as in our own; coordinating committees existed in only a few, and only France had a statutory marine body, born just weeks previously.

Conversations with ministry officials swiftly warmed up upon distribution of the President's first annual report with the Johnson and Humphrey messages attesting to high-level support. They seemed further disarmed by candid discussion of problems within our government of coordination, agency rivalry, and bureaucratic games, recognizable problems that elicited sympathy for any governmental body charged with extracting a harmonious, unified, and spirited program out of so many subdivisions. They were also captivated by the freshness of ocean issues, and quickly got down to brass tacks on their practical importance. Most officials, incidentally, had a pragmatic view toward science and little acquaintanceship with oceanography. Not surprisingly, conversations turned to money. Publicly they inquired about how much the United States planned to invest in marine research and development. Privately, they were asking how much it might cost their countries to participate. At that stage, our only message was one of direction, not explicit targets, but we strongly emphasized our hope that whatever was done at sea would not be a unilateral U.S. thrust. The contact with France, quietly pursued behind DeGaullist antipathy for America, was to flower over the next three years in a fruitful bilateral association through Yves La Prairie, director of CNEXO, Centre National pour l'Exploration des Océans, and his able associates. This cordiality, however, was not equally offered by foreign scientists. My credentials as a presidential appointee might even have been a handicap. In these European capitals, the cleft between pure scientists and their governments was even more heavily polarized than in the United States. Nevertheless, they welcomed this American initiative as a possible lever to pry open their governments' penurious treasuries just a trifle more.

The mission in opening intergovernmental communication on marine affairs, according to the wires from our embassies, was fruitful. And we discovered that all countries saw the need for more marine information, initiatives for exploration were generally dormant, support by other countries required prior demonstration of "payoff," and leadership by the United States was essential for progress.

GOVERNMENTAL PLANNING MACHINERY

Once soundings had been taken in both domestic and international waters, we decided to move on three fronts: to improve the status of science activities at home, to brighten the international mood, and to tighten up the existing in-house machinery. In these formative years, the Council rather than the State Department took the lead in international cooperation, deriving authority from Section 6 of the Marine Sciences Act: "The Council, under the foreign policy guidance of the President and as he may request, shall coordinate a program of international cooperation in work done pursuant to this Act, pursuant to agreements made by the President with the advice and consent of the Senate."

The substantive reason for Council enterprise was the need for an active focus of leadership that did not exist in the State Department. Until 1966, the State Department's primary continuing concern with the oceans had stemmed from international conflicts over fishing rights. After a peak of activity in 1958 that led to four salient conventions on law of the sea, State's interest in marine affairs had atrophied. Spread thinly, the State Department staff was usually obliged to function as a fire brigade, cooling off crises but only rarely able to sustain continuity on issues. In that atmosphere, any creative effort to undertake new initiatives, much less to generate policy alternatives, was neither feasible nor encouraged.[9] Small wonder that presidents stoked the foreign policy fires from the White House.

In 1966 State's modest interest in marine affairs was subdivided among numerous offices. There was the post of Special Assistant to the Secretary for Fish and Wildlife to which Donald L. McKernan had been appointed in the fall of 1966. Additionally there were the Bureau of International Scientific and Technological Affairs headed by Herman Pollack, several functions under the Assistant Secretary for International Organization Affairs, Joseph J. Sisco, and the Office of the Legal Advisor under Leonard C. Meeker.

Pollack's role as science advisor to the Secretary proved vital to marine affairs. His post had been created some twelve years earlier in response to recommendations of the Hoover Commission, later reinforced by a 1950 study for the State Department by consultant Lloyd V. Berkner on "Science and Foreign Relations."[10]

As Eugene B. Skolnikoff writes, prospects were remote for dealing with or influencing policy,[11] but in my view under Pollack the operation began to pick up steam.

Sisco's shop, dubbed "IO," was responsible for preparing U.S. positions at all United Nations meetings, through several subordinate offices. The first unit led by James Simsarian was responsible for U.S. positions at, for example, the Intergovernmental Oceanographic Commission.[12] The second office dealing with the UN was headed by David H. Popper, who had attended the 1958 Law of the Sea Conference. He became re-involved in the fall of 1966 to steer U.S. representative James Roosevelt's proposal before the General Assembly on the general importance of the oceans to the world community, into a resolution asking the Secretary General to survey international marine science activities, and to examine mechanisms for preventing duplication and encouraging cooperation.[13]

The Legal Advisor became involved in the spring of 1967 when Humphrey asked him to chair an advisory committee, composed of legal specialists of the concerned federal agencies, to guide Council funded studies on law of the sea in response to Section 4(a)5 of the Act.

State's internal fragmentation and rivalries soon became apparent at many Council activities, where observers from all five sectors, attending to fulfill their responsibilities, became a source of delay and frustration. And while all five offices in State nibbled at kernels of interests, none displayed initiatives to meet the opportunities ahead.

To reinforce Council capabilities, it seemed necessary to recruit staff with scientific and foreign policy credentials who would be comfortable with the Council's venturesome style—a most unlikely combination. Glenn E. Schweitzer, who had been Pollack's science attaché in Moscow and of great assistance to me during conferences with Soviet science policy officials in June 1966, was now in Washington and fretting over State's slow pace. Recognizing Schweitzer's ample qualifications, I obtained his transfer from Secretary Rusk with the notion of his providing liaison with Pollack. One of his first projects was preparation of a five-volume inventory of marine science activities of some ninety-nine nations, the first time that the economic and social importance of the sea was catalogued.[14] For several years following he was to assist in foreign policy aspects of marine affairs.

Our next step reflected the knowledge that whatever international plans might be conceived and nurtured in the Council, they would ultimately have to be reared in the agencies with funds for implementation. Moreover, we felt that State should have a far stronger role of leadership in pulling together fragmented agency interests than they had previously displayed. In February I proposed that Humphrey discuss State's role at a sufficiently high level to include all internal and competing State cognizances and obtain a commitment. A meeting was arranged by Humphrey on February 10, 1967, with Foy D. Kohler, Deputy Undersecretary of State, and the outcome was creation by Rusk of an ad hoc interagency committee with State Department chairmanship. This was later to become a permanent Committee on International Policy in the Marine Environment (CIPME). Kohler was appointed Chairman by Rusk, and Pollack was made executive secretary, a position of leadership that afforded an opportunity for his shop to enhance its uncertain internal status.[15] Staffing was funded by the Council. Pollack created a panel infrastructure that generated a rich agenda reinforced when necessary by the Council Secretariat. The relationships between our two offices prospered.

In September, Secretary Rusk wrote to the Vice President that he was upgrading the ad hoc committee to a full-fledged State Department committee, jointly serving the Council. This move to make CIPME clearly State's baby, we subsequently discovered, had been inspired by Pollack to counter any threat of takeover by the Council when the Interagency Committee on Oceanography was replaced with a Council committee structure. While the Department of State's jurisdiction was based on presidential delegation rather than on statute, Rusk acted diplomatically but firmly to nail down their role. But he acknowledged to Humphrey: "The deep interest which you have taken in the international aspects of United States policy and programs regarding the oceans has had the beneficial effect of stimulating a considerable concentration of senior-level attention on them." Any threat by the Council had the welcome result of jolting State into assuming a responsibilty that Pollack had difficulty promoting otherwise. I was privately pleased, because the Council's primary objective was to help agencies to do their job, not to do it for them. State's territorial gambit was to be repeated early in the Nixon administration,[16] but as happened with numerous agencies, the play for power and position was not necessarily accompanied by action.

Following Humphrey's European odyssey and overseas encouragement, I requested that CIPME broaden its study of regional exploration to consider a major and different enterprise in international cooperation. In June, Edward Farrington, an Army Corps of Engineers colonel on detached duty to staff CIPME, came up with the appellation that was to last—a "Decade" of exploration. Enthusiasm within the Committee began to grow and Schweitzer gave this highest priority, along with issues on the international legal regime, to provide President Johnson an opportunity to expand on his earlier policy statement when he addressed the opening of the 22nd session of the United Nations General Assembly. By June, overseas reaction to the President's first marine affairs report was beginning to be heard, virtually all favorable. Among other factors, in playing up international cooperation and by playing down the military component of the U.S. program, we had headed off unwarranted foreign suspicion that we were pushing the marine program for a military advantage.

About then came the shock that gave an important fillip to the still embryonic Decade concept. Ambassador Arvid Pardo of Malta made a surprise proposal to the United Nations General Assembly on August 18, 1967, that the seabed resources be internationalized and demilitarized in the interest of mankind.[17] Refinements in CIPME had already taken on enough form to accelerate it through the Council machinery. The plan was to follow the previous year's successful pattern of new initiatives, but by now Roy Dillon in the Secretariat had detected on his sensitive budgetary antenna that the fiscal situation was so gloomy that such a mode of operation was risky; in fact, we were more likely to be asked by the Budget Bureau to cut rather than reinforce last year's requests. At the August 31 Council meeting, Foy Kohler reported that he would have preferred to develop international policies in CIPME at a more orderly pace and regretted that "our hand has now been forced by the Malta initiative." Expressing the view that the Malta proposal was headed in the wrong direction, Kohler then outlined what had been previously under CIPME study: sovereignty of the seabed; a resolution at the UN to create a Committee on the Oceans; a "decade of cooperative exploration of the resources of the seas beginning in 1970"; and an outline of principles as a basis for U.S. speeches by members of our UN delegation. While he called for no action on the Decade or legal regime, the Council approved three lower-powered internationally related proposals: (1) That the United States invite foreign aquanauts and doctors, including Soviet specialists, to participate in Sea Lab III, a man-in-the-sea experiment by the Navy; (2) that the United States invite the USSR and other interested nations to cooperate in studies of the marine geology of the Bering Sea; and (3) that the United States introduce the concept that intensive, long-term investigations of limited ocean areas (ocean acres) be incorporated in international ocean surveys and encourage establishment of the first ocean acre in a South Pacific international marine preserve.

At the October 16 Council meeting, the Decade was one of a series of nominations before the Council for new initiatives. Kohler, who had the responsibility for presenting it, adopted a cautious State Department stance. According to the minutes,

Kohler noted that while the concept of a decade of exploration of the oceans or ocean floor would be useful, the main thrust of our counter approach at the UN in an effort to be constructive would be a proposal to set up a Committee on the Oceans. The product of this year's ocean debate, including the concept of the Decade would be developed by that Committee. The Decade, however, would *not* be introduced with the idea that it would be passed on by this General Assembly, but rather would be one of the projects referred to the Committee.

Coached by McKernan, Kohler was also emphatic that the United States should not include living resources within the Decade proposal, leaving that extension in scope from seabed resources to be raised by other nations.

The uneven, weak support for the Decade, even from State which stood to gain foreign policy goals at little cost, was continued at the November 18 session, where an intervention from the Chairman nailed it down. But the prospects of throwing the Decade concept into the uncertain UN political crucible via a new and yet to be established UN committee was deeply disturbing, for survival there was most unlikely.

At this point, a second channel for gaining high-level support that had opened up in May for generating new ideas for the President's 1968 State of the Union message became unexpectedly available—the White House office of Joseph A. Califano, Jr.

JOHNSON'S CALIFANO DRAGNET

It is ironic that while technicians in the bureaucracy bite their fingernails in frustration because of upper-level roadblocks to their ideas, the White House damns the same bureaucracy as "uncreative" because it is unable to produce instant or politically dramatic solutions to break through the enduring pile of unsolved, complex problems. Kennedy and Johnson both called the bureaucratic apparatus ponderous and unimaginative. Nixon questioned its loyalty and endeavored to fill vacancies with recruits who passed political screenings. In 1964, Bill Moyers adopted a technique to link the White House, the bureaucracy, and the world of expertise outside of government through a series of fourteen ad hoc "search and distill" task groups to inject new vitality into lagging domestic programs.[18] Moyers lost enthusiasm when he recognized the limitations of top-of-the-head pontificating by experts and the debilitating effect on governmental operation if all innovations were injected from the top down.

By 1966, with Joseph Califano, the White House dragnet approach was again endemic. According to Harold Seidman,[19] "Forty-five task forces were organized. . . . Papers were circulated on an 'eyes only' basis, and when agency people were included on the task forces they were reluctant to tell even their bosses about what they were doing. The task force operation bred a miasma of suspicion and distrust without producing very much that was usable." Seidman's pungent diagnosis may in part reflect the BOB's dismay in 1966 when their previous role as task force leaders was suppressed by White House staff.

In mid-1967, Califano's exercise was revived, and my assistance was solicited May 17. Our May 28 reply included eight suggestions for marine programs:

1. Develop a technological capability for the Navy to operate anywhere in the oceans, at any depth, at any time;

2. Develop by 1980 a world-wide map of fish resources and plans to utilize these in combatting protein deficiencies;

3. Institute a world-wide buoy network coupled with observations from spacecraft by 1975 to improve weather forecasting and to aid studies of hurricane modification;

4. Begin a program to rehabilitate harbors and ports; to anticipate use of artificial islands for commercial activities, power plants, and landing fields for supersonic aircraft; and to utilize ship unloading systems more effectively via containers, barge lighters, and hovercraft;

5. In conjunction with states, establish multidisciplinary regional institutes to study problems of seashore and estuarine use for aquaculture, recreation, and industrial development, to obtain the optimum use of these limited resources;

6. Develop a new treaty in the sea paralleling those in outer space and on arms control that might call for advance warning of research submarines, aid to stricken craft, etc. and demilitarize the seabed and progressively demilitarize the other parts of the deep ocean itself;

7. Foster development of a new ocean industry that would achieve critical mass by combining considerations of shipping, coastal engineering, and offshore engineering;

8. Emphasize the concept of international oceanic exploration.

Only silence emanated from Califano after these proposals. But on July 13, Presidential Science Advisor Donald F. Hornig asked for marine affairs topics that he could send to Califano. I responded directly to Califano, with copies to Hornig and Humphrey (who was not initially included by Califano in the exercise). Entitled "Potential Areas of Increased Emphasis in Marine Sciences," the memo now urged the President to:

1. Strengthen food-from-the-sea programs to aid the war on hunger;

2. Improve world-wide navigation "to reduce hazards of collision" by making available latest technology;

3. Initiate grant-in-aid programs for state or regional groups to resolve conflicting uses of the shoreline;

4. Initiate harbor development to rehabilitate urban waterfronts;

5. Develop U.S. assistance to international regional programs of oceanic exploration to encompass U.S. training of foreign oceanographers, lease U.S. research ships, and accelerate data dissemination;

6. Initiate a world-wide buoy and satellite network correlating oceanic interests with the already proposed World Weather Watch;

7. Develop a treaty on the seabed paralleling that in outer space to head off conflicts;

8. Advance ocean-based industry to aid the private sector;

9. Develop federal safety standards for offshore operations "as the population of drilling rigs offshore increases" (anticipating Santa Barbara by two years);

10. Strengthen the regional approach to marine sciences in Latin America—to implement the President's commitment at Punte del Este;[20]

11. Standardize procedures and allocation of radio frequencies to improve anticipated traffic in ocean communication;

12. Establish a cooperative international program for sharing oceanographic technology;

13. Accelerate Arctic research;

14. Strengthen the handling of ocean data;

15. Develop administrative procedures to ease contributory funding for projects in marine science, by superposing the requirements of individual agencies to effect a composite threshold of justification that might warrant support.

That shopping list triggered an exciting response. On August 17 Califano addressed a joint memo to William D. Carey, Assistant Budget Director, and to me requesting elaboration, with details as to each proposal, the problems attached, inadequacies of existing programs, cost-benefits anticipating "people whom the programs reach," and alternatives considered and rejected. By then Council staff had been examining implications of earlier proposals that facilitated a more detailed reply. A proposal to inaugurate an "International Decade of Exploration" was explicitly included, probably the flash point of inspiration when the Decade was born. Through August and September these proposals for presidential initiatives were further refined, sharpened by Council staff and then exposed to the strongest group of outside experts I could assemble.[21]

The price tag of all new initiatives came to $78.8 million. Confronting any proposals for new money was a stern guideline from the President for budgetary discipline. Reflecting on that warning, the consultant panel wrote on October 7 that "the marine affairs program is small compared with others designated as national programs; indeed it has not yet reached 'critical mass.' Guidelines established by the President permit little if any growth in individual agency budgets. Therefore, we believe it will be essential this year for the Council to intervene in the normal budgetary process if momentum is to be maintained toward declared national goals." So that the panel should not simply appear to be a source of pressure for more funding, I insisted that they bite the bullet by assigning priorities within their list and by specifying which investments should be made if only $30 or $50 million were available. Reluctantly, they responded.

On October 16, I carried to the Council the combined recommendations of the Secretariat, its consultants, and CIPME, including proposals on the Decade. Our objective was to gain ammunition for dual approaches to the President—via the Vice President as Council Chairman and via Califano—with the force of consensus by three separate sources: insiders at a policy level, inside experts, and outside experts.

The State Department representative, Foy Kohler, reminded the group not to be too optimistic about evolution of the Decade concept, inasmuch as our resources, mainly dollars, might be limited. UN Ambassador Arthur J. Goldberg had already been instructed to go slow.[22] That cool and halting treatment of the Decade concept revealed State's way of buying time to deal with Pardo's proposal. To advocates of the Decade who saw much deeper merit, it appeared as a cloud on the horizon.

Robert M. White, then Director of ESSA and Chairman of the Council's Committee on Ocean Exploration and Environmental Services, supported the concept as sound but warned that for his part-time committee to develop it further would take at least another six months. The second cloud.

At that point, I urged at least a Council commitment to the Decade with relatively modest funding provided in FY 1969 (roughly $200,-000) for a full-time planning staff to elaborate the concept in ways not now possible with White's committee of part-time volunteers. The issue was left for later action. Dissatisfied with that impasse and gaining Humphrey's support, I requested each member to vote within a week on how he stood on each issue, with resolution scheduled at the November 15 meeting. Our aide memoire to Humphrey urged "body English" on the outcome. For the Decade initiative I added that "there is some controversy as to whether to include fishery resources as well as minerals. . . . I believe fisheries should be included. . . . State seems nervous . . . because it might upset existing international fisheries research agreements. . . . I recommend endorsement."

By the date of the meeting Vietnam was heating up, and all cabinet officers were painfully aware of the budget crunch. The Vice President opened with a frank approach to the problem:

Especially in a year with tight budgets, the Council actions assume special importance in laying before the President additional options for allocation of limited resources. . . . In the competition for funds, the Council must also assume a unique role as an advocate for marine science budgets. Otherwise, in the absence of any one agency now having a prime responsibility for oceanic activities, there is the hazard that the President could end up with a program that is either weak or unbalanced and thereby fail [to help the President] to meet the legislative mandate.

He went on to note, "Many of the proposed initiatives require very small funding levels for fiscal year 1969; but even these cases reflect the determination of the government to begin planning for more intensified activity that could earn substantially greater support in subsequent years." Everyone hoped for relaxation of the budgetary constraints after Vietnam.

Humphrey opened the Decade issue by unequivocally stating his support and hope for Council concurrence. John D. Young of the Bureau of the Budget added momentum by endorsing the assignment to the Council of funds for planning, on grounds that Council staff neutrality would overcome jurisdictional problems and maintain surveillance over all activities underway to minimize waste. Kohler again raised State's objection to inclusion of fisheries, but Humphrey intervened and urged the Council to keep the initial scope as broad as possible, subject to possible later contraction if necessitated by tactics of international negotiation.

The Council supported the Vice President's view, provisionally supported the concept and voted funding for planning staff through the Council budget. That modest victory was sweet, but the prospects of another long delay in gaining the Council's full endorsement of the plan itself were disturbing. The alternative route via Califano's dragnet now offered greater promise.

Between the October and November Council sessions, Califano took a further step to winnow down the avalanche of proposals for presidential support. On November 8 he convened a task group on marine affairs, including Budget Director Charles Schultze and his aide, John D. Young; Secretary Ignatius and Assistant Secretary Frosch from the Navy; Stewart Udall; William Gaud; Undersecretary of State Nicholas DeB. Katzenbach; Donald Hornig; and myself. Some sent alternates. Unexpectedly, I was requested to open the meeting with a background statement on the issues and the proposals. Nervous about muffing this critical opportunity to gain support, I began with a brief history of presidential defaults for ocean affairs prior to 1966. Pointing out subsequent legislative developments and the opportunities for President Johnson, I then outlined the proposals having greatest potential impact. I decided to play my best card first, the Decade. It proved to be the only one challenged by the others present; the earlier clouds had become thunderheads.

Pollack, representing Katzenbach, was frank and steadfast in his opposition. Apart from the same caution that Kohler had earlier expressed, he felt no announcement should be made by the President until other nations had been fully consulted. Moreover, any underwriting of the project should be publicly guaranteed by advance commitment of funding because in other cases when presidential support was motivated primarily for quick political effect the commitment of funds had been withheld and U.S. credibility had suffered. Pollack's arguments seemed to be gaining credence with Califano until his final point—that projects such as the Decade should be woven into the UN debate as a tool for the State Department but not employed by the President as political instruments. Califano gagged. Subsequently, he instructed Katzenbach to develop this initiative personnally, even if over the protests of professional diplomats who felt the need to use this ploy in UN negotiation.

Katzenbach's November 30 response was a masterpiece of balanced persuasion: "I think the identification of the President with the proposal for a major international cooperative program for the exploration of the oceans would have both domestic and international advantages. It would be a mistake for the President to make any statement on this subject until the current round of debate at the UNGA is completed, hopefully sometime in December." And in reflection of Pollack's point: ". . . it would be preferable if the Executive Branch had a clearer picture of the exploration program and a firmer view of what the dollar cost would be. It would be possible to give the President's statement more punch and to make it more precise and meaningful to both the domestic and international maritime community if the statement were to be made after this information is in hand, i.e. the end of February 1968."

From the Califano soiree came a green light to move on five marine affairs issues. Katzenbach was instructed to develop a proposal for a U.S. initiative through the UN on "A Decade of Exploration of the Seas," with my assistance. Udall was similarly asked to frame a five-year program for fish-protein concentrate. The Council Secretariat was asked to handle estuarine problems, considering regional authorities modeled on the River Basin Commission and on safety at sea. A fifth group would expand on port and harbor development.

With inputs to Califano completed, we returned to gaining support for marine affairs through the "initiatives" path that the Council had hacked through the budget jungle the year before. On December 11, Humphrey transmitted Council recommendations to the President, especially noting unanimous support by Council members. After taking note of the burden then on the President to balance the budget, and mindful of the administration's broad objectives, Humphrey wrote:

In all cases, the Council determined that these proposals (1) contribute to Great Society goals by providing clear economic benefits and enhanced quality of living to the American people, and (2) respond to foreseeable challenges of the 1970s: expanding world population, maritime threats to world order; waterfront deterioration in coastal cities; increased pollution and conflict in cases of the coastal margin; expanding requirements for maritime sources of oil, gas, and minerals; technological changes in maritime shipping; and increasing needs for recreational opportunities. . . . These proposals would thus demonstrate U.S. vision and leadership in the world community, and Presidential initiative to implement the mandate of the Marine Resources and Engineering Development Act by prudent investments having long-term multiple benefits.

The twelve priority elements began with "Food from the Sea," an initiative begun one year before and threatened with infant mortality. The International Decade of Ocean Exploration was second. The President thus had two separate recommendations for the Decade on his desk. The outcome was carefully held in absolute secrecy, a custom dictated by presidential experiences with leaks in the Washington sieve. Until the State of the Union and other special messages were released none of us, including the Vice President, knew which, if any, of these proposals had been accepted.

PRESIDENTIAL RESPONSE—JOHNSON LAUNCHES THE DECADE

The gratifying finale flashed on January 17, 1968. The Council had won a brass ring on the presidential merry-go-round. In his State of the Union Message, President Johnson stated: "This year, I shall propose that we launch with other nations an exploration of the ocean depths to tap its wealth and its energy and its abundance. . . ." As befitting a necessarily generalized overview, that proposal was sufficiently vague that everyone listened for the second shoe to drop. The second annual report that had been at the printer before the President's decision was released in early March, and while embroidering some details, it, too, preserved an enticing generality. Said Lyndon Johnson:

Other nations are also seeking to exploit the promise of the sea. We invite and encourage their interest. . . . For our part, we will:
—Work to strengthen international law to reaffirm the traditional freedom of the seas.
—Encourage mutual restraint among nations so that the oceans do not become the basis for military conflict.
—Seek international arrangements to insure that ocean resources are harvested in an equitable manner, and in a way that will assure their continued abundance.
Lack of knowledge about the extent and distribution of the living and

mineral resources of the sea limits their use by all nations and inhibits sound decisions as to rights of exploitation. I have therefore asked the Secretary of State to explore with other nations their interest in joining together in long-term ocean exploration.[23]

The second shoe dropped on March 8 when, in his Conservation Message, Johnson proclaimed his wholehearted support for the concept.

Even in the Age of Space, the sea remains our greatest mystery. But we know that in its sunless depths, a richness is still locked which holds vast promise for the improvement of men's lives—in all nations.

Those ocean roads, which so often have been the path of conquest, can now be turned to the search for enduring peace.

The task of exploring the ocean's depth for its potential wealth—food, minerals, resources—is as vast as the seas themselves. No one nation can undertake that task alone. As we have learned from prior ventures in ocean exploration, cooperation is the only answer.

I have instructed the Secretary of State to consult with other nations on the steps that could be taken to launch an historic and unprecedented ad-venure—an International Decade of Ocean Exploration for the 1970's,

. . . .

We hope that those nations will join in this exciting and important work.[24]

State immediately began informal elaboration to some forty-six foreign missions in Washington, laying the groundwork for more formal discussions and, they hoped, for sympathetic response. Justifiably dissatisfied with mere rhetoric, the press needled the Secretariat for more details. Our response was a press conference, on the record. Almost the first question I was nailed with was, "How much money?" If funded at the levels consistent with goals, I stated that it would involve $3 to $5 billion by the United States over the decade of the seventies. On a tapered upward rate of expenditure from present annual U.S. levels of roughly $100 million for exploration, funding levels would triple by 1980.[25] The total international investment would be twice that level. Finally at the press conference we promised early release of a "white paper" with full details. Budget Bureau officials sitting in noticeably winced.

Meanwhile, in-house planning was stubbornly pushed by the Council Secretariat but with limited success. For one thing, agency capabilities and experience for such planning proved severely limited, with White's committee mainly able to muster linear extensions of on-going activities that inevitably reflected the egg crate of agency interests and specialized disciplines. By May 1968 plans had been manicured and polished sufficiently by Council staff to be released by the Vice President as a "white paper,"[26] distributed throughout the United States and abroad as a basis for discussion of the concept of the Decade at international and national meetings. Its tone epitomized some golden words of those on the *Challenger* expedition a century before who took the first important steps to explore the sea: ". . . the science of abyssal research cannot, from its nature, advance slowly and gradually; it must proceed by strides."[27]

That momentum was to carry through in President Johnson's commencement talk at Glassboro College in June 1968, where he met with Soviet Premier Aleksei N. Kosygin: "Finally, I can suggest other opportunities for cooperation between the United States, the Soviet Union, and other nations—cooperation to extend our knowledge, cooperation to develop our resources that man has scarcely touched. There is the problem of exploring the deep-ocean floor. There is the American proposal for an international decade of undersea exploration."

THE INTERNATIONAL ARENA

Indeed, the most essential Decade partner was the Soviet Union. Their oceanographic capabilities approximately matched ours, so that together we accounted for roughly 70 percent of the global capability. By good fortune, communication between American and Soviet oceanographers had been reinforced during IGY and subsequent oceanographic exercises. Operation of the IOC had critically depended on United States–Soviet cooperation, and the first and second International Oceanographic Congresses, in New York and Moscow, fostered information exchange even further. The Decade concept, however, anticipated a far deeper involvement and, it was to be hoped, commitment, by the Soviet government.

When the UN progeny of the Pardo proposal—an ad hoc Committee on Peaceful Uses of the Seabed discussed in Chapter 6 first met in New York in March 1968, I sought an informal conference with one of the senior representatives of the Soviet delegation, K. V. Ananichev, whom I had met in Moscow in 1966. Completely uncommitted as far as his government was concerned, he was intensely curious as to U.S. motivation for the Decade. From the dialogue that ensued I felt he would have been better satisfied if I had responded that the Decade was entirely a ploy to delay hasty decisions at the UN in the wake of the Pardo proposal, about which both governments were apprehensive. Any altruistic motives were suspect as concealing yet some other military or commercial interest, but the door was opened for a trip to Moscow to explore the proposition further.

On April 30 I wrote to D. M. Gvishiani, Deputy Chairman of the State Committee for Science and Technology, proposing an early visit to explain the Decade and taking note of suggestions for expanding scientific exploration made earlier by Soviet officials. Since the Decade proposition might appear on the agenda of imminent planning sessions of IOC and of the UN Ad Hoc Committee, a prior private exchange might promote harmony at these international forums. His May 23 reply was affirmative, although he would not be personally available. With the conference in Moscow arranged, I wove an itinerary around it to include stops in London, Bonn, and Oslo to elicit their support at the forthcoming IOC session, with a return via La Jolla, California, to meet with SCOR at its annual meeting.

In early June, with Schweitzer as aide and translator, I departed for Moscow with the blessings of the Vice President. In Moscow, five meetings were held with Soviet officials of the Foreign Ministry, State Committee for Science and Technology (comparable to *OST*), Fisheries Ministry, Hydro-Meteorological Service, and Academy of Sciences. After explaining the President's Decade proposal and its relevance to pending UN Resolutions 2172 and 2340, I urged the Soviets to participate at forthcoming meetings of IOC, SCOR, and the UN Seabed Committee, and to work toward establishment of appropriate international planning and coordinating mechanisms for the Decade, with IOC in a leading role. Sessions were cordial, correct, but candid, with cold-war rhetoric silenced by the tragic news of Robert Kennedy's assassination.

Inevitably, questions turned to the common denominator—money. I proposed and they readily concurred that our two contributions be equal —so the question was, how large? Was this to be the U.S. response to the Soviet initiative in space? To some extent they were relieved at the relative modesty of funding anticipated, especially that this was to be far less than in space; space expenditures had clearly been a drain on their resources as well as ours. No commitment was explicitly requested (nor given) on adoption of specific programs, the starting date, and duration. On this score they implied a need for high-level Soviet decisions concerning oceanographic budgets, which at that time did not seem to be accorded priority emphasis, a problem not unknown in the United States.

In the course of discussion with O. N. Khlestov, head of the legal department of the Foreign Ministry, the Soviets broadened the agenda to include legal, arms control, and economic aspects of ocean activities because they recognized a role of the Marine Council that went beyond science. The Soviets were interested in close consultation with the United States during forthcoming discussions on seabed and territorial water questions as well as on the Decade. They accorded high priority to the prospective third Conference on Law of the Sea out of concern for proliferation of extensive unilateral claims to territorial waters, resolving the seabed regime was of less importance because they felt its exploitation to be far ahead. Their highest priority on the UN Ad Hoc Committee agenda was the proposal already tabled to reserve the seabed for peaceful purposes, and they urged approaches similar to those for outer space and Antarctic treaties.[28]

Support was forthcoming from Academicians Ye. K. Fedorov of Hydro-Met, who was curious about the international focal point because of rivalries among international bodies, and Ishokov of Fisheries, who was worried about the possibilities of the new technologies leading to overfishing and about extension of territorial claims, primarily in the Southern Hemisphere.

Ye. I. Sklyarov and K. V. Ananichev, representing the Executive Board of the State Committee on Science and Technology, which bears responsibility for the coordination of all marine affairs, concurred on all substantive points, felt experts could work out details, and favored IOC as focal point with a new planning staff to bring in viewpoints of other specialized agencies as well. Ananichev remarked that he realized that the Decade was not solely a ploy to fend off the Malta proposal and that it could proceed in parallel with development of a legal regime, even as a logical prelude to exploitation. He hoped the United States and the U.S.S.R. could work together at the UN to convince developing nations that the Malta proposal was based on false expectations of early economic return and to head off any new international body to manage the seabed. I returned with some confidence that we would not have a roadblock there, provided we sought cooperation through the sanitary shield of multilateral rather than bilateral activities; the Soviets clearly felt that too-open rapport with the United States would fuel the Red Chinese fire. Our embassy reported to Washington that the Soviets understood the character of the Decade and U.S. motivations for its inception, and had reacted very favorably to all substantive elements of the proposal.

In the United Kingdom, meetings were held separately with R. H. J. Beverton, Secretary of the Natural Environmental Research Council (NERC), which had coordinating responsibilities in oceanography; with officials of the Ministry of Technology, and of the Foreign Office; George E. R. Deacon, Director of the National Institute of Oceanography; Sir Solly Zuckerman, Chief Scientific Advisor to the government, and L. J. Lighthill, Secretary of the Royal Society and Chairman of NERC's Committee on Fisheries and Oceanography.

The British had a spate of concerns. The Foreign Office worried about possible proliferation of international bodies, competition within national priorities for limited funds, and difficulties with internal co-ordination. As to domestic coordination, they were taking steps to correct fragmentation through two new committees besides NERC, one under Zuckerman, one under Mintech, but their bureaucracy seemed impregnable. And despite funding increases, marine sciences had been accorded neither leadership nor priority emphasis at a high governmental level. Others wondered about private industry participation, a puzzle still to me except for opportunities to develop hardware. Afterward, our embassy euphemistically reported that the British seemed "unrushed" about developing any position on the seabed regime. Some time later, I was to receive a warm personal note from Lighthill congratulating the U.S. on its initiative.

In Oslo I met with Robert Majors, Director of the Council for Scientific and Industrial Research, and others interested in marine biology, offshore exploration, fisheries, and foreign affairs. All were curious about costs, and worried about proliferation of international activities that required for a small nation like Norway a substantial fraction of staff time at meetings. They favored a regional approach to exploration to surmount national jealousies.

In Bonn, Minister of Science Gerhard Stoltenberg and other F.R.G. Science Ministry officials were enthusiastic and willing to move, unequivocally supporting goals of the Decade but asking about costs and procedural weak points, especially as to the possibility of international coordination without creation of a new body. One result of the talks was to give Stoltenberg ammunition in support of quadrupling his own small oceanographic budget by 1972.

After the dust of that journey settled, it became striking how much all of these governments had in common: relatively low priority for marine affairs, fragmentation of internal jurisdiction and rivalry among ministries, a lack of internal coordination and interest in U.S. procedures, limited funds for new projects and requirements that increases be justified on grounds of expected "pay off" rather than more science, opposition to the creation of new international bureaucracy, the desire for international comity. How familiar these issues sounded.

But there were distinctions, almost as though each nation revealed its personality through this small sample: West Germany alert to new opportunities; the United Kingdom uncomfortable when the boat was rocked; Norway agonizing over its small size; the U.S.S.R. eager to match the competition.

At the IOC's planning session on June 13 the Decade received a mixed reception, but a formal recommendation, jointly drafted by the U.S. and U.S.S.R. representatives, was adopted supporting it. The U.K. was skeptical, stating it was an ineffective device to get more money from member governments; Japan, Germany, France, and Brazil gave solid support.

To suppress U.S. authorship, which seemed curiously necessary in international political circles, the IOC recommendation "considered" the U.S. Decade proposal and went on to endorse "the concept of an expanded, accelerated, long-term and sustained program of exploration of the oceans and their resources [LEPOR], including international programs, planned and coordinated on a world-wide basis."

On June 17 after my return, Humphrey held a press conference reporting on the favorable response by the Soviet Union and other nations to the proposal.[29] Later, I was to visit France, Italy, Korea, Japan, Brazil, Venezuela, Mexico, and Canada in a quiet campaign for support. In all conversations, it was evident that nations were ambivalent about participation; each desired assurance that others would play, yet each was apprehensive about another gaining political advantage. Not only was this a chicken and egg enterprise, it meant dealing with an entire hatchery. But in 1968 prospects were bright for a portentous international exploration of the sea.

A NEW THEME OF THE UNITED NATIONS

The sea had long been an arena for international cooperation, but over the centuries, primarily in the development of public and admiralty law to protect property rights and preserve order. There were innumerable conventions, treaties, bilateral and multilateral agreements. International machinery had also evolved quietly and unsystematically, function by function, and geographic sector by sector.[30]

Not surprisingly, within the UN the question of fragmentation had arisen in connection with a growing interest in the opportunities offered by the sea for world development. ECOSOC (Economic and Social Council) Resolution 1112(XL) of March 7, 1966, had requested the Secretary General to make a survey of the present state of knowledge of the resources of the sea beyond the continental shelf excluding fish, and of techniques for exploiting this research. UNGA Resolution 2172 (XXI) of December 6, 1966, requested the Secretary General to survey marine science and technology activities by members of the UN family of organizations, member states, intergovernmental and nongovernmental bodies, and to formulate proposals as to the most effective arrangements for an expanded program of international cooperation. The focus of attention was scientific and technical. Then came the Pardo proposal to enliven the 22nd session of the UNGA, with stirrings of interest by lesser developed countries suddenly switched to broader political questions on the potential of marine resources for accelerating their economic growth. Proposals were tabled both to study the questions of benefits and to buy time to rationalize the associated legal regime for the seabed. Twin issues of cooperative exploration and an updated legal regime were thus developing side by side. (The second is treated in the next chapter.) For both, any advance beyond these UN fact-finding enterprises inevitably hinged on the questions: What stake do nations have in the sea, individually or collectively? And how will this be revealed in geopolitical dynamics?

The General Assembly at its twenty-second session did, however, begin consideration of the general question of jurisdiction over the deep ocean seabed and undertook an "Examination of the question of the reservation exclusively for peaceful purposes of the seabed and the ocean floor, and the subsoil thereof, underlying the high seas beyond the limits of present national jurisdiction, and the use of their resources in the interests of mankind."

With CIPME and Council assistance, the State Department evolved a position to support careful study by the General Assembly and proposed a Committee on the Oceans that would be competent to examine all marine questions brought before the Assembly, to encourage international cooperation in exploration of the oceans, and to assist the General Assembly in considering questions of law and of arms control. As a further step, we suggested it might develop a set of principles to govern states in the exploration and use of the seabed.

At the conclusion of debate on December 18, 1967, the General Assembly unanimously adopted Resolution 2340, establishing an Ad Hoc Committee of thirty-five states to prepare a study on various aspects of the seabed beyond national jurisdiction for consideration by the Assembly the following fall. The study would examine (1) activities of the United Nations and its specialized agencies related to the seabed; (2) relevant international agreements; (3) scientific, technical, economic, legal, and other aspects of the question; and (4) suggestions regarding practical ways of promoting international cooperation in the exploration, conservation, and use of the seabed and its resources. Here was the primary, albeit untried, political theater to consider the Decade, to complement the IOC that was identified in the UNGA action as one source of technical assistance.

Considering the number of different international entities having interests, the Council Secretariat encouraged U.S. representation to seek support for the Decade in all; we succeeded with the UN Economic and Social Council, the Executive Committee of the World Meterological Organization and its Commission for Maritime Meterology, and the Council of the Food and Agriculture Organization, as well as IOC.

Meeting three times during 1968, the Seabed Committee considered background papers prepared for it by the UN Secretariat, the IOC, and other UN specialized agencies. The United States brought before it a special proposal for the Decade. Meanwhile, pursuant to the UNGA Resolution 2172, an international group of experts (including Henry A. Arnold from Council staff) outlined existing programs and suggested steps to the Ad Hoc Committee to strengthen international cooperation in the field.

The Committee's final report illuminated emerging conflicts of interest and heightened awareness of technical and legal problems associated with exploiting the deep ocean floor. Especially it tempered with realism the early excited expectations of immediate and large returns from resources of the deep seabed. During the fall of 1968 the General

Assembly reviewed that report and after extensive debate in the Political Committee of the General Assembly adopted four resolutions, three cosponsored by the United States. The four propositions packaged as UNGA Resolution 2467 resolved to:

1. Replace the ad hoc arrangement with a 42-member standing Committee on the Peaceful Uses of the Seabed and the Ocean Floor Beyond the Limits of National Jurisdiction, to expand the studies carried out earlier by the Ad Hoc Committee;[31]

2. Urge measures to prevent pollution of the oceans;[32]

3. Support the U.S. proposal for an International Decade of Ocean Exploration within the framework of a comprehensive long-term program of scientific investigation and call on the IOC to play a leading role in coordinating the program;[33] and

4. Request the Secretary General to study the question of establishing international machinery to promote exploration and exploitation of seabed resources and their use.[34] (The United States considered this proposal premature and therefore abstained.)

Broad interest was revealed in the cosponsorship by twenty-eight nations of the UN Resolution endorsing the Decade. Many were well aware of the need for careful preparatory planning, including the identification of specific areas for investigation. All wanted to avoid a proliferation of new international bodies for its implementation, expanding bilateral and regional arrangements within its over-all context.

The Decade resolution had singular importance in giving substance and visibility to a long-standing, unfulfilled ambition of the IOC for a "long-term and expanded program of oceanographic research." That rather vague objective was suddenly spotlighted by UN recognition of the fragmented state of ocean activities, not just scientific oceanography. No single organization existed to overview the many national projects. No machinery existed to integrate research and exploration, much less to focus on their practical objectives. No forum existed where various international bodies such as FAO and WMO, concerned respectively with fisheries and meteorological studies, could air their separate findings or collaborate with a weaker IOC dealing with oceanography as a whole. Added to this fragmentation internationally as well as domestically, the ocean scientists lived apart from fisheries and atmospheric scientists.

Central to the recommendations of the Secretary General was the need for political adrenalin in IOC to enable it to serve as a focal point for coordinating international marine science activities, in cooperation with other interested international organizations, particularly FAO and WMO. The United States and other nations then made specific suggestions for increasing the resources available to the IOC and for restructuring the organization, some of which were adopted in 1969.

During this time the IOC was becoming increasingly nervous that the new UN Seabed Committee would usurp its jurisdiction—a serious matter substantively because the new UN committee had no competence to deal with highly technical details of research, data exchange, and so forth. The scientific traditions of IOC were also threatened by possible assignment of research priorities on a political basis. Thus the U.S. preference for IOC's lead role in the Decade was a step toward improving coordination machinery and subduing their apprehensions. Subsequently, however, failure to support the IOC in this expanded role because of internal UN politics revealed that the question of coordination was unsolved. Maritime affairs were still splintered, and the issue was to erupt again in UNGA Resolution 2580(XXIV) and especially ECOSOC Resolution 1537(XLIX) of June 30, 1970, which called attention to the haphazard, piecemeal, and random attack being made on problems of ocean space.

In 1968, important steps were required internationally to implement the Decade concept: (1) development of a multinational organizational framework for planning and cooperation; (2) identification of specific projects to be carried out; (3) commitments by individual nations of their individual exploratory capabilities to cooperative endeavors.

Response to these questions, especially the third, would devolve from a world still operating on the basis of nation-state self interest.

Most nations placed priorities on exploring the contiguous continental shelf and coastal fishery stocks, and these programs received the bulk of national financial support. In the near term, much of the world's ocean exploration will probably continue to be of this nature. However, the sharing of experiences and data even from such regionalized efforts could be of general benefit. The greatest returns will, of course, derive from the pooling of resources and the sharing of responsibilities in specific collaborative projects on a regional basis. In the deep oceans, where, because of the high cost of operations international collaboration is essential, only areas with the highest potential interest to the major users would be given priority attention. So actions were taken.

The international commitment to an International Decade of Ocean Exploration was realized in 1968 by UN General Assembly Resolution 2467D(XXIII), and the General Assembly Resolution 2414(XXIII) which endorsed the concept of a long-term and expanded program of oceanographic research including the Decade. The UN Seabeds Committee favorably considered the comprehensive outline in August, and in September it was formally approved by the IOC. It was considered favorably by ECOSOC and the 24th session of the UN General Assembly in 1969.[35]

As difficult as had been enlistment of international support, we were to find our domestic brethren would have to be as assiduously wooed and won.

CONFLICTS BETWEEN GOVERNMENT AND SCIENCE—THE FAILURE OF SYMBIOSIS

Would the scientific community support the proposal? Moreover, would American scientists seek support from scientists abroad who would be consulted by their governments regarding merits of the enterprise? When the Decade proposal was publicy launched by President Johnson in 1968, it failed to prove as spontaneously alluring to scientific cooperation as we had hoped. So as to enlist their support, the Secretariat had previously convened a large group of representatives from most of the academic guilds, NASCO, the National Security Industrial Association, and the Marine Technology Society in Washington in January 1968, for a one-day briefing. We found unqualified backing from such individuals as Warren S. Wooster, then back at Scripps, but it became apparent that wider support among scientists would have to be sought amidst hostile mutterings about federal domination.

By the summer of 1968 it was clear that systematic collaboration with the scientific community was needed, to engage both their ideas and their support. Leaders who had given private encouragement began under pressure from their own constituencies to snipe at the project. The chairman of NASCO, John Calhoun, declared that support would be conditional on the turning over of full planning responsibility—and funding—to the Academy. To do so would have guaranteed disinheritance by Johnson. But worse, few governmental officials were convinced that NASCO could do the job.

Thus began a nervous gavotte that never resulted in a mutual embrace. There was a deep-seated suspicion that the flag- and arm-waving of the Decade announcement was a political maneuver by a President who was considered increasingly hostile to science and that the initiative would not be backed by sufficient funds. Bad as that was domestically, it appeared to the scientists even worse that the UN should be involved. Its newly established Ad Hoc Committee to Study the Peaceful Uses of the Seabed and the Ocean Floor Beyond Limits of National Jurisdiction was regarded as a threat to the IOC, which the scientific community controlled. In that atmosphere, cooperation by the oceanographic community could be hoped for only after money was available. Few scientists understood our budgetary obstacle course and the need for their help in developing a rationale that would earn presidential endorsement. Few cared to get embroiled in the politics. It was easier to take potshots from outside.

Federal agencies were similarly concerned over adequacy of funding, but their apprehension also reflected the cleft with the scientists and the difficulty in joint planning of expeditions that would have some relevance to the problems the agencies were expected by statute to solve. Moreover, there was clear recollection that in cases where the United States had been prompt in offering to help fund an international scientific enterprise, it had frequently ended up as Uncle Sam's child, with little funding from beyond our borders. To be sure, since late summer 1967 the international climate for marine affairs issues seemed increasingly hospitable, but the level and persistence of response were highly uncertain.

Aware of their uneasy interdependence, the government and the scientific community could not live apart but they never found a way to live together. All of the characteristics of symbiosis wherein each partner derives strength and sustenance from the other were obviously present: the government's advancement of the Decade and achievement of its varied goals critically rested on a sound core of scientific exploration; and the fiscal support wanted by scientists for research chosen on the basis of spontaneous curiosity depended on availability of federal funds. The variety of courtship styles among birds, animals, and humans is a source of constant amusement to the uninvolved spectator, and to the observer of relationships between the Council and the National Academy on the issue of planning the Decade, the ballet of approach and rejection must have been equally entertaining. The result, however, was never fully successful.

Our initiatives failed to negotiate any long-term *entente cordiale* for planning and conduct of the Decade enterprise, principally because of disagreement about the control of funds. We had hoped to establish a detente through a shared planning mechanism. If the Decade were to differ from its precursors in coherent, continuous exploration devoted to clearly enunciated goals, a unified planning staff was needed to synthesize the objectives of science and to delineate exploration of resources subject to private development or for public purposes. The same staff would be required to develop details, milestones, priorities, and timing and would be a clearinghouse for matching goals with resources of manpower, instrumentation, shore- and sea-based facilities, ships, submarines, buoys, together with the funds requisite to supply them. They also would be responsible for the end products of the Decade: a varied family of maps, atlases, research reports—counterparts of the nautical charts of sixteenth-century expeditions.

To accomplish these goals, we anticipated a staff of perhaps ten full-time people, with a balanced division between insiders and outsiders, but representing a wide range of specialties including instrumentation, systems analysis, data processing, economics, and industrial technology.

Delineation of government goals was to be achieved by the existing Council committees concerned with ocean exploration. External goals were to be generated by a part-time advisory board composed of individuals appointed by the two Academies and by commercial and industrial interests. Both sets would then be evaluated and blended by the planning staff, their product forming a basis for U.S. positions in international negotiation, for generating U.S. funds, and for guidance to the actual participants.

The problem was, who would be in charge. For this post we hoped to recruit a scientist who combined the necessary stature and the collateral skills of planning, leadership, and negotiation. We contended that he would have to be a full-time government official, to represent the government when defending budgets internally and before Congress or internationally when dealing with private participants and other nations. The academicians insisted he be chosen from outside the government. I offered a compromise that an outside planning leader with acceptable qualifications be nominated by the Academies, but that he occupy a government post. We could offer him the government's highest civil service rating, so that salary should be no barrier. Offers to head a new office for Decade planning were extended to several leading scientists, but with negative results. Some had other commitments; some admitted concern about retaining their professional status if they assumed a bureaucrat's mantle.

To follow up the 1968 presidential announcement regarding the Decade, the Council hoped to develop sufficient detail by the fall to justify presidential funding of specific items having long lead time. This would put chips in our pocket for entering the poker game at international conferences. But our impasse with the Academy threatened to bring the Decade to a screeching halt.

We chose to defer the planning apparatus and instead to invite the National Academy of Science and the National Academy of Engineering to formulate their independent proposals to the Council.[36] This unique contract with both academies was announced by Humphrey July 23, 1968. While each academy personified a quite different set of disciplines, cultures, life styles, and affiliations, close collaboration between them was required. NASCO was largely composed of university-based scientific research specialists and the NAE of industry-based engineering designers and managers; together they presented a unique blend of experience, intellect and point of view. But first they had to learn to communicate and live together. Under chairmanship of Warren Wooster and with William E. Shoupp, Vice President of Westinghouse, as Vice Chairman, both committees met for three weeks at Woods Hole. Also participating were guest specialists and observers from federal agencies. Their varied proposals were filtered by a steering committee, and a first draft of their report was delivered soon after. It was of immediate value.

Meanwhile, the Council had been approached to fund a second Decade-related study by a consortium of institutions called the Gulf Universities Research Corporation (GURC) which had been endeavoring to get a handle on a marine program in the Gulf of Mexico, but between problems of institutional rivalry, part-time leadership, and shortfall in federal funds for a proposed Gulf Buoy project, it rocked along slowly until about 1968. Their proposal for a "Gulf Science Year" to be articulated with the Decade was proposed by John C. Calhoun representing GURC, and the Council responded with the conviction that study carried out in that geographical sector would guide broader planning.

By fall of 1968, despite a renewed effort to elicit support from Califano, no major commitment of new funding had been made by President Johnson. Partly he was alarmed at budget deficits, but he also deliberately deferred new starts that would appear to tie his successor's hands. Actually, the government was loaded with underfunded new starts from the prior seven years because of the burgeoning Vietnam distraction. The incoming President would obviously not welcome more demands, and Johnson's decision may in the long run have increased the hospitality subsequently shown by Nixon to the Decade.

With the change in the administration, the NAS/NAE study presented several new problems. First, the Academies let it be known that their report would press hard for a major increase in funding, without which the Decade should be abandoned. Second, recommendations were to be drawn in broad terms, without indications as to what could be done at alternative levels of funding or what internal priorities should be set. Advice simply to spend more money was never welcomed by government and I visualized such a report as inflaming opposition from the new President. Not able to persuade the NAS/NAE group to temper their over-all demands or to select priorities, and still shepherding the Stratton report and the backlog of unfinished business in marine affairs left from the Johnson era, we were nervous that premature release of the report before the Council under Vice President Agnew had examined the issue would lose the whole ball game. The Academies did agree to delay publication until after the May request by the President that the Council move ahead on Stratton's substantive, if not organizational, proposals.

In June the Academies' report, *An Oceanic Quest*, was released. It was an excellent job of craftsmanship, fully sympathetic to the Decade concept including its emphasis on uses as well as study of the sea. Moreover, the report took unequivocal note of national goals: to benefit directly the growth of the national economy; to obtain information required for management and conservation of resources, for improving the effectiveness of nonextractive uses; for prediction, control and improvement of the marine environment; and for the making of social, political, legal and socio-economic decisions related thereto; to provide a technical basis for the reduction of international conflicts in the oceans; and technology for future ocean research and utilization. The report related ocean uses to Decade program elements in mineral resources, living resources, waste disposal and ocean transportation, and suggested research goals in geology and nonliving resources, biology and living resources, physics and environmental protection, and in geochemistry and environmental change. After all these years, their viewpoint toward pragmatic goals had substantially changed.[37]

On national program management, the NAS/NAE report was ambiguous. It asserted that "it should be an early task for a central planning staff to relate the concepts and proposals in this report to the on-going and planned programs of Federal, state and private institutions . . . ," but it never said where the central group should be located or how composed. It also said that "a new composite body is required to coordinate federal activities with those of state or private organizations (including industry)"[38]—which sounded like tacit acceptance of our proposal but without details.

As to funding, they were insistent that "if much less than $100 million of new money per year (averaged over the Decade) can be made available for the U.S. program of ocean exploration as described in this report, it would be undesirable to identify the set of programs as IDOE."[39] A "possible spending pattern" called for $150 million the first year, peaking to $500 million by 1975, and dropping to $250 million by 1980. The intent, of course, was to goad the government treasury with an implied threat to withdraw support if the full commitment were not made—preconditions we felt unrealistic, both as to level and rate of increase of funding, and as to a future commitment that no President ever makes if avoidable. The inpasse on planning organization and funding level stood firm. But our courtship continued.

THE NIXON FUNDING DECISION

After the 1968 elections and before the new President was inaugurated, a huge official question mark hung over the Decade. Would it win fiscal support from the outgoing or the incoming administration? And would worry by several conservative Congressmen about the seabed legal regime spike our guns? They alleged that the United States was giving away its birthright to adjacent seabed, and since the Decade was seen as intimately related, some members questioned the Executive Secretary on whether we would also be giving away data.[40] Responding to that argument, I contended we were getting three units for the price of one: that is, assuming we were to fund one-third of the total international activity, the Soviets one-third, and all other nations one-third, and assuming equitable data exchange, we would be receiving twice the volume of data that we collected. Senate Concurrent Resolution 72 supporting the Decade proposed by President Johnson passed the Senate. Congressman Lennon introduced its House counterpart,[41] but never brought it to the floor for fear of debate and defeat by conservative colleagues who had been innoculated by the National Oceanography Association, as discussed on page 149, and telegraphed their anxieties about a giveaway in their spate of cautionary resolutions in 1967.[42]

With far less gusto than previously, Califano turned again to his "eyes only" task group exercises—first a dragnet for ideas, then a follow-up to pan political gold from the voluminous offerings sluicing through from feasibility studies. My first response of October 14 on marine science options opened with an appeal to a President whose 1966 and 1968 statements on international aspects of the oceans were portentous steps toward peaceful use, to follow through by credible implementation. I proposed that the President: (1) start the International Decade of Exploration in 1970; (2) submit presidential proposals on law of the sea to the UN at this session; (3) revitalize the lagging Fish Protein Concentrate Program by establishing goals to develop a (global) FPC capacity to meet needs of 200 million people by 1980; (4) propose legislation on the coastal zone to promote orderly development through grant assistance to states; (5) strengthen the politically popular Sea Grant College and Program Act; (6) improve U.S. fisheries; (7) proclaim an Arctic policy; (8) expand capabilities for exploration and weather/climate prediction by a northern hemisphere buoy network by 1980.

On December 4 I sent Califano a status report to notify him that three of these proposals—the Decade, FPC, and Sea Grant—were in trouble; participation by six agencies was contingent upon approval of some $36 million over budget allowances. On December 5, the Vice President submitted to the President recommendations for FY 1970, to "continue the steady and widely recognized progress generated under your Administration, selectively advanced within prudently managed budgets, to enhance economic growth, coastal development, national security and international understanding." The Decade was the first of a list of nine items that included domestic fisheries, FPC, Arctic policy, Sea Grant, a data buoy system, reduction of oil spills, standardizing of instruments, and promoting optimal use of the coastline.

On December 16 Califano convened a task group comprising Udall, Boyd, Charles J. Zwick, the new Budget Director, Ignatius, Matthew Nimitz (his assistant), and myself to discuss recommendations. The President's response was in his January 17, 1969, marine affairs message and budgets for FY 1970. The Decade was listed first among a catalog of priorities by the outgoing administration. But with marine science increases restricted to 12 percent over the prior year, only token funding was earmarked to start the Decade. The key governmental decision was left for incoming President Nixon.

The Nixon administration did move swiftly to consider the unfinished business of its inheritance and at the February, 1969, session under

Vice President Agnew, the Decade was accepted for priority study and instructions for early review. As detailed in chapter 3, the Decade was on the May 23 agenda for action. Trotted out on the Council stage for its key audition, the Decade failed to gain approval. The rationale for the Decade advanced by Robert White on behalf of his committee, albeit reinforced by *An Oceanic Quest*, was rejected. The Executive Secretary was admonished to take it back to the drawing board if we were to try again at all. Moreover, with the creation of the Committee for Policy Review (CPR) and with Agnew's appointment of the Executive Secretary as its chairman, I was doubly responsible for the Decade's resuscitation. Never had I come from a Council meeting so depressed.

Moreover, I reluctantly came to the conclusion that with no outside scientist willing to take over governmental leadership of Decade planning, I would have to do it myself. In June I formed a Decade task group: Glenn Schweitzer, William H. Mansfield III, Bill L. Long, and other Council staff; several from agencies; and one outside scientist, Garth Murphy from the University of Hawaii. The task group met in plenary session almost daily, breaking up into panels but reassembling periodically to begin the planning job the Secretariat had visualized a year before. Now at least some power tools were in hand: the Academies' *Oceanic Quest;* the paper produced by White's committee concerning governmental interests; an eloquent document formulated by scientific experts serving various interested international bodies meeting at Ponza, Italy;[43] and program proposals generated at the IOC.

But completely lacking was a clear set of program goals or criteria for setting priorities. The first thing needed was a plan for planning, and the second was a commitment to that plan by all concerned agencies, so that the final product could not be undercut by a recalcitrant participant as having been directed "at the wrong question," a favorite bureaucratic device.

A plan for a plan emerged, followed by contentious acceptance by the Committee for Policy Review.

The next step was to muster additional planning tools: a base line of Decade-related current activities representing perhaps 30 percent of the federal marine sciences budget; a list of marine-related national goals and explicit steps for their achievement via the Decade; a master list of Decade proposals; and finally, the framing of a U.S. plan at different funding levels to be defended in the internal budget process and advanced as the U.S. proposal before international bodies.

Our own deadline was the fall budget season. Facing so short a lead time we hoped to extract from the task group creative proposals that had been consigned to dusty desk drawers after rejection in prior budget reviews, but which still had enough luster to warrant a second go-round. We also had several earlier planning documents. In recognition of the

growing importance of programs to observe and predict oceanic conditions, the Council in 1968 had begun drafting a five-year federal plan
for Marine Environmental Prediction (MAREP) prepared by its Committee on Ocean Exploration and Environmental Services as a guide to
agencies for the development of future federal programs in that area.
(MAREP was released separately in 1970 to encourage discussion of
national efforts in these fields and to assist with technical matters related
to planning U.S. participation in the Integrated Global Ocean Station
System.[44])

The Council's Committee on Ocean Exploration and Environmental
Services was also preparing a Ten-Year Plan for Ocean Exploration
(TYPOE). Its goal was to provide a framework for considering national needs for and benefits from ocean exploration, goals, and priorities in developing federal programs and in determining when
exploration should be conducted, by whom, and with what milestones.
Both reports were useful.

Of greatest disappointment in the summer of 1969 was the paucity
of fresh ideas from the agencies. They essentially proposed more spending for old but "underfunded" program elements. Even these proposals
were devoid of convincing rationale. Only the Navy had been through
drills of this kind before, and then always in relation to national security
objectives that could be couched in quantitative terms. No other agency
appeared to have experience in planning for civilian goals.

Our only course was now to apply straightforward management
techniques to develop a Decade plan. By late August, the planning
group had a draft. Agencies were pessimistic about its future, even
alarmed that if the plan succeeded it would be at the expense of old
favorites. What lifted the whole enterprise from an exercise in utter
futility was the unexpected September 9 request from the White House
to the Executive Secretary stating that the President wanted interim
marine science proposals immediately. The pent-up disappointment and
frustrations on Capitol Hill about inaction in marine affairs had finally
but quietly exploded when Congressman Rogers Morton had a private
chat with the President at a leaders' breakfast.

Although Murphy and several individual scientists had participated
in the planning exercise, we now felt the need to establish broad support by the scientific community. First the plan was tested on a strong
panel of outside consultants, then on a special meeting with the NAS/
NAE committees, aided by an elaborate assortment of explanatory
color charts which would pale even a good military briefing. Generally,
the reaction was good, the criticism being curiously leveled largely at
the enthusiastic style of presentation by Council staff. But once again, the
NAS/NAE group came back with its high price of admission. They
strongly criticized the limits for planning purposes set by the Council as

too small. Ironically, that same day, a director of a major oceanographic laboratory was telling the President's Science Advisor, Lee DuBridge, that the Council was requesting more funds for the Decade than could be fruitfully employed by the scientific community. Could that community be schizophrenic?

Review within the CPR was far more substantive. Out of a full and frank discussion came a drive to increase emphasis on environmental quality goals and to suppress emphasis on resource development, including fisheries, a position strongly espoused by James R. Schlesinger, representing the Budget Bureau. The major difficulty lay in fleshing out environmental plans because none of the agencies and hardly anyone in the scientific community had been concerned with ocean pollution.

A second iteration brought consensus. The plan was ready to send to the White House, together with the other nominations for presidential support. At that point, we simultaneously sought concurrence from the Council members themselves, with Agnew writing each a letter of encouragement. With reservations only about lead agency assignments, again consensus!

On October 16, we recaptured the brass ring. Speaking for the President, Agnew announced on October 19 the five initiatives in marine science of the Nixon Administration.

With the help of Lee DuBridge, the Council had obtained a commitment of $25 million of new money to get started in 1971. Included was support for the National Buoy Project that had been initiated by the Council in 1967 and was now ripe for expansion.

As elaborated subsequently in the fourth Council report, the Decade would be intended to:

(1) Preserve the ocean environment by accelerating scientific observations of the natural state of the ocean and its intentions with the coastal margin—to provide a basis for (a) assessing and predicting man-induced and natural modifications of the character of the oceans; (b) identifying damaging or irreversible effects of waste disposal at sea; and (c) comprehending the interaction of various levels of marine life to permit steps to prevent depletion or extinction of valuable species as a result of man's activities;

(2) Improve environmental forecasting to help reduce hazards to life and property and permit more efficient use of marine resources—by improving physical and mathematical models of the ocean and atmosphere which will provide the basis for increased accuracy, timeliness, and geographic precision of environmental forecasts;

(3) Expand seabed assessment activities to permit better management —domestically and internationally—of marine mineral exploration and exploitation by acquiring needed knowledge of seabed topography, structure, physical and dynamic properties, and resource potential, and to assist industry in planning more detailed investigations;

(4) Develop an ocean monitoring system to facilitate prediction of oceanographic and atmospheric conditions—through design and deployment of oceanographic data buoys and other remote sensing platforms;

(5) Improve worldwide data exchange through modernizing and standardizing national and international marine data collection, processing, and distribution; and

(6) Accelerate Decade planning to increase opportunities for international sharing of responsibilities and costs for ocean exploration, and to assure better use of limited exploration capabilities.[45]

The National Science Foundation was assigned responsibility for planning, coordinating, and funding the Decade.

The Nixon commitment to the Decade reappeared in April 1971, when in his message forwarding the final report of the moribund Marine Sciences Council he stressed the Decade among several components of his enhanced budget. The report brought up to date the status of Decade planning in NSF, its funding, and its progress under IOC:

In fiscal year 1971, the distribution of the $15,000,000 appropriated for the Decade among the three areas of emphasis will be approximately as follows: 20% environmental quality, 45% environmental forecasting, and 35% seabed assessment. Expenditures in the environmental quality program are being deferred this year during the development of carefully designed program for a major Decade effort in pollution studies, which should be completed by 1972. Forty percent of the research in fiscal year 1971 will be accomplished by Government agencies with the remaining 60% being carried out by academic and non-profit institutions and industry; in succeeding years the Federal agency involvement will be reduced.[46]

FY 1972 funding to NSF of $20 million dropped in FY 1973 to $18 million.[47] The auspicious thrust of initial funding may have been the summit. Further growth had been promised for the following year, but despite pleas by NSF's new director, William D. McElroy, funding remained about level; high-level advocacy seemed muted and the planning capability never assembled; the Decade was fast becoming another pipeline for project funding. Like a plane launched from a carrier, the Decade was off the flight deck but dipping slightly above the water. Whether it would gain the essential altitude remained to be seen. Other countries also were slow to commit funds with the same enthusiasm with which they had endorsed the concept and seemed prepared to support two years before. The U.S. program for the Decade was crystallizing, but it read like a domestic rather than a multinational enterprise.[48]

By 1971, the entire international arena had been trampled to a quagmire by other events that would dampen the opportunities for the oceans to unify rather than divide. The UN was developing serious internal strains, partly the consequence of rapid growth in membership of newly independent nations who could control the vote but who were limited in economic or political capital to influence world affairs. Ocean issues could hardly remain immune. The Decade became temporarily sidetracked; its blooms of cooperation were wilting. The vision of an international enterprise to benefit all nations seemed forgotten.

CHAPTER 6

One Sea and One World

Roots of Maritime Law—Who Owns the Deep Seabed

At roughly the same time that the International Decade of Ocean Exploration was being advanced through federal and UN apparatus, three other issues were about to be staged in the global arena—on sovereignty over seabed resources and their exploitation beyond limits of national jurisdiction (including definition of those limits), on seabed arms control, and on the detrimental effects of marine pollution. At first glance, each issue was distinct; the Decade appeared as a scientific question, the seabed regime a legal and political question, pollution a social and economic question, and arms control a military question. Moreover, all emerged amidst vast ignorance about the nature of the seas and a pervasive miasma of indifference to their importance; the international setting, albeit on a grander scale, resembled the domestic. By 1971, in the short span of five years, the scientific, economic, social, legal, military, and political questions were uncritically homogenized and being examined energetically in every possible international forum.

Around 1966 and following close behind the new U.S. marine policy implemented under President Johnson, international indifference warmed to enthusiasm as speculation spread about benefits from seabed minerals, oil, and gas. The same factors that had subtly converged to turn U.S. attention to the oceans were ripening everywhere. Demands for resources were growing. More powerful technology facilitated more distant petroleum extraction and fishing. And initiatives by American entrepreneurs as well as the government added momentum to interest by the world community. Isolated events then coalesced and ignited world-wide attention, some to elicit man's noblest desires for collectively seeking a tranquil, humane society, and some to trigger a savage territorial imperative based on short-term self-interest, unenlightened by study or perspective of longer-term benefits. The seductive promise of early economic return transformed opportunities for international comity in exploration and development into contentious debate. In an unwitting scramble for riches, Pandora's box was opened in terms of such questions as who owns the sea and seabed. The initiatives and confrontations stimulated by these legal issues, the moves and countermoves by different interests within the United States and by individual nations and their voting blocs, constitute at a planetary scale another chapter in the politics of the oceans.

What made these internationally tinctured issues unique was that 85 percent of the marine territory involved was beyond national sovereignty, in the areas where resources must be considered as held in common. By no means did this neutral sovereignty dispel conflict. Some countries asserted jurisdiction hundreds of miles seaward, thus eroding the traditional freedom of the seas. Long-range fishing fleets began to threaten fisheries historically harvested by coastal states. The possibility of stationing weapons of mass destruction on the sea floor heightened the specter of an arms race in the sea. And the less developed countries viewed the deep seabed as an inequitable technological preserve of more advanced nations and the call for abatement of pollution an added cost that only the developed could afford. The ancient maritime customs relating primarily to freedom of navigation on the ocean's surface were challenged by a new experience regarding submarine resources, disarmament, pollution, and research, for all of which legal arrangements were in a relatively embryonic state.

This law of the sea had been rooted since Roman times in the common law of property rights. But while dry-land boundaries were subject to quantitative survey and describable in deeds and covenants, the seaward extension of these boundaries was limited by difficulties in marking a fluid medium and in contributing to its military defense. What we know as the "three-mile limit" over territorial waters, generally attributed to a famous Dutch jurist, Cornelius von Bynkershoek (1673–1743), was determined solely by the range of shore-based cannon. Adding to this complication in defining boundaries was the mobile sovereignty that surrounds each ship at sea and on which special protection had been conferred. From agreements born in the maritime commerce of Tyre and Sidon, a decentralized random code of conduct slowly developed, consolidating customs, common practice, and tacit consent of merchants and shipowners.

This pragmatic development of sea law concerned property rights of a very special and elite group, the commercial or trading class who emerged in Europe beginning in the thirteenth century and their political protectors. Out of a feudal culture, a commercial renascence nourished the artistic and literary era about which so much more is known. As instruments of empire, the traders required protection, and in extending their mercantile life styles to the marine domain, established principles that have since been frozen into the statutes of maritime law to protect property (and to a lesser degree persons) against perils of the sea. Some of these risks arose from a hostile environment, some from mutiny, others from piracy, and still others from inept seamanship that threatened the safety of ships and cargo. The distinguished scholar Myres S. McDougal contends that most international law has been generated "by people creating expectations in each other about the requirement of future decision by simply cooperating or engaging in collaborative activities. . . . The great bulk of our inherited prescriptions in the law of the sea had their origin in this way, and when the community achieves legislative prescription of this kind there is some guarantee of its rationality."[1]

Maritime law then evolved from ancient admiralty laws into a new branch, now to secure the common interests of different communities over claims of special interests. This public order of the oceans is a far more recent invention, examined in 1930 but not codified until 1958. While this authority was a necessary condition, it was not sufficient. The dynamics of decision-making and subsequent enforcement complete the process and bring into play the social and political interdependencies that transfer ocean activities to the broader geopolitical theater. In the world today no state is free to do what it wants to do, not even the United States, the Soviet Union, or China. All nations are imbedded in the web of a closed system so crowded that any action creates a ripple of reaction. Given the atomistic, pluralistic quality of the international community and the almost universal consensus against a monolithic and centralized world government, a minimum, if not an optimum, order has been exercised by diplomatic instruments of accommodation and retaliation, supported by implicit economic and military measures of coercion. Primarily we rely on voluntary commitments from the various states.

The historical roots of greatest relevance to contemporary marine issues lay in attempts by the world community to deal with freedom of the seas for the purposes of navigation. By 1966, this concept also encompassed freedom of access, freedom of innocent passage, freedom of exploitation and use—and freedom of abuse—of the oceans and the seabed beyond predetermined sovereign limits. This evolution reflected the impact of technology not only in new relationships between nations afforded by modern transportation, communication and techniques of resource extraction, but also in interaction and conflicts among the different functions themselves—fishing, transportation, mining, research, and environmental conservancy. Technology extended the two-dimensional oceanic arena for navigation to a three-dimensional medium involving fish and seabed resources that jarred customs, institutions, and law. It became apparent that since the law is by its practice conservative, and technology in its capacity to induce change is radical, the two tend to advance at vastly different rates. At a global scale, in 1966, this dilemma was scarcely perceived.

The contemporary legal framework was also being influenced by a new political phenomenon—the emergence of new nation-states that, never having had a merchant class, viewed uses of the sea from quite a different perspective than did Western Europe and the United States. For the new states, self-interest dictated exploitation to achieve swift economic parity.[2] At the same time, the Soviet Union, which also lacked a history in international trade and had only limited access to the adjacent sea and seabed, was inclined to cast its growing maritime ambitions in the aristocratic tradition of naval dominion as much as utilization.

As cited in chapter 3, the United States sought to implement its 1966 Marine Sciences Act in the context of multinational cooperation toward practical as well as altruistic objectives of a community of nations at peace. The liberal philosophy espoused by President Johnson in his July 13, 1966, *Oceanographer* address set the course. Council Chairman Humphrey set the pace. The Council and its member agencies had an assignment to develop proposals against the backdrop of international law that comprised the rules of the game. At the very time the U.S. government began to re-examine the maritime game, the laws that governed the uninhabited regions of the planet were beginning to be subject to the same political forces and changes in values that were shaking all human society.

THREATS TO MARITIME ORDER

By 1966 a sequence of technological circumstances concerning energy and mineral extraction, fishing, and military operations had engaged legal interest because all posed subtle changes to the maritime status quo. Oil, gas, and minerals were known to exist on the submerged terrace that surrounds the continent, exhaustion of on-shore reserves made such off-shore exploration attractive, and technology was making it feasible. In the climate of economic reconstruction after World War II, assertion of mineral rights of the underwater landscape first called into question the adequacy of maritime law to protect entrepreneurs interested in submarine resources, but whose huge capital investments required protection by assignment of exclusive rights to exploitation.

On September 28, 1945, President Harry S. Truman issued two

surprising policy directives on ocean affairs. Intended to assist the domestic oil industry by clarifying federal versus state jurisdiction of offshore resources, they shook the international community. Proclamation Number 2667 unilaterally extended American rights by stating that "the government of the United States regards the natural resources of the subsoil and the seabed of the continental shelf beneath the high seas but contiguous to the coasts of the United States as appertaining to the United States, subject to its jurisdiction and control."[3] While the proclamation was silent on the exact boundary of jurisdiction, the accompanying White House news release interpreted the area covered as "contiguous to the continent and covered by no more than 100 fathoms of water." The other proclamation, Number 2668, dealt with coastal fisheries in relation to the high seas and served to carefully differentiate jurisdiction over conservation zones in the water column from that over resources on the seabed beneath. Not surprisingly, other nations soon exercised their prerogatives by making unilateral claims, but often blurring distinctions between seabed, subsoil, and fishing rights. At that point, several nations on the west coast of South America that had narrow continental shelves staked their claim at not less than 200 miles,[4] an arbitrary limit that Senator Magnuson reports was based on the 200-mile neutrality zone mandated by President Roosevelt to protect the western part of Latin America during World War II.[5]

The next factors impinging on maritime law developed in the early 1950s in relation to undersea warfare. The United States had successfully brought together nuclear power, the submarine vehicle, and the missile, creating a technological troika. Military strategy rapidly embraced the marine environment because of its advantage of masking naval forces and thus providing the advantage of surprise in attack or survivability when under attack. The U.S. Navy skillfully developed this strategic capability. Aware that competitors could do the same, they also developed defensive measures of passive listening arrays on the seabed to detect and track hostile craft. At the same time, with our world-wide commitments to contain communism, our presence on the seas was demanded everywhere. Freedom of transit for both surface and subsurface operations became vital to our national security, best guaranteed by a discipline of narrow limits to territorial seas that would afford transit through international straits that were regarded as partaking of law for the high seas.

In the 1950s a third set of events began to upset tranquility at sea: far-ranging tuna boats were being impounded by the South American countries exercising their 200-mile limit claims, and were being ransomed by the State Department rather than permitting our fishermen to buy licenses and thus inadvertently legitimizing these claims that abrogated traditional freedom of the seas. Meanwhile, long-range foreign craft were making huge catches off our own shores, much to the chagrin of domestic fishermen, and triggering cries for legal protection. By 1966 they were to obtain shelter in a 12-mile but not 200-mile zone, with our State Department aware of the national security argument.

Next in this sequence of developments stimulating interest in law of the sea was the growing murmur of potential seabed wealth beyond boundaries of national sovereignty. Undersea deposits of nodules containing manganese, cobalt, nickel, and copper were known to pave wide stretches of the seabed. Because of their commercial importance, this untapped resource was given widespread publicity as motivation for accelerated ocean research.[6] The Mohole project began to confirm feasibility of drilling for oil in deep water (its successor, the deep drilling ship *Glomar Challenger* indeed having extracted 400-foot-long cores with oil indications in 11,720 feet of water).[7] While speculation of commercial exploitability dates back to 1959, surprisingly it was neither nation-states nor industrial entrepreneurs who uncorked the economic prospects of the nodules and other seabed resources. Rather it was private individuals and organizations who idealistically coupled marine benefits to world order and institutions.

Along with these technology-induced considerations, a new threat to public order of the oceans was paradoxically arising due to conflicting interpretations of the Convention on the Continental Shelf that had been negotiated for the very purpose of heading off conflict. Motivated largely by the Truman Proclamation, that convention was framed at the United Nations Conference on the Law of the Sea convened at Geneva on February 24, 1958, and concluded on April 29, with eighty-six nations represented (not, of course, including those whose independence has been won subsequently). Three other conventions emerged from that initial conference, and a concluding session in 1960, concerning the territorial sea and contiguous zone, the high seas, and fishing and conservation of living resources.[8]

A great deal was accomplished at this Conference, for which participants accorded much importance to preparatory work of the International Law Commission. That group of specialists, which first met in 1949 and worked in relative freedom from purely political considerations, stirred the waters sufficiently to develop extensions of the traditional law of the sea—and compromises—that were later adopted at the Geneva sessions. The preparatory work took eight years! As one analyst stated, documentation on the 1958 conference

. . . reflected the potent moves, countermoves and positions of individual states and blocs. There was a constant attempt to lobby delegates concerning certain points. There was lesson number one: That the real reason for a state voting in a certain fashion must be learned if one would become proficient in the given game of estimating how that state would vote on hypothetical proposals. . . . There were the blocs: the European, the Latin Americans, the Africans. There were the peacemakers, principally the Indians, who sought harmony for the sake of harmony. The international lawyers of renown outnumbered the scientists, and many important decisions were . . . grounded on national pride. Not enough . . . were truly based on science; yet it must be admitted that the sciences of fisheries and oceanography are infants.[9]

Policy analyst Robert L. Friedheim undertook a statistical correlation of voting records to demonstrate a North and West struggle against South and East on territorial sea and contiguous zone; North versus South on fishing rights of coastal states; an East-West struggle over cold-war issues; even a muted quarrel over how to solve the problem of rights of landlocked countries.[10]

Perhaps no new doctrine of international law has ever before developed so rapidly, with so little precedent and with such potency to marine affairs as the Convention on the Continental Shelf. Within ten years it was to be shredded with controversy.

Article 1 defined the continental shelf "as referring to the seabed and subsoil of submarine areas adjacent to the coast but outside the area of the territorial sea, to a depth of 200 meters or, beyond that limit, to where the depth of the superjacent waters admits of the exploitation of the natural resources of the said area." The continental shelf was thus defined quite differently from common geologic usage, and ignored its physiographic characteristics. Unprecedented was the establishment of a flexible limit of sovereign rights that depended on the state of the art of relevant technology. However, it was the inherent ambiguity and conflict between vague notions of "exploitability" that was rationalized as the basis for extension of claims, and of "adjacency," that in 1967 lit the fires of international debates at the UN.

Despite the highly elastic connotation of Article 1, Article 6, was explicit in stating that "where the same Continental Shelf is adjacent to the territories of two or more states where coasts are opposite each other . . . in the absence of agreement [the boundary between the two is] a median line, every point of which is equidistant from the nearest points of the baselines from which the breadth of the territorial sea of each state is measured." State Department geographer G. E. Pearcy in 1959 interpreted that statement as suggesting that when the depth of exploitation is not limited by technology, the ocean bottom everywhere is sliced up and defined as belonging to one or another coastal state.[11] Full implications of that interpretation were sketched by resource economist Francis T. Christy, Jr., who delineated median lines crisscrossing Atlantic and Pacific ocean beds to represent a wild new map of the globe:[12] boundaries extended out halfway across ocean beds, so that such small islands as the Azores and Bermuda gained fantastically greater territorial claims below the waterline compared to land above. Today, few nations or sea lawyers accept Pearcy's thesis. Deliberately or not, the language of the Continental Shelf Convention had the effect of postponing the decision as to exclusive rights for exploitation. It was almost as though a time bomb had been ignited, with a fuse about eight to ten years long. For in the matter of sovereignty lay the new dreams of wealth, position, and power to be derived from the seabed, all in hibernation. In 1959, the *Christian Science Monitor* speculated on whether "an international agency, perhaps within the UN [might] take over jurisdiction."[13] While that cogent thought must have intrigued many, it was to lie dormant for some years. The U.S. Senate had casually ratified the sea law treaty on May 26, 1960, with scarcely a thought about ambiguities in text or other implications.[14] Discussion on the Senate floor took less than half an hour.

Between 1960 and 1964, the issue was almost completely quiescent, and in the succeeding two years only a few mild questions bubbled inconspicuously to the surface. Soon after the Continental Shelf Convention took effect in 1964 through ratification by the required twenty-two signatory powers, Congressman Richard Hanna proposed a bill authorizing study of legal problems arising out of the management, use, and control of natural resources of oceans and ocean beds. Because of the accident of his designating the National Science Foundation as study agent, the bill was referred to a rival congressional committee. Soon after, however, a similar bill introduced by Congressman Lennon authorizing studies by the Coast Guard assured referral to his and Hanna's Merchant Marine and Fisheries Committee.[15] The proposal died, but this charge later showed up as Section 4(a)5 of the Marine Sciences Act, with responsibility for such legal studies assigned to the President with advice and assistance of the Marine Sciences Council.

The relationship between technology and continental shelf development was highlighted as a legislative issue in a 1964 address by the author, who called attention to the neglected role of engineering as the connective tissue between scientific understanding of the sea and its practical use.[16] Referring to that speech on February 10, 1965, Senator E. L. "Bob" Bartlett from Alaska introduced S. 1091 to establish policy to accelerate exploration and development of the continental shelf, to encourage private investment, to develop an engineering capability to operate on the shelf, all through a new operating agency patterned after the AEC. While the bill died, several provisions regarding the role of technology later found their way into the Marine Sciences Act. Even then, no one stumbled on the issue of Continental Shelf jurisdiction.

On March 7, 1966, the Economic and Social Council of the UN (ECOSOC), which had a long-standing interest in natural resources and their implications for the less developed countries, passed an unheralded Resolution 1112(XL) requesting the Secretary General to make a survey of the present state of knowledge of seabed resources and of the techniques for their exploitation.

Then, on July 13, came President Johnson's eloquent warning:

> Truly great accomplishments in oceanography will require the cooperation of all the maritime nations of the world. And so today I send our voice out from this platform calling for such cooperation, requesting it, and urging it.
>
> . . . .
>
> We greatly welcome this type of international participation. Because under no circumstances, we believe, must we ever allow the prospects of rich harvests and mineral wealth to create a new form of colonial competition among the maritime nations. We must be careful to avoid a race to grab and to hold the lands under the high seas. We must ensure that the deep seas and the ocean bottoms are, and remain, the legacy of all human beings.[17]

This milestone concept had been nursed by a small group of policy analysts, including Naval Captain William Behrens, who saw the possibilities of coastal states setting a pattern inimical to U.S. security interests and capable of inciting maritime conflicts that would resemble the seventeenth-century battles over terrestrial frontiers. While such notions were floating about with hazy visibility and uncrystallized advocacy in rarified policy circles, President Johnson captured its essence in the *Oceanographer* address, which was to serve as potent guidance for policy to come. Of key importance was Johnson's reference to the sea as a "legacy of all human beings," a phrase that was to become of critical importance in dealing with the later concept of "a common heritage of mankind." Although the report on *Effective Use of the Sea* prepared by the President's Scientific Advisory Committee was released by him the same day, it was relatively silent on international cooperation or collaboration on legal issues involving the seabed.

President Johnson's speech received widespread press coverage and favorable comment. Within the State Department there was anxiety that this major plank of foreign policy had been drafted the night before. The Defense Department was privately pleased because the position was well aligned with its necessary concern for freedom of the seas, favoring minimal areas under national sovereignty that could inhibit naval operations. But none of the federal agencies had yet developed positions; what was later to evolve as a family quarrel was initially only an exercise in ideology.

When these questions of seabed development innocently burst upon the various agencies of the U.S. government, the coastal nation states, international bodies, private industrial interests and concerned citizens, the legal regimes of the sea, once thought to be an esoteric body of jurisprudence reserved for specialists, were found to be a potent issue of unexpected ramifications.

PROPOSALS FOR INTERNATIONALIZATION: THE COMMON HERITAGE OF MANKIND

Geopolitical implications of seabed sovereignty had begun to warm up when the Commission to Study the Organization of Peace in 1965 recommended that title to the entire ocean (beyond a twelve-mile limit for fish and a specifically defined continental shelf for minerals) "be vested in the international community through its agency, the United Nations," with benefits assigned to the UN.[18] The Commission, then with Clark M. Eichelberger as Chairman, is a research affiliate of the United Nations Association that was formed in 1939 to study what should take the place of the defunct League of Nations, and has been a stubborn advocate for a strong UN. The Commission proposal was largely motivated by a desire to provide independent income from seabed resources for a fiscally embarrassed United Nations, but it was also rationalized as a policy initiative to avoid controversy arising from competing claims, assure economically effective use of ocean resources, prevent military uses, avoid ocean contamination, and assure "equitable" allocation of profits from ocean exploration. To implement its title, the UN should establish a Marine Resources Agency.

Others picked up this theme. In November 1965, a Citizens Commission on Natural Resources also urged at the White House Conference on International Cooperation that a United Nations agency be vested with responsibility for international marine resources. Congressman Paul G. Rogers, a member of the House Merchant Marine and Fisheries Committee and one of the most active members of the Subcommittee on Oceanography, grasped these implications in his June 29, 1966, address to the fledgling Marine Technology Society. Taking note of the Eichelberger report, he warned of the need for new law of the sea as nations grow more conscious of the ocean's riches and inclined toward "international pilfering."[19] The question of title to the seabed also intrigued Senator Frank Church. While serving as a U.S. delegate to the UN he prepared a report for the Senate Foreign Relations Committee, released February 20, 1967, suggesting that the UN be made financially independent through ownership of the oceans' mineral resources.[20] The World Peace through Law Conference representing 2,500 lawyers from 100 countries, adopted a resolution submitted by Aaron L. Danzig on July 13, 1967, declaring that the resources of the high seas beyond the continental shelf were the province of the United Nations.[21]

The concept generated by Eichelberger's Commission was being widely discussed, but largely in forums predisposed to strengthen world government. All were based on the "untold wealth" of the seabed, notwithstanding the relatively limited information on mineral distribution and on extraction technology, a gap that the UN began quietly to fill by a request for a study by the Secretary General in Resolution 2172(XXI), discussed on page 262. But even more important to be resolved were the finer grain questions of defining functions for a new UN agency, creating administrative apparatus for management and control, setting criteria for disbursing proceeds, and integrating operations into the world market place. When this Commission proposal was discussed at the second Law of the Sea Institute in June 1967, William T. Burke, a scholar in maritime law, questioned whether the utopian goals could survive genesis in the General Assembly. He contended that the recommendation would have to be reconstituted so that its decision-making process, especially "those disposing of the new source of wealth to be placed in its control [would] faithfully reflect the present distribution of power, wealth and skill among the members," and that the "gargantuan puzzle of disarmament" be separated from the resource exploitation issue.[22] In rebuttal, Eichelberger recalled how the competence of the General Assembly was similarly questioned on two other complex technological issues, first when it took over creation of an International Atomic Energy Agency despite the proposition of the United States and Great Britain that it was "too involved to be sent to the General Assembly," and when the General Assembly proclaimed in 1961 that celestial bodies were not subject to appropriation and that space ships could not carry weapons of mass destruction. Eichelberger urged lifting our sights, and recalled Johnson's plea to prevent anarchy on the seabed. Burke's pessimism about the General Assembly was later to be confirmed by events.

THE PARDO PROPOSAL

The most powerful stimulus to international debate on the continental shelf was detonated by UN Ambassador Arvid Pardo of Malta on August 17, 1967. The substance of his proposition was concisely projected in the request by Malta to inscribe on the agenda of the General Assembly at its 22nd session an item entitled "Declaration and Treaty Concerning the Reservation Exclusively for Peaceful Purposes of the Sea-Bed and of the Ocean Floor, Underlying the Seas Beyond the Limits of Present National Jurisdiction, and the Use of their Resources in the Interests of Mankind."[23] The General Assembly was electrified by his elaborate, scholarly assertion of riches on the seabed to be readily harvested with immediate profit. He proposed that the UN should internationalize the seabed beyond some narrow limit of national jurisdiction by a particular interpretation of the Convention on the Continental Shelf or, if necessary, by its amendment, and that it create a new UN organ to administer this internationalized seabed. Proceeds would be disbursed to the less developed countries. In these propositions, the definition of seabed geography lying beyond national sovereignty was linked to institutions for its management. Clearly, the narrower the band of jurisdiction by coastal states, the greater the area of seabed resources, especially those just adjacent to the continental shelf, would be available to be shared by the international community.

This Item 92 on the General Assembly Agenda was tabled, later followed by the Secretary General's note urging separation of issues between peaceful uses, scientific activities, and resource exploitation, so as to concentrate on questions of the legal status of deep-sea resources and ways to convey benefits of exploitation to developing countries.[24] The fifty-eight nations who participated in the debate advocated positions that ranged from vesting title to the seabed in the United Nations to a moratorium on unilateral exploitation of seabed resources. While some opposed any present consideration by the General Assembly, on December 18, 1967, Resolution 2340(XXII) was adopted, creating a thirty-five member Ad Hoc Committee to Study the Peaceful Uses of the Seabed and the Ocean Floor Beyond the Limits of National Jurisdiction. Chaired by Ceylon's Ambassador to the UN, Hamilton Shirley Amerasinghe, the Committee was to prepare a study for the 23rd session that would examine activities of the UN and its specialized agencies dealing with the seabed; relevant international agreements; scientific, technical, economic, legal, and other aspects of the question; and suggestions on promotion of international cooperation in exploration, conservation, and use of the seabed and its resources.[25] By this act, the UN was now squarely involved in ocean affairs.

The less developed countries quickly and uncritically hopped on the bandwagon but few were prepared with facts on the issues. Only during the following year of inquiry was a modicum of rationality restored as to the prospects of early reward, which up until then had been based more on fantasy than on fact. By then, however, the aroused hopes were to find political expression through another route, less promising in the distribution of benefits Pardo had sought.

That the original proposition should come from Pardo of Malta was initially puzzling, but his own background suggests a logical basis for his initiative. Unlike most UN ambassadors, Pardo was not a professional diplomat. He was first a scholar and second a public servant of the international civil service. A native of Malta, he had received a degree in international law in Rome, but was then interned for five years during World War II in Italy and Germany. He entered the UN service immediately after and served in a variety of posts until his appointment as ambassador in 1964. His island nation of Malta, located at the crossroads of Mediterranean traffic, had gained its independence only in 1964 and was still finding its role tangential to the majestic issues being debated at the UN. It is not surprising that the seabed issue that had been percolating quietly in scholarly discussions of new routes to world order should furnish a natural theme for Pardo's careful research and initiative.[26]

Pardo has told me, incidentally, that he developed his brief without consultation. Just before his famous address, he discussed it with the Israeli and the U.S. representatives; the latter was sufficiently shocked to return the next day with a plea to demur. Back in Washington, consternation followed the speech, but by good fortune, unlike so many other situations when the United States seems caught by surprise, some preparation had already gone into the development of a position that several of us had hoped would be offered at the UN that autumn by President Johnson.

Evolving U.S. Policy

Since the fall of 1966, ramifications of the seabed question had been receiving attention in the U.S. government, but at a low priority everywhere except in the Marine Sciences Council. The tiny Council staff, endeavoring to respond to Johnson's request for initiatives by January, saw there was too short a time to complete any major study of foreign policy implications. In fact, we scarcely comprehended the complexity of the interaction of marine technology and law. Eichelberger had wired Humphrey expressing hope that Johnson's concept would not be transformed by bureaucratic legerdemain or disappear completely by neglect. We responded with assurances that the Council intended to follow Johnson's policy guidance. When the Council's first annual report was being prepared, we were obliged to be silent on seabed issues, except for reporting Johnson's speech. The State Department objected to our restating the "heritage of mankind" concept, and Johnson concluded his marine affairs message with a weaker commitment to his initial position rather than a stronger one we had drafted, simply stating: "We shall bring to the challenge of the ocean depths . . . as we have brought to the challenge of outer space . . . a determination to work with all nations to develop the seas for the benefit of mankind."[27]

The opportunity for the Council to become involved in policy formulation had opened accidentally in late fall, 1966, when our representative to the UN, James Roosevelt, unexpectedly communicated to Washington his intention to address the General Assembly on marine affairs. The Council Secretariat was pleased that some initiative was being asserted in marine affairs in the international arena and at the same time concerned to head off any premature or ill-considered proposal. Roosevelt's contribution led quickly to UN Resolution 2172(XXI), requesting the UN Secretary General to make two surveys—one called for by ECOSOC and one "of activities in marine science and technology, including that relating to mineral resource development, undertaken by . . . United Nations . . . organizations, various member states and intergovernmental organizations concerned, as well as by . . . other interested organizations." The Resolution also requested the Secretary General, in cooperation with specialized UN organs to formulate proposals for "ensuring the most effective arrangements for . . . cooperation . . . through science [and in] the exploitation and development of marine resources, with due regard to the conservation of fish stocks." So fishery and mineral resources were both put on the agenda for study.[28] While seemingly innocuous, this action dealing with scientific and technical questions laid the groundwork for UN self-education on the political seabed issue subsequently ignited by Pardo.

That fall, the Council Secretariat put its first priority on the issue of seabed disarmament. Senior officials of the Arms Control and Disarmament Agency (ACDA) were asked to report on the status of research on opportunities to extend to the ocean the already eight-year-old arms limitations discussion with the Soviets. ACDA replied that they were "studying" the matter. After encouragement from our office, reinforced by indications of Humphrey's interest, ACDA began to move; by February 1967, we felt satisfied they were giving the issue serious attention.

The Council Secretariat then turned to the legal regime, in recognition that technology had outpaced the law. To be sure, technology had proved since World War II to be a potent instrument of prestige and power in world affairs, both military and economic. In the case of the oceans, however, its influence was initially not to trigger a competition for technological supremacy but rather to heighten territorialism. The question was whether any steps lay ahead wherein all nations could benefit, and thus head off the debilitating race Johnson had warned about. While aware of Eichelberger's proposals, we felt a problem lay in implementing valid long-range goals with operationally viable apparatus. So we began our own study of the deep seabed.

The issue as we saw it initially was to determine what next steps of leadership should be taken in the national interest to extend the law of the sea, for the long as well as the short run. National security led the list of considerations, especially in preserving freedom of transit through straits that was essential to naval operations. But that concept had favorable implications in regard to merchant shipping whose freedom from uncertainty of goodwill of nations bordering straits was important to trade of the entire world community. And the concept of freedom of the seas sustained a moral principle. Our security also depended upon greater stability in international relations, so a question naturally followed as to what were the interests of other nations that could be better satisfied by new arrangements that would be compatible with the premise of freedom of the seas. Besides national security, what were the other economic, scientific, and political objectives of our own nation and how would these be served? In short, we were looking for the unusual solution that would satisfy interests of different nations—developed and undeveloped—and our own. At the same time, we were no doubt indulging a smidgen of idealism.

Every nation was obliged to examine its interests regarding the seabed regime. Some would be slowed by long traditions. Some would have neither tradition nor a multidisciplinary staff available to undertake study. Some would stake claims by preliminary positions to provide maneuvering room for subsequent negotiation and compromise. Some were involved in or anticipated resource development along their coasts that seemed to overshadow all other considerations; some would want to project distant water fisheries. Most would place more emphasis on exploitation than on conservation or environmental protection. At that stage, we concentrated on the seabed resources.

As we saw it, the latter issue initially rested on rejecting any belief that the 1958 Convention on the Continental Shelf was unambiguous in defining boundaries of national sovereignty. This opened up the question of a range of alternatives of the boundary of national sovereignty that was colloquially referred to as "wide" or "narrow" shelf. We inquired how each option would affect various military, economic, and political interests, both in the United States and abroad. By this time, Council staff were in consultation with individual State and Interior Department staff concerning innovative proposals by George Miron for a buffer zone. Such a zone would separate a narrow coastal band of national jurisdiction from the heart of the ocean bed under universal jurisdiction. The zone would have mixed jurisdiction to balance interests for different shelf widths, domestically, and particularly between developed and developing countries which especially hoped to gain from the exploitation of resources beyond national jurisdiction. As a corollary to that step, following the intent of the Marine Sciences Act, we contracted for studies in depth by a number of the nation's most distinguished legal experts, William Burke, Louis Henkin, and Paul M. Dodyk, hoping to flush out alternatives, and the pros and cons of each, as grist for the governmental policy mill. Not only was the Council Secretariat intent on ferreting out implications of change in the legal regime of the seabed; we also sought to examine legal problems dealing with fisheries and with the little discussed issue of freedom of research. When Secretariat discussions with State, Defense, and Interior officials revealed these questions to be of little immediate concern, we proposed that Humphrey take steps (as recalled in chapter 5) to establish two high-level committees—the Committee on International Policy in the Marine Environment (CIPME), and a task group under State's Legal Advisor to oversee studies contracted by the Council.

By the spring of 1967, Council staff were also in close consultation with the Stratton Commission on this issue. Three of their members had a keen interest in and expertise on these matters: Carl Auerbach, a professor of law at the University of Minnesota; Jacob Blaustein, founder of the American Oil Company, largest stockholder in Standard Oil of Indiana, and widely respected for his individual efforts toward world peace; and Leon Jaworski, a distinguished partner in the largest law firm in Houston (subsequently elected to presidency of the American Bar Association), and well acquainted with the oil business. Formed into a subcommittee by Stratton, these three were well aware after the Pardo proposal that the development of a U.S. position would not await completion of their report in January 1969. They undertook to prepare an advance position for the Commission, found themselves in substantial accord, argued it vigorously before the full Commission to gain their endorsement, then frankly urged adoption within the government.

By the fall of 1967 the Council contract studies were completed, and while we decided against premature commitment to a position, their findings were released in the Council's annual report:

—There should be deliberate policy decisions on the extent of the Continental Shelf; a precise definition of its seaward boundary seems desirable. A buffer zone might be established to bridge the boundary between the Shelf and the seabed with the coastal states' interests in the ocean floor given special protection in the Zone.
—The U.S. should seek an international legal framework which promotes freedom of oceanographic research within waters subject to national control and on the Continental Shelf.
—Consideration with regard to living resources might be given to establishment of a global conservation authority which would strive to extend and improve existing international regulation of high seas exploitation in the interest of conservation and efficiency.[29]

All three were to prove prescient in illuminating the key issues involved as a backdrop for an enlightened U.S. policy.

Henkin's rationale for a buffer zone between the coastal band under coastal-nation jurisdiction and the common property beyond advanced a concept discussed earlier in Washington circles, and was immediately embraced within the Commission; the Council Secretariat, while publicly reserving its position, privately found it attractive and compatible with notions it had derived in rudimentary form the year before. In our appraisal, foreign policy in relation to lesser developed countries and to promotion of world order clearly favored a narrow shelf, as did U.S. defense considerations. Domestic economic considerations favored neither (on which most of the petroleum industry disagreed). Yet nations with a narrow shelf would be nervous about operations too near their shore. The buffer zone seemed like a way to have your cake and eat it too, in balancing conflicting interests.

As to any government-wide position on the seabed issue, CIPME had made relatively little progress by late summer, despite the efforts by

Herman Pollack and constant prodding from the Council staff. No one had yet cut through the maze of interconnecting issues that pivoted on the boundary question, but we were still hoping to have some positive proposal to lay before the UN. It was Pardo's speech that electrified the State Department; everyone agreed that while its idealism was constructive, his proposition went too far, too fast, with insufficient hard data. The State Department's response was almost immediately subdued when the Congress was hastily pushed into the act by the fledgling National Oceanography Association. Playing on conservative biases of selected members of Congress, the NOA staff claimed that our State Department was adopting Pardo's proposal and in so doing was giving away a birthright of America in its offshore treasures at the edge of the continental shelf. Although the petroleum interests had substantial influence in NOA, there is reason to believe that at that time they had not studied the Pardo proposal, developed a position, or maneuvered the NOA initiative. Rather, it was a knee-jerk reaction on NOA's part. The rumor mill carried further innuendoes of intrigue—that Pardo's speech had been prepared in the Council Secretariat!

The result was a blizzard of resolutions in both House and Senate, mostly in opposition to vesting authority over deep ocean resources in the United Nations.[30] That NOA had catalyzed this hasty response was acknowledged publicly by Senator Norris Cotton when he backed away from his own resolution (S.J.Res. 111): "I would say very frankly to the Committee that the first draft of this resolution was prepared for me by representatives of the National Oceanography Association. As far as I am concerned, this matter in my resolution of directing American representatives in the United Nations to oppose action or to take any particular attitude, I think might well be deleted."[31]

Nevertheless, there were deep-seated congressional concerns that the United States was relinquishing claim to valuable assets of yet unknown scope, and that the United Nations was unqualified to assume such broad responsibilities. In that same vein, Congressman Paul Rogers urged that for purposes of exploitation, "The United States should have the right to occupy the ocean floor to the Mid-Atlantic Ridge and assume the responsibility to defend it."[32]

On the other side of the issue, Senator Claiborne Pell, who as an unabashed advocate for an ocean program had sponsored the Sea Grant proposal and coauthored *Challenge of the Seven Seas*,[33] saw an opportunity to excite interest in the future of international ocean development and plunged into the fray with speeches and resolutions in September 1967, in support of the Pardo thesis.[34] There was little collegiate response.

Along with State Department officials, I met off the record with both Congressmen and Senators in efforts to reassure them that we had absolutely no commitment to Pardo's or Pell's proposals, later reinforcing this in a colloquy during a hearing before Congressman Lennon.[35]

In the fall of 1967, the State Department, justifiably wary of hasty innovation and gunshy of Congress, chose to back into a position of doing nothing until publicity cooled down. Reminded by the Council Secretariat and Humphrey that something had to be said after the Pardo bombshell and the repeated assertions by the President in support of international cooperation, State carefully structured a position. On grounds that information essential to rational development of seabed resources was acutely deficient, and that the United States felt that to avoid precipitous action a very careful study must be made of legal ramifications, our UN Ambassador Arthur J. Goldberg proposed that the General Assembly establish a Committee on the Oceans to examine all marine questions brought before it, to stimulate international co-operation in exploration (the Decade proposal was being protected for presidential announcement), and to assist the General Assembly in considering questions of both law and arms control. As a first step Goldberg proposed that a "set of principles" be developed to underlie a code for exploration and use of the seabed. A new doctrine was being launched.

As the U.S. position evolved, the Council Secretariat endeavored unsuccessfully to gain presidential support for elaboration of the "common heritage of mankind" theme. The State Department argued against it on grounds that a few Congressmen, alerted by the NOA initiative, would interpret the statement as selling out to Pardo and intervene so effectively as to upset any development at all. In fact, during the UN debates there was apprehension that since the congressional reaction was predominately negative, Goldberg would be undercut. We thus undertook to gain support on the floor by Senators Mansfield and Pell. Mansfield responded immediately by backing Goldberg,[36] echoed by Senator Norris Cotton. Pell, however, took a much bigger, independent leap, embroidering on Pardo's proposal, and in March 1968 he proposed an elaborate treaty in the Senate to govern exploration and exploitation of ocean space and to ban all underwater military activities except passive detection.[37] The treaty would apply the same freedom-of-the-seas principles to the seabed that govern surface operations, create an international licensing body to oversee commercial exploitation, establish a Sea Guard to police the treaty, and ban weaponry on the ocean floor.[38] He favored the narrow shelf, expressed both in depth (550 meter isobath) and in distance (50 nautical miles). Pell's resolution had a cool reception in the State Department and vitriolic criticism from the oil industry.

Early in 1968 the United States' position was strengthened and clarified with the President's support of international exploration in his 1968 State of the Union message, with his environmental message proposing the Decade, and with the seven-part initiative elaborated in the Council's second annual report transmitted by the President to Congress. CIPME's hard work was beginning to bear fruit. In the March 1968 report of the Council, the U.S. position began to unfold on two major questions: "What should be the seaward limit of the Continental Shelf?" and "What resources are there beyond the Continental Shelf and who should control them?" Picking up the administration's policy theme, the Council said:

A desirable early step . . . would be international accord on certain general principles . . . the seabed should not become a stage for a new form of colonial rivalry and should not be subjected to claims of national sovereignty. Rather, the seabed should be open to exploration and use by all states without discrimination. International standards should be set to foster orderly exploration and use. . . . Cooperative scientific research . . . should be encouraged together with broad dissemination of results. Activities on the seabed should be conducted with reasonable regard for the activities of other states. Pollution and interference with the traditional freedoms of the seas should be avoided.

The Council went on to lay the groundwork for the Decade and related the proposal to the seabed issue: "Our lack of knowledge of the scale and location of ocean resources also hampers the making of sound policy decisions, domestically and internationally. . . . The United States is thus encouraging the international development of long-range plans for intensified cooperative exploration of the oceans." Then the Council made two proposals; first, for "unmodified ocean habitats for research and education . . . as ecological baselines . . . characteristic marine features such as a deep ocean trench, a group of sea mounts, and an uninhabited coral atoll might be set aside." Second, we proposed that "international ocean acres, i.e., limited ocean areas designated for intensive research over a period of many years . . . might be established in the vicinity of marine preserves."[39]

With the new ad hoc seabed apparatus then clanking louder at the United Nations, the United States government was under increasing pressure to develop a position on the legal regime. The first issue to be resolved was the limit of national sovereignty, a wide shelf or a narrow shelf. All parties in the U.S. government were willing to consider the options open, initially not constrained by the existing Continental Shelf Convention. But the more intensively the major federal agencies engaged in study, the more evident was the irreconcilability of their viewpoints. The impasse blanketed CIPME. The Department of Defense, ably staffed for mounting policy studies, developed its position quite soon. It favored a narrow jurisdiction, recognizing that the Navy would have far greater access to the seas within the scope of treaty obligations for operating submarines and emplanting listening devices essential to detection for antisubmarine warfare if all but a narrow band of shelf were beyond national sovereignty. The Departments of Commerce and Interior were obliged to consult with affected interests in the oil industry as to how the disposition of continental shelf resources, especially off our own shores, would stimulate private investment and affect company profits, the stability of U.S. sources, and balance of payments.

Before governmental decisions hardened into flint likely to strike sparks, consultation was sought January 14, 1968, with the National Petroleum Council (NPC), an industrial advisory body to the Secretary of Interior (created June 18, 1946) to provide guidance from petroleum and natural gas interests. The oil industry wanted a wide shelf. By July, 1968, the NPC had developed an interim position through a Committee on Petroleum Resources under the Ocean Floor chaired by E. D. Brockett, Chairman of the Board, Gulf Oil Corporation. Based on their interpretation of the 1958 Convention on the Continental Shelf, NPC contended that "The United States, in common with other coastal nations, now has exclusive jurisdiction over the natural resources of the submerged continental mass seaward to where the submerged portion of that mass meets the abyssal ocean floor and that it should declare its rights accordingly."[40] This was the "wide shelf" position—argued on the geologic grounds that the continent over which nations have sovereignty extends submerged to where its toe intersects the deep seabed, perhaps at a depth of 12,000 feet. The NPC also claimed that "this was the intent of the framers and delegates who composed the 1958 Geneva Convention," and the Congress and President who ratified it. On this point, there was to be sharp and acrimonious disagreement. And while not buying the extreme position of a median line boundary, the NPC and its legal advisors such as Northcutt Ely, believed that the birthright of the United States as protected by the 1958 Convention was being given away. Apart from its legal interpretations, the NPC argued that the wide shelf would help the United States maintain supplies of energy for national security, and help an expanding domestic petroleum industry to meet declining reserves by looking to offshore sources. The NPC contended that seabed and high seas status of overlying waters could be separated, that the existing international laws were adequate and that long-term arrangements should reflect stable policies that "will encourage, not deter, exploration, recovery and use of deep-ocean minerals." Soon after, the American Branch of the International Law Association, under its Committee on Deep Sea Mineral Resources chaired by Northcutt Ely, espoused the same position as NPC, but slightly hedged on "exploitability" provisions as not yet technologically advanced to validate the wide shelf advocated by the NPC.

Mindful of the shock waves beginning to ripple through the petroleum industry and their influence in governmental affairs, the State Department had encouraged solicitation of NPC's advice before a possible impasse could develop. If this had been thought of as smoking out that position, the officials were probably surprised and hurt when they were promptly stung by the angered wasps. The United States wanted to crystallize its position in the UN Seabed Committee by June 1968, but the effect was to rush the NPC to a decision based less on logical analysis than on gut feelings. At the NPC's plenary session to consider the draft report, the approval by representatives of all the major and independent oil companies was 95 to 1. The single opponent to the wide shelf position was eloquently and heroically argued by Jacob Blaustein, who by then had the rationale of the Stratton panel on which he served, and of Henkin's study. His efforts were beaten back, but he advocated the task group's position privately to oil interests and to State Department officials. Even Julius Stratton, who served on the Board of Directors of Standard Oil of New Jersey and was personally convinced of the wisdom of his panel's views, became embroiled, but he too failed to persuade industrial associates. While the industry was then gaining a foothold to lobby, the government rapidly lost the collateral chance to explain or seek adherence to a counter view. A tragic polarization occurred that was ameliorated only by NPC's pro forma caveat that its position was based solely on economic considerations, not on national security or foreign policy consideration.

So by June 1968 the Departments of Commerce and Interior, feeling obliged to support NPC, were at loggerheads with Defense. For almost the first time, the Navy and the petroleum industry found themselves at odds. The State Department did not press foreign policy considerations and saw their primary role as a moderator between Defense and Interior Departments. The Secretariat's stance at CIPME meetings was to attempt to gain some innovative position that would demonstrate U.S. foreign policy leadership, but we carefully refrained from publicly arguing either side of the issue. By then, however, we found confirmation of our initial, albeit tentative position. Staff and consultant analysis indicated that the United States did not have to depend on deeper submarine sources for reserves in the interest of national security, and that the legislative history of the 1958 Conference did not support NPC legal interpretations. Even the self-interest of the oil companies seemed uncertain in that many were drilling offshore in foreign waters in profit-sharing partnership with other coastal nations, and their proposed extension of coastal jurisdictions would perpetuate a patchwork of agreements that were regularly shaken as unstable foreign governments came and went and demanded greater shares of the profit. They might benefit from a narrow shelf.

Considering the dissension so rampant at the UN, Humphrey and the

Council Secretariat saw an opportunity to advance one unifying theme with the seabed issue. The roots of intergovernmental amity could arise from a concept of the common heritage of mankind, beneficial to the third world, to the United States, and to world order. We also made two assumptions that proved wrong: that the seabed issue could be dealt with in isolation from other sensitive maritime legal questions, and that if a proposal were uniquely able to serve a variety of foreign as well as domestic interests, they would perceive the benefits and jump on a common bandwagon.

The State Department, chairing CIPME, extracted a temporizing compromise out of the domestic impasse in the form of a draft resolution of principles,[41] although the commitment to dedicate proceeds from deep seabed exploitation to world community purposes was a mark of some progress. President Johnson authorized their introduction by the United States at the second session of the UN's Ad Hoc Committee, thus following up Goldberg's commitment. Issued June 28, 1968, they were intended as building blocks for international accord:

. . . there is, and will remain, an area of the seabed and deep ocean floor beyond national jurisdiction. . . . There should be established, as soon as practicable, an internationally agreed precise boundary delineating the area beyond national jurisdiction. Exploitation and use of the natural resources of the seabed and ocean floor prior to the establishment of a boundary should not prejudice the location of that boundary. . . . Beyond national jurisdiction . . . no State should claim or exercise sovereignty or sovereign rights over any part of the area. . . . There should be no discrimination in the availability of the area for exploration, scientific research . . . internationally agreed arrangements governing the exploitation of resources of the area should be established as soon as practicable. Such arrangements should include provision for the orderly development of resources; conditions conducive to making of investments; preserving the integrity of investments in the area made prior to agreement on such arrangements; the dedication . . . of a portion of the value of the resources recovered from the deep ocean floor to world and regional community purposes; and accommodation among the commercial and other uses of the deep ocean floor and marine environment.
. . . The seabed and deep ocean floor beyond national jurisdiction should be used exclusively for peaceful purposes in accordance with the U.N. Charter. Since this does not preclude military activities generally, specific limitations on designated military activities will require the negotiation of a detailed arms control agreement. Military activites not precluded . . . would continue to be conducted in accordance with freedom of the seas and exclusively for peaceful purposes.[42]

This latter question on seabed arms control had already been split off for resolution in a different forum than the General Assembly of the UN, as explained at the end of this chapter.

Among other provisions, the United States advocated prompt negotiation of international arrangements, concerned that delay would unleash the nationalistic dragons. (With three years of delay, by 1971 they were out of the cage.) Again in retrospect, perhaps there was no way to hasten the UN process, which was beginning to bog down as the number of new voting members increased and aligned themselves to ignore the major powers or else to block them, as violations of the Charter increased, and as suspicion arose that the ocean issue was being manipulated to serve the maritime powers.

By the winter of 1968, having received a number of reports by the Secretary General[43] and the Ad Hoc Committee, the UN moved to higher ground by promoting the Ad Hoc Committee to permanent status. In the 2467A-D series of resolutions, discussed on page 238, the new body was given a rich agenda for study.

Within the U.S. government, it was clear the CIPME was not able to resolve the boundary issue. I appealed to the President via Califano to develop a clear position for debate at the projected March 1969 UN Seabed Committee session; I failed. The last chance of the Johnson administration to take an explicit step occurred at the January 10, 1969 meeting of the Council. Amidst the other bruising issues on governmental reorganization and coastal management, CIPME brought to the Council a minimal proposal for an interim position of a moratorium on unilateral claims, to head off any further deterioration in the already faltering UN position on a seabed regime.

Secretary of State Dean Rusk, after ringing support to such a concept, warned that if we were to continue to exhort consideration of the "heritage of mankind," we should mean what we said. This was the first indication of an unequivocal position by the State Department itself, and a welcome thrust. But the full range of possibilities was not debated by the Council. The Council acted affirmatively to support the moratorium position, but for the government the dilemma remained on how to harmonize the various national interests.

As the baton was handed from Johnson to Nixon in a cloud of uncertainty, the seabed issue could not be ignored because of its presence on the imminent UN agenda. Under a new secretary, William P. Rogers, the State Department was obliged to generate a government-wide position on the seabed regime, and he moved to sustain CIPME's role and responsibility. Similarly, the seabed disarmament question had progressed apace in the UN's Eighteen Nation Disarmament Committee (ENDC). Henry A. Kissinger, the President's Special Assistant for National Security Affairs, swiftly picked up the ball on disarmament; later he was also to give effective assistance in moving the seabed issue.

By now, the Stratton Commission report had been released, baldly challenging the NPC position on the boundary of national jurisdiction as "not warranted either by the language of the definition of the 'continental shelf' or its history," and rejecting their proposal for exercise by coastal nations of sovereign rights over the submerged continental block "as contrary to the best interests of the United States." Among other factors, they warned of the danger that coastal nations without important mineral deposits might feel justified in claiming exclusive access to the superjacent waters, a practice already troublesome in the southern hemisphere. The Commission proposed as a limit to the shelf the 200-meter isobath, or 50 nautical miles, whichever gives the coastal state greater area. Then they recommended new international agreements for mineral resources underlying the deep seas—an International Registry Authority and an International Fund to disburse proceeds from a tax on deep sea exploitation through the World Bank UN Development Program, which in turn would finance research and exploration and food-from-the-sea programs (to develop protein supplements). The Commission proposed creation of an intermediate zone between their narrow shelf boundary and a 2,500-meter/100-mile limit, but "only the coastal nation or its licensees, which may or may not be its nationals, should be authorized to explore or exploit the mineral resources of the intermediate zone. In all other respects, exploration and exploitation in the intermediate zone should be governed by the framework recommended for the areas of the deep seas beyond the intermediate zone."[44] While the buffer or intermediate zone was of mixed jurisdiction, benefits of its exploitation were intended to be shared internationally.

Not surprisingly, the new administration felt unencumbered by the Council's last desperate efforts at consensus. Now chaired by Deputy Undersecretary U. Alexis Johnson, CIPME met February 28, 1969, to reopen the debate, with the Stratton proposals squarely on the table. It promptly became evident that they were back to Adam and Eve. Representatives of Defense, Interior, and Commerce almost instantly assumed the identical, uncompromising positions that they had declared all through the prior year's development in the Johnson administration, and felt no commitment to the January 1969 action by the Council. More than that, Interior and Commerce claimed that since their policy level officials were new on the scene, they would require far more time to develop positions. The consensus had fallen apart. The Council Secretariat, with this priority issue now staffed by William Mansfield, kept up a steady pressure for action.

Alexis Johnson endeavored to exert leadership as CIPME chairman by a round of private discussions with the contending agencies and representatives of industry. Apparently he came away with the opinion that no position was the best position, a matter the Senate was to bring into the open.

As the UN committee wrestled with the seabed issue during the summer, the United States gingerly put forward a position on international machinery, stating that such machinery, as part of an international regime, could include an international registry of claims governed by agreed criteria and supplemented by appropriate procedures. Under such machinery, governments would be responsible for adherence by their nationals to internationally established criteria, and the system would require adequate procedures for verifying compliance. The U.S. representative cautiously pointed out that "no more and no less machinery should be created than would be required." Without clear orientation, the UN Committee was fraught with dissension. Many individual nations asserted unilateral claims.

That missed opportunity for U.S. leadership in 1969 seemed serious. By November, I had been in communication with the White House, urging them to intervene to break the CIPME deadlock. One result was an upgrading of discussion to the National Security Council. The Undersecretary's Committee of the NSC under Elliott L. Richardson was charged with reviewing the impasse and recommending actions. I was asked to represent the Council Secretariat; in that role, however, I was careful to point out that I could not commit the Council Chairman.

Richardson requested his legal advisor, John R. Stevenson, to endeavor to reach a compromise position. This was exposed to the Committee on January 29, 1970, my last day in office. After careful study, I concluded that the State Department draft proposal reflected a compromise between the Defense and Interior positions, more for the sake of compromise rather than for a position that could be rationalized on its own. Moreover, if it was intended to straddle the Defense-Interior rift, it had failed to achieve a balance and was clearly tilted toward the NPC wide-shelf view.

At a session conducted in solemn decorum, State's proposal was explained and discussed to ensure that its implications were fully understood. At that stage, I decided to throw caution to the winds, to reveal for the first time my own position on the seabed regime, and to argue for junking the State proposal. First, I felt obliged to challenge Interior's position for a wide shelf argued by then Undersecretary Russell Train, a matter I deeply regretted because of my high personal regard for Train. Among other factors, there was no evidence to support his contention that profits to U.S. oil interests would in the long run be greater in exploiting submarine oil sources under wide-shelf provision (even assuming that a valid objective). With development expected worldwide, exploitation off our own shores was but a single source of profit. Frequently, U.S. companies were whipsawed in their negotiations, a situation less likely to occur under a single international regime.

Second, I questioned the validity of the proposal because it failed to assert any position by the State Department with regard to foreign policy interests of the United States, either in support of Stevenson's proposition or in relation to alternatives. Seabed strife from unrestrained exploitation and conflicting claims was inimical to world order, and our leadership in ocean affairs could be a factor in bolstering our prestige generally in seeking peace amid antagonism over our involvement in Vietnam. We had an interest and a policy in helping the less-developed countries to help themselves.

Third, our security interests were not served by the wide shelf; even the Navy admitted that the NPC argument for protecting deep sea oil reserves off our shores was specious.

Fourth, few alternatives were considered, the most conspicuously absent being the one proposed initially by Henkin and later refined and supported by the Stratton Commission.

Fifth, the risk of "creeping jurisdiction" was increasing whereby individual countries might try to expand vertically through the water column their jurisdiction over freedom of the seas, research, and fishing over a band as wide as the seabed. Ample evidence of this trend had already appeared in Brazil's objection to geophysical surveying off their coast, in more frequent capture of our fishing vessels, and in a growing chorus of individual nations unilaterally asserting claims.

Finally, I reminded the group of the opportunity for the President to assert a leadership in ocean affairs that had so far been lacking in the Nixon administration.

Representing the Joint Chiefs, General Earle G. Wheeler rose to support my position as did the Department of Justice representative Richard G. Kliendienst, each for different reasons. The consensus was back to the drawing board. The positions outlined at the meeting were carefully weighed by Richardson who, after wide consultation in and outside of government, shaped his synthesis of the issue and made proposals through NSC to the President.

These actions culminated in the May 23, 1970, release by President Nixon of a major unequivocal policy.[45] The President proposed that all nations adopt as soon as possible a treaty under which they would renounce all national claims over the natural resources of the seabed beyond the point where the high seas reach a depth of 200 meters and would agree to regard these resources as the common heritage of mankind.

This important concept that the Council had tried so hard to advance from Johnson's 1966 springboard was finally a matter of presidential doctrine. In elaborating Nixon's proposal, the Council explained:

This treaty would establish an international regime for the exploitation of seabed resources beyond this limit. The regime would provide for the collection of mineral revenues for international community purposes, particularly assistance to developing countries. It would also establish rules to prevent unreasonable interference with other uses of the ocean, protect the ocean from pollution, assure the integrity of investment necessary for such exploitation, and provide for peaceful and compulsory settlement of disputes.

Two types of machinery would authorize exploitation of seabed resources beyond a depth of 200 meters. Coastal nations would act as trustees for the international community in an International Trusteeship Area comprised of the continental margins beyond a depth of 200 meters off their coasts. Agreed international machinery would authorize and regulate such exploitations beyond the continental margins.[46]

The machinery, incidentally, would be composed of a balanced representation of interests, rather than one-nation, one-vote that was hampering the UN.

These principles were developed into a draft treaty by Louis Sohn, then on leave from Harvard University and advising State, and subsequently introduced into the UN Seabed Committee as the first U.S. position on the Continental Shelf. Here was a version of Henkin and the Commission buffer-zone concept, now labeled, "Trusteeship Area," a term that found some unfortunate opprobrium internationally because it recalled colonial protectionist policies. Nevertheless, this was the first proposal to be advanced by a major power that could appeal to all parties and suppress a grab for seabed resources.

In August, the United States presented the UN Seabed Committee with a draft UN Convention on the International Seabed Area, in the form of a working paper, elaborating on the President's policy statement. It was to gain warm editorial praise as being in the direction of world order and away from national advantage or economic privilege[47] but it did not prove the miracle drug many had hoped for the cure of maritime nationalism.

During this time the initial response by the 91st Congress to Pardo's treaty subsided, although there was keen interest in response by the Executive to the Stratton proposal. The chairman of the Senate Committee on Foreign Relations created a Subcommittee on Ocean Space and made Senator Pell chairman. Hearings were opened in July, 1969 and government witnesses freely confessed that they had developed no position. In the colloquy that ensued Pell accused the administration of a "no-policy policy," with which Alexis Johnson felt obliged to concur.

To head off jurisdictional collisions, the Senate Commerce Committee, which had historical cognizance in all ocean affairs, established a Special Study on United Nations Subocean Lands Policy, chaired by Senator Ernest F. Hollings, a symbolic move agreed upon by all parties to keep the Commerce Committee in the picture while giving Pell maneuvering room. While not pursuing the international issue, the Hollings study panel was later elevated to subcommittee status to attack a wide range of marine issues of more direct interest. Meanwhile, the Senate Interior Committee's Special Subcommittee on Outer Continental Shelf, chaired by Senator Lee Metcalf, began to hold hearings and to assert its jurisdictional claim.[48]

Out of an elaborate set of hearings, the Metcalf Committee adopted the interpretation of the Continental Shelf Convention espoused by the NPC, arguing against the narrow shelf and upholding the 1958 Geneva Convention as "positive, reliable and adequate." Then, while endorsing the general framework of Nixon's May 23 declaration, it strongly objected to the boundary provisions on grounds that "to offer to renounce its inherent sovereign rights to the mineral estate of its continental margin in the hope that these few recalcitrant nations would mend their ways and begin to adhere to the freedom of the seas doctrine is like offering to pay ransom to bandits in order to encourage them to stop stealing."[49] Rather it wanted measures "to insure the protection of investors who desire to exercise present high seas rights to explore and exploit the wealth at the deep seabed beyond the limits of the submerged land continent." Curiously, the report was inconsistent in saying it was a "serious misconstruction" to interpret the Convention as extending present U.S. rights. Perhaps, then, the NPC view was not a valid but a desired position.[50]

Interior Committee Chairman Henry M. Jackson had earlier communicated these views to the President, with the result that the August 3, 1970, draft treaty based on the Nixon statement was actually downgraded to a working paper and controversial points were muted for the UN Committee.[51] The NPC, having been defeated in holding the line in the executive branch, clearly had friends on Capitol Hill. In 1971, S. 2801 sought to protect deep-sea exploitation of hard minerals.

While this debate was raging in the United States among legal specialists,[52] parties at interest, and in government offices, the public was generally indifferent, even bored by the technical details. One exception lay in continuing reports of the Commission to Study the Organization of Peace and in a new project by the Center for the Study of Democratic Institutions. In an informal climate intended to relax usual obligations to adopt traditional postures amidst political rhetoric, the project director Elisabeth Mann Borgese was determined to provide a forum as to international law and apparatus that could lead official thinking. Initially, her objective was to move the United States off dead center, later to foster interest at the UN, but these objectives proved ephemeral because of the march of events.

In the summers of 1970 and 1971, under the Center's sponsorship, a Pacem in Maribus convocation on the island of Malta developed a number of propositions of majestic proportions.[53] While that initiative may not have catalyzed progress, it furnished an important stage for examining the issues in a much longer time perspective. By that time the author, free to participate as a private individual, proposed a systemic approach to management of the marine environment that would take note of ecological integrity, interactions of resource development, conservation, pollution, and scientific exploration, and would perhaps counter the compulsive focus on territorial sovereignty becoming outmoded in a technological age.[54] If deep seabed resources were to be thought of as the common heritage of mankind, so should the entire marine environment beyond national jurisdiction and all knowledge about it. On that basis, new international machinery would be needed for gathering and processing information about that common heritage, and for participatory decision-making by the world community on goals and priorities. A far broader issue was in the making of alternative marine futures, discussed in chapter 10, toward which U.S. policy could be more clearly and broadly focused.

In December 1970, the UN decided to convene a new Law of the Sea Conference in 1973. The United States first opposed that move but, when clearly outnumbered, supported it. Fishing rights, territorial sea limits, freedom of scientific research were intertwined, as creeping jurisdictions were expressed not as initiatives of individual nations but of power blocs. We entered the lists with a draft convention on territorial seas, passage through straits, and fisheries on August 3, 1971, exactly one year after introduction of a draft treaty on the seabed. But at the UN Seabed Committee, none of the U.S. proposals were gaining acceptance.

In retrospect, some critics of the U.S. actions contend that the wrong policy was advanced initially, that action was premature because few nations were prepared to respond, that for the U.S. to do anything but drag its feet was a major blunder. Perhaps. But it is also possible that the United States was right in its substantive proposals and its idealism, but so dilatory and vacillating in its tactics at a time the UN was metamorphosing, that an opportunity for a major step toward order was lost. It is also possible that the conditions arising in the UN after 1969 made any U.S. approach unlikely to succeed.

PAINSTAKING PROGRESS AT THE UNITED NATIONS

Apart from questions of international law, various facets of marine affairs had been under surveillance by intergovernmental organizations for decades.[55] The International Hydrographic Bureau (IHB—now International Hydrographic Organization) was created in 1921 to deal with the application of oceanographic data to safe navigation. The International Council for the Exploration of the Seas (ICES) had long dealt with fishery statistics for the North Atlantic. Intergovernmental fisheries commissions specialized by region or species harvested had been created to deal with conservation problems.[56]

Within the UN family of specialized agencies, six treaty organizations created with broad mandates had within their ambit of functions cognizance over particular sectors of marine interests: the UN Educational, Scientific, and Cultural Organization (UNESCO) (in which the Intergovernmental Oceanographic Commission was located), the Food and Agriculture Organization (FAO), the World Meteorological Organization (WMO), the Intergovernmental Maritime Consultative Organization (IMCO), the International Atomic Energy Agency (IAEA), and the World Health Organization (WHO).

The FAO, while predominantly concerned with terrestrial agriculture, has within its structure a major Fisheries Department concerned with fisheries productivity, environmental influences, harvesting, management, overfishing, economics and marketing, resource assessment, and utilization. It also serves as executive agent for the UN Development Fund.

The WMO functions as a clearinghouse to facilitate intergovernmental exchange of meteorological data. Because of the reciprocal interaction of ocean and atmosphere, of the role of the atmosphere in transporting pollutants to the ocean, and of the requirement for ocean sites for sea-surface as well as weather observations, WMO has necessarily been interested in oceanography.

IMCO was established as an information as well a rule-making body to deal with safety at sea of life and property, largely focused on ship operations. In recent years, they have included pollution of the marine environment from oil and hazardous cargo on their agenda.

The IAEA, concerned with the development of peaceful application of atomic energy, has placed emphasis on radioactive safety associated with accumulation of radionuclides at sea from atmospheric fallout, and discharge or disposal of materials directly at sea.

Within its terms of reference, WHO has been interested in health questions of waste disposal and pollution in estuarine and coastal areas that would affect bathing, water quality, and seafood contamination.

The Economic and Social Council (ECOSOC) serves the General Assembly as the informational funnel and coordinating agent for all of these affiliated specialized agencies. While ECOSOC expects program initiatives to be generated and implemented by subsidiary specialized agencies, it has endeavored to unify ocean interests by recommendations for leadership by the Secretary General, and by creation of an inter-organizational committee for exchange of plans and program coordination, discussed in chapter 10. ECOSOC's performance and leadership have been criticized widely. And in the aggregate, all international bodies dealing with marine affairs were found subcritical in size, of low visibility and modest prestige in international political circles.

Straddling this mosaic of functions and jurisdictions is the UN General Assembly and its Secretariat. The Seabed Committee created in 1967 was its most conspicuous instrument in dealing with legal and policy issues, correlated through various resolutions with technical competences of the specialized agencies. After its founding, the Ad Hoc Committee on the Seabed met three times in 1968 and considered background papers prepared for it by the UN Secretariat, the IOC, and other specialized UN agencies. Its final report reflected a heightened awareness of the complex technical and legal problems uncorked by the issue of seabed exploitation, eventually tempering with realism the inflated expectations of early sumptuous returns from the new frontier.

The Committee debates had exposed a variety of positions represented by individual nation-states and blocs voting their self-interest. One problem arose because the data on seabed resources were so sparse that, even given the narrow motivation of economic gain, the route to satisfaction of self-interest was uncertain. Narrow versus wide sovereignty, national versus international control, freedom of the seas versus constraints were at issue, and steps toward world order were at stake.

Finding a common ground was perplexing. Power blocs manifested generally at the UN were reproduced at diminutive scale in the Seabed Committee: technologically advanced haves versus have-nots; Latin America versus the North. The same divisions that had plagued the 1958 Geneva Convention were now augmented by additional factions, although the East versus West contest was subordinated by mutual interests of the two superpowers in freedom for naval operations. Underlying these positions were alleged suspicions by the LDC's that the U.S. position was invariably conditioned by the self-serving colonialism of a major power protecting its oil industry. Publication of the NPC report lent credence to these views. The smaller nations, initially beguiled by Pardo's arguments for an international agency to disburse seabed profits selectively for their benefit, had begun a campaign for a narrow shelf. But years later when indications that the United States might favor a narrow shelf were first privately leaked, and later confirmed by the Nixon proposal, our motives were again suspect, now in terms of military ambitions through freedom of passage through straits. Either way, the United States was caught in a bind. Smaller nations were also curious about the extent to which the United States and the Soviet Union might agree and were more apprehensive than pleased with such prospects because of anxieties that alliances between the two superpowers would suppress all other interests.

In August 1968, acting independently on reports it had received in response to earlier resolutions, the ECOSOC adopted three resolutions:

Resolution 1380(XLV) invited the Secretary General to continue to promote further systematic investigation of mineral and biological resources, and specifically invited FAO to review food resources beyond the continental shelf.

1381(XLV) called on the General Assembly to endorse the concept of a coordinated long-term program of scientific research including such initiatives as the Decade.

1382(XLV) pushed for UNESCO to "study appropriate means of imparting a wider knowledge of the sea and its resources as part of secondary education programs."

During the fall of 1968 the General Assembly reviewed the Seabed Committee's report (mindful of the Secretary Generals's report on seabed resources which among other things cautioned against hopes of quick wealth)[57] and ECOSOC resolutions. After extensive debate in the Political Committee, the UNGA adopted on December 21 four resolutions, three cosponsored by the United States and one on which we abstained.

2467A(XXIII), passing unanimously, established a 42-member Standing Committee on Peaceful Uses of the Seabed and the Ocean Floor Beyond the Limits of National Jurisdiction and empowered it to expand on studies by its progenitor. This was a quantum jump in providing a permanent forum for marine issues, little more than one year after the topic first surfaced.

2467B(XXIII) urged measures to prevent pollution of the oceans.

2467D(XXIII) discussed proposals previously supported by the United States for the Decade and, within the IOC framework of a comprehensive and long-term program, gave IOC a leading role in coordination.[58]

2467C(XXIII) requested the Secretary General to study the question of establishing international machinery to promote exploration and exploitation of seabed resources, an initiative the United States was then obliged to declare premature, among other reasons because of controversy back home in CIPME and nervousness on Capitol Hill, and concern that such machinery might be too readily controlled by smaller states that outnumbered the advanced countries but lacked technical expertise.

Ceylon's Ambassador to the United Nations, Hamilton Shirley Amerasinghe, was re-elected chairman of the new UN Seabed Committee and it went to work, meeting three times in 1969.

The legal subcommittee of the Seabed Committee devoted primary attention to carrying out the mandate of the 23d General Assembly: To formulate legal principles that would promote international cooperation in the exploration and use of the seabed.

The economic and technical subcommittee of the UN Seabed Committee focused its attention in 1969 on a report submitted by the UN Secretary General setting forth an analysis of possible types of international organizational machinery that might govern exploitation of the deep seabeds. While a number of developing nations suggested the need to create an international agency that would regulate and control seabed exploitation, more and more countries suggested that the international authority should itself engage in exploitation.

In the fall of 1969 the 24th UN General Assembly reviewed the Seabed Committee's report and debated it at length in the Political Committee. In December, the Assembly adopted four resolutions 2574A-D(XXIV) on seabeds:

(1) Requesting the Secretary General to ascertain the views of member states on the desirability of convening at an early date a Law of the Sea Conference dealing with the full range of law-of-the-sea issues, including the seabed;

(2) Referring substantive seabed issues entrusted to it back to the Seabeds Committee and asking the Committee to prepare a comprehensive, balanced set of principles in time for the 25th General Assembly;

(3) Requesting the Secretary General to prepare a further study on various types of international machinery, particularly a study on machinery with extensive powers; and

(4) Declaring that states and persons, physical or juridical, are bound to refrain from all activities of exploitation of the resources of the deep seabeds, pending the establishment of an international regime.[59]

The United States opposed the first of these resolutions because it believed that treating all legal issues, including seabeds, at a single Law of the Sea Conference would only increase the difficulty of making progress on any. We saw the problem of creeping jurisdiction and trade-offs that might occur at the expense of freedom of the seas because the door was open to reviewing and junking all the tenets of the 1958 conference. Especially, the United States was worried about limits for the territorial seas. In the absence of any international agreement, we had long espoused a three-mile limit. Most countries adopted a twelve-mile limit, the furthest anyone visualized in 1960 when the U.S. proposal for a three-mile limit was defeated by a single vote. And when in 1970 the United States said it was willing to seek a treaty under the UN to fix the limit at twelve provided that narrow straits would continue unencumbered,[60] others were moving to two hundred. The hazard was all the greater in the absence of preparatory work by experts out of the glare of the political stage, as had been accomplished before the 1958 Conference. This trend toward homogenizing issues had been raised by several scholars in the field, but the United States had hoped the seabed issue could be dealt with on its own merits. The United States strenuously opposed the fourth resolution, which amounted to a call for the prohibition of all further exploitation of the deep seabeds, pointing out that such a freeze could spur initiatives for unilateral extensions of jurisdictions over the seabeds. The Council position at the end of the Johnson administration would have supported this moratorium.

In this second year of debate, further progress was made toward

adoption of basic principles, including consideration of those drafted
and advocated by the United States, but there was little progress on the
gut question of revisions to the Continental Shelf Convention and international management of the deep seabed. Nations had come to see the
significance of the issues and were being assiduously careful.

The UN Seabeds Committee held three more sessions in 1970, and
while its Legal Subcommittee made some progress, disagreement was
so strong at the end of these sessions that no declaration of principles
was in fact agreed upon. The dramatic U.S. proposal on the seabed
failed to gain support, and paradoxically triggered fourteen Latin
American and Caribbean nations to assert the right to set any limits of
jurisdiction. For the first time in the life of the Seabed Committee, the
Soviet Union backed away from any principles on an international regime, probably as a sign that Moscow would accept only the most rudimentary form of international apparatus, if any at all.[61] So the upsurge
of nationalism and the Soviet's chronic apprehension over controls by any
international body administered a serious setback.

In December 1970, however, the United Nations General Assembly
adopted resolution 2749(XXV) on Seabed Principles, following two
years of negotiation. They represented some accomplishment, especially
in recognizing that the seabed beyond the limits of national jurisdiction,
as well as the resources of the area, are the common heritage of mankind.
The declaration also provided that states shall pay due regard to the
rights and interests of other states that may be affected by seabed activities and that the principles shall not affect the legal status of the
water and air space superjacent to the area.

Resolution 2750 A–C(XXV) on Law of the Sea was also passed:
requesting the Secretary-General to study problems arising from the
production of minerals from the deep seabed; requesting the Secretary-
General to prepare an up-to-date study on the question of free access of
landlocked countries to the sea; and deciding to convene in 1973 a Conference on the Law of the Sea that would deal with the establishment of
an equitable international regime for the seabed beyond the limits of
national jurisdiction, a precise definition of the area, and related issues
concerning the regimes of the high seas, the continental shelf, the territorial sea and contiguous zone (including the question of international
straits), fishing and conservation of the living resources of the high seas
(including the question of the preferential rights of coastal States); the
preservation of the marine environment; and scientific research.[62]

Once more, the critical decisions were deferred and the debates
gravitated to the politics of the issues rather than the issues themselves.

The developing countries became apprehensive lest their lack of capital and technology deprive them of an equitable basis for participating
in benefits from seabed exploitation. To strengthen their bargaining

position, they doggedly introduced consideration of other maritime legal regimes related to the breadth of territorial sea, rights of passage through straits and fishing practices, all of which added not only new dimensions of complexity but also of controversy. One hundred eight members of the UN voted at the General Assembly for the 1973 Law of the Sea Conference. Only seven opposed and two abstained, mainly the Soviet Union and its associates. That apparent rush for an early Conference reflected two opposite schools of thought. One group wanted swift change, albeit through due legal process. The other wanted no change but felt that international agreement would be the only deterrent to unilateral extensions of sovereignty, and thus complete anarchy.

The Seabed Committee was given the task of preparing for 1973, with its scope of cognizance considerably extended beyond geographical limits of the ocean floor outside of national jurisdiction. Membership in the committee was increased from forty-two to eighty-six. By the summer of 1971, its agenda was nourished by additional draft treaties, including one by Pardo that encompassed over 150 articles, distinguished by his idealism, but now with a 200-mile resource and pollution control zone under coastal states, which represented a substantial retreat from his initial position. Others were submitted by the Soviet Union, Tanzania, the Latin states, Ceylon, and India.

With the Seabed Committee now examining all legal aspects of the sea, the United States on August 3, 1971, submitted three draft articles, on the territorial sea, on passage through straits, and on fisheries. Article I would limit the width of territorial sea or combined territorial sea and contiguous fisheries zone to twelve miles. Article II would preserve high seas freedom of navigation and overflight through straits. Article III on fisheries would provide for regulation by appropriate international or regional organizations that, for purposes of conservation, would set quotas and ensure compulsory enforcement. Coastal states would have a special preference in quotas determined by their potential for harvesting, except that existing percentages by other states would not be reallocated to coastal states. Special provisions would be made for anadromous species, for cases where international organizations were not operating and for resolution of disputes, although trial and punishment would be administered by the government whose flag was flown by the transgressor.

Article I also clarified the ambiguities of rights of innocent passage provided for in existing conventions because the question of what was "innocent" left extensive discretion and unpredictable interpretation to coastal states; and by the U.S. proposal, the right of free transit would be inseparable from freedom of navigation on the high seas.

The U.S. argument for fisheries regulation was based on the need to protect fisheries as a major source of protein for three-fourths of the world's population, even though it represents only about 20 percent of total protein consumed. The fisheries provisions aroused complaints from U.S. fishermen who felt that traditional fishing by foreigners off U.S. shores should not be condoned, and from many coastal nations who wanted exclusive unilateral rights to regulate contiguous fisheries. The United States was not adamant on any of its proposals except for the territorial sea and straits.

At the August Seabed Committee meeting, many other countries set forth their positions, and it was clear that there were wide differences that could not be resolved if each stood squarely on its interpretation of interests and rights. Any legal evolution that would permit nations to partition off areas of seabed, annex wide bands of contiguous seas subject to coastal jurisdiction, and restrict navigation through straits could hardly promote harmony.

There are serious questions as to whether the large number of fishing organizations implicit in the U.S. proposal together with the International Seabed Resources Authority would not encumber the situation with too many functionally separate but legally interconnected organizations. In fact, there is much skepticism about creation of any new international body, given the unsteady performance of many.

Nevertheless, if fisheries are to serve the world community, as pointed out in chapter 7, it is essential that management be practiced by limiting entry and reducing the number of undisciplined, submarginal, fishing units. International rights must be assumed to exist, and delegated to national agents.

There is, therefore, a basis for considering the marine environment and its resources beyond national jurisdiction as the common heritage of mankind. This is a fundamental concept that, unfortunately, is all too often employed as a tactical device to define what is left over after territorialism is satisfied. It may be a more potent concept than freedom of the seas. But even that concept is in need of defense by all nations, not simply those who are alleged to be its beneficiaries. If a right cannot be enjoyed because of lack of material means, its remedy does not lie in its abolition, but by technical assistance. On this issue, the United States has been silent. Thus, the common denominator organizationally is data and analysis capability to serve all interests, without access to which the LDC's would retain their suspicion that only the advanced nations who control the collection of data will benefit from it.

Two issues remain at the starting gate: protection of the seas from pollution, including hazards of spillage from supertankers, and freedom of scientific research. Yet they may provide the lowest common denominator of possible consensus. 1969 and 1970 reflected a growing concern by coastal nations over offshore pollution. The disastrous foundering of the *Torrey Canyon* and later tanker casualties led IMCO to formulate the International Convention relating to Intervention on the High Seas in cases of Oil Pollution Casualties and the International Convention of Civil Liability for Oil Pollution. Public furor over the disposal of obsolete nerve gas rockets off the Atlantic Coast produced a series of initiatives in Congress and ultimately a major review and statement of policy by the President to restrict indiscriminate disposal of waste at sea. With a concerned awareness of the experimental voyage of the U.S.S. *Manhattan* through the Northwest Passage and the prospect of vessel casualties in the ecologically sensitive Canadian Arctic, the government of Canada took a bold step for a coastal nation to protect its shores in 1970: a pollution control zone to extend one hundred miles from its territorial sea to prevent pollution from shipping, established unilaterally by the Arctic Waters Pollution Prevention Act. In late 1971, proposals to ban ocean dumping were submitted to preparatory sessions of the UN Conference on the Human Environment. A start was made in preparation for the 1973 LOS on establishing international principles to protect and preserve the oceans from continental sources of pollution, to establish international standards, to provide for cooperative data acquisition and dissemination, to provide for warnings and liabilities in event of damage, to assist the LDCs in meeting standards without economic penalties, and to permit actions in event of grave and imminent dangers.

Over the six years since the UN engaged questions about the sea, world-wide interest widened and deepened. By then, one analyst had concluded that taxable revenues from exploitation of deep sea minerals would only amount to $80 million annually two decades hence, thus helping to squash earlier excitement over quick riches.[63] But the treasure of oil and fish nearer the coastlines remained a viable prize. Resounding idealism had deteriorated to contests of bald self-interest. And while the Secretary General was repeatedly called upon to study the questions, the opportunity for his stronger leadership was sadly aborted. The superpower's position stemmed from desires to minimize constraints on freedom of action; smaller states wanted minimum encroachment by the technologically precocious. Some coastal states bargained possible economic gains off their shores for favorable political alignments. Western Europe was cautious generally; Latin American nations opted for broad and extensive national sovereignty, an anathema to the United States, but in a way it was chickens from the Truman Proclamation coming home to roost. Most lesser developed countries, however, favored a narrow shelf and strong international law and controls. They developed a bloc of 77, which constituted the most powerful voice in the UN, short by 7 of the needed 84 to gain a two-thirds majority to swing any issue. Ever more coastal nations pressed for advantage by membership on the

Seabed Committee; by 1972 it had swollen to 91, almost the size of the UN itself and equally hampered by sheer numbers.

At least by the end of 1971, all of the different positions were on the bargaining table. But whether a consensus could be martialled that would serve the world of the future was of grave concern. Considering the novelty of the seabed regime, the gaps in technical knowledge of the sea, problems in interweaving technology with statecraft in the tapestry of legal, scientific, economic, political, and social questions involved, the national prestige, security, and territorial claims at stake, a long gestation period was to be expected. The year 1973 was destined to become the year of decision. But the initial quest for international comity and management was delayed so long that all the primitive territorial instincts were given the chance to hatch. Surely, if nations approach 1973 with the same views they held in 1971, the results of the conference will reflect a maritime outlook of the nineteenth century rather than the twenty-first.

In the meantime, two other issues encumbered the negotiations—on freedom to conduct scientific research near the shores of coastal states, and on protection of the sea against pollution. Both issues were grasped as chips in the international poker game without regard for long-term implications. On the other hand, only a common threat to the closed environment of our home planet could bring unity of purpose and a spirit of cooperation to the nation-states and even to rival international bodies. And ironically some nations were reported as trading away the principle of freedom of research on which information essential to preserving the health of the environment would have to be based. The superpowers were especially apprehensive about support by so many nations for not only a wide shelf but also wide territorial seas above. Both legal operations of intelligence-gathering ships and navigation through straits (traditionally permitted as innocent passage not prejudicial to peace, good order, or security to the coastal state) were at stake. To gain 12-mile limits to territorial seas and exemptions for straits, the United States appeared inclined to bargain away other legal conditions, and by March 1972 the Nixon doctrine on the seabed and 1971 fisheries proposals seemed abandoned in favor of a 200-mile exclusive contiguous zone for living and nonliving resources. Official U.S. indifference to freedom for research appeared to be checked, however, largely as a result of objections from the scientific community and the National Advisory Committee on Oceans and Atmosphere. To be sure, there was great difficulty in separating genuine research from military snooping and commercial reconnaissance. But to prohibit exploration closer than 200 miles, as many nations proposed, or to require complex clearance procedures, could be fatal to scientific study in the most critical reaches of the sea.

Principles were being discussed concerning freedom for research (espoused by the United States in 1968) that involved advance notification, cooperation by the coastal state, sharing of all data and open publication, and protection of rights of coastal states. Without such agreements, resource development, environmental forecasting, and pollution warnings could all be severely inhibited, and the cultural ethic and common heritage themes of international comity eroded.

The years 1972–73 with a UN Conference on the Human Environment and on Law of the Sea will be portentous years of international decision. Some of the challenges that must be faced both as to substantive issues and the decision-making process are discussed in chapter 10. If these challenges are not met effectively, the world may find the consequences by the year 2000 far more difficult to resolve.

Seabed Arms Control

Paralleling interests by the international community for utilization of the seas for peaceful purposes were direct measures to minimize their use in warfare. The threat of nuclear calamity lurked behind the thinking of all individuals and nations alike. Above the striving by the superpowers to increase the potency and number of weapons of mass destruction could be heard the first notes of a new theme—that strategies of national security based on increases in nuclear armament with spiraling costs contributed to the balance of terror without necessarily adding to security. This recognition, coupled with the hazards of nuclear accidents, revived in 1962 fresh initiatives for arms control. Treaties toward limitation of conventional weapons had been attempted in the 1920s and 1930s—and had failed.[64] The nuclear age gave urgent motivation to try again.

Undersea vehicles had already demonstrated that clandestine armaments and occult weaponry could be masked by the marine environment. The oceans were more and more attractive to military planners. But from the point of view of national security in 1966, Vice President Humphrey and the Council Secretariat saw an oceanic analogue to the Non-Proliferation Treaty (NPT). The nuclear powers were already seeking ways to head off the spread of nuclear weapons to other countries; let them consider at the same time how to bar their entrance in sectors of the environment not yet actively regarded as potential launching sites. In the upper waters of the ocean, four nations, the United States, U.S.S.R., the United Kingdom, and France, were currently engaged in the use or development of submarines to carry nuclear-tipped ballistic missiles. That particular milieu thus could not qualify as nuclear-free. But how about the seabed? One could visualize underwater silos for missiles, or bottom-crawling tractors. Neither was yet a reality. Nevertheless, their prohibition was not generically a new idea.

Comparable steps had been taken in a 1959 treaty declaring that Antarctica shall be "used for peaceful purposes only," and in a 1967 treaty similarly reserving the moon and other celestial bodies. But so far as the Council staff knew, no corresponding initiative had been taken at a policy level regarding the seabed.

In November 1966, Schweitzer and I had a long talk with the Director of Research for ACDA, and found to our dismay that no staff studies had been undertaken on this possibility of seabed arms control. We urged attention to nonproliferation on the seafloor and also in deeper waters well beyond the depths of contemporary submarine operations but potentially of great potency for running and hiding amid hilly seabed landscape. The vacuum of policy research was brought to Humphrey's attention and he promptly encouraged staff to push ACDA while he put this question on his February 1967 agenda for discussion with Deputy Under Secretary of State Kohler. We also felt it necessary to prepare intelligence estimates on what the Soviets were already doing or preparing to do in that arena.

The first hint of Council interest in the seabed disarmament question was reflected in the 1967 annual report. The first chapter on "International Cooperation in the National Interest" concluded with a section that dealt with not only swords but also plowshares in the world community. But on the point of seabed disarmament all that could be said explicitly was that "The United States is continuing its efforts to encourage all nations to become partners to the 1963 treaty prohibiting nuclear testing in the seas, atmosphere and outer space."[65]

While the ACDA studies were underway, the concept of demilitarizing the seabed broke loose in several quarters, at the UN in Pardo's proposal, and in the U.S. Congress. In the spring of 1968 at the Ad Hoc Seabed Committee, the Soviets proposed full disarmament of the oceans, a propaganda stratagem that they knew we could not accept because of existing submarine deployment of Polaris missiles on which we heavily relied for a second strike capability. Since they were building a similar capability it is certain they would have been equally unwilling to sacrifice their investment.

By the end of 1967, the concept of seabed arms control had sufficient promise that Humphrey and I had hoped to offer President Johnson an opportunity for an initiative at the opening of the UN in 1968. In that we failed, but at least we were somewhat prepared to respond to the Soviet proposal. Parenthetically, the United States had long sustained a position at the UN that steps toward nuclear arms control were so sensitive and technically complex that some forum other than the political stage of the General Assembly was necessary for progress. Toward that end, a special committee of eighteen nations called the Eighteen-Nation Disarmament Committee (ENDC), including both nuclear and non-nuclear powers, had been created in December 1961, and had been hard at work in Geneva on numerous issues. In light of that history, the United States responded to the Soviets with a draft resolution to the G.A. and the Ad Hoc Seabed Committee that would refer to the ENDC the question of arms limitation on the ocean floor alone, rather than on the initially improbable range of all ocean-based nuclear weapons. Our objective was a workable, verifiable, and effective international agreement that would prevent use of this new environment for the emplacement of weapons of mass destruction. The U.S. position asserted that the seabed and the deep ocean floor beyond national jurisdiction should be used exclusively for peaceful purposes, but a significant caveat was appended that this be in accordance with the UN Charter which does not preclude general military activities consistent with freedom of the seas and exclusively for peaceful purposes. Specific limitation on designated military activities thus required negotiation of a detailed arms control agreement.

The ENDC put this issue on their agenda in July 1968. By the fall of 1968 the issue was so far developed within the U.S. government that it was referred to the Committee on Principals that served the National Security Council as an Under-Secretary-level policy-planning committee on arms control. By then, ACDA had developed a powerful comprehensive paper outlining implications to be tested at a staff level one step below the Committee. I was asked to represent the Marine Sciences Council. The issue was developed in terms of gains and penalties to security to be anticipated in taking such a step. The major argument that ensued was largely between Chairman of the Joint Chiefs, General Earle Wheeler and myself. Since some advantage of seabed deployment might conceivably be discovered at a later date, he argued that we should not now cut off any of our options by arms limitations. Based on facts developed by the ACDA, Navy and CIA, and from my own experience in deep ocean technology, I argued that we had not even begun feasibility research, much less hardware research, on such a system. Even if presently supported, it was a good ten years off, would be costly and without much expectation of support in the present budget squeeze. More important, it appeared to offer few strategic advantages since any fixed site would be far more vulnerable to detection than a mobile Polaris.[66] Our deep ocean technology was far ahead that of the Soviets, and at a time when the cold war had been chilled even further over Vietnam and we were eager to gain support for the NPT from non-nuclear nations who felt they were yielding their options, even small steps such as proposed here in agreements with the Soviets might have virtue in relation to our over-all security. Some highly technical questions remained as to what launching devices were or were not included, what areas away from coastlines would be declared nuclear free, and, finally, how inspection for adherence to treaty obligation might be conducted. Privately, Assistant Secretary of the Navy Robert Frosch confirmed that Navy studies supported my views.

President Johnson was advised of the debate, but following his instructions to defer new issues, a decision was still pending when his administration expired. It is not certain that Johnson would have approved even the generalities of our resolution for referral to the ENDC in view of objections of the Joint Chiefs, had it not been for pressure to respond to the Soviet propaganda ploy, but by no means did referral to ENDC commit the United States to a detailed agreement.

In the meanwhile ENDC was enlarged to twenty-six nations and was renamed the Conference of the Committee on Disarmament. It was making progress of its own and a key session was scheduled for March 1969, just a few short months after Nixon assumed office. This was a hot potato dropped into the hands of his security advisor, Henry Kissinger. Again the issue was prepared for a meeting of the NSC, with Kissinger enjoining his staff to draft a major paper citing options and their implications for the President's consideration.

In February 1969, Kissinger presided over a preparatory session attended by representatives of the interested agencies. The staff work was a model of objective analysis. By then Gerard Smith was heading ACDA and was well briefed by his own staff on the merits of both the substance and the strategy of a positive U.S. position on the seabed disarmament. The confrontation between the Joint Chiefs on one side, ACDA and the Council Secretariat on the other, erupted a second time. Smith was eloquent in his arguments, liberally salted with humor reminiscent of an adroit horse trader. My arguments were along technical lines. The telling question that hung unanswered was: If the prospects for seabed deployment were so promising as to warrant maintaining our options at any cost, why was not a penny of Navy money being invested for necessary research? Kissinger listened but did not reveal his position.

President Nixon had to shoulder the burden of decision, and he would certainly be considering pros and cons of the issue as expounded by Henry Kissinger and based in part on the debate that Smith and I carried at the preparatory session.

On March 18, 1969, President Nixon wrote a letter of instruction to Ambassador Gerard Smith, head of the U.S. delegation, outlining the U.S. position:

First, in order to assure that the seabed, man's latest frontier, remains free from the nuclear arms race, the United States delegation should indicate that the United States is interested in working out an international agreement that would prohibit the emplacement or fixing of nuclear weapons or other weapons of mass destruction on the seabed. . . .

Such an agreement would, like the Antarctic Treaty, and the treaty on outer space which are already in effect, prevent an arms race before it had a chance to start. It would insure that this potentially useful area of the world remained available for peaceful purposes.[67]

Simultaneously, the Soviets submitted to the Committee a draft convention calling for the prohibition of nuclear weapons and military installations of any kind on the ocean floor outside a twelve-mile limit. Clearly with our intensive seabed arrays to listen for submarines, we would have to oppose this extreme measure. On May 22, the United States countered with its own version, focused on fixed-weapon emplacements only.

By September the Soviets had backed down to meet U.S. reservations, and the United States agreed to the twelve-mile limit for the contiguous zone exempt from nuclear prohibitions instead of the three miles we had supported as the outside limit from shore. In October a joint U.S. and U.S.S.R. draft was laid before the disarmament conferees, comprising a draft treaty proposal under which parties would not implant or emplace on the seabed beyond the maximum contiguous zone any objects of mass destruction as well as facilities specifically designed for storing, testing, or using such weapons.

In the fall the United States-Soviet draft was considered by the Conference of the Committee on Disarmament (CCD) the UN Seabed Committee, and the UN General Assembly. Many smaller nations were critical of the two superpowers, fearful of their combined nuclear and technological superiority. While the outpouring of resentment was fired by other frustrations having little to do with the oceans, Sweden's Alva Myrdal led with the curious attack that superpower collaboration was designed to keep smaller nations from progressing in peaceful applications of nuclear technology. Suggested changes proposed by many nations were considered, and the General Assembly, not satisfied with the draft treaty, referred it back to the Disarmament Conference for further work and completion if possible.[68]

The original draft treaty jointly agreed upon by the U.S.S.R. and the United States brought further growls of disapproval from the Disarmament Conference, partly from suspicion of collusion between the two superpowers, partly from a view that the enterprise was much ado about nothing because it was believed that neither power had begun or was about to begin such a strategy and thus was giving up nothing, and finally because most countries lacked inspection capabilities that would enable them to challenge a suspected offender. Consequently, some changes were made in the original draft, especially in Article III to provide that "in the event that consultation and cooperation have not removed doubts and there is serious question concerning the fulfillment of the [treaty] obligation . . . parties to this Treaty may . . . refer the matter to the Security Council."

On December 7, 1970, the UN General Assembly adopted a resolution by a vote of 104 in favor, 2 against and 2 not voting, commending the treaty as drafted by the CCD and requesting that it be opened for signature and ratification at the earliest possible date. On February 11, 1971, the treaty was placed open for signature and was signed by 62 nations including the three depository governments—the United States, United Kingdom and Soviet Union. The treaty was to enter into force following ratification by 22 nations, including the three depository nations (see Appendix 30 for the text of the treaty).[69]

From one of those present at the Washington, D.C., ceremony, the author received a private note saying, "Some of us remember where it all started."

As suggested in chapter 10, it was only a start.

CHAPTER 7

Marine Resources: Government-Industry Synergism

GOVERNMENTAL ROLES

Men have sought fish and extracted salt from the ocean for thousands of years, but only within the decade of the 1960s have three separate developments converged to impel systematic exploration and exploitation of the full potential of marine resources. First, scientific oceanography has generated new knowledge of what is in and under the sea. Second, new technologies have overcome the obstacles of a hostile environment to facilitate access and extraction of resources at a competitive market price. And, third, new demands for every kind of raw material have followed the growth and industrialization of world populations.

In 1966 the United States advertised by legislative mandate its new awareness of the seas' potential. At that time, the government tacitly assumed that exploration and extraction were largely the province of private enterprise. Federal administrators well understood, however, that satisfaction of mixed, overlapping, public and private goals required a delicate blend of mutual involvement by government and industry. In fact, over the past three decades technology has moved these sectors closer together, and the emerging partnership in developing both living and nonliving marine resources has been enforced by several unique conditions that attend virtually all uses of the sea. Even the Marine Sciences Act of 1966 consolidated resource recovery goals with considerations of requisite exploration, technological development, and public/private management, although details were left to the implementing agent—the President of the United States, with advice and assistance from the Council.

While the marine policy was unprecedented, other public policies concerning full employment, economic growth, reductions in trade barriers, encouragement to private investment, and stable and ready access to strategic materials had long provided a backdrop to industrial activities in the marine environment.

In 1966 the marine resource of highest value, commercially extracted world-wide, was seafood. Close behind and growing rapidly were petroleum and natural gas. Salt, magnesium, and bromine were being extracted from the seawater; sulfur, coal, iron, phosphates, sand and gravel, tin, and even diamonds were recovered offshore in shallow water; and prospects were brightening for exploitation of manganese, copper, cobalt, and nickel from the legendary nodules which paved thousands of square miles of the sea floor. For millenia, most of these resources had been inaccessible; fish had been taken largely by small coastal craft and minerals had been recovered only from shallow water by primitive techniques. New technologies extending man's reach and muscle now offered both positive and negative consequences. Oil could be drilled in deeper water, and all physical resources throughout the marine environment appeared to be within reach. Simultaneously, however, unconstrained access to fish stocks and to whales threatened witless depletion or extermination of some species.

An additional economic and legal enigma then surfaced. Whereas resource recovery is most likely to be developed by private-venture capital and entrepreneurship following a traditional profit incentive, the marine resources involved are in the main subject to public ownership and management. Moreover, their development may be motivated jointly by private profit and by traditional public economic goals as well as goals of national security benefiting from prestige. Since our national concepts of sovereignty, economic and legal principles, and cultural and social organization have historically been based on private rather than public property, the new situation demanded clarification of the rules and ultimately a more enlightened relationship between government and industry. The situation is further blurred because 85 percent of the marine territory involved lies beyond jurisdiction of coastal nations. This no-man's land (or everyman's land) is subject to a variety of territorial claims by nation-states, as well as regional and intergovernmental agreements.

The economics of marine exploitation are yet another matter. Fuel and mineral commodities of marine origin must enter markets already fed by terrestrial sources, wherein prices, processing sites, and transportation patterns have been previously established. Exploitation agree-

ments between private interests in one country and governments of another, and the emergence of multinational corporations, add further
complexity. Because of the heavy capital investments involved and the
low mobility of mining, processing, and refining equipment, any disruption of the status quo understandably sends shock waves through
the entire industry. Yet demand for most of the resources offered by the
marine environment is growing faster than new sources can be discovered on land. In some degree all sectors—industry, the U.S. government, other governments, and academicians—recognize the inevitable
fact that an industrialized society will eventually turn to the sea to help
meet its needs. But they hold differing views as to how and when. The
Marine Sciences Act listed as third in its array of policy objectives
"encouragement of private investment enterprise in exploration, technological development, marine commerce and economic utilization of
the resources of the marine environment." But just what this meant in
operational terms had to be deciphered.

Early in the Marine Council's life, the Secretariat endeavored to open
a dialogue with industry in terms of what industry wanted, what it
needed, and what federal encouragement it should have in the broad
perspective of national interest. Since "industry" was such a broad institutional category, we separately examined petro-chemicals, hard minerals, fisheries, marine services, and aerospace.

The initial reaction of the oil interests, who were investing about $200
million annually for offshore surveys and resource development, was to
preserve the status quo. The industry had high confidence that its
previous offshore activities would expand rapidly, protected by depletion
allowances and import quotas. Like most industry, it had traditionally
opposed governmental "tampering" with the private sector. It was eager,
however, for the Navy to unlock some of its classified technology, especially related to submerged oil operations; and it strongly encouraged
pre-investment reconnaissance of our own continental shelves, but only
on a scale so broad that the presence of oil pockets would not be publicly
revealed; this latter intelligence would be developed by its own finer
grain proprietary surveys. The industry urged better weather services
to improve offshore safety and grumbled about the unpredictable schedule of governmental auctions of oil lease sales in the Gulf of Mexico. On
the legal status of deep-sea resources beyond the continental shelf, an
issue over which the industry and government were later to collide, there
was ambivalence. By no means were the industry's views monolithic;
small independents differed from the majors, and those in offshore or
foreign development differed from others confined to U.S. land sources
and markets.

Mineral interests were more conservative in estimating the prospects
for marine sources and were similarly opposed to governmental genera-

tion of requisite technology. They objected to being bound by the practices of offshore oil leasing, citing the slower exploitation of minerals and the low mobility of mining equipment. Moreover, they were to differ with the oil industry on a preferred international legal regime.

The fisheries sector articulated highly differentiated views depending on whether their quest, for example, was for tuna, shrimp, Northeast Atlantic haddock, Chesapeake Bay oysters, or Pacific salmon. Except for tuna and shrimp, most harvesting sectors of the industry, in contrast to the marketing sectors, were in dire straits and tended to seek short-term remedies. All wanted a stronger program of research, yet complained that the Bureau of Commercial Fisheries, established for that explicit purpose, had been too academically oriented and did not relate to problems confronting the industry. Most of all, they wanted protection both from seizure of our boats in Latin American waters and from "invasion" of Japanese and Soviet craft into our own coastal waters.

The aerospace, marine services, and instrumentation industries' approach was simpler. They had long been used to a symbiotic détente with government in applying skills of private technological management to achieve public goals. Historically, these goals had been oriented toward national security (the Navy was a familiar customer); and the instruments of cooperation had been federal contracts for research, development, and procurement, with government–industry partnership resembling administrative agreements rather than buyer–seller relationships in a free market.[1] Since they had been supplying means that met someone else's requirements, they had accumulated limited experience in dealing with "ends," especially in the civilian sector. So the oceans market involved two sets of unfamiliar products—tools for scientists who wanted better information about the marine environment, and equipment for commercial resource extractors who sought to use that environment. The aerospace interests primarily hoped to apply their expertise in conventional modes, through larger federal budgets for marine components to meet civilian needs.

Around the periphery of these highly involved sectors were the banking and investment interests. They, too, reflected a dichotomy of attitudes—some were conservative, others highly adventurous. Some promoters of new enterprises were starry-eyed about the appeal of exotic marine development for the bored investor still looking for pathways to instant wealth. One investment house, having opened seminars on marine opportunities,[2] later nervously backed away from its early optimism.

All sectors of industry complained about the multitude of federal agencies, but some who had good communication to bureau chiefs were unwilling to advocate reorganization that might threaten their channels or displace government personnel familiar with their problems.

Faced with this conglomeration of private activities, the Council first sought to clarify the complementary roles of industry and government. A major study was sponsored with the National Planning Association to catalogue different instruments previously employed to meet public and private goals concurrently, as a reference framework of alternative measures by which government might stimulate private enterprise to invest in marine resource developments. The report listed fifty-seven such measures, arranged in five broad categories: government sponsorship of research and development; financial measures; legal, regulatory, and administrative measures; social overheads and other social measures; and international measures. Included were certain devices such as taxes, antitrust regulations, and production allowables or quotas, which the business community may have considered discouraging rather than stimulating to accelerated investment, although NPA felt the proposals would ameliorate certain of these practices and improve the overall investment climate. The list included:

Government Sponsorship of Research and Development: 1. Exploration of potential resources. 2. Resource delineation, including *in situ* characteristics. 3. Technology of resource extraction (e.g., mining). 4. Resource processing technology. 5. Weather research and development. 6. Oceanographic, geological, or geophysical data. 7. Communications technology. 8. Technology of navigation and location fixes. 9. Transportation technology. 10. Sensors and instruments of measurement. 11. Materials research. 12. Deep submergence systems (rescue and salvage, bottom work stations). 13. Man-in-the-Sea research. 14. Marine biology and ecological research (food, drugs, control of fouling organisms, water pollution). 15. Other scientific and engineering research.

Financial Measures: 1. Sharing of research or capital facilities costs. 2. Sharing of government research facilities. 3. Direct subsidies of production costs. 4. Government stockpiling of strategic materials. 5. Government purchases of other goods and services (procurement and RDT&E). 6. Special encouragement programs, such as small business loans, land grants, foreign aid, loan guarantees, and reinsurance. 7. Taxes (depletion allowances, fast tax writeoffs, tax abatement). 8. Tariffs and import quotas. 9. Retirement of inefficient units from production.

Legal, Regulatory, and Administrative Measures: 1. Laws of ownership and property rights. 2. Leasing: agreements, regulations, and administration of lease sales. 3. Patent system. 4. Antitrust laws and administration. 5. Public utilities. 6. Quasi-public corporations (e.g., COMSAT). 7. Production allowables or quotas. 8. Forecasts, special economic and planning studies, budgetary proposals. 9. Setting of national and international standards on equipment design or component design.

Social Overheads and Other Social Measures: 1. Development of human resources through education assistance programs such as the Sea-Grant College Program, research grants to universities, contracted university research. 2. Provision of statistical and other information services. 3. Ports, harbors, waterways, navigation aids. 4. Related transportation systems (rail, highway, air). 5. Rescue service (e.g., Coast Guard, Navy, Air Force). 6. Police protection and law enforcement system. 7. Recreation preserves and facilities. 8. Pollution control systems (water, land, and air). 9. Tsunami and earthquake warning systems. 10. Shark-repellent measures or other marine life repellent measures. 11. Safety standards and inspection of equipment and facilities. 12. Weather services: reporting of weather data, monitoring hurricanes and other potentially severe storms, weather forecasting, and weather control. 13. Mapping services (pilot charts, bathymetric, current data). 14. Catastrophe assistance. 15. Public health services. 16. War on poverty, depressed area programs, urban redevelopment programs, and so forth, in maritime regions. 17. Providing for land and space requirements for economic and social activities.

International Measures: 1. Projects of international cooperation in gathering oceanic data, weather data, geophysical and geological data. 2. International cooperation in certain government-sponsored R&D items. 3. Cooperative binational ventures in delineating and exploiting ocean resources. 4. Help in developing foreign markets for equipment manufacturers in the United States for overseas marine applications. 5. Assistance to U.S. business firms to develop overseas ocean resources to serve foreign or U.S. markets. 6. Ports and harbors. 7. Law of the sea, national jurisdiction issues, regulations in use of ocean resources.[3]

At the same time, we were mindful that the Commission on Marine Science, Engineering and Resources (COMSER), which enjoyed the benefit of heavy representation from the private sector, had been charged to recommend "an adequate national marine science program that will meet the present and future national needs. . . ." In going beyond merely a federal program, it was obliged to examine the goals and activities of the private sector in intimate detail; one of the panels had been explicitly created to deal with "Industry and Private Investment."[4]

The Council thus chose to defer over-all policy consideration of resource development and of public/private collaboration until the complex array of issues and relationships was better understood. Nevertheless, as public goals were considered—international tranquility, economic growth, improved social benefits domestically, and even such international humanitarian concerns as the war on hunger—it was clear that none was attainable without major private participation. Within a few months we were to add environmental protection to this list.

Initiatives adopted by the Council in the fall of 1966 (described in chapter 3) thus carried significant, albeit indirect, implications for the private sector. Item 1 on international cooperation stressed peaceful uses of the oceans; beyond pure altruism it was based on the recognition, already underscored by costs of our involvement in Vietnam, that relaxation of world tension would potentially release added funds for those marine-related projects that had long waited for budget support. Item 2 on food from the sea to meet protein malnutrition abroad anticipated that federal seed capital would aid private venture in developing both the technology and market intelligence for this humanitarian project. Item 3 on Sea Grant emphasized the federal training of specialized manpower and held that industry would benefit both directly and indirectly from improved dissemination of scientific information and from collaborative research between universities and local marine industry. Item 4 on data systems was intended to improve delivery of environmental information to all customers, industry included, by application of modern systems techniques for the collection, storage, and retrieval of vast quantities of raw observations.[5] Items 6 and 7 concerned mineral surveys of the continental shelf to identify potential new sources, and the strengthening of ocean-based observation networks to improve prediction of near-shore weather, severe storms, and ocean conditions—matters of key importance to the offshore industries. Item 8 concerned expansion of the Navy's capability for recovery of lost equipment and for deep-ocean engineering where industry would be directly involved as contractors or would indirectly benefit from research fall-out. None of these proposals was designed explicitly to aid industry in the direct fashion they desired, but their conspicuous potential for serendipity reinforced our initial perception that public and private activities must be separate but closely related.

The legislation stated that the first obvious requisite of the federal role was leadership, since it appeared that no ocean activity could be advanced by traditional private initiative alone. However, to allay industry's fear that government leadership might usurp the private sector, we felt obliged to spell out limits of federal involvement. In the first annual report in 1967, the Council delineated the federal role to:

—sustain a naval science and technology to meet national security needs;
—support foreign policy objectives by fostering international agreements, understanding and cooperation, and by supporting technical assistance to developing nations on the principle of self-help;
—enhance capabilities for describing and predicting the state of the oceans and the weather and provide services to marine interests;
—explore and foster exploitation of fish, minerals, and energy resources by mapping, by appropriate development of technological capabilities, by formulation of means for public and private collaboration, and by encouragement of private investment;
—aid abatement and prevention of pollution and assist in the conservation

and improved utilization of recreational, esthetic, and economic resources of our sea coast and Great Lakes;
—protect life and property at sea, and along the coast;
—nourish basic knowledge and develop scientific facilities and manpower.

The report dutifully added: "At the same time, the Government recognizes that the vitality of our industrial organizations and the creativity of our scientists have been major factors in our progress toward better understanding and use of the seas. Indeed, the bulk of Federal funds devoted to marine efforts has been expended through grants and contracts with private industrial and academic organizations."[6]

In 1969 the Stratton Commission outlined its view of federal participation, adding to the Council's list requirements to: (1) assist in planning for optimum use of limited public resources including the resolution of conflicts among users of the sea which cannot otherwise be adjudicated; (2) adopt regulatory policies which will not discourage private investment; (3) negotiate acceptable international arrangements to conduct marine industrial and scientific activity, to conserve marine resources, and to prevent pollution of the seas.[7]

The availability of venture capital was seen as a major criterion of future development. On the one hand was a growing market demand for marine resources, but on the other was a series of unknowns. As against competing land deposits the marine resources were of indeterminate richness and distribution, the costs and the technology of extraction were uncertain, and the legal safeguards of investment were shadowy. Thus the most useful federal role appeared to be the reduction of unknowns and of risks. The Council publicly inquired: "How should marine technology required for industrial development be supported when advances of the activity are in the public interest but the returns on investment are too long deferred or otherwise less attractive than alternate ventures available to private capital?"[8]

The question elicited few answers. Even industries affected volunteered few suggestions. This issue was of such concern to the Council that in its second annual report it devoted an entire chapter to discussing "Partnership with Non-Federal Institutions." By the time the third report was completed, openly entitled "A Year of Broadened Participation," the Council had completed studies, taken initiatives, received rebuffs, encountered frustrations, and concluded with a more explicit definition of government's role, at least regarding minerals, to:

(1) Insure an adequate, dependable, diverse supply of raw materials . . . ;
(2) Have available mineral supplies at lowest cost consistent with the satisfaction of other national objectives;
(3) Maintain a sufficient resource base for national security;
(4) Conserve the Nation's mineral resources by using them wisely and efficiently;

(5) Preserve the quality of the environment—air, water, and land—while obtaining the needed mineral resources;

(6) Maintain safe and healthful working conditions . . . ;

(7) Manage the mineral resources of Federal lands in accordance with sound business principles;

(8) Manage U.S. mineral resources so as to assist in maintaining a favorable balance of payments;

(9) Provide supporting services such as mapping and charting, weather and sea forecasting, aids to navigation, ice breaking and channel maintenance, and rescue-at-sea activities; and

(10) Provide a climate for industry to produce efficiently under competitive conditions the minerals required for the domestic economy and foreign trade.[9]

As its main theme of support the Council called for a "balanced, civilian marine technology as a specific assistance to an emerging marine industry . . . in those research and development areas where investment involved a high initial cost or long deferred return in investments,"[10] a conclusion based on the wide experience that innovative research and engineering has the capacity not only to improve products and to spawn new ones, but also to generate entirely new markets and industries. This generalized view was later reinforced by the Commission. One question ducked by the Council but addressed by the Commission concerned the vehicle of that governmental support. Their answer was NOAA. But as to policies, criteria of support and modes of subvention, the Commission report was silent.

Throughout its life, the Council recognized that effective utilization of the sea would heavily involve the private sector. Moreover, the government in the public interest might have to become a full, risk-sharing partner. Toward this end the Council Chairman (representing the full herd of federal agencies) and the staff were uniformly sympathetic. No philosophical polarization dictated that government and business should eye each other as perpetual adversaries. But it was also evident that general abstractions were worthless; each sector would have to be considered separately and in substantive detail.

FOOD FROM THE SEA

The industrial maritime sector consisted of four elements: the mercantile shipping interests, the vendors to the Navy (increasingly underemployed), the fuel/mineral industries, and the fishermen. Since the last group was the most continuously articulate in protesting past failure of governmental policy initiatives, their plight led Congress to highlight "Rehabilitation of our Commercial Fisheries" in the declaration of policy of the 1966 Act. Fishing and whaling had been key industries at the founding of the Union, and men of the sea had been actively represented in the Congress ever since. Despite a fast return of the fishermen to their nets after World War II and a sharp upturn in the catch to prewar levels, the domestic fishing industry was in trouble. Congress had for twenty years engaged in studies of their problems and in palliative measures, but had failed to arrest further decline. The Act now gave the President a chance to remodel the entire system.

Unfamiliar with the symptoms of the problem, much less its causes and possible remedies, the Council Secretariat first set out to educate itself. To bolster staff expertise, attempts were made, without success, to recruit such authorities as Wilbert M. Chapman, a fisheries biologist of enormous physical stature equally matched by intellect, experience, and boldness, or Dayton Lee Alverson, one of the Bureau of Commercial Fisheries' most perceptive and talented scientists. While we later filled that slot with competent staff, we also learned to lean heavily on distinguished consultants from both industry and academia—Chapman, Francis T. Christy, Jr., William C. Herrington, Jake Dykstra, and James A. Crutchfield—distinguished scholars in resource economics and leaders in the industry itself. Utterly candid in outlining the predicament of a sick industry, they contributed briefings and position papers to supplement BCF studies and enlighten the Council. With generalizations hazardous because of fragmentation of the harvesting sector of the industry, we learned that the situation was roughly this: world markets for fishing products were expanding; the U.S. market was similarly expanding; the U.S. fish production after a sharp postwar revival had remained almost static while others surged ahead; some fishery stocks were endangered by overfishing; the U.S. had neglected to act in the alleviation of protein malnutrition with fish.

In 1956 the United States had been second in production only to Japan, but by 1968 our position had dropped to sixth and volume had fallen to 4.1 billion pounds, down from a 1962 peak of 5.4 billion pounds. Our fraction of 13.3 percent of total world catch in 1950 was down to a minuscule 4 percent, and represented about 5 percent of the $9.1 billion value of the 1968 world catch; yet some of the most productive fishing grounds in the world were near our own shores. On the other hand, U.S. consumption of fishery products had grown at a faster rate (7.1 percent annually) than that of any other country, some being used for pet food and protein supplement in animal feed. To meet this demand, U.S. imports had skyrocketed from 33.2 percent in 1957 to 76 percent in 1968, valued at $823 million. Some of the resulting balance-of-payment deficits had, however, been offset by repatriation of profits from U.S. business firms operating from overseas bases.

Foreign fleets tended to be newer, better equipped, and generally more innovative in technology. The Soviet Union had dramatically advanced in both size and catch efficiency of its fishery fleet, the result of a capital investment of over $4 billion since World War II. Considering that their coastline was limited and their yet-inefficient agriculture did not meet the domestic imperative for protein, it was not surprising that they led in distant-water fisheries.

That the number of U.S. fishermen and shore workers, as well as the number of fishing craft, had declined in the 1950s and 1960s was paradoxically discovered to carry favorable as well as unfavorable implications, in that inefficient units were dropping out. Nevertheless, the productivity of U.S. fishermen in terms of volume per capita had increased only about 1 percent annually, substantially less than the almost 2½ percent of other sectors of our economy, excluding agriculture, which was growing at 5 percent. Finally, except for the tuna and shrimp fisheries and processors of imported frozen chunks who were basking in relative prosperity, most of our fisheries were economically depressed, with fishermen receiving far less than the national average income.

What were the causes? Our consultants listed many: unlimited access to common property resources; archaic institutional constraints and legal measures ostensibly for conservation but in reality handicaps to technological innovation to protect the status quo; fragmentation of jurisdiction at state, federal, and international levels; lethargy and force of tradition among entrepreneurs; lack of vertical economic integration of the industry (a discontinuity between fish catching and marketing); vigorous international competition coupled with lower foreign wage scales and lack of U.S. tariffs on imports; inefficient practices with superabundance of small, obsolete fishing units, overcapitalization, and overfishing; entrance of new countries into the fish business (40 of them nonexistent when the 1958 Fisheries Convention was negotiated); low rate of return on capital investment resulting in slow capital formation; prohibition of purchase of cheaper fishing vessels built abroad if operated out of U.S. ports (based on a 1793 law).[11]

So much for the trip hammer of serious economic, legal, institutional, and social factors in fishery management. The next questions concerned the biological factors that underlie the entire enterprise—such matters as ecological influences on the biomass; density and distribution of species; population dynamics; fish predators, parasites, and disease; artificial genetic improvement; effects of pollution; competition of sport fishing with commercial fishing; overfishing. Despite years of energetic research, estimates of total biomass available for annual harvesting without depletion ranged wildly from two to forty times the present level.[12] Equally uncertain was forecasting of abundance of both exploited and unexploited stocks in determining the relationship between spawning stocks and their subsequent progeny, and defining the relationship between fishing effort and yield.[13] No wonder arguments blazed on fishery stock management, on how to extract the maximum yield without stock depletion. Although there had been some technological advances on fish location and harvesting (the latter including fish attractants and new techniques of straining or trapping with nets), with few exceptions fishing was still a primitive food-gathering activity based on hunting for the prey. And of 40,000 species of fish, only about 150 were commercially harvested, with 55 of these accounting for 97 percent of the catch.[14] Most were found in coastal rather than in deep water.

Amid the great debate on optimizing economic versus resource yield, decisions had to be made regarding fishery management. Lacking sound information as a bulwark, decision-makers found themselves buffeted by a mixture of rational pressure and emotion from apprehensive vested interests. Who was making these decisions? And on what basis? Domestically, the burden of responsibility for conservation and management of living resources was clearly that of individual states out to the three-mile limit; but some states had de facto jurisdiction over all fishing units operating out of their ports, even to deep-sea operators. Thus, in the North Pacific, fleets from California, Oregon, Washington, and Alaska operated under highly disparate rules. Limitations on bait, fishing gear, boat sizes, and duration of season, as well as on species harvested and catch limit, were developed state by state, ostensibly as conservation measures. Closer examination revealed that the great majority of "scientifically based" regulations promulgated on the pretext of conserving stocks were in fact politically expedient devices to allocate the resource among competing fishermen, simply to reduce political pressures and resolve economic conflicts. As Crutchfield has said in remarking on the one-man gill-net boats catching salmon in the Northwest, "The vessels would be larger and the industry more capital intensive were it not for the quaint custom of reducing excessive fishing pressure by eliminating the more efficient methods of catching salmon," i.e., with large ships concentrating on schools at river mouths. It was shocking to discover that modern methods had been excluded more to protect inefficient traditional fishermen than to conserve stocks.

To be sure, instances of overfishing by U.S. fishermen of coastal stocks had been documented as caused by too rapid extraction or poor timing that reduced parental stock. But even in such cases, the methods of protection adopted at one time or another frequently failed. Internationally, fishing nations had been mindful for decades of the need to manage stocks jointly and had devised various arrangements for this purpose. Multilateral agreements were policed by each participating nation or a multinational commission with its own extra-national technical and administrative staff and responsibilities, but the tradition of adopting policy or making decisions by unanimous agreement had led to debilitating compromises. Moreover, conservation measures adopted were seldom monitored. It was not surprising to find some fisheries subject to participation and constraints of three or more individual states, three or more nations, and several international bodies. Paradoxically, coastal nation preferences provided for in the 1958 Convention on Living Resources had never been exercised, so abuses had gone uncorrected even when unilateral policing by a coastal nation would have been possible. Meanwhile, developing nations had begun worrying about whether resources to which they were "entitled" were being denied them as a result of grandfather clauses that protected the rights of those who fished first and who now had more advanced technical information and capabilities.

From this maze of predicaments swirled not only confusion but also window-rattling lament. U.S. coastal fishermen in fifty-foot craft were frequently within earshot (if not gunshot) of large, handsome foreign vessels, especially Soviet and Japanese. The charge: invasion of U.S. territory, followed by pressure on Congress and the State Department to force the rascals out. The invaders claimed to be in international waters and not fishing the same stocks as immobilized coastal fishermen, but lacking close inspection until 1971 no one believed them. More important, no one had the courage to tell the American fishermen that their economic misery was partially self-generated. Those fisheries experts who had vision enough to understand the problem, those knowledgeable bureaucrats in federal and state governments who hesitated to express unpalatable opinions to their clientele, those politicians who did not want to risk the wrath of voters, all had stood silently aside watching the confrontation. And at the polls each fishing unit equals one vote. The domestic response was calculated to extend the old patterns of too many small boats manned by too many harvesters of fish.

It now became obvious to the Council that more uniform and rational federal regulations were essential to foster both conservation and economic vitality of the fisheries, to transform fragmented state jurisdictions that had been perpetuated by the pressures of vested interests. Moreover, wider over-all policy was necessitated by the biological facts of fish migration. The Truman Doctrine had already extended resource management beyond the three-mile limit for fisheries as well as for seabed minerals, and the Fish and Wildlife Act of 1956 gave the Secretary of Interior authority to manage high seas fisheries, but this authority had been exercised primarily in international negotiations. In domestic concerns the federal government had defaulted.

Since creation of the Fish Commission in 1871, federal intervention of one sort or another had evolved as piecemeal, spasmodic responses to symptoms. No one had endeavored to diagnose, much less to treat, the malignancy. Even when policy had been coordinated at the federal level for purposes of international negotiation, the national interests in the sea had proved to have many dimensions, not usually compatible. First and foremost was the Navy's claim to freedom of the seas, already threatened by unilateral extension of sovereignty. Next, national policy was to bail out U.S. tuna vessels impounded by Latin Americans rather than to have our fishermen purchase licenses and tacitly sanction the two-hundred-mile claim. More would be involved in such creeping jurisdiction than fishing rights alone: a coastal nation acceded the right to license the fishing boats of other nations within some seaward extension of its national boundaries might eventually claim the same sovereignty over innocent passage through critical narrow straits, or control over scientific research and of ocean pollution. Thus, our fishery negotiations had always been constrained by subliminal considerations.

Another discovery by the Council was that world-wide competition coupled with unlimited entry had resulted in an enormous overcapitalization in proportion to fishing targets, so that the problem extended beyond our own shores. A breakthrough to obtain information and develop markets for uncaught species that would keep vessels profitably employed more of the year might alleviate the problem, but could still not help our existing fleet, hampered by its limited range.

As the Secretariat began to consider policy initiatives, it became clear that our straitjacket might be loosened by a national subsidy to replace the overabundance of small ships with fewer but larger and more modern ones, thus offering greater potential return on investment. Individual fishermen, however, are ambivalent about this step toward improved profits; they treasure the freedom to work for themselves even in submarginal enterpreneurial units, and those concerned about being squeezed out as surplus were not receptive to the idea of adapting to other occupations. Another roadblock developed: the Budget Bureau steadfastly opposed subsidy, contending that common fishery resources with unlimited entry should continue to be harvested by whatever nation could do it most efficiently. This argument for world-wide economic optimization brushed aside other considerations of national interest, including the need to buttress our position at the international conference table in the interest of responsible fisheries management. In fact, it may have even contributed to confusion and did not induce rational development of resources on a world-wide basis.

Another alternative would be to limit entries. This would bring legitimate complaints of lower returns on the industry's capital investment unless a means were found for retiring older craft. When the "tragedy of the commons"[15] became evident in the case of whaling, the response by the nations involved was not voluntary restraint but even further capitalization to grab a larger slice of a shrinking pie. The demise of the blue whale is a monument to man's greed.

One solution that suggested itself spontaneously was to consider each fishery as a unified entity, to be managed by one responsible party, under rules of limited entry or a quota system based on some consistent, albeit incomplete, set of fishery facts. To the fishing industry, this and other institutionally radical alternatives, such as federal rather than state regulation of fisheries, were pure heresy.

Given this baffling array of dilemmas, the preferred government solution had been to strengthen our competitive position mainly by support of fisheries research via the social-overhead route. A family of laboratories had been organized around the coasts, each oriented to study the problems of its local sector of the industry but all, most unfortunately, subcritical in size. BCF programs embraced resource development and management programs (62 percent), processing and marketing (13 percent), advanced technology of fishing gear (8 percent), economic research (2 percent), and international activities.[16]

The Bureau of Commercial Fisheries, organized in 1956, increased their budgets from $9.5 million to $38.8 million in 1966, and were engaged in a host of activities. But the economic facts established that fifteen years of accelerated fishery research that provided information on resource management had not paid dividends in arresting decline of the industry.

No quick pragmatic remedy could be expected for these enormously complex ills. Rehabilitation of the domestic fishery would require careful therapy and a long convalescence. As a starter, the Council gave strong encouragement to the Bureau of Commercial Fisheries to expand their range of studies to encompass broad policy and ways to improve fishing productivity, and to kick the habit of reacting to local pressures by providing individual palliatives. BCF management welcomed this thrust but seemed constrained by internal handicaps, frowns from the Budget Bureau, and the merely casual interest of their Secretary. We hoped that the Commission by sheer competence and prestige could break down the reluctance of the patient, but we did not wait for their prescription.

Cooperative programs by government, industry, and the scientific community could significantly increase the proportion of fishery products supplied to the U.S. market by the domestic fleet, simultaneously benefiting the consumer, the fisherman, and the general economy. Also, our position to negotiate international agreements could be greatly strengthened through a more potent U.S. presence in fishing grounds. The Council tried to spotlight the situation by unusually frank discussion in the Council reports of problems, if not of remedies. Each report peeled back layers of misinformation, but we realized they were probably not read at the grassroots where policy was politically influenced. The diagnosis set forth in the 1969 report, after citing the superannuated condition of the U.S. fishing fleet, pointed out some of the "interrelated factors" contributing to the problems. They included the tangled regulations of multiple federal, state, and local authorities that increased costs and reduced efficiency; the overabundance of fishermen organized in small entrepreneurial units, operating in suboptimal craft; the high cost of building vessels in U.S. shipyards and high insurance rates; the separation of harvesting and marketing interests; and inadequate biological and technological data. It also pointed unequivocally to the alarming symptoms: an absolute reduction in the average output per U.S. fisherman in the same period (1957–67) within which the average output per worker in the general U.S. economy had risen over 30 percent and in the agricultural sector had soared by 67 percent.[17]

Numerous studies were conducted within and outside the government to seek ways to stimulate the harvesting sector. Of particular importance was a March 1968 conference of 266 representatives of all parts of the industry at the University of Washington in Seattle, partially supported by the Council. The conference underscored the need for the federal government to re-examine fishing restrictions that reduce efficiency of the fisherman, to consider limiting the number of fishermen exploiting stocks already being fully harvested, to improve statistics on fisheries, to strengthen U.S. policies concerning fishery management, and to take into account food shortages on a world-wide scale in promoting development of the domestic industry.[18] Private industry was urged to develop and expand the seafood market and to provide better education and training for its members. Unfortunately, industry ignored its own leaders.

By 1968, BCF Director Harold E. Crowther found strong support from Council staff and consultants for proposals to reorient BCF management and programs. These were instantly quashed by the Budget Bureau in the winter of 1968, rationalizing their opposition by a contention that it would be foolish to build up an industry they felt could never compete. They also attacked BCF management as incompetent while denying them the opportunity to cast off any past inadequacies. Council staff argued that the game was worth the candle if only because of our balance-of-payments deficit and our need to sustain a maritime presence to bolster influence in the world community. In addition, we hoped to pique the President's interest in aiding the deprived sectors of our people, which surely included many fisherman. If BCF needed reorganization, which that agency itself had under voluntary study, it would find incentive in a policy-level budget push. We lost, partly because in the final crunch we were unsupported by Secretary of Interior Udall, who had too many other fish to fry in too small a budget pan.

In the meanwhile, a perceptive study had been conducted by Lee Alverson, temporarily brought by BCF to Washington, D.C. His study was not completed before the change in administration; when that occurred, the Democratic head of BCF was abruptly shunted aside and

Alverson, whose partisan lineage was inconspicuous, was offered the position as chief. By then, however, illness in the family required that he return to Seattle. But he had learned that his report was to be impounded by the new officers of the Interior Department (even the one copy released to the Council Secretariat was withdrawn), a casualty of the reductions Nixon was relentlessly extracting from the Johnson budget. Despite their lip service to fisheries, the new Secretary of Interior and his immediate lieutenants, apparently under instruction from the Budget Bureau and unwilling to fight, refused to support any expanded program of national fisheries. Alverson has in fact since told me that regardless of family problems he would have declined because the Department of Interior was not inclined to resolve fishery problems.

At the end of the Johnson administration, the Council, closer to prescription for cure, urged that:

1. Steps should be taken to:
   —provide uniform Federal guidelines for fishery conservation and management that would eliminate unnecessary restrictions and inconsistencies and foster selective revisions of State and local laws;
   —encourage the development of competitive ships and equipment in the United States by revising the fishing vessel subsidy program; and
   —encourage increased opportunities for the sale of the catch of U.S. fishermen and for vertical integration of the U.S. industry.
2. The Bureau of Commercial Fisheries should give increased attention to:
   —estuarine ecology, aquaculture, the technology of extracting concentrated fish proteins, expanded knowledge of fish populations on the Continental Shelf, and development of fish locating and catching equipment and methodology;
   —an improved Federal program of statistics, data processing, and information dissemination to the fisherman through extension programs; and
   —assistance to industry to produce clean, safe, and wholesome fishing products with a level of graded products comparable to those in meats and poultry.[19]

The Council Secretariat was also impressed through its survey of world-wide aquaculture activities that "farming of the sea" could make significant contributions to the domestic economy as well as meet protein malnutrition, and hoped its 1967 contract study by Ryther and Bardach would excite private as well as BCF interest.

Of special import, the Bureau of Commercial Fisheries agreed to review the federal, state, and local laws that by restricting use of efficient gear fostered economically irrational conservation measures, and to recommend modification to permit adoption of technological innovations leading to higher economic yields. The gut question of putting more powerful technological tools in the hands of American industry had not been answered. The Council, through its initial contract studies, had opened the issue in the hope that a marriage might be consummated between the innovative high-technology sector and the retarded fish-harvesting sector, but the matchmaking proved ineffective.[20]

In January 1969, the Stratton Commission released their report with essentially the same diagnosis as the Council. Their recommendations, though softly worded, would have given guidance to the renewed quest for a fishery policy.

In 1969, the Council Secretariat seized on these Commission proposals as priority targets for study by the incoming administration. By the fall of 1969 our attempt to strengthen fisheries activities via the presidential initiative route (productive of the five October 1969 actions described on page 165) had failed, largely due to violent Budget Bureau objections in the White House task force and in the absence of any support from the Vice President or OST. Unfortunately, an article in *Science* pessimistic about future growth in yield was publicized in Washington and used indiscriminantly as a club against more support.[21] By the time the article was refuted by other fisheries specialists, much damage had been done. Budget Bureau representatives later remarked about their surprise that the Council's Secretary was so tenacious, because all the other initiatives were threatened by this one. But I firmly believed a strengthened fishing industry, while contributing to economic development especially in some coastal communities, could also improve our balance of trade, broaden our stature as a marine power, and enhance U.S. capability to assist the world community in optimizing fish production. The industry was clearly not about to help itself; and the government still had a hangup on fishery policy.

From 1966 to 1971, two Presidents took note of fisheries interests in some eleven messages to Congress or in statements ranging in subject from fish protein concentrate and international conservation to improved purity of fish products.[22] Except for Johnson's support of fish protein initiatives, none reflected any significant change in policy; and except for sporadic outcries against foreign fishermen close to U.S. shores, Congress was equally ineffective. In this single area the Council most conspicuously failed to ignite the presidential fuse.

Nevertheless, the Council was not reluctant to start a battle on another front—a program for development of fish protein concentrate to aid in combating malnutrition.

THE SAGA OF FISH PROTEIN CONCENTRATE

In September 1966, the Council had extracted nominations for action regarding fisheries from the report of the President's Science Advisory Committee and from the agencies, but no new approaches were suggested. Crowther was brand new as McKernan's successor as chief of BCF, and his shakedown cruise had only begun. Study of the complexity of fishery ailments had quickly ended the Secretariat's age of innocence and left us with a melancholy recognition that very little of merit could be mounted, much less formulated, in the first cycle of presidential initiatives to aid rehabilitation of the industry.

We tried another route. The program on the sea's potential to meet protein malnutrition had already been examined by PSAC and had received pats on the head by Congress. The President himself had stated, "Next to the pursuit of peace, the greatest challenge to the human family is the race between food supply and population increase. That race is now being lost."[23] Two-thirds of the grain, milk, and meat sources of protein were consumed by one-quarter of the earth's two billion peoples. Three-quarters of that population derived half their protein from fish. The sea offered an attractive source for meeting the unsatisfied needs.

In a number of nations of Asia, Africa, the Middle East, and Latin America, citizens of all ages—but especially children—suffer from malnutrition resulting usually from protein deficiency. The outcome is not only personal misery; it is now well documented that children who sustain severe protein shortages through the age of five often suffer permanent disability.[24] Shadow photographs taken with high intensity illumination dramatically reveal that the brain of such an unfortunate never develops sufficiently to fill the skull. If diet is supplemented at an earlier age the child will catch up; but not after five years. Often in sight of their homes are coastal waters rich in fish, some not caught at all, some harvested by their own fathers for export, some by other nations. That too little of the harvest finds its way into indigenous diets is explicit in the adage of the fishermen's hungry children. The paradox is that most of these fishermen today fish for money, not food.

This cruel pattern as well as the tragic and intractable disparity between world food supply and expanding population provided a natural target when, during its formative months, the Council endeavored to respond to President Johnson's request for recommendations to be fitted into his immediate program.

Fish are largely protein. Moreover, industrial processes had long been used to extract protein from low-cost fish for animal feed, and at least one visionary entrepreneur, Ezra Levin, had endeavored since 1950 to refine a fish protein concentrate (FPC) for human consumption. The BCF had initiated process research in 1962, and in 1966 had persuaded Congress to authorize pilot plants of sufficient capacity to debug production problems. The dried powdered protein is manufactured by grinding up whole fish and extracting fats and water with a solvent such as isopropyl alcohol. The resultant flour has some instantly attractive properties. Nutritionally balanced[25] and easily digestible, FPC was also bacteriologically and biochemically safe with proper handling; was relatively stable without refrigeration or other special processing; was readily incorporated as an enrichment agent at a 5–10 percent level in cereal products, soups, opaque drinks; and it had no detectable "fishy" flavor or odor. Ten grams per day met minimum daily protein requirements of a child. Finally, it was readily available and its eventual cost of 20–25 cents per pound projected in 1966 would be competitive with other protein sources such as dried nonfat milk, soy or peanut flour.

The problems? Cost in 1966 was still too high for widespread use in diets of the economically depressed; only less fatty species of fish could be utilized at that time; industrial scale plants had not been designed or tested at pilot scale; marketing methods were lacking, since an unfamiliar additive was not likely to be used spontaneously in indigenous diets. Equally serious, the Food and Drug Administration had initially barred fish protein concentrate from the market on grounds that it was "filthy" because it originated from whole, unprocessed fish. When this objection was overcome, FDA raised a question of its health hazard because of the fluoride content present from powdered fish bones. Until FDA changed its rules, marketing here or abroad was unthinkable, and the U.S. food processing industry was understandably bearish.

In September 1966 the Council decided that world-wide malnutrition warranted priority status for an initiative regarding fish concentrate in recommendations to be made to the President. The Secretariat then undertook two parallel exercises: to develop a viable FPC program focused on world hunger, and to remove the FDA albatross. On the latter, BCF with Council staff support arranged a crash set of testing and human-feeding experiments with advice from the Marine Protein Resource Development Committee of the National Academy of Sciences. Humphrey exerted gentle but steady pressure on HEW Secretary John Gardner, and the combined efforts precipitated favorable FDA action February 2, 1967. In that process, we were to learn of the backdoor influence of the dairy industry, fearful that their market for dried milk solids was threatened by FPC.[26] The FDA action, however, had subtle reservations that were later to prove a continuing if not fatal impediment to FPC development. They prohibited its use in formulated foods, the only reasonable market for the product; permitted sale only in one-pound consumer packages (as if consumers were to sprinkle FPC on breakfast cereal); limited the fluoride content, although relaxing this later; and limited use to one species of fish (also later expanded). Since many other processed foods that listed a full chemical factory on the wrapper had FDA approval, such quibbling was a nonsensical obstacle to domestic production on a commercial basis.

The second thrust was development of a credible demonstration project. Following a crash study for the Council by Donald F. McKernan with his former BCF colleagues, then conceptual elaboration by Council consultants, the program was taken over and structured by an interagency task group ramrodded by Henry Arnold of our staff.

The project was explained as follows:

On a cooperative bilateral basis, it is proposed to select and survey the fishing potential and market feasibility of FPC products in three less developed countries; one of these countries will be selected as the place in which to foster development of a local capability to produce and distribute FPC. The specific objective would be to demonstrate that—consistent with local needs, fish supplies, people, and customs—it is feasible to meet animal protein needs of a large number of pre-school children and pregnant mothers promptly and economically. An initial goal for a small country would be to provide by 1971, ten grams of animal protein daily to each of one million people. This can be expanded as experience or circumstances dictate.

The main elements of the proposed FPC demonstration program . . . are to:

—*develop commercial process for producing FPC*, including research to improve the present process and make it more economical and suitable for other species; also included: research on food technology, research on problems of toxic fish, and development of appropriate guidelines to foster stringent quality control. Design, construction, leasing, and operation of the authorized pilot plants are part of this development;
—*improve the fish catching*, *landing*, *and processing capabilities of three protein-deficient countries;*
—*develop markets for FPC in at least one protein-deficient country*. Local fishing potential, analysis of local eating customs, as well as market and distribution patterns for appropriate food forms, would be evaluated;
—*establish a viable commercial FPC system in at least one protein-deficient country;*
—*encourage other nations and private interests to establish commercial fishing industries wherever feasible.*

The concept of utilizing FPC to help meet protein deficiencies was a "technology exporting" concept. Rather than shipping fish protein to protein-deficient countries, it would help those countries, through the importation of U.S. technological capability, produce the fish protein themselves.[27]

This rationale for FPC was thus a sharp departure from earlier interests that had been focused solely on research and with different motivation. BCF and Congress had jumped on the original bandwagon in 1965 with the primary objective of helping the domestic industry find a market for trash fish caught incidentally to expensive species, or to widen the market for inexpensive species already being harvested for fertilizer and fish oil. Legislative initiatives to that effect had been taken in the 89th Congress by Senators Bartlett and Magnuson and Congressman Hastings Keith to authorize construction and lease two demonstration plants, at politically expedient sites on each coast.[28]

Over-all program management of the demonstration project and lead-agency responsibility for all government participation were assigned to the Agency for International Development. Technical support was to be provided by BCF and HEW. Special attention was given to establishing and maintaining world-wide quality and sanitary standards through such world organizations as the Food and Agriculture Organization. Inasmuch as success depended critically upon development of a low-cost product, consumer acceptance, and effective marketing, U.S. industry was encouraged to take a major role in planning and developing the program.

That assignment to AID put the Council in something of a quandary. On the one hand, the AID history included evidences of fluctuating leadership and policy, generally slow-moving management, and a running battle with Congress. On the other hand, the Agency had the responsibility and experience for overseas technical assistance and was already staffed abroad by representatives who could bridge the gap to direct recipients. Moreover, AID agreed to establish a special office that the Council Secretariat would help staff and advise.

Political support for the new approach was superb. Humphrey was lyrical; Administrator William S. Gaud of AID was enthusiastic; President Johnson mentioned this program in several speeches and messages. In his 1968 message "To Build the Peace" he said:

> We must also tap the vast storehouse of food in the oceans which cover three-fourths of the earth's surface. I have directed the Administrator of the Agency for International Development and the Secretary of Interior to launch a five-year program to:
> Perfect low-cost commercial processes for the production of Fish Protein Concentrate. . . .
> Develop new protein-rich products that will fit in a variety of local diets. . . .
> Encourage private investment in Fish Protein Concentrate production and marketing, as well as better fishing methods.[29]

Not only did the 90th Congress endorse this new approach but it specifically amended the Foreign Assistance Act and urged the President and AID to increase emphasis on the more effective use of fish protein and other concentrates in the war on hunger. It was the freshest of Council initiatives. And the amount of money required was so small that even BOB interposed no objections.

During the first year, surveys of FPC opportunities based both on local need and available supply of fish were conducted in a number of Latin American, Asian, and African nations, with Chile selected for more intensified study of commercial marketing of FPC fortified foods. AID advertised to purchase about $1 million of domestically produced FPC to enrich grains provided by the Commodity Credit Corporation for distribution in AID food programs primarily for children, pregnant women, or nursing mothers. By then, a number of countries were displaying interest as partners in the program; Canada and Sweden were pushing their own FPC initiatives; numerous U.S. food distributors were examining the opportunity to participate or to devel their own processes; and BCF was accelerating its research.

By 1968 the Council was sufficiently encouraged to state in the annual report that initial goals were to provide adequate FPC to meet protein needs of one million children by 1971, on which basis we anticipated a pilot operation capable of producing enough for seven to ten million by 1975. Hoping for a self-sustaining industrial base, we forecast that needs of up to 200 million could be met by 1980, using only a small fraction of the projected world fish catch.[30] By 1969 Korea and Morocco were added to the countries for trial marketing. Domestically, AID contracted with Alpine Marine Products Corporation in New Bedford, Massachusetts, to supply FPC at 42 cents a pound, using an approved adaptation of the VIOBIN process developed by Ezra Levin, a technique differing from that initially approved by FDA but which subsequently gained concurrence.

At that point the first of a flock of unexpected problems began to flutter around our shoulders. The BCF could not seem to muster the in-house management to uphold its technical assignments, and BOB monitors put limits on future funds unless the Council Secretariat could elicit guarantees of improved BCF performance. In a conference with Secretary of Interior Stewart Udall, I urged that the project be pulled out from BCF and, highlighted by the prestige of his own office, be assigned to a special director. Enthusiastic about FPC, he consented, but when the promised increase in funds was cut by BOB, he felt he should withdraw from his commitment. BCF then set up its own special office.

Next, the estimates of cost of the pilot plant exceeded initial appropriations and required a recycling through congressional mills. While this plant was in Aberdeen, Washington, the district of the chairman of the House Interior Appropriations Subcommittee, opposition was triggered when Levin claimed that industry rather than government should undertake the project. At another round of hearings (March 26 and June 26 and 27, 1968), my testimony, along with others, claimed that the government initiative had already been cleared by industry, pointing out that a representative of VIOBIN had been present when twenty-six different companies had met in January 1966, in support of the government's pilot plant. We also cited the industry's lack of motivation for research because of uncertain markets. As to Levin's claim that the pilot plant would preempt industrial production, we countered that the projected plant was too small for production runs; on the contrary it could provide technical experience for future industrial exploitation by helping to reduce the cost of the product to the 20 cents per pound that had been estimated as the target price that would gain widespread market acceptance. To offset the increased estimates for plant construction, we now proposed to build a single plant instead of two, thus requiring no new authorization. Congressman Lennon helped lubricate the legislative process by the euphemism that the action was "reducing the number of fish protein concentrate experimental plants." I spent many hours with reluctant congressmen, finally eliciting support from Dingell, who held a key committee position. Congress acted with P.L. 90–549, signed on October 4, 1968. Then a new roadblock appeared. The new administration imposed a budget cut accepted by Secretary Hickel that excised the plant. That move had to be overcome by more congressional intervention. By then, the Stratton Commission had lent its weight to FPC.

Ironically, the third misfortune to strike the program was industry's own failure to produce the contracted protein flour. The Alpine operators had apparently misjudged availability of hake close to the New England coast at the proper season and their ability to induce fishermen to go further offshore at the prices they were willing to pay. Furthermore, because the pilot plant using the VIOBIN process for extraction was delayed in construction, then pushed into production before it was tested, a substantial fraction of the initial output failed a protein-richness test, probably due to excessive heating in the process. AID accepted delivery of 172 tons, rejecting the remainder of a total 430 tons produced, an amount which still would have fallen far short of the 1,071 tons contracted for.

That year, Korea was found an unsuitable base for an FPC industry because it lacked an inexpensive source of fish. Meanwhile, however, Brazil, Pakistan, and Peru were undertaking studies of their own.

By late 1968 the question of domestic protein deficiency had been opened, yet Council inquiries directed to federal agencies repeatedly brought denials and rejoinders that, if such were indeed found, existing sources of protein would be adequate. Again, congressional inquiries finally uncorked the facts that led to an exposé by Senator Ernest F. Hollings[31] through his assignment on the Senate Labor and Public Welfare Committee. By May 1969 a Council proposal to consider including FPC in school lunch programs was accepted by all agencies concerned, including a reluctant Department of Agriculture.

To top off the dismal sequence of tribulations, the AID Food from the Sea office showed signs of running out of steam. The agency had failed to elicit support from its foreign missions, and its energies to coordinate action through a ponderous superstructure were being absorbed in a perennial reorganization.

As two items on the bright side, research was proceeding on alternative methods of protein extraction from whole fish, and late in 1971 the pilot plant was dedicated and open for business. But by then, it was evident that the project had bogged down. On the questionable assumption that FPC had to find acceptance in a free market, economists were taking potshots at the concentrate because of its continued higher cost than alternative sources. This overlooked the facts that the new pilot plant had not yet established production costs and that in any event the cost would be only one factor in public favor. Ironically enough, one such economic study sponsored at MIT by the Marine Sciences Council[32] reached the negative conclusion that in less developed countries cost would probably be the controlling factor in marketability and acceptance; in better nourished countries, attractiveness of FPC food products, rather than price, would control.

In retrospect, the cost issue was far less instrumental than was FDA policy in arresting the project. FDA claimed that, apart from health, they must take into account cultural and esthetic preferences of the American public. According to an account in *Science*, "Virgil Wodicka, director of the Bureau of Foods, agrees that gelatin made from hooves or sausage made with ears and snouts might also be psychologically repellent, but says these products have been around for a long time and are culturally acceptable. The idea of eating whole fish, though, is new, and the FDA as the 'technical representative of the consumer' believes in protecting its charges from surprises."[33]

Could it also be that the dairy lobby was still influential in blocking this competition? In comprehensive hearings by both House and Senate on the pilot plant authorization in 1966, strong support for FPC development came from General Foods Corporation and a number of other food processors. The only objection came from the National Milk Producers Federation, which stated, "Surely it would be unwise to recover protein from fish waste [*sic*] and leave unused present stocks of high-quality food the acceptance of which both at home and abroad is unquestioned. . . . We oppose the use of public funds to develop a food product from fish material ordinarily considered filthy and inedible."[34] Whole fish and shellfish are, of course, eaten all the time—sardines, oysters, clams, and lobsters, to mention a few. But FDA remained adamant. The canard about filth had already been effectively speared by a NAS committee that included several distinguished food scientists, industrial toxicologists, fisheries economists, and marine biologists and which concluded that a "wholesome, safe and nutritious" product could be manufactured. H. M. Burgess of General Foods said that the brier patch of FDA restrictions hedging the market completely discouraged private enterprise from entering the field. The failure of Alpine did nothing to restore confidence. Commenting on FDA restrictions, the MIT study asserted that "economically they are as restrictive as outright prohibition . . . without justification and contrary to the public interest."

A 1971 Cornell University report estimated a U.S. domestic market for protein supplement at 3,100 million pounds annually, of which only 1,100 million pounds is being met.[35] Unfortunately, in comparing costs of protein sources, the report used MIT estimates of FPC costs only in relation to prices of raw fish without recognizing higher costs at an early stage of product development and without considering the important contribution made by government agricultural subsidies in having reduced costs of soy flour and nonfat dried milk.

FPC was blocked from the urban ghettos and Indian reservations where protein shortage is known to exist, at incalculable cost to the nation. It is paradoxical that a premium concentrate, aimed at the affluent citizen, is now being developed by a coalition of American and Swedish firms, Nabisco and Astra Pharmaceuticals. If this gilt-edged product opens the market for fish protein concentrate, the original intent to help American fishermen may finally be realized and a new avenue opened for attacking the problem of hunger, here and abroad.

DAVY JONES'S MINERAL LOCKER

The policy issues concerned with fisheries differ totally from those of nonliving marine resources. Such fossil fuels, ores, and minerals are basic sources of the energy, construction materials, metals, chemicals, and fertilizers required by any advanced industrial economy. Over the years, industry in the United States had developed a dependable and economic supply of raw materials from terrestrial sources to meet demands of an expanding population with a rising standard of living. In the process the twin incentives of economic development and national security had induced a wide range of federal initiatives to aid industry in such areas as developing supplies of minerals at lowest cost, exploring for domestic sources to assist in maintaining a favorable balance of payments, providing a climate for industry to produce efficiently under competitive conditions, conserving the nation's mineral resources by using them efficiently. These projects, however, had little relation to resources of marine origin.

To meet its needs in the late 1960s the United States was importing 21 percent of its minerals and 13.5 percent of its petroleum; the percentage was considerably higher for certain strategic metals such as manganese. Looking ahead, the U.S. consumption of nonfuel minerals was expected to double by 1980 and that of petroleum products to increase by about 50 percent. The world-wide demand for minerals also was projected to increase because of industrialization and rising levels of consumption. Such projections created incentives both for technological advances to permit use of lower-grade sources or substitute materials and for geological exploration of new reserves. In recent years, however, the richness of terrestrial mineral bodies had decreased and production costs risen; with the aid of new technology, industry was now ready to look seaward for oil, gas, sulfur, and for new sources of aggregates and other materials. Until recently this quest had been pursued largely in the relatively shallow waters of the continental shelf. Although little of that shelf had yet been systematically surveyed to determine the presence, distribution, and richness of seabed deposits, this new frontier was to be a potent source.

Gulf of Mexico prospecting for oil and gas first took place offshore about one mile from the shoreline in 1938; the first discovery well in shallow water was completed in 1947; the first beyond sight of land in 1964. By 1968 over 9,000 wells had been completed at an investment exceeding $12 billion. In 1967 offshore production accounted for 12 percent of the U.S. crude, valued at $1 billion, almost double that in 1965. The value of petroleum production through 1967 from lands of the outer continental shelf under federal jurisdiction totaled about $4 billion, and if state lands inshore were included, over $5 billion. This had contributed a windfall to both federal and state treasuries, from leases to explore and exploit and from royalties on production. Following the first federal offshore lease sale in 1954, the transaction of 1,276 oil and gas agreements had resulted in a total bonus income of almost $4 billion. Federal royalties from production of oil, gas, and sulfur from the outer continental shelf had totaled more than $700 million through 1967, and the federal government was then receiving more than $12 million per month in royalties.[36] Overseas, by 1970, exploration was being pushed off the coasts of 78 countries; 28 countries were producing 2 million barrels per day, double that off the U.S. coasts.

In some ways, it was the Truman Proclamation of 1945, followed by the Outer Continental Shelf Lands Act of 1953, that launched the domestic enterprise. Once industry had a stable claim to title, it was willing to take risks. Bills initially passed by Congress to quitclaim offshore resources in favor of state jurisdiction had been vetoed by President Truman but subsequently were upheld by a June 5, 1953, Supreme Court decision. In that same year a law was enacted quitclaiming submerged lands to the states out to three miles or to the historical limit, if greater.[37] The 1958 Geneva Convention, ratified by the United States March 24, 1960, further clarified legal positions.

Notwithstanding this intense burst of activity, industry had both complaints and problems. The spasmodic timing of state and federal lease sales sharply affected their decision-making that required systematic geophysical surveying and lead time for financing. They were dismayed that the highest bid had on occasion been rejected by the Bureau of Land Management. And they were all too aware of the high costs of offshore drilling. Hurricanes in the Gulf of Mexico posed a constant and expensive risk. Drilling in Alaska's Cook Inlet confronted tides of twenty-three to thirty foot variance, currents of four to eight knots, and ice floes six feet deep.

On the plus side, slant drilling now enabled up to sixty wells to be sunk from a single platform, although this bonus was to prove a handicap when fires on offshore platforms spread to adjacent wells. Moreover, offshore enterprise offered greater opportunities than the diminishing returns landside and a high ratio of success coupled with low exploration cost. On a cost basis alone, industry would ostensibly have preferred cheaper foreign sources except for their potential instability; on this score, offshore sources provided a lever in negotiations. (Some sectors, however, have been found to profit in sustaining high crude-oil prices by keeping out cheaper foreign products because their depletion allowance is computed on the basis of crude prices.)

In the entire mix of considerations, government had been intimately involved: providing a favorable financial climate through import quotas, domestic production allowables, depletion allowance (27½ percent reduced in 1969 to 22 percent), and capital gains tax. The ability to write off as current expenses costs that other industries must capitalize was probably the most valuable tax incentive to the industry. Government had contributed as well through such social overhead measures as weather mapping, navigational aids, marine search and rescue, port and harbor construction and maintenance, statistical reporting, and transfer of such Navy technology as saturation diving techniques, underwater cameras and communications, corrosion-resistant metals, and TRANSIT satellites for precise location. Notwithstanding the substantial income from leasing offshore rights, government was investing a much smaller amount for marine research and development. The Budget Bureau uniformly opposed earmarking of income for special purposes while subtly encouraging those policies conducive to offshore revenues.

The Council Secretariat soon recognized that the primary considerations of policy regarding the energy industry were woven into an intricate tapestry of almost invisible agreements and relationships with federal and state government. National energy requirements had led industry to propose higher allowables related to offshore development; larger lease size; reduced royalties in some blocs including unleasable areas; longer production time than the current five-year expiration; and freedom to exit from marginal areas. Within the Council's jurisdiction, we felt that our role was to assure that the social overhead services worked well. Mild pressure from industry to accelerate these measures met with little encouragement in the Johnson White House and an even cooler reception in the 1969–71 interval under Nixon.

Primary involvement of the Council with the offshore oil question came unpredictably from another direction when the *Torrey Canyon* and

Santa Barbara oil spills blazed into headlines and demanded action to protect the marine environment from oil pollution. The *Torrey Canyon* disaster, occurring on the eve of my trip with Humphrey to Europe early in 1967, gave a chance to sound out British opinion firsthand. During that interval, in daily touch with Washington, we sought advice from American science and industry to recommend to our U.K. friends. None was forthcoming because little attention had been given by U.S. industry to this emergency, and our primary assistance lay in U.S. troops ignominiously cleaning up beaches by hand. It did develop that the British government had been in a flap over which ministry, if any, was responsible; this red tape had absorbed valuable days when containment would have been easier. With that lesson in mind, the Council Secretariat aided the White House in developing a government-wide contingency plan for what response should be made, and by whom, should the same accident strike U.S. waters. Since the Coast Guard had the command and control capability for quick on-the-scene action, but the Department of Interior was responsible for environmental preservation and rules for offshore oil production, a blunt confrontation met the question of who should be in charge. With the Secretariat serving as mediator to the contending parties, compromise resulted in the first publication of a plan in 1968.[38]

That step proved a lifesaver at Santa Barbara in 1969. While the moves and countermoves by Secretary Hickel and the industry, by the Geological Survey and others, constitute a fascinating epic beyond the scope of this book, it was amply apparent that the oil industry had previously assumed little responsibility for environmental protection and endeavored to keep the issue out of the Council's hands. The Department of Interior was pushed as lead agency to deal with oil spills from the Hill and the White House, not only because it was the home of the federal water pollution control agency but because it could also be expected, from its traditional rapport with industry clientele, to be more sympathetic to industry's problems. For a while Hickel's uninhibited initiatives spiked that tradition. Then came a spate of more oil stains spreading in the marine environment from Gulf of Mexico platforms, from colliding tankers, and in Puget Sound from the ridiculous experience of pumping a large diesel barge "full" while the overboard drain was open.

The oil companies began to read the signposts. For the first time individual companies and the American Petroleum Institute sought countermeasures to the oil contamination problem, rather than leaving the quest for containment and cleanup entirely to the federal government.[39] But during this interval, facing the issues of the international legal regime and of environmental pollution, the industry and the Council stayed at arms' length. The Secretariat was determined to keep its independence.

As to nonfuel minerals, ore bodies or placer concentrations on land might logically be expected to extend seaward, making this submarine landscape a rational target for mineral exploration. The continental shelves of the world were estimated to provide sources covering some ten million square miles, larger than the North American continent itself; those around the United States totaled one-fourth of the land area. The economics naturally promoted sharpest interest in the higher-priced minerals in the shallowest water—gold and diamonds in submerged beaches or river valleys, both of which have attracted investment with some degree of success.[40] Marine phosphates had been extracted but not commercially exploited because deposits compare unfavorably with land sources. Yet, since shipping costs represent a high fraction of the delivered price, proximity of marine sources to markets could be a major factor in future development (in respect to Japan, for example). Public leases in 1961 to Collier Carbon and Chemical Company were unused, and commercial attempts off the West Coast in 1961–67 were unsuccessful.

A completely different situation existed regarding deep-sea nodules. What had been largely a curiosity since their discovery in 1875 by the H.M.S. *Challenger* began in the 1960s to arouse commercial interest. Thought of originally as a source of their primary ingredient, manganese (of which the United States imports 99 percent of its requirements), these nuggets came subsequently with changing market conditions to be regarded with additional interest because they contained copper, nickel, and cobalt. Moreover, the copper averaged 1.04 percent in some parts of the Pacific compared to 0.63 percent in U.S. land sources. The economics, however, were blurred because of such factors as proximity of source to existing heavy mining investment, available processes for beneficiation or extraction, and proximity to markets. Spangler[41] argues that because the United States imports so much of its needs, an offshore source would be of national interest to improve balance of payments, and the accompanying table 13 underscores his point. However, he admits the sudden opening of a new source could seriously disrupt world markets. An alternate strategy might be to bring in such sources only gradually so as to meet growing world demand while simultaneously keeping prices from rising exorbitantly.

Venture capital of existing mining companies was slow to enter this field, but by 1968 the pace accelerated with six companies involved. Subsequently came an announcement by Deep Sea Ventures of Tenneco of a new extraction process expected to make the nodules competitive.[42] Once this has been established, other mining firms may enter the competition as a defensive measure, although many uncertainties remain regarding markets, jurisdiction, and federal assistance.

Mindful of these developments, the Council continued bearish about

TABLE 13
TRENDS IN U.S. IMPORTS AND EXPORTS OF SELECTED MARINE
GOODS AND SERVICES, 1950–70
(in millions of current dollars)

Imports	1950	1955	1960	1965	1970	Total 1950–70
Fish[a]	198	259	363	601	1040	9735
Petroleum[b]	592	1026	1537	2093	2770	33158
Manganese[c]	42	72	82	110	34	1465
Copper[d]	243	455	353	425	529	8928
Nickel[e]	72	183	147	250	423	3848
Tin[f]	202	179	119	180	202	3530
Transport[g]	643	946	1333	1808	2404	29586
Total Imports	1992	3120	3934	5467	7402	90250
Exports						
Fish[a]	28	40	41	70	118	1089
Petroleum[b]	499	646	468	418	487	11904
Copper[h]	88	218	291	293	358	4571
Nickel[i]	3	14	10	4	13	237
Magnesium[j]	1	5	4	13	24	143
Phosphate Rock	15	20	38	11	10	492
Sulfur	33	51	42	66	34	999
Transport[k]	859	1137	1337	1720	2265	31002
Total Exports	1526	2131	2231	2595	3309	50437

[a] Includes edible and nonedible fishery products.
[b] Includes crude petroleum and refined products.
[c] Manganese ore.
[d] Copper metal and manufactures.
[e] Nickel metal, incl. scrap.
[f] Includes alloys in chief value of tin, ores, and concentrates.
[g] Includes freight on foreign-carrier imports and U.S.-carrier port expenditures abroad.
[h] Copper metal and manufactured goods.
[i] Nickel metal, incl. alloys and scrap.
[j] Includes metal, alloys, scrap, semifabricated forms, and powder.
[k] Includes freight on exports carried by U.S.-operated carriers and foreign carrier expenditures in U.S. ports.
Source: *Statistical Abstract of the United States*, 1950–71.

prospects, influenced by the negative report on seabed mineral prospects by its consultant, Economic Associates,[43] and noting no special pressure from the industry itself. Mineral recovery was considered of lower urgency than other programs, although the public interest involved in the balance-of-payment aspect was never adequately explored. Partly, this ho-hum attitude stemmed from industry itself, but even the Department of Commerce, queried on this point, was not concerned. And the Budget Bureau adopted the same point of view as with fisheries, holding that global economic efficiency should be the criterion. If other countries could mine more cheaply, they should do so at whatever cost in gold flow. The Council of Economic Advisors sang the same song in 1968, and no change in policy appeared with the change in administration.

In the case of mining, however, the embryonic state of technology for exploration and extraction led the Council to advocate stronger government assistance. The Bureau of Mines was encouraged to expand its activities at Tiburon, California, and made a fitful start, only to be shot down by the Budget Bureau.

In a nutshell, the Council saw that 1966–71 interval was not the time to launch an elaborate effort to extract minerals from the sea. But industry–government partnerships for that thrust could well be set up in advance, before some critical urgency permitted the politics of the issue to dominate deliberations.

Muscles for Marine Technology—Ocean Engineering

Given these experiences with the fishing, petroleum, and mining industries, the Council Secretariat took a long look at those functional aids to industry necessary to national interests and decided that the support of research and development in ocean engineering, which had been perennially arrested in the federal armory, was a proper role of government. As early as 1964, speaking as an individual, the author had endeavored to point out that engineering, as the functional bridge between science and national goals, warranted a stronger role in marine affairs.[44] This rationale was adopted as a major new objective in Sections 2b(3) and 2b(7) of the Marine Sciences Act.

Engineering was regarded not as an end but rather a means; the development of marine tools, techniques, and facilities could be of service both in everyday maritime chores and in research. Such historical examples of successful ocean engineering as ships and submarines, undersea cables and tunnels, coastal protection, and offshore oil operations all have evolved from classical engineering principles concerning propulsion, materials, sources of energy, structures, and communications, and have been shaped by practical considerations of cost, safety, reliability, and ease of maintenance. But these principles have had to be extended to accommodate unique marine forces. What was new about ocean engineering was the maritime environment itself: sea-surface motion; tides and currents; wave impact and wind loading; hydrostatic pressures at depth; buoyant forces; opacity of sea water to electromagnetic energy; high attenuation and scattering of light energy; high conductivity of sounds; lack of gaseous oxygen for man or for chemical combustion; presence of the common elements in sea water; the variable two-phase nature of water-bottom interface; and severe corrosion and fouling.

The Navy in its long need to cope with these factors had evolved effective methods for living with the sea. But for civilian applications except for shipbuilding, marine technology was in its infancy. In this recognition, beginning about 1965, such organizations as the Marine Technology Society were founded, while the National Security Industrial Association and the older societies of mechanical, civil, electrical, and naval engineers added new vitamins to their program activities. The newly formed National Academy of Engineering put ocean engineering on the front burner. All received encouragement from the Council Secretariat and were invited to comment on federal programs.

In the meantime, the Council sought justification from the agencies for stronger federal funding of engineering research and development, in the belief that enhanced funding would provide further incentive to industrial participation. From the outset, the Council placed emphasis on a deep-ocean search and retrieval capability for military purposes. The *Thresher* catastrophe in 1963 dramatized the compelling reasons for finding and recovering objects from the deep-ocean bottom; the nation was then frustratingly incapable of accomplishing such a rescue in water more than 400 feet deep. Three years later, following the collision of two U.S. aircraft over Spain, the recovery of an unarmed nuclear weapon from 2,850 feet of water near Palomares demonstrated that some embryonic progress had been made. That task, however, required three months, dozens of ships and aircraft, thousands of people, and millions of dollars. Through the Navy's own initiatives, with Council endorsement, funding was subsequently increased slowly, and the Navy invested primarily in a submarine-rescue capability of small, highly maneuverable submersibles, a system recommended for development after *Thresher* and initially operational in September 1971. The second project was development of NR-1, a small nuclear-propelled research submarine that was placed and retained in a super-secret category, but which had the potential of confirming a major technological breakthrough in power plants. Pushed by Rickover, it needed no assist from the Council.

To add versatility to the hydrospace tool kit, the Council next considered the merits of buoy networks, instruments to observe oceanic phenomena from spacecraft, inexpensive power sources, reliable underwater instruments and communications, large ultra-stable floating platforms, unmanned vehicles, deep submersibles, and techniques for saturation diving. The Council distributed a report at the 1967 IOC meeting that included vibrant color photographs to enlist interest in the new era of space-ocean technology.[45] The Council also extracted a list of priorities through a contract study with Southwest Research Institute.[46]

Included on the President's 1966 selection of marine initiatives was a buoy program that had been sketched out by the Interagency Committee on Oceanography then converted to reality by Council staff. Responsibility for its development was assigned by the Council to the Coast Guard, which had long experience at sea with navigation aids, was newly authorized to engage in oceanographic research, and would enjoy the chance to work with sexy technological tools.[47] By then, however, another factor entered the picture. The elaborate Mississippi Test Facility that NASA had erected to proof-test propulsion of Apollo boosters before transport to the Kennedy Space Center was running out of mission because of the truncated Apollo program. Looking desperately for some follow-on, the Center's director moved with great alacrity to point out to his influential senior Senator, John Stennis, that their under-utilized magnificent computer was conveniently close to the Gulf Coast and the facility could direct some of its talents to the oceanographic area. Before long, the Coast Guard's buoy program was moved lock, stock, and barrel to Mississippi. The first initiative in civilian marine technology looked promising. By FY 1972 the program had grown to $13 million—far less than capabilities and needs merited, but the best that could be done under budget constraints. Industry was participating heavily in contracts.

A different chronology attended the diving project. Industry was gung ho in 1966 on the grounds that when divers could escape the constraints of umbilical air hoses and hard hats by operating as free swimmers with breathing apparatus, their increased manual dexterity, maneuverability, and sensing would immeasurably aid underwater exploration and development. Manned diving had already become a half-billion-dollar a year industry.[48] It was foreseen that such free agents moving about under the sea could serve a multitude of purposes. In exploitation of minerals they could make visual surveys and perform such tasks as evaluation of deposits, placement and inspection of drills and cores, maintenance and repair. They could aid in monitoring, controlling, and correcting beach erosion and pollution. Their skills could assist underwater construction, tunneling, harbor development, installation of sewer outfalls, and aids to navigation. And certainly, in their traditional role, they would be better able to act in salvage operations and disasters at sea. Men living in the sea, working in submerged pressurized laboratories, and equipped with self-contained breathing apparatus and lockout minisubs, could open whole new worlds to scientific investigation of marine ecology, geology and geological mapping, aquaculture, archeology, human adaptability, and all facets of marine biology.

Outside enthusiasm ran high for such possibilities. Technical advances were made in deeper safe operation (600-feet became routine), better diver communication and comfort. The Navy's Sea Lab experiments to 600-foot depths were already underway and seemed destined to advance the technology by developing safe mixtures of helium and oxygen gases for prolonged deep exposure and protocols of descent.

When this project stumbled in 1968, the Council Secretariat began a special exercise that was carried to completion in 1969. Under a staff aide, John Billingham, borrowed from NASA's manned-flight research, a comprehensive program plan was developed, and help to ease it through the interagency mill was enlisted from John Craven, chief scientist of the Navy's deep submergence systems project. By that time, however, industry had completely switched goals, dropping the emphasis on saturation diving in favor of manned undersea operations from fixed or mobile shells, resistant to outside pressure and affording a natural atmosphere. Council consultants, injecting industry's viewpoint, pulled the rug out from under the accelerated man-in-the-sea program by mid-1969, and the government applied a *coup de grâce* when a Council panel chaired by Assistant Secretary of Commerce Myron Tribus could find little justification for its advancement. Biomedical research continued in the Navy and at such centers as the University of Pennsylvania, Duke University, and Virginia Mason Hospital in Seattle, but not at the desired pace. Project TEKTITE,[49] covering a 60-day operation in shallow water (50 feet), was a highly successful enterprise, but even that evolved painfully through two successive phases because of a budget drought. With prospects gone for the accelerated man-in-the-sea program, only more generalized proposals remained from the Council's fourth report. These centered largely on safety, health, coordination, and communication in the development of manned undersea activities.[50]

Other new marine technological initiatives were twinkling here and there throughout the government: the Advanced Research Projects Agency in DOD began advocating the development of large stable platforms, the Navy and ARPA pushed air-cushion vehicles, and the Maritime Administration made a fresh attack on technological modernization of the merchant marine. But the aggregate ocean engineering effort was dragging. The Navy was by far the best equipped to conduct ocean engineering, yet Navy funding for ocean engineering remained level from 1967 to 1971. Considering inflation, this meant in real dollars a sharp decline. Moreover, the support was sifted thinly over a large number of projects, and leadership was widely distributed in specialized Navy labs that were in sharp internal competition for a lead role.

The Council, however, appreciated that the specialized manpower required to perform ocean engineering that would couple scientific oceanography to practical application was in short supply. Sea Grant was its chosen instrument to build up a long lead-time, national ocean engineering capability in the academic institutions.[51] By 1969, the Sea Grant Office also began to realize the need for others than engineers in the technological enterprise—"specialists in many other disciplines [are] needed to turn their energies and intellect to the sea—from economics, law, business, public administration and foreign affairs." NSF established as a goal "by 1975 [to] be instrumental in turning out annually 100–300 ocean engineers at the graduate level, 100–300 ocean engineers at the bachelor's level and 500–900 ocean technicians."

On the civilian side, the Council Secretariat had lit the kindling and stoked the fire, but the activity never heated up. In the first instance, no single operating agency could muster the necessary complement of staff and funds. It is a melancholy fact that when the man-in-the-sea initiative showed temporary promise, Tribus's interagency committee felt that, apart from the Navy, the only agency that had sufficient in-house capability to manage a civilian marine technology program was the National Aeronautics and Space Administration.

Without the government impetus, contributions by aerospace and marine services industries also sputtered. Combining the saltwater experience of offshore oil firms with the high technology of aerospace industry seemed a promising approach and was attempted. The effort foundered on the lack of motivation, imagination, and political sophistication of both industries—the maritime and the aerospace—to undertake on their own initiative or to rationalize for government support the base of technology needed to master the sea. In 1967[52] when invited to address each field separately, I challenged both to specify how the seas could further national interests, aided and abetted by new technology. Curiously, I was criticized by the trade press as hostile to industry. Except for NSIA, industry groups generally did not respond. Rather sadly, in 1971, officials of NOAA were making the same charge of inadequate industrial response.[53]

In recognition of the problem, the Stratton Commission put its growth-promoting bets on a major new program of federal technological projects, to:

Achieve capability to occupy the bed and subsoil of U.S. territorial sea and learn to utilize continental shelf and slope to 2,000 feet.
Achieve capability to explore depths to 20,000 feet by 1980 and utilize the depths by the year 2000.
Initiate a comprehensive fundamental technology program (NOAA).
Establish National Projects to focus marine effort on specific areas of opportunity and need (NOAA).
Establish a National Project of test facilities for undersea systems (NOAA).
Involve private industry in planning and conducting National Projects (NOAA).
Plan and administer programs to advance marine technology so that industry can assume early responsibility for development.
Utilize Navy development capabilities for fundamental technology through cooperative arrangements with NOAA.[54]

The Commission's prescription for this retarded stepchild was a massive injection of federal funds to underwrite a new function to supplement the consolidation of existing functions in their proposed new agency. From a modest funding level of slightly under $20 million in FY 1968, the Commission recommended that support for fundamental marine technology should be expanded more than sixfold to $130 million annually during 1971–75 and tenfold to more than $200 million by 1976–80. Nearly half of the funds would be devoted to the advance of a comprehensive technological capability to operate on the continental shelves and deep ocean. Other funds would be devoted to National Projects that included: (1) test facilities and ocean ranges, (2) Great Lakes restoration feasibility tests, (3) continental shelf laboratories, (4) pilot continental shelf nuclear plant, (5) deep exploration submersible systems, (6) pilot buoy network (already under way), (7) pilot harbor development, (8) deep ocean stations, (9) seamount stations, (10) mobile undersea support laboratory, and (11) a large stable ocean platform.

The Commission's focus of marine technology was largely on hardware, on means enhanced by a generalized capability, not on ends. No distinctions were drawn between engineering research and development and technology. The latter would include the software of information transfer, capital, resources, and organization that is mobilized to deliver a desired good or service to the marketplace to meet a prescribed objective, and for which scientific understanding, invention, or craft are inputs. But there was limited discussion of the coupling process, including the role of private entrepreneurship and the use of indirect federal incentives to facilitate its expression. The legislative charge to the Commission to produce a national (not just federal) program remained a baffling puzzle which their proposed National Advisory Committee for the Oceans was expected to unravel in guiding their proposed marine technology program.

Nevertheless, in the spring of 1969, this marine technology proposal of the Commission was deemed separable from the organizational issue and ordered the subject of Council study along with the other Commission proposals, as recounted in chapter 3. Assigned under the Council's Committee for Policy Review to the task group chaired by Tribus, the proposals fared poorly; they were tagged as unconvincing by all the agencies and were heavily criticized by Tribus himself. Already laboring under severe budget constraints, agency personnel might have been expected to be skittish about a new competitor for funds. But even the Council consultants gave little support to the Stratton technology proposals, which in their view seemed too much like a solution looking for a problem.

When the opportunity for political intervention arose in September 1969 with the White House request for presidential initiatives, the Secretariat's proposals for support of general-purpose marine technology were instantly rejected, notwithstanding recognition that, apart from marine objectives, an underutilized aerospace industry would be the immediate beneficiary. The Nixon administration, seemingly determined to allow no inroads on the federal purse, was paradoxically indifferent to the needs of industry and to the underemployment of highly talented scientists and engineers previously trained with these public funds. Not until 1971 would that point of view be reversed.

Throughout this quest for a civilian marine technology, the problem had never been formulated in a way that the goals would justify the investments. And in the absence of crisis and of an influential pressure group outside of government, no civilian operating arm inside government had the fortitude to push this high-cost, cutting edge of marine affairs through the budgetary resistance. By the spring of 1971 the Council had been dismembered, and NOAA, on which the Commission bet its chips, had yet to add engineering to its top agenda.

On May 26, 1971, Senator Hollings introduced S. 1963 to give a burst of support to marine technology, now rationalized by the unemployment issue. It faltered in the absence of public support. While it appeared that the initiative, like a yo-yo, would return now to Congress, neither Congress nor the Executive, Council nor Commission, industry nor academic experts, had cracked the stubborn enigma. The intractible nature of the marine technology issue suggested that something more fundamental might be wrong. A further diagnosis is attempted in chapter 10.

CHAPTER 8

Organizing Federal Efforts

THE QUEST FOR ORGANIZATIONAL SOLUTIONS

Examination of the political brew in marine affairs over a ten-year period shows one ingredient continuously bubbling to the surface and attracting attention from congressional cooks: demands for federal reorganization. This was, of course, in accord with the time-honored Washington cult that sweeping out old structures is tantamount to reform and progress. As Harold Seidman has acidly remarked, reorganization "is prescribed to purify the bureaucratic blood and prevent stagnation."[1]

The congressmen who examined the riddle of oceanography became true believers in that dogma. In 1960 the House Science and Astronautics Committee sponsored the first in-depth diagnosis of federal marine ailments. Facing the evidence of disarray, they posed as the question to be answered, "How can this nation best mobilize for a concerted, long-range program of research and exploration of and in the sea?"[2]

The committee report identified as the prime objective, "Federal organization and appropriations to support the proposed effort."[3] It then cited three complex obstacles to achieving the maritime goals: the current fragmentation of the program among eleven different agencies; the low rating of marine affairs in the agenda of national priorities; and the gaps in certain civilian functions among existing agencies.

Six years of dogged congressional effort were to go by before even an interim solution to these problems could be reached with the creation of the Marine Sciences Council; four more years would be needed before a new agency would emerge on behalf of civilian maritime affairs.[4]

Congress began by addressing itself to three persistent challenges. Their initial proposals to draw together the fragmented program were variations on the single theme of improved coordination machinery. Although the executive branch took a number of steps to strengthen its mid-echelon Interagency Committee on Oceanography, a skeptical Congress conferred its bicameral blessing of statutory responsibility for coordination on the Office of Science and Technology. That thrust was pocket vetoed.[5]

The second phase of inquiry emphasized the need for high-level leadership to harmonize disparate goals and programs of existing agencies, but also for more powerful apparatus to raise marine priorities, integrate and reinforce the aggregate of soft low-echelon voices. Through the 1966 Marine Sciences Act, the President was given that responsibility, backed up by the interim Marine Sciences Council, chaired by the Vice President.

The third phase of the legislative quest for a still more influential and permanent organizational device seized on centralization rather than coordination. This theme first appeared on the congressional list of remedies through proposals of Congressman Bob Wilson and Senator Edmund Muskie in 1964.[6] Such measures wilted as premature. Aware of the limited size of the program and of the lack of sufficient external injunctions, old hands in the Congress, including Warren Magnuson and his staff, were dubious whether the field could earn a viable organizational identity. Seldom had agencies been given birth simply from the desire for administrative tidiness. In those few reorganizations where external pressure was lacking, only strong initiative by the President had carried the day, as when the Department of Transportation was created in 1966 over both congressional and interest-group resistance. The gambit of marine consolidation had the strength of logic but no political propulsion.

Marine partisans outside of government nevertheless coveted a central, highly visible agency and cited NASA's demonstrated power to mount a powerful space program and to tease funds from both Congress and the President, perhaps forgetting that NASA's success was largely due to the exhilarating climate in which it was prospering. The clamor for marine action and for a new agency emanated largely from the scientific community and the aerospace industry. Because fishing, shipping, offshore oil, and other industrial users of the sea remained silent, the campaign unfortunately appeared to be motivated by self-interest of the potential technological armorers rather than by dedication to maritime interests of the nation per se.

Within government, marine specialists who would also have stood to benefit from such an agency feared their own components might be lost in the vortex of consolidation. Congressional hearings from 1959 to 1965 contained fulsome assurances from individual agencies and from representatives of three administrations that fragmentation resulted naturally from diverse statutory assignments and could be accommodated by whatever coordinating machinery was then in vogue.

That many congressmen were eager to produce some new federal structure strong enough to fight for more funds became clearly evident during hearings in 1965 and 1966. Notwithstanding this expectant atmosphere, Senator Magnuson's cool appraisal of the immaturity of the field, and a lack of Senate staff enthusiasm for any measure that would dilute the growing impetus for a Council, left initiatives for reorganization to the House committee. Congressmen Lennon, Mosher, et al. had misty perceptions of an ultimate goal, but little to suggest in the way of either a clear target or a strategy that would achieve organizational goals without stirring up jurisdictional rivalries among other House committees. They reluctantly concluded that consolidation would be accepted only if the notion originated outside of both legislative and executive branches.

The vehicle for this major new thrust, to be generated by a quiet coalition of House members interested in the sea and those interested in the aerospace industry, was to become the Commission on Marine Science, Engineering and Resources.[7]

Cynics often regard the creation of a prestigious commission as a ploy to delay or even block action on some urgent controversial issue. It provides a decision-maker handling a political hot potato with a legitimate excuse for inaction and with maneuvering room to preserve options. Sometimes commissions are expected to build support for controversial courses of action by legitimizing rather than advising, thus to tranquilize public or congressional agitation. But both industry and Congress knew what they wanted from this child of the House Oceanography Subcommittee, and it was not delay. In their eyes, the Commission was an instrument of progress toward more funding for marine affairs, probably through proposals for a new marine agency.[8] Legislative history that led to authorization of the Commission in the Marine Resources and Engineering Development Act was detailed in chapter 2.

It is difficult to know what expectations President Johnson had for the Commission, either when he signed the bill into law or when he later appointed what was to prove a virile and hard-working band. By coincidence, about the same time he signed the Marine Sciences Act authorizing a commission, Johnson released a report by his own Science Advisory Committee that proposed a new operating agency composed of the Environmental Science Services Administration, Geological Survey, Bureau of Commercial Fisheries, and parts of the Bureau of Mines and Coast

Guard. While this centralization was to improve marine advocacy as well as coordination, PSAC's rationale was based more on unifying the earth sciences than interrelated governmental functions.[9] The similarity between the PSAC proposal and the subsequent Commission recommendations proved remarkable, but by the time the Commission issued its findings, some members of the PSAC panel had changed their minds.

APPOINTMENT OF A PRESIDENTIAL COMMISSION

The Marine Sciences Act provided for a Commission appointed by the President of 15 members from federal and state governments, industry, and academia, augmented by four congressional advisors. It was empowered to conduct a "comprehensive investigation and study of all aspects of marine science in order to recommend an overall plan for an adequate [*sic*] national [not just federal] oceanographic program that will meet the present and future national needs"; recommend a governmental organizational plan with estimated cost; and submit a report "to the President, via the Council, and to the Congress not later than eighteen months after the establishment of the Commission [which] shall cease to exist thirty days after it has submitted its final report."[10]

After President Johnson signed the measure into law, only one entity within the executive branch shared congressional aspirations for a Commission powerful enough to hoist marine affairs up the ladder. That entity, fortunately occupying a key position to give that boost, was the Council Secretariat, which outsiders and insiders alike had wrongly adjudged as competitive to the Commission.

Immediately after the Council was activated, August 17, 1966, Vice President Humphrey and the Executive Secretary examined the Commission's role and agreed on its importance as a major spur to marine reorganization. It could not serve this purpose, however, if used for repayment of political obligations or artfully loaded with special interests. Rather, it should be composed of distinguished individuals primarily outside of oceanography and aerospace so that findings would not be criticized as self-serving. The opportunity to convert that philosophy into practice fell to Humphrey himself when the White House personnel office, responsible for preparing nominations for all policy-level presidential appointments, and aware of the sensitive legislative coupling of Commission and Council, turned the matter over to him. With general instructions, he bucked the assignment to the Council Secretariat. By informal arrangement, nominations were subject to review and possible veto by John Macy, head of the personnel office, or by White House political advisors, after which any of their counterproposals were subject to veto by Humphrey or the Secretariat before submission to the President.

It was decided to balance membership on the basis of the legislative instructions, industrial uses of the sea, such scholarly disciplines involved as economics, law, foreign affairs, science and engineering, and geographic distribution. Among the qualified candidates we also sought a few who by their past activities had gained recognition by interested congressmen, or who had access to the President himself. Although the legislation permitted five federal government members, we decided to increase the number of outsiders to the extent practicable, by earmarking slots for Navy, Interior, and Commerce, but in the process aroused NSF's objection because it had an important oceanography role yet could not be represented.

Meanwhile, 900 nominations for the Commission cascaded on the White House. Many came from congressmen who had a jurisdictional interest in marine affairs—so many, in fact, that we decided any direct reflection of their preferences would make more enemies than friends. All had to be considered, but a number of the finalists were new names injected from the Secretariat. Congressional representation was fortunately simple—the chairman and the ranking minority member from subcommittees of the two Houses having jurisdiction.[11]

By October 1966 the research was completed, our nominations had been cleared with Humphrey and had gained concurrence from Macy's office. From Johnson, however, came only silence. By December the more volatile members of the trade press were angrily accusing the Council of blocking appointment of the Commission because we wanted no competition. The concern of these ocean-oriented journalists was stirred by more than the legitimate role of gadfly to bureaucratic lethargy. Vendors in the ocean business, especially the high-technology purveyors of instruments and equipment, bought advertising in the trade press. The longer the federal budget drought continued, the greater the danger of small ocean enterprises dropping out of sight, along with their advertising. Even the larger firms, which had been looking for growth through diversification, were beginning to cool to marine markets. So some of the trade press, sincerely concerned about the oceans, were also nervous about staying in business; some of their frustration was understandably vented in the Council's direction. Neither Humphrey nor the Council chose to reveal that Johnson was sitting on the nomination list. In desperation, however, I asked Humphrey in December to intervene. Action came about one month later, when a few changes in the list of appointees were negotiated and the President gave his approval.

We were especially pleased that the President had accepted the recommendation for Julius A. Stratton, the distinguished president emeritus of Massachusetts Institute of Technology, as chairman. Since saying "No" to the Vice President of the United States would be considerably tougher than refusing the invitation of a presidential aide, Humphrey agreed to approach Stratton about the chairmanship of the

Commission. He succeeded in pleading the President's case; and this acceptance by the then Chairman of the Board of the Ford Foundation was perhaps as important as any development that followed; for the success of the Commission depended heavily on the prestigious, objective, and effective chairmanship it received.[12]

On January 9 the White House announced the appointments. Within two weeks Stratton was receiving other messages: strong encouragement to get moving and offers of assistance even from some disappointed candidates; complaints that more oceanographers and aerospace representatives were not appointed; warnings that the Council had obstructed prompt appointment and should be watched carefully. Some "old hands," not appointed to the Commission, gossiped to the press that "the Commission runs a risk of being not much more than a minor advisory appendage to the Marine Council," and further fanned competition by mentioning that the Council had "a half-year head start on the Commission. Catching up may not be easy." After lamenting that "there is concern that the highest level of expertise is not represented," an analysis in *Science* magazine concluded, "Some eighteen months hence one perhaps will learn whether the Stratton Commission has risen to the challenge or has simply confirmed the cynical view that the true role of government advisory bodies is to support whatever views are arrived at by those responsible for their appointment."[13] That question might have been equally applied to the 1966 MacDonald PSAC panel, the 1969 Wakelin task force, the Ash Council mentioned later, all involved in marine reorganization and each entailing a different objective by the appointing authority.

Stratton was able to avoid the usual commission frustration of having to wait six to twelve months for appropriations, since the Council Secretariat had already defended funds for the Commission's launching at the Budget Bureau and on the Hill. He promptly called for an organizational meeting, and scheduled the inaugural session so that Humphrey could be present when presidential certificates of appointment were awarded. The press was invited. Both Humphrey and Stratton read opening statements on the importance of the oceans and on the determination of the Council and Commission to work as complementary, not competing, instruments. Both were sincere; but over the following two years, deliberate as well as inadvertent actions pitted one against the other, with proponents of the marine field curiously unmindful that such warfare could be harmful. Both principals were conscious of the dangers of such territorial dissension, however, and while steadfastly maintaining independence in viewpoint, reciprocal invitations were extended for each to attend all the other's formal meetings, except executive sessions.[14]

In his initial search for staff Stratton received an offer of assistance from the naval aide to Assistant Secretary of the Navy Frosch, offering his personal help and that of the ICO staff that he commanded; but, aware of the probable loyalty of staff to the organization that pays their salary, Stratton was determined to sustain the independence of the Commission and courteously turned the Navy's offer aside. By April, Stratton had chosen as staff director Samuel A. Lawrence, a former examiner for the Bureau of the Budget.

While deferring the question of government reorganization for consideration by the entire Commission, Stratton established panels on basic science; environmental monitoring; management and development of the coastal zone; manpower, education, and training; industry and private investment; marine engineering and technology; marine resources; and international law. The panels were designed for assessing the status of marine activities, for identifying opportunities and problems, and for proposing measures to be taken; their reports constituted the primary source material upon which the Commission based its own final conclusions. That logical subdivision almost undid the Commission at one point, however. Each panel unilaterally headed in its own direction of inquiry with the implicit understanding that when it agreed internally on recommendations in its specific area, the full Commission should accept them. Such a procedure could have led to a patchwork of sectorial considerations, lacking the coherence and objectivity available from the Commission as a whole. Stratton's diplomatic skills saved the day, and during its last months the Commission hammered out a consensus to support all proposals unanimously, as one body. There was no minority report.[15] By virtue of his leadership he was also able to suppress any tendencies by individuals to protect special interests with which they might be identified, noting that the Commission was not to be another pipeline of pressure on the President for parochial concerns.

Unanimity was all the more remarkable because proposals from two panels had initially triggered heated dissent. The Panel on International Law, feverishly examining the international legal regime prior to the United States' negotiations in the United Nations General Assembly, had adopted and articulated a "buffer zone" and narrow continental shelf concept that was an anathema to those Commission members who favored wide U.S. jurisdiction over its adjacent seabed. Eventually, the legal panel made its case and won support by the entire Commission. Coincidentally, most of those initially opposing the "buffer zone" position were members of the panel on marine technology, which also failed to get widespread support for its proposals. That panel's recommendation called for massive undersea technological development, by way of a glittering array of devices, submarines, undersea facilities, and demonstration projects that would have gladdened the heart of all aerospace advocates of accelerated spending. Since these initial proposals were not rationalized by explicit social or economic gains to the nation, the Commission trimmed them down. Nevertheless, the Commission supported a major effort to build general technological capabilities and to undertake pilot developments, with a substantial price tag.

Although the Commission ultimately adopted a comprehensive view of marine affairs, it decided to exclude from their scope of inquiry questions of national security or of the U.S. Merchant Marine. The first was bypassed to limit the scale of inquiry and to avoid confrontation with powerful defense interests; the second, as with the Council, to steer clear of a hornets' nest of maritime issues within the purview of the marine science legislation but seemingly beyond reconciliation.

COMMISSION DYNAMICS

The Commission began with a vast fact-finding set of field hearings, contacting well over a thousand individuals, and pleasing but inadvertently inflaming aspirations of hungry marine interests. By December 1967 the Commission was bustling with activity but had reached relatively little agreement on where it wanted to go. Not that some of its members lacked ideas. The Marine Engineering and Technology Panel had set out an exciting menu of technological dishes. Other panels were moving in separate directions at varying speeds. After soul searching, the Commission found its bearings and elevated its dialogue from a program to a policy level of considerations.

Stratton now recognized that the size and pace of activity required an enlarged staff; at first this was underwritten by transfers of limited Council funds. But Lawrence, sensing a need for staffing beyond the scope of direct congressional appropriations, borrowed staff from government agencies. At that stage, Commission member Taylor Pryor also came to the rescue. Long a zealous entrepreneur in Hawaiian marine activities, he generously volunteered financial assistance for the technology panel staff through his private Oceanic Foundation, and encouraged his board chairman, James Wakelin, to participate. Soon, Commission staff totaled thirty-five, three times the Budget Bureau's initial allowance.

By October, the Commission realized that the eighteen months provided by law for completion of their report would prove inadequate. Quiet probing on the Hill and at the White House indicated hospitality for an extension, and legislation toward that end was formulated by the Budget Bureau and transmitted as an administration request. Congressional approval was prompt.[16] This delay in reporting reawakened grumbling among the outsiders. But later events revealed a distinct likelihood that if the report had reached Johnson on time—July 1968 rather than January 1969—the President might have rejected the report, and the possibility of a new agency might have disintegrated.

The Commission had agreed to Humphrey's request that it serve as a sounding board for Council initiatives, since its members represented a broad range of national (not just federal) interests and were free of the burdens and constraints of operating responsibility. Commission members therefore sat in on Council advisory panels so their advice at preliminary stages of policy formulation might forestall torpedoing in the ultimate Commission report. In turn, the Commission was to realize as they completed their task that a friendly nod from the Council could expedite or even prove crucial to presidential adoption of their recommendations. Not only was their report required by law to undergo a Council critique en route to the President, the only continuity in government for advancing their efforts once they retired rested in the Council Secretariat.

The need for extending the deadline of the Commission report reopened an old issue regarding life of the Council. Both Houses had originally agreed that it should be a temporary body, pending analysis by the Commission and action on their recommendations. The House was especially insistent that the Council's life span was to be so restricted that it could in no way inhibit Commission findings, especially any recommending the Council's demise. Extension of the Commission's life now meant extension of the Council life. By the time this came up for deliberation, House members Lennon and Mosher were strong Council supporters. By then, too, it was apparent that if the Council went out of business in the short four months originally provided after submission of the Commission report, the interval would be far too brief to allow response from the executive and legislative branches and a serious hiatus could occur.

Thus both House and Senate were pleased at a provision of the Commission extension that disconnected the life of the two, pegging the Council at June 30, 1969, regardless of when the Commission reported. Their final date was January 9, 1969. Floor statements rationalized the extension in terms of the political calendar: that the original Commission curfew of July 9, 1968, and the November 9, 1968, Council termination would both have landed in the middle of a hot election campaign and the waning days of a lame-duck Congress. The extension went through without debate.

During its two-year life, the Commission met almost monthly; attendance by members was high; participation was intense as members rather than staff delineated issues, postulated alternatives and selected targets and directions for federal action. The result was a total of 122 major and 52 detailed recommendations covering such diverse issues as described in preceeding chapters on coastal management, law of the sea and national technological projects.[17]

PROPOSALS FOR REORGANIZATION: THE "WET NASA"

Stratton left until almost last the most baffling and potentially controversial question of federal reorganization. While the issue lurked in the wings from the beginning, it was not marched onstage until the spring of 1968. To be sure, candidate proposals for reoganization had been stacking up. One of the first was a brief from Navy staff to its Commission member proposing that the Navy serve as government-wide focal point for civilian as well as military development, especially regarding technology for which they were undisputed leaders among federal agencies. By law the Navy had long provided hydrographic information beyond our shores for civilian as well as military purposes, and now a simple legislative mandate could authorize a Navy-civilian counterpart to the Army's Corps of Engineers. The proponents wheeled up such big guns as John Foster, Director of Defense Research and Engineering, to back their proposition. Swept under the rug in these arguments were Navy difficulties in acquiring budgetary support for even its military-oriented research; congressional appropriations committees were unlikely to approve an additional drive for nonmilitary activities. Yet to be unleashed was the Mansfield amendment to defense appropriations, section 203, which in 1969 explicitly prohibited funding for research and development not having clear military application.[18] Initially, the Commission's technology panel echoed the Navy song, but before deliberations were through, that close harmony had fallen silent.

The second organizational proposal came from an informal, hard-working subcommittee of the three government representatives on the Commission—Robert M. White from Commerce, Frank C. DiLuzio from Interior, and Charles F. Baird from Navy.[19] Free of coaching by their host agencies, they concluded that consolidation of closely related existing agencies within one of the cabinet-level departments was the answer; they proposed Interior. Commission members promptly shelved that suggestion, disenchanted by the historical management disarray of that department, its longstanding failure to support marine affairs, and its historic orientation to land resources of the West. By June 1968 Stratton, becoming concerned about organizational alternatives, invited suggestions from Council staff. At his urging the Secretariat made available its comprehensive study of administrative impediments to marine policy that could potentially be remedied by restructuring of functional components. Of thirteen alternatives, that of consolidating necessary functions into an independent organization looked most promising. Many of these functions already existed, but the full array bore no resemblance to any existing agency. If contemporary organizational entities were to be consolidated, the study suggested, they would have to be put in a mixer to whip up such a change in internal structure as would homogenize their past histories and rivalries and open up the opportunity for fresh leadership. Moreover, the organizational result would be expected to represent both a commitment in the national interest and a service to marine clientele.

The potential for action associated with such an independent agency began to titillate the Commission. Although some of its members looked at reorganization for the usual reasons of economy and efficiency, and some put emphasis on form rather than substance by a sterile organization chart shuffling, others recognized the power dynamics of federal structure and its ligaments to a constituency to be served. To gain for a prospective agency the necessary "critical mass," they now began thinking in terms of including the whole Coast Guard, not just the marine science and engineering components that had been proposed as far back as 1966 by PSAC. The Coast Guard's relatively large budget and manpower pool were a great temptation; consolidated with other elements they would bring the total marine-related budget close to a billion dollars, and that level of funding would surely gain attention.

The final proposal was tentatively crystallized by August. Said the Commission in its final report:

Because of the importance of the seas to this Nation and the world, our Federal organization of marine affairs must be put in order. . . . Present Federal marine activities have grown over the years largely without plans to meet specific situations and problems and are scattered among many Federal agencies . . . [and] relate only marginally to the central mission of the department. . . . [They] have not acquired the visibility and attendant support necessary to be effective. Many of the scattered marine programs are too small to have impact. Equally important, their isolation from each other, . . . has caused an inevitable degree of insularity, overlap and competition. But perhaps most significant, their isolation has made it very difficult to launch a comprehensive and integrated program to remove the obstacles that stand in the way of full utilization of the oceans and their resources.[20]

The definition and diagnosis of that same complaint had troubled Congress for fully nine years. But now, unequivocally, and with authority, the Commission stated as its solution: "Creation of a major new civilian agency, which might be called the National Oceanic and Atmospheric Agency, to be the principal instrumentality . . . for administration of the nation's civil marine and atmospheric programs." Further, it was to be "established as an independent agency reporting directly to the President." It was to be composed of "the U.S. Coast Guard, the Environmental Science Services Administration, the Bureau of Commercial Fisheries (augmented by the marine and anadromous fisheries function of the Bureau of Sport Fisheries and Wildlife), the U.S. Lake Survey and the National Oceanographic Data Center." It was also to include the transferred Sea Grant Program, the government-funded, privately managed National Center for Atmospheric Research, and the NSF programs of Antarctic Research. An essential component of the proposed assemblage would be the weather bureau function of ESSA. This, it turned out, was substantially larger than the ocean functions and well beyond the statutory assignment to and inquiry of the Commission. No one, however, wanted to unbolt the ESSA anatomy, especially since there were logical grounds for integrating ocean and atmosphere observations, research and services.[21]

Thus the Commission felt that it had met its own criteria that "the national ocean program must be of a size and scope commensurate with the magnitude, importance and complexity of the problems it seeks to solve, the services it seeks to render, and its potential contribution to the well being of society." Recognizing, however, that existing marine components did not add up to the full spectrum of needed functions, they recommended that NOAA assume responsibility for: institutional support for university-related laboratories and coastal-zone laboratories; development of fundamental marine technology; formulation and implementation of national projects; grants to states for coastal zone management and research laboratories; and development and coordination of weather modification.

In addition the Commission visualized the executive branch beyond NOAA as engaging in: government-wide planning "to articulate objectives and develop plans for their orderly attainment, including the delineation of responsibility among the various participants"; advocacy to "promote action to advance the national ocean program"; evaluation to "assess the progress of the nation in meeting objectives and inform the nation thereof"; coordination of "policies and basic procedures to assure consistent actions in meeting common objectives"; and communication to "facilitate cooperation among the various marine interests, including groups within the federal government, by insuring effective communication."

For these purposes, some functions would "best be provided by staff agencies within the executive office," while others would be carried by outside advisory machinery. That the Council fulfilled these functions seemed to be overlooked, and the Commission recommended continuation of the Council only until NOAA was formed.

On the matter of involving outside interests, the Commission stated that "participation by principal elements of the [marine] community should be part of the process for formulating major programs and evaluating progress" through "establishment of a committee which might be designated the National Advisory Committee for the Oceans . . . drawn from outside the federal government . . . broadly representative of the Nation's marine and atmospheric interests . . . appointed by the President with advice and consent of the Senate."[22] The Committee was to include observers from each of the federal agencies concerned, and was to have a small full-time staff. Its principal function would be to (1) advise all U.S. government marine-related agencies; (2) inform the Congress; (3) assist the states and private interests; (4) guide the fundamental marine technology development program; and (5) submit a periodic report assessing the national ocean programs in such areas as national goals and long-range plans, facilities, manpower, national projects, scientific investigation and oceanographic operations.[23]

As to in-house, government-wide coordination, the Commission observed that the 1966 Act vested continuing responsibility in the President for planning and coordinating federal marine activities across the board. They also took note that the Council and its staff, established to advise and assist, "have responded with vigor and imagination to the challenge of giving coordination and direction to the present fragmented marine activities." They went on to acknowledge that "issues have been raised and action set in motion which would have been delayed or overlooked in the absence of the Council and its capable and dedicated staff." But they then endeavored to head off potential conflicts in the power structure by recommending that the Council be continued only "until decisions are reached on the Commission's organization plan." To some extent the Commission confused the steering role of a council and the operating role of NOAA and seemed to consider upper-level policy planning and coordination among Navy, State, Interior, NSF, Smithsonian, AEC, and NASA as secondary. And they missed the advantage of NOAA having the Council as an ally rather than a competitor.

Looking to the future, they asserted that "functions of leadership and control remain [apart from NOAA] that can be exercised only within the President's own office." Wistfully they said: "Presidential staff groups will, of course, intercede as necessary on the President's behalf to identify problems not being addressed, to mediate issues, and to exercise leverage in getting agencies to work together on matters of common concern."

No explicit remedy was thus proposed for the endemic disease of departmentalism diagnosed first in 1959 and temporarily controlled by the systems approach and authority of the Council. Retirement of the Council in 1971 required re-establishment of a coordinating committee under the Federal Council for Science and Technology, recalling the earlier era of the ICO, its orientation toward means rather than ends, its tribulations, structural weaknesses, and limited influence.

Apart from federal reorganization, section 5(b)5 of the Act called on the Commission to analyze the full spectrum of marine activities, private as well as public, "including the economic factors involved, and recommend an adequate national [not just federal] marine science program that will meet the present and future national needs without unnecessary duplication of effort." If indeed it were to address non-federal activities, the Commission would have had to examine public/ private goals, roles and interactions. It chose to touch this enigmatic task only lightly.

Having agreed on its findings, the Commission and its staff began the massive job of drafting as its report a full set of informative, attractive, persuasive documents. For selling its conclusions, two groups were to be approached: interested members of Congress (beginning with the four who had been appointed as advisors) and key officials in the executive branch. The Commission report, by law, was to go to the President via the Council and (separately) to the Congress. The first provision was a reasonable routing for the chief executive to gain advice from his in-house marine advisors; the second was a precaution by suspicious House members to ensure that the report would reach Congress in its pristine form, free of any tampering and possible subversion in the presidential office.

Anticipating its responsibilities, the Council on February 28, 1968, had authorized an ad hoc committee to undertake detailed review and to submit comments to the Council; Humphrey asked the Council's Executive Secretary to preside over the committee, appointed at the level of assistant secretary.[24] Stratton, who was present at the meeting, accepted our invitation to try out his draft in a private dry run, calling on the technical skills and government-wide viewpoints of that committee before submitting the report to formal review by the full Council.

The group met for a week in isolation at Woods Hole, Massachusetts, the last two days with Stratton and Lawrence present to receive informal comments. A consensus of the ad hoc committee was sought but not insisted upon; primarily there was a threat that the organizational issue could tear apart the Council itself if members were to adopt parochial positions on reorganization affecting any of their subdivisions. That expectation was later to be realized in bizarre ways. But at Woods Hole participants gradually shed their bureaucratic cloaks and, in a candid collegiate effort, got down to the gut issues on their own merits. Most participants, as presidential appointees unlikely to continue regardless of who won in November, entered debate in an uninhibited spirit. A draft of the entire Commission report, replete with 122 major recommendations, had been circulated in advance and comments were solicited on all.

Unanimity among committee members developed on issue after issue, with the following views emerging as to strengths and weaknesses of the organizational recommendations:

"Federal reorganization is the most important . . . Commission recommendation. It is the best of all alternatives considered and pro-

vides a critical mass for federal marine affairs [with three dissents].
. . . Any adverse effect on the Department of Transportation must be
minimized. . . . More efficient operation of fleets and facilities will be
provided. Ocean exploration and general purpose technology missions
would be provided a civilian home, necessary to future growth." However, "Presidential and Executive Office attention to operational marine
matters will be reduced considerably. . . . The components transferred
should be integrated for more efficient operation when combined in the
new agency. . . . The new agency should have a strong in-house competence to guide its contract program [in marine technology]. . . .
The technology program should be better balanced with social and
economic needs." More elaborate discussion of the ad hoc report was
given by the Executive Secretary in subsequent congressional testimony.[25]

In supporting the critique on agency reorganization, Assistant Secretary of the Navy Frosch contributed significantly to the analysis, ignoring
the power play previously made by his staff. And corresponding officers
of Commerce, Interior, and Transportation agreed on the merits of a
new independent agency. Although far from hostile, representatives for
OST and BOB were more coy in refusing to show their hands.

Our advice to Stratton on many of his proposals included ways to
render the report yet more salable, but, unfortunately, the Commission's
time was ebbing too fast for any of the suggestions to be adopted.

The Woods Hole analysis was tested in a further two-day meeting
of the ad hoc committee at Airlie House November 15–17, 1968. By
then, the Department of Transportation representative had apparently
briefed his boss, Secretary Alan Boyd, and all hell broke loose. The Department, having been strenuously assembled just two years before, was
still vulnerable to fracture, especially since civil airline interests still
preferred FAA's independence, and disaffected marine interests, seeking
an independent Maritime Administration, had blocked the logical in-
corporation of that body into DOT. Boyd had labored diligently to make
the new department work, and he greatly feared that dislodging even
the politically feeble Coast Guard would cause the rest of the structure
to come tumbling down.

Assistant Secretary of Transportation Frank W. Lehan was faced
with the awkward predicament of withdrawing his previous support of
the ad hoc committee's draft. Angered by the threat posed by the imminent report, Boyd was later to write President Johnson a scathing attack on the Council and especially its Executive Secretary for meddling
in cabinet-level operations. While that criticism was checkmated by a
letter of praise from Dean Rusk to Johnson, the fuse of presidential
opposition had been lit. DOT was regarded by President Johnson and
his assistant Joseph Califano as their brainchild. It was a major step

ahead toward improved federal management, and they were determined that no commission on less important issues would undermine what they had worked so hard to accomplish at the cost of so many bruises from vested interests.[26]

In late fall, 1968, Stratton visited several cabinet officers to share the contents of his report directly, partly to inform, partly to persuade. Boyd's antagonism was again apparent. Another impediment appeared when Magnuson could find no time to receive a briefing; Stratton was to discover that this response possibly resulted from Commission staff's deference to the House over Senate interests in the mistaken belief that the Commission owed its allegiance primarily to the side that had been its sponsor.

By November Stratton was keenly aware of President Johnson's adverse reaction. Commission member George Reedy (a long-time staffer for Johnson, most recently as White House press secretary) had sounded out Califano and was sharply rebuffed. And neither Reedy nor Commissioner Leon Jaworski (a Texas associate of Johnson) could pave the way for Stratton to meet with the President. As the administration was coming to a close, efforts seemed academic.

The Commission was comforted by the reflection that conspicuously hospitable reception of their report by Johnson might have led to its rejection by his Republican successor, who would bear the responsibility for implementation. That at least one Commissioner had a direct entree to incoming President Nixon was reassuring.

PRESIDENTIAL RESPONSES

At the Council's last meeting under Humphrey, the ad hoc committee submitted its report on the Stratton findings, and proposed they be adopted as the Council's views in forwarding the Stratton report to the President. The merits of the organizational proposal were emphasized, albeit with caveats, including those insisted upon by Lehan. The response by cabinet officers, however, was far different. Most opposed the reorganization move. Boyd was stoutest in opposition but he was joined by other representatives from Interior, Commerce, and the National Science Foundation; even Secretary Rusk expressed a caution. Only the Secretary of the Navy supported the move.

No action was requested of the Council, but their generally cool reception was reported to a President occupied in clearing his desk for a successor.[27]

One new function conspicuously lacking among existing civilian agencies had been recommended by the Commission as a new engine for NOAA—the buildup of marine technology. This was the muscle that had long been essential in moving toward new maritime goals. More than that, it was the power plant for the whole enterprise, offering the functional analogue to NASA. This feature naturally raised the greatest hope among those industries accustomed to dining with NASA and DOD but now threatened with a low-calorie diet. There had been disclosed, moreover, a treasure trove of projects potent enough both to stir the hopes of NOAA's advocates and to alarm the Budget Bureau into efforts to bury the whole proposal before its riches could be spread out in presidential view. With Johnson, this containment had been unnecessary because of his animosity to dislocation of the Coast Guard.

With the nation poised for the January 20 inauguration of a new president, the January 9 release of the report was not likely to arouse headlines. The *New York Times* nevertheless accorded it a warm reception.[28] *Science* magazine headlined NOAA, "Refitting of Nation's Ark," and resurrected the "wet NASA" appellation that Commission members had hoped would be suppressed because of its negative connotations for budget-minded policy makers. Concluded *Science*, "Although the commission has attempted to launch its report with something of a splash, many of its major recommendations may sink without even a ripple unless the new President gives them his backing."[29]

The Commission recommendation most breathlessly awaited and almost universally applauded by marine interests was that to create NOAA. The trade press was lavish in its praise, but such fireworks from this earnest band of enthusiasts had not previously carried much weight at the Eisenhower, Kennedy, or Johnson doors. Would a Republican president expected to be interested in pleasing industry be different?

In a campaign address at the major oceanographic center of Miami on October 30, 1968, Nixon had supported the marine field but ducked commitment to imminent Commission proposals:

Making full use of the 1966 Act, it will be a first priority of my Administration to present to the Congress an integrated and comprehensive program in oceanography. The purposes of this program will be to:
—Establish national goals for our oceanographic effort.
—Provide a more effective framework for coordinating the work of private industry, of the universities, and of government at all levels. Consideration should be given to establishing a Sea Exploration Agency, with responsibility for coordinating our activities in Oceanography.[30]

No one was quite sure what SEA comprised, especially if a coordinating rather than operating agency. However, campaign rhetoric in America is an art form and not often a serious commitment. Both Democratic and Republican platforms had contained dutiful, even thoughtful, references to future marine development, and Humphrey had issued a strong statement of marine futures toward the end of his campaign.[31]

After Nixon's victory, his exchange of correspondence with Congressmen Mosher and Lennon and letter to me was a hopeful clue, at least as to preventing a hiatus in Council operations. But there was grave uncertainty as to NOAA's fate.

Several business friends of the Nixon administration began communicating their support of the Stratton proposals. Among these was Commissioner John Perry, a newspaper publisher from Florida, whose ardor for ocean affairs had led him to establish a corporation to manufacture small submersibles for the underwater jet set. While a Republican, Perry had supported Johnson editorially in 1964; in 1968 he was a Nixon rooter, and his yacht had often been the site for private strategy meetings by the Nixon high command during the Miami convention. So Perry, feeling his access to the Chief Executive would open doors for Stratton, was optimistic about the outcome. The primary lobby for NOAA was the aerospace industry. Ultimately, it was to gain support from the U.S. Chamber of Commerce and several state governors. But generally, support from users of the sea was feeble, and relatively little emanated from the scientific community that had started the enterprise on this decade-long evolution.

By good fortune, presidential advisor Arthur Burns had placed the Commission report on the Nixon agenda along with an extensive shopping list of major business for the new administration. Feeling every avenue should be explored to gain presidential recognition, the Council Secretariat sought additional communication with various echelons of the new White House staff. As mentioned in chapter 3, the outcome was a presidential request to Vice President Agnew to have the Council review the Stratton report. After briefing new Council members on the scope and content of the government's marine program, Council Chairman Agnew requested them to examine Stratton's key proposals and to reply to him for preparation of Council recommendations to the President. At his further request, the Secretariat personally briefed each new cabinet officer on the role of the Council, on its business carried over from the Johnson administration, and thrust of the Commission proposals. It developed that President Nixon, endeavoring to counter departmentalism from the beginning, had already appointed each of these new and harried cabinet officers to a series of interagency councils invented to deal with a variety of domestic issues, and the Marine Sciences Council looked like just another distraction. Although suffering under the shock wave of new responsibilities, each was respectful of the Vice President's request and found time for a tête-à-tête.

Many were well prepared for the marine responsibilities of their departments and were possibly marine supporters. HEW Secretary Robert Finch was highly knowledgeable about the marine issues from having served while Lieutenant Governor of California as liaison with the Governors Advisory Committee on Ocean Resources. John Volpe, Secretary of Transportation, who as governor had hosted a Conference on the Sea in Massachusetts, was also familiar with the general issues. Secretary of Interior Walter Hickel, surprisingly relaxed after bruising sessions of the Senate confirming committee, was well acquainted as former governor of Alaska with maritime issues on fisheries, eager to consolidate his balkanized Interior Department under new leadership.

Agnew's note to Council members to analyze the Stratton proposals was shunted down into the bureaucracy for advice, frequently from the same staff who had prepared memoranda for their predecessors. Considering that each cabinet newcomer felt he must demonstrate his departmental loyalty in the coming wars with bureaucratic rivals, it was not too surprising that on the NOAA issue the three cabinet officers most concerned should each reply, in effect, "The consolidation makes sense, provided NOAA is not made an independent agency but is lodged in my bailiwick!" One twist of fate found Robert M. White, ESSA's Director carried over by the new administration and advising Commerce Secretary Stans. White was in an awkward spot when consulted on the Department of Commerce position. Although a member of the Commission and fully in support of its recommendations for an independent agency, he was obliged to help his boss in a hard game of territorial chess. Neutral Council members such as Lee Haworth carefully avoided taking sides, but Secretary of the Navy Chaffee (coached by the only DOD assistant secretary carried over, Robert Frosch) took a constructive view that NOAA would serve a broader national interest.

Responses from all Council members on the Stratton recommendations were to be reconciled by an Executive Office committee, representing the White House, Budget Bureau, OST, the office of the Vice President, and the Council Secretariat, charged with formulating a report for the Vice President to transmit to the President. With conflicting advice from cabinet officers, it became evident that the standoff on NOAA was not to be broken. The Vice President's personal representative on the task group, Jerome Wolff, proposed backing out of any further consideration of NOAA because of the political problems it could occasion for his boss. The Budget Bureau and OST representatives promptly endorsed that view, but for other reasons. The Bureau still fretted that NOAA would be a NASA-like sponge on the budget. The OST staff, reversing its prior support of its 1966 PSAC panel, still objected to enhancing marine priorities. The drafting committee requested me to prepare Agnew's response to the President. Unwilling to abandon the NOAA proposal, I stubbornly tested version after version that in some fashion kept alive the opportunity to evaluate NOAA. Finally—a consensus. Six program areas were recommended March 27 to the President by Agnew for consideration by agencies when developing their FY 1971 and future programs, including some Commission proposals that could be implemented without any major reorganization.[32]

Also included was a temporizing recommendation to keep the NOAA issue alive, with its merits to be assessed at the same time as those of other prospective organizational proposals related to the environment, natural resources, and maritime transportation.

President Nixon's response to Agnew's recommendations was announced May 19 with two directives. The first, to Agnew, requested the Marine Sciences Council to consider certain Stratton proposals. The second directive, to examine NOAA, was addressed to the recently created President's Advisory Council on Executive Organization, under the Litton Industry president, Roy L. Ash.[33] At that stage, the Ash Council was a new and little understood creature, born by presidential decree to examine the structure of staff offices surrounding the President himself and such brave and radical ventures to improve federal management as the superdepartments that Nixon submitted to Congress in 1971. Except for John B. Connally, none of the members had any government experience, having made their reputations in the business world. Progress depended on full-time staff assembled from the Bureau of the Budget, many loaded with pet ideas they now had the opportunity to sell.

By this time, incidentally, it was clear that the Commission was to be denied a chance to make its own case directly. Perry was unable to arrange a presidential audience for Stratton. Stratton's personal initiatives with Lee DuBridge and members of the Ash Council were turned aside. NOAA advocacy thus resided with a vocal but small chorus from outside government, with certain persistent Congressmen, and with the Marine Council Secretariat.

Congressional interest had erupted immediately after the Commission report was released, from such supporters as Senator Hiram Fong of Hawaii whose staff, reflecting that state's growing awareness of its maritime geography, had kept track of marine affairs during the long arduous months of report gestation. Interest, however, was largely concentrated in the House. The Lennon-Mosher team signaled its impatience to act, and announced hearings to receive the administration's response to the Commission. They were to be smitten by a dual misfortune. Lennon was temporarily indisposed with a detached retina, and when he returned, he was confronted with the sudden loss of his most effective staff aide, John Drewry, who after 20 years as Merchant Marine Committee Counsel was forced into early retirement by its chairman, partly because he was felt to contribute too much of his time to helping the Oceanography Subcommittee. In September those hearings were opened. To gain some idea about the status of NOAA, Lennon asked the Ash Council to report on its progress. An inept staff response invited the jibe from Lennon that they were just beginning to consider whether to consider NOAA.

In the meanwhile, unexpected pressure descended on the White House from Congress. When the Council was up for its third extension, members of both parties drew attention to the dormant NOAA proposal. Then such influential Republican members as John Anderson and George Bush, who were incidentally not on the Merchant Marine Committee, pushed ever harder from the floor.[34] Joining a group under Mosher's leadership that included Bob Wilson and Rogers C. B. Morton of the Republican National Committee, they repeatedly sought an audience with Vice President Agnew to ask about the administration's priorities for marine affairs generally and NOAA specifically. Without an administration position, he put off this encounter. Finally, armed with the five Nixon-approved initiatives on marine affairs, Agnew agreed.

The session, held on October 13 in the Vice President's ceremonial room in the Capitol, opened with his somewhat vague abstract on the importance of the oceans. Respectfully silent during this analysis, the visitors then opened with questions. Agnew asked aide Jerome Wolff rather than the Council Secretary to reply on NOAA, probably because my views were well known not to accord with the administration's. Ducking the NOAA issue entirely, Wolff put the responsibility on Ash's shoulders. The delegation left, unhappy about both the administration's foot dragging and the Vice President's reluctance to move the enterprise off dead center. All were suspicious that the five-point program was a delaying tactic.

Throughout the Commission study the Council's Executive Secretary had studiously avoided a public position on organization, although a suggestion was made in March 1967 that a public-private organization similar to COMSAT should be considered.[35] By December 1968, in the expectation of being replaced, I supported the idea of an independent agency in a Washington, D.C., speech before the American Oceanic Organization and in an article for *Oceans*[36] magazine. The continuation of my appointment by President Nixon compelled me to return to the discipline of presidential staff and to forgo further public advocacy until the President had made a decision. That silence was eventually penetrated.

In testimony before Lennon September 16, 1969, I was asked to present the administration's position, and my own. I supported NOAA while duly acknowledging several reservations about the Commission report that the earlier ad hoc committee had generated. My support for an independent NOAA at that time was rationalized "on the expectation of better management, greater efficiency and improved performance . . . in providing a host agency for unmet civilian needs such as ocean engineering and technology, and coastal management." New structure was visualized to embrace various functions spread over government, but without explicitly phrasing the support as simply welding together existing agencies.

In my view, all of the previously listed functions would be better performed in NOAA than at present, based on the following criteria: fulfilling public need; contributing to improved presidential decision-making; affording a stable organization of stature that could attract high talent; providing better performance at same or lower cost; subject to more responsive and innovative management; and helping Congress in its oversight function. Most important of all, I suggested that, if the nation were to lead the world in rational use of the oceans and coastal zones, a high level spokesman was needed in the executive branch. No single person had both the position and authority over use of funds to meet these needs.

Testimony in support of NOAA came also from numerous spokesmen, most advocating NOAA as proposed by the Commission.[37] Included were statements from Ed Reinecke, then Lieutenant Governor of California, W. M. Chapman, Jacob Blaustein, Thomas D. Barrow, Hubert H. Humphrey, F. Ward Paine, Norman J. Padelford, and Douglas L. Brooks.

Once the Nixon administration had adopted the five-point program, it soon became apparent that leadership of the Council Chairman was being drained away. Evidently under instructions as a political spokesman, the Vice President was busy hitting the hustings with controversial speeches. Also, stung by his own maladroitness in leading the wild Council herd, he withdrew from that responsibility and left all decisions on major issues to the Secretariat. Without the leverage of an interested Chairman, the future influence of the Council Secretariat was cast into shadow. Perhaps another Executive Secretary could succeed better in enticing Agnew's personal dedication, and that prospect firmed up my own long-postponed decision to resign.

With NOAA not in sight, it seemed almost certain that one more year-long extension of the Council was in order, to June 30, 1971. Resignation on January 31, 1970, would permit fulfillment of my responsibility for seeing the fourth annual report completed, and still give a successor at least eighteen months to get his feet on the ground and gain the satisfaction of accomplishment. The resignation was discussed amicably with presidential aides Peter M. Flanigan and John C. Whitaker, and I was asked to nominate possible successors. Departure was publicly announced on November 12. As the suggested January 31 deadline approached, no replacement had been appointed, so I recommended temporarily filling the post with the most experienced staff aide in the Secretariat who had been my invaluable right arm throughout the Council's life, Roy Dillon. This recommendation was accepted, and

Roy's pro tem assignment carried through to the demise of the Council itself in April 1971.

THE NIXON DECISION—NOAA AT LAST

By late November 1969, despite the Ash Council's tight secrecy lid, the Council Secretariat learned that their draft analysis of NOAA would veto any step toward consolidation. They cited two grounds: the oceans were not important enough to the federal interest to warrant this conspicuous symbol of priority; and any such step should await some larger scale reorganization to deal with all sectors of the environment, including natural resources. Apparently the cautious strategy of the Ash Council was to test draft proposals in the privacy of White House operations, revising as necessary so that the version for public release coincided with presidential endorsement.

Aware that their recommendation could sound the death knell of NOAA, I decided to intervene with the most direct move—a letter to the President. In it I tried to muster all the arguments for NOAA, in a sequence that might attract the attention of the President in his current political frame of mind. The December 4 communication took note of the widespread bipartisan support of his five-point marine sciences program, but stated that another step remained to assure that these initiatives were effectively implemented. Such management problems were enumerated as the independent operation of three civilian fleets, each with idle or poorly utilized ships, the scattered data collection, storage and retrieval capability of ten individual agencies, and the lack of a civilian ocean engineering capability to shepherd our "man-in-the-sea program" and to improve reliability of equipment and thus reduce wasted time at sea. Difficulties of the President in fulfilling responsibilities under P.L. 89–454 for managing this program of civilian marine affairs were outlined as the result of rivalries, lack of agency breadth of view, the low priority in civilian agencies, the vulnerability of cooperative agreements forged by the Council to centrifugal forces of departmentalism, and the piecemeal adjustments in congressional appropriations.

A high-level spokesman and manager was needed via a new, independent agency. Nine criteria were then listed for evaluating improvement in marine functions through creating an agency such as NOAA or other alternatives proposed by cabinet officers.[38] The letter closed with the contention that "the growing problems facing the President in providing desired leadership and direct program management for marine science affairs in the Executive Branch requires Federal reorganization along the lines of the proposed NOAA . . . to carry out the National mandate to use the oceans more wisely and to assist the non-Federal marine community in working toward National goals."

That letter was later mentioned by White House aides as a major contributing factor in the President's rejection of the initial Ash proposals to dump NOAA. Its impact was reinforced by congressmen concerned over the administration's vulnerability on this issue.

Meanwhile, a separate project in the White House had opened a new backdoor to presidential attention. In the fall of 1969, a series of task forces similar to the Moyers-Califano groups had been organized, now largely of distinguished outsiders to develop key issues for the President to consider in his forthcoming State of the Union address and other messages. When consulted about adding a task group on marine issues I concurred, feeling that a favorable recommendation on the Stratton report by a prestigious group might lead to the Nixon imprimatur on proposals previously ignored as stepchildren from the Democratic administration.

In setting up the task force, however, OST grabbed the ball and chose to ignore Council staff in subsequent developments. On October 10, James H. Wakelin, Jr., was named chairman of a group largely from the oceanographic and aerospace communities.[39] The Committee also included Chalmers G. Kirkbride and other NOAA advocates; and since the chairman was the author of the proposal leading to the Commission legislation and had worked with Stratton, everyone had high hopes for the product. They were to be bitterly disappointed. The group met in plenary session several times but evidenced confusion about its assignment. Wakelin was apparently receiving private signals as to what was wanted from at least three sources: the Office of the Secretary of the Interior, OST staff saying that they represented White House views, and from the Ash Council staff. Only Wakelin and Paul Fye met with the White House staff, and they never fully revealed their instructions nor why the study was so politically constrained. However, the inputs were interpreted, or misinterpreted, as instructions to kill NOAA. The smothering was attempted via proposals for a mini-NOAA combining Sea Grant, the National Oceanographic Data Center, and the National Oceanographic Instrumentation Center in an independent agency. The National Marine Agency, as it was called, would "accomplish its mission primarily through existing [private and public bodies,] and be responsible for the coordination of marine affairs among the various [federal] agencies," a role any experienced policy analyst knew could not be achieved, even if the Director of NMA were to chair a coordinating committee for that purpose. Ash apparently endorsed all but the independent status for NMA, planning instead to put it in Interior, and his staff began to lobby on the Hill for support. The Wakelin report was wrapped up December 18 in a form compatible with the second draft of the Ash recommendations, and leaked to numerous congressmen. They transmitted their violent objection both to Wakelin and to the White House. Kirkbride, who was on the task force, joined the protest in letters to the White House.[40]

Congressman Mosher, writing the White House directly to inquire about action on NOAA, received promises from staff that by January 1970 there would be a decision and permission to publicize that schedule. Ash was told that the President wanted some concession to congressional pressure. But January came and went, and all was quiet on the Potomac. The annual Council report was drafted with holes left for a possible last-minute decision, but none was forthcoming.

Administration torpor was an obvious target for Congress. The question was—who would take the initiative? The Merchant Marine Committee had legislative oversight of many of the agencies that would be housed in NOAA, but others were under cognizance of different committees. To push for agency consolidation through H.R. 13247, introduced as a symbol of support of Commission proposals, would inevitably open a Pandora's box of disputes. Thomas A. Clingan who was to replace Drewry was not yet fully on board; Lennon was badly handicapped to take initiative. Thus the House members fervently hoped that the President would bell the cat by creation of NOAA through executive branch initiative, preferably by a reorganization plan that could not be nibbled away by Congress.

On the Senate side, John G. Tower, George Murphy, and Mark Hatfield had taken public stands in support of a new independent agency and were looking to Magnuson to act. Senator Ernest F. Hollings, newly appointed chairman of an oceanography subcommittee, was educating himself rapidly on the issues and was eager to assert his leadership role. Careful not to telegraph his punch in advance, Hollings on March 5 blasted not only the administration but also the President directly, charging indifference. He was supported on the floor by his committee chairman, Magnuson, and by Senate Majority Leader Mansfield. His powerful speech signaled new leadership in marine affairs in the Senate.[41]

White House reaction was cold silence; but the staff was furious, highly sensitive to criticism of the President. Tolerance to dissent was low, with the belief that opposition to the President's position was opposition to the President.

In April 1970 the Ash and Wakelin proposals were on the President's desk, Hollings' speech had been ignored, and the grapevine carried word that presidential action was imminent. Urgent calls were reaching me in Seattle from Stratton and other members of the Commission, and from others fearful of a calamity. At that point, I decided to use a pending address on marine affairs before the National Association of Broadcasters as a podium to publicize the dilatory response to the Commission, and again to press for an independent agency. I was hopeful of rousing support on the issue. There was none.

The President faced a painful dilemma, since any reorganization step required expenditure of political capital. The Congress wanted action, and delay could precipitate a return of initiative in marine affairs to the legislative branch. Temporizing with a mini-NOAA would please few. On the other hand, the Budget Bureau advisors wanted inaction, and their control of the Ash Council led in that direction. Still worse, any moving about of marine agencies might please the Cabinet officer recipient, but many more bruised and indignant losers would be left thrashing about in the wake.

Time and tactics to refloat the grounded NOAA were both running out. But one small possibility appeared when Senator Hollings realized that he had access to one of the most powerful figures in the administration and one to whom the President regularly turned for advice: Attorney General John N. Mitchell.[42] Hollings proposed that he, Stratton, and I should privately meet with Mr. Mitchell and lay our cards on the table, in hopes of gaining his support over the increasingly ominous momentum of the Ash proposals, one of which, to convert the Budget Bureau into a stronger Office of Management and Budget (OMB), had already been successfully lobbied on the Hill by Ash Council member John B. Connally.

All three of us carried the argument. Mitchell was clearly unhappy about the Ash Council's general proposals and was willing to transmit contrary propositions to the President. However, he did not favor an independent agency, on the grounds that too many already reported directly to the President. He saw the possibility of putting NOAA in an existing department, endowing it with sufficient strength to act as though it were independent. He felt that some of those in the Department of Justice worked well on that basis, the FBI for example. Whitaker who attended from the White House listened attentively.

Mitchell did speak to the President, and it soon became clear that Nixon was amenable to this mode of reorganization. The only question was, which department? Volpe, Stans, and Hickel all wanted the package, and all objected to their losses if it went elsewhere. Stans, close to the President after years of loyal fund raisings while Nixon was between public jobs, was the most vocal in his arguments; he was amply buttressed by White's rationale for a Commerce home because ESSA, the largest civilian oceanographic agency, was already there. Hickel was already in a precarious position with the President, although his forced resignation was still ahead. DOT was the least logical recipient, but Volpe, like his predecessor, was persuasive that loss of the Coast Guard would open up the Department to further gutting by those interests who had disliked its original creation.

On July 9, 1970, President Nixon announced his decision by transmitting to Congress Reorganization Plan 4. Said the President:

The oceans and the atmosphere are interacting parts of the total environmental system upon which we depend not only for the quality of our lives, but for life itself.

We face immediate and compelling needs for better protection of life and property from natural hazards, and for a better understanding of the total environment—an understanding which will enable us more effectively to monitor and predict its actions, and ultimately, perhaps to exercise some degree of control over them.

We also face a compelling need for exploration and development leading to the intelligent use of our marine resources. The global oceans, which constitute nearly three-fourths of the surface of our planet, are today the least-understood, the least-developed, and the least-protected part of our earth. Food from the oceans will increasingly be a key element in the world's fight against hunger. The mineral resources of the ocean beds and of the oceans themselves, are being increasingly tapped to meet the growing world demand. We must understand the nature of these resources, and assure their development without either contaminating the marine environment or upsetting its balance.

Establishment of the National Oceanic and Atmospheric Administration —NOAA—within the Department of Commerce would enable us to approach these tasks in a coordinated way. By employing a unified approach to the problems of the oceans and atmosphere, we can increase our knowledge and expand our opportunities not only in those areas, but in the third major component of our environment, the solid earth, as well.

. . . .

Under terms of Reorganization Plan No 4, the programs of the following organizations would be moved into NOAA:
   —The Environmental Science Services Administration (from within the Department of Commerce).
   —Elements of the Bureau of Commercial Fisheries (from the Department of the Interior).
   —The marine sport fish program of the Bureau of Sport Fisheries and Wildlife (from the Department of the Interior).
   —The Marine Minerals Technology Center of the Bureau of Mines (from the Department of the Interior).
   —The Office of Sea Grant Programs (from the National Science Foundation).
   —Elements of the United States Lake Survey (from the Department of the Army).
In addition, by executive action, the programs of the following organizations would be transferred to NOAA:
   —The National Oceanographic Data Center (from the Department of the Navy).
   —The National Oceanographic Instrumentation Center (from the Department of the Navy).
   —The National Data Buoy Project (from the Department of Transportation).[43]

The National Oceanic and Atmospheric Administration thus proposed would consolidate in the Department of Commerce all of Stratton's major components except the Coast Guard; however, the Coast Guard's buoy program would be added to the closely related functions. After pulling and hauling for a decade, a new agency was about to be born. Joy resounded on Capitol Hill and the plan looked like smooth sailing. It was not to be quite that simple: opposition through a resolution of disapproval,[44] H.R. 1210, unexpectedly developed from Congressman John Dingell, who years before as head of the House Subcommittee on Oceanography had found his initiatives sunk by the Kennedy veto. Primarily he saw NOAA as the handmaiden of a Department of Commerce so dominated by industrial interests as to be incapable of objectivity on issues of the marine environment, and the conservationists gave him considerable ammunition.

Nevertheless, the small but determined band of NOAA advocates was vastly relieved at the President's transmittal of the reorganization plan, and none took a critical look into the mouth of the gift horse. Drums were again beating for congressional concurrence although the rhythm from outside was slowed from the exhaustion of too much delay. Many of us consulted by both Congress and the White House on the acceptability of NOAA in Commerce had supported the arrangement as a political compromise.[45]

Disapproval of the reorganization plan was rejected by Congress on September 28. With NOAA certain, speculation began to pop out names of candidates to head the agency. My earlier resignation was correctly interpreted as a firm decision not to have my hat in the ring, entreaties of old friends notwithstanding. Many people were consulted on possible candidates by both committees of Congress and by Will E. Kriegsman on the White House staff, who was earnestly endeavoring to ensure the success of an enterprise that he had piloted through the final months of family debate. As to qualifications, any NOAA leader would have to fight hard for status, especially in the budget competition, before his Cabinet officer, the Budget Bureau, the President, and Congress. The same clutch of the President's advisors who had lost their strenuous battle against the birth of NOAA would now regroup to keep it from acquiring either the funds or the leadership to match its mandate. Success would depend on a NOAA chief's ability to relate marine issues to the compelling social concerns of our time. Moreover, he would need close rapport both with his bosses and with the yet-to-be mobilized constituency from industry and the coastal states who would be users of the sea, rather than generators of information about it. He would be expected to transform the building blocks of NOAA into a homogeneous structure, meeting functional needs and displaying a totally new architecture. Top-flight talent would have to be attracted, to bring the charisma of renewal to an enterprise largely led by the same small cast of competent individuals who comprised the ICO in 1960. This eclectic, interdisciplinary man would also have to be acceptable to a Republican administration.

Such questions were tempered by another consideration: How much leadership did the administration really want? The desired speed of the chariot would clearly depend on the choice of its driver.

NOAA was born October 3, without a selection. Robert White, administrator of ESSA, the largest component of the new enterprise, was named acting head of NOAA; but this hesitation to appoint a permanent chief seriously hampered a prompt start. White was already fighting a hard but losing battle for budget; his initial package of proposals totaled about $450 million, but the Office of Management and Budget, exploiting a weak justification of old ESSA and BCF favorites, cut the allowance to the prior year's level of about $300 million. White and Stans were desperate. Such a drastic reduction undercut any growth potential for NOAA. Worse, it was a blow to prestige that could seriously set back the program itself. In a last desperate move, Stans put his own prestige on the line and traded on his personal friendship with the President. The President agreed to a $30 million increase.

The early months of NOAA found it hobbled, lacking a permanent administrator as well as permanent staff in the other five positions provided by statute. As an intermediate measure, White assigned most of the organizationally significant positions to loyal hands from ESSA and BCF. On January 28, 1971, a presidential announcement confirmed White's selection as head. Number 2 position went to Howard W. Pollack, an energetic, talented former Congressman from Alaska. Number 3 was Robert Townsend, White's ESSA deputy. Numbers 4 and 5 were yet to be recruited. The delay had not been fatal, but the slow start was symbolic.

Other ominous signs appeared. None of the five initiatives announced by the administration in October 1969 was assigned to NOAA. The other agencies received fatter increases in budget. NOAA itself announced no new targets. Coastal management legislation that the administration had supported in the fall of 1969 was now languishing. And marine technology was given not even a nod.

THE TECHNOLOGY GAP AND OTHER UNFINISHED BUSINESS

Nixon's NOAA differed from the Commission's in three major respects: it had been denied independent status; it did not incorporate the Coast Guard; and it was not blessed with new functions, in particular, that of marine technology. Considering the powerful opposition within the Nixon administration, it was perhaps remarkable that NOAA should have emerged at all. Clearly the machinery was bolted together on the basis of what was politically feasible, given both the institutional problems and the personalities, rather than what was desirable. But the omission of a marine technology function was a curious paradox, since closing of the technological gap was a prime objective of many proponents, especially from industry.

During the arduous ten years of policy planning for a truly national effort in marine affairs, the role of technology had been the last to emerge; but by the time the Marine Sciences Act was hammered out its importance had been fully recognized. Section 2(b)6 stated the national policy implemented by the President: "should be conducted so as to contribute to . . . the development and improvement of the capabilities, performance, use and efficiency of vehicles, equipment and instruments for use in exploration, research, surveys, the recovery of resources, and the transmission of energy in the marine environment." Included were tools both for research and for exploitation. In Section 8, marine science was defined to apply to "oceanographic and scientific endeavors and disciplines, and engineering and technology in and with relation to the marine environment." The term "ocean engineering" was, however, so new and its meaning so obscure that the Council in its first annual report felt obliged to devote a full chapter to an interpretation.

The tactics for Council implementation of those provisions had uncorked several difficulties. By far the largest fraction of federal support for ocean engineering came from the Navy, reflected in its complex, varied, but specialized equipment to transit, probe, and utilize the sea for defense. But the state-of-the-art generally was primitive, even as compared to the far newer field of space exploration. And no one had responsibility to correct that deficiency. As the Council had said: "To carry out many of the explicit foreseeable tasks requires a versatile ocean engineering that is not yet in being. No existing agency has specific responsibility for developing a reservoir of general ocean technology."[46] The ambition of the Navy to be assigned that role was not assuaged by the Council in its 1967 set of initiatives, because it was felt that in the face of budget cuts the Navy would not be able to justify funds for such a generalized project designed largely to aid the civilian marine capability. Apart from agency jurisdiction, the Council questioned, "How can the government support multipurpose marine technology serving several public interests, yet lying below a threshold of budget justification for any single agency?"[47] A third set of problems appeared at the interface of government and the private sector, as discussed in chapter 7. Industry could be expected to undertake marine technological development on its own initiative only where the ratio of return to investment compared favorably in amount and timeliness to alternative ventures on land. Between such purely private efforts and governmental military programs lay grey areas in which the federal government might have to sponsor additional research and development to meet objectives of the Marine Sciences Act.

Despite Council support of new technology to extract fish protein concentrate, to develop improved buoys, to prevent oil spillage or alleviate pollution, to improve spacecraft observations, and to advance naval deep ocean technology, over-all funding had not grown to match the opportunities. Serious diagnostic questions arose in 1968 when Council staff failed to elicit from the civilian agencies either clear definitions of engineering steps necessary to achieve their missions or initiatives to defend budget increments for strengthening the existing but feeble engineering capabilities. Many of the agencies were found to share identical engineering objectives, yet in none did they achieve adequate visibility or support. The first step toward federal marine technology was thus to discover how one existing agency could perhaps be given lead responsibility for developing ocean engineering for the entire civilian sector. The smell of NOAA was in the air.

Nevertheless, by 1970 ocean industry was withering from the drought in federal funds for marine technology. Growth scarcely was keeping pace with inflation, and prophesies by the few optimists were rejected. Into this unpromising milieu, NOAA was born, but without the technological glands visualized by its advocates.

Meantime, with attention riveted on the birth of NOAA, other marine legislative issues were stacked on a back shelf. Chief among these were coastal management legislation. Congress was delaying action, partly to determine the relationship of a new program for support of state-managed coastal authorities to the President's February 8, 1971, proposal for land-planning reform. Both House and Senate wanted to lodge that coastal management and research responsibility with NOAA, but they faced a double-barreled shotgun.

First, Nixon had previously assigned such responsibility to the Department of Interior, although it had never been actively carried out. Under its new Secretary, Rogers C. B. Morton, the Department could be expected to repel rivals; Morton was basking in the prospects albeit politically remote, that within Interior a superagency on the environment might be constructed to which NOAA would be transferred. Second, the administration had withdrawn its initial support for coastal zoning to imbed that function in a far-reaching measure of land-use planning that would most certainly not be assigned to NOAA. As discussed in chapter 4, Senator Hollings restarted engines in February 1971 with a bill unequivocally putting coastal management into NOAA's hands.[48] He hoped to weld a new constituency in the field of marine affairs, but by May of 1971 it was evident that the bill was winning only modest support.

Hollings then perceived another issue of growing national importance and far more potent political leverage: unemployment in the aerospace industry. He tied that concern to underfunded marine technology in a new bill intended to inject sizable budgeting support into NOAA,[49] thus bringing the technology issues raised by Stratton back into view. His first hearings were in Seattle, perhaps the most depressed community in the country, with unemployment running close to 15 percent. Invited to serve as his lead witness, the author stressed the need to buttress marine technology with socially relevant goals:

. . . despite some major gains, [the issues reveal] the repeated cycle of governmental indifference to the importance of the oceans; a gap between promises and performance, confusion as to priorities, and a mismatch of resources with goals. . . . NOAA has not met expectations of its supporters, largely because the Administration has not supported it fully . . . it is as though we finally have a new, glittering automobile but with an underpowered engine beneath the hood and with watered, low-octane gas in the tank. The weak motive power . . . is inadequate marine technological capabilities in NOAA to mount a vigorous program of civilian marine affairs. . . . It is through technology that our navy modernized its capabilities to fulfill its functions. . . . The civilian counterpart of naval technology did not previously exist in the agencies that were consolidated into NOAA, and it is lacking today. . . . Paradoxically, in [carrying out the marine sciences mandate] we might help solve three other problems: (1) arrest decay of treasured engineering and scientific capabilities built at public expense over the last twenty years; (2) meet growing threats to our marine environment; and (3) utilize marine technology in stimulating our economy and reversing our growing failure to compete in international markets, a matter that recently led Secretary of Commerce Stans to note that for the first time in 75 years we face a trade deficit.

Examples were then cited of social and economic needs that could be fulfilled by improved marine technology—development and conservation of domestic fisheries; production of protein concentrate to help meet world-wide malnutrition; aquaculture to farm the inshore waters, which promise to be more productive in pounds of protein per acre than any agriculture on land; preservation of our marine environment with better information collected by ships, platforms, buoys, satellites, then stored with modern data processing techniques, structured and made readily available to all clientele; data for better weather forecasting; technologies to deal with the inshore environment by improved dredging technology; steps to prevent beach erosion; pilot development of large platforms for offshore nuclear power stations so as to overcome problems of waste heat dumped into shallow estuaries and to increase radiation safety of nearby cities; rehabilitation of the waterfront in all coastal cities, anticipating containerization and other modern practices; dealing with oil spills, both containment and prevention by harbor radar for surveillance and ship control; and developing new tools for research, including improvement in reliability of instruments.

Emphasis was also given to updating our merchant marine and ship-yards and to the Navy whose funding for naval technology had re-mained roughly constant for the previous six years. (In fact, the gen-eral purpose technology component of the marine affairs budget that had grown from $14.8 million in FY 1967 to $29.4 million in FY 1971 had leveled off for FY 1972 and was to remain static in FY 1973.)

Three points were emphasized: that support for technology was stalled; that it should be more goal oriented; and that NOAA should take the initiative, if necessary impelled by new legislation.

Apart from NOAA's technological horsepower, other questions were being asked in Washington bureaucratic circles about its performance.

The NOAA Odyssey

Ever since 1807, when the first marine agency was established, new functions and new agencies have been created helter-skelter in response to isolated needs—to maintain naval defense, to foster safe navigation, to collect customs, and, in recent years, to aid the domestic fishing industry, control pollution, create harbors, and protect the shorelines. While they shared the common denominator of the sea, these functions were not coordinated, and beginning in 1956, a variety of mechanisms was in-vented to interrelate these diverse elements. In the 1950s an informal Coordinating Committee on Oceanography enabled scientists in dif-ferent federal agencies to exchange data about the environment and re-search intelligence. In 1959, the Federal Council for Science and Tech-nology was created by executive order, and a short time later it activated and then shepherded an Interagency Committee on Oceanog-raphy to coordinate programs and prevent duplication. In 1966 the Congress superseded this informal arrangement with the statutory Marine Sciences Council. Each of these mechanisms of increasing authority met the needs of its time, but improved steering machinery was no sustitute for a stronger propulsion unit. NOAA was designed for that purpose.

Its creation was based on these premises: this nation's dependence on the oceans will increase and require stronger governmental leadership and priority to maintain the quality of the marine environment; to foster extraction of food, energy, and minerals for a growing industrialized population; to assure an effective voice at the international conference table as other nations facing the sea assert their claims; and to focus energies of our states, our industry and academia on common problems. Moreover, a variety of marine activities is supported by a common base of environmental science, exploration, and technology that must grow to meet needs outlined above and must be effectively coordinated as between federal agencies, federal with state, and federal with private.

NOAA was assembled from a number of existing components and

missions. Albeit with new titles, most kept their functional identity.

How well has NOAA succeeded? According to a reporter for *Science* magazine:

Nine months after its creation, NOAA is an anemic agency without clear identity . . . it suffers from a lack of administration support, budget stringencies, and the absence of a constituency. Under NOAA's present leaders, who seem more committed to remaining members in good standing of the Nixon team than championing the nation's marine effort, it is unlikely that the agency will steer the national program on an independent course . . . almost three-fifths of the agency's current $283 million budget goes for atmospheric activities and nearly two-thirds of its 11,300 employees work in atmospherics. . . . Proponents of a strong national oceanographic program express the fear that White (a meteorologist) may tend to give preferential treatment to atmospherics. . . . [Moreover] few innovative changes have been made.[50]

According to Townsend, Number Two man, "NOAA is essentially the old ESSA with some new parts fitted in." Even White was quoted as saying he is "the first to admit NOAA is more like a collection of the groups that came into it than a coherent agency." Further said *Science:* "Conservationists interviewed have little faith that NOAA will fulfill its environmental function as long as it remains under the Department of Commerce," mindful of the initiative by Maurice H. Stans to personally summon Delaware's Governor Russell Peterson (a Republican) to argue against Peterson's strong coastal management legislation. The author was quoted as saying that "NOAA's prospects for extending its authority also hinge on whether it gets control over coastal affairs. . . . To subdivide the marine environment artificially and assign the coastal function to another agency would renew wasteful splintering that NOAA was intended to correct." Yet White could not very well undercut the administration's position of subsuming coastal management under land-use planning, with authority assigned to Interior. Sometimes White's full range of talents were hamstrung.

Very little was said about whether NOAA would or could pick up functions of the Council. Under P.L. 89–454, the President had clear statutory responsibility for coordination and leadership. Advice and assistance were available from the Council, at least until April 30, 1971. Yet it was permitted to lapse without a presidentially appointed Secretariat, and under a Vice President more inclined to partisan politics than to federal management. As a consequence the House Appropriations Committee cut Council funds so severely that it could not even complete its full last year of authorized life, and its supporters in Congress, dismayed, felt they could do nothing.

On the matter of coordination, since important marine functions yet resided in the Coast Guard, Interior, State, NSF, Smithsonian, NASA, and AEC, some unity in approach was obviously required. In the absence of the Council, a new committee was appointed early in 1971 under FCST, chaired by Robert White, the Interagency Committee on Marine Science and Engineering (ICMSE). Gilman Blake was added to OST staff to oversee marine sciences. But by its terms of reference as discussed in chapter 10, and with the uncertain support from its convening authority, OST, there are major questions as to whether the new committee can perform better than its precursor, ICO, in meeting the needs cited earlier by the Commission for "functions of leadership and control . . . in the President's own office."

While the Nixon administration had taken the critical step of generating NOAA, it was Congress that moved to establish the advisory committee recommended by the Commission. Several versions were dropped in the legislative hopper, some echoing the hope of the Commission's technology panel for an advisory body with administrative and budgetary power. But H.R. 2587, which passed the House of Representatives on May 17, 1971, and its slightly different twin, which passed the Senate on August 2, were very close to the Commission's final proposals. The twenty-five-member National Advisory Committee on the Oceans and Atmosphere (NACOA) was to be appointed by the President to review progress of U.S. programs, advise the Secretary of Commerce as to NOAA's performance, and prepare an annual report to the President and Congress on an over-all assessment of the nation's marine and atmospheric activities. The bill, P.L. 92–125, was signed into law August 15, and appointments were announced October 19.[51] William A. Nierenberg was selected as chairman, and William J. Hargis, Jr., as vice chairman. The question then to be answered was, "Just what does the convening authority expect to be accomplished?" No clue was available from the President in the appointment announcement. Nor did he send any message or emissary of intent when NACOA met first on December 11, 1971. Congressmen Lennon and Mosher were present, however, emphasizing the role they foresaw for NOAA and expressing hope that it would move vigorously to assess the state of the government's marine affairs program and not pull any punches when it was first due to report, because 1972 was an election year. In its April 28, 1972, issue, *Ocean Science News*, after an interview with NACOA's staff and chairman, reported that the first report "won't try to shake the establishment," taking note that this is an election year. Federal staff reported that NACOA had taken a poll concerning implementation of Stratton's proposals to advance government programs and had found little progress three years after release of the study. NACOA would be faced with a serious dilemma as to its candor in its first report.

In the struggles to attain and activate NOAA and NACOA, one fact has become clear. The setting of public policy in an effervescent, cantankerous democracy requires more than a tacit presidential nod. To mount and carry out a national policy as vital as the study and development of marine affairs requires leadership. Yet, since 1970, there have been almost no statements as to government-wide goals and plans by NOAA's administrator, his boss, or any high-ranking policy officials. One symbol of emphasis was detail on marine affairs in the special appendixes of the President's Budget; it disappeared after the FY 1971 edition. In 1972, the annual marine affairs report of the President, due by law on January 31 although often delayed until March, had not been released by June. Yet, marine affairs leadership can be given by just one man in the United States: the President, himself. And no matter how loud the Congress fusses and fumes, no matter what words are set to legislative music, the subsequent implementation depends on the operations of the bureaucracy, including those dealing with science and technology in every operating agency and in advisory roles to the President. Portions of the next two chapters suggest what may be expected or required, on the basis of a decade of past experience in marine science affairs.

CHAPTER 9

Policy Implications from a

Decade of Marine Development

Warriors in the Bureaucracy

Marine affairs over the last decade have been so episodic and composed of such disparate issues and institutions as to frustrate description, much less evaluation. While all activities were imbedded in a common saltwater environment, that unity of nature was far less relevant than the unifying—or divisive—values and influences in human affairs that shaped political choice. Marine activities did not take place in isolation, but were intertwined with dozens of other nonmarine themes and shared the same stage. Events completely foreign to maritime issues both fostered and deterred goals. Some of these events were contemporary; some transpired years previously and hibernated until new circumstances set them on collision course with marine-related events. Some developments occurred accidentally and spontaneously, but others were deliberately extracted from the governmental apparatus that traditionally processed issues slowly, sequentially, and sectorially. Marine affairs thus derived much of its texture from the broader envelope of public policy and its generating processes, and from special interactions with science and technology.

Reciprocally, experiences from marine affairs may provide clues to the reduction of broad dilemmas in science policy and in American government, to help guide science and technology to a rendezvous with unmet social goals; to foster a systems approach to the competing, divergent interests in government, and to expose propositions regarding

368

management of federal programs that warrant special acceleration but which defy comfortable departmental assignments and tend to languish in orthodoxy. This chapter is devoted to such generalizations, and in the case of science policy, to some prescriptions.

Like all other sectors of American government, marine affairs have been strenuously challenged to respond to common aspirations of a free people, to satisfy varying demands and reconcile conflicting interests of a pluralistic society, to accommodate new and changing values, life styles, and institutions, and to expand individual opportunities. In efforts toward purification, imperfections and inequalities have been identified and reforms advanced, although not always with prompt or unqualified success.

We find that the President of the United States occupies the central position of power to set national priorities of public purpose and public purse, to harness the energy of a variegated people with a delicate balance among different interest groups so that none can unduly dominate the others. That exquisite balance, however, is not a matter of static equilibrium; it is constantly shifting with changes in national life, in our culture and counterculture. As one result, the incumbent President is obliged to examine a constant parade of issues and their effects on the nation, his office, his party, and his place in history. Simultaneously, the exercise of presidential power has become more difficult because political maneuvering room seems to have shrunk and consensus is ever harder to mobilize.

The primary countervailing force to excessive presidential ambition or inertia and the corrective to inequitable priorities is the Congress. The ratchets of progress and self-renewal seem to depend on a healthy level of tension between the two branches, and there is ample evidence from marine affairs that Congress can take effective legislative initiatives and follow through. There is no automatic guarantee, however, that Congress can fulfill its function. In its choice of issues, Congress often displays a randomness forced by short-term expediency, steered by preferences of powerful committee chairmen, and limited both by time available to members and by size of staff and research capabilities. Because one set of committees controls policy and another appropriations, and because executive coordination has no congressional counterpart, the legislative process suffers from incongruity.

Sometimes the governmental system displays a virtually Newtonian system of checks and balances, so equilibrated and interlocked that the President, Congress, and the bureaucracy are inhibited from individual enterprise to deal directly with an issue. But it does react. The two most telling external stimulants are crisis and political pressure from special interests or articulate citizen coalitions. But the typical, most universal reflex aroused by such threats is a retreat toward the status quo and territorial defenses. The government tends to do best tomorrow what it did yesterday.

While the executive branch is often treated as a monolith, government agencies are as highly varied in their origins and evolution, ideology and biases, subcultures and styles, as people. They may be bold or timid, well organized or confused, creative or parochial, reliable or unpredictable, responsive or lethargic. The bureaucracy is a microcosm of our pluralistic society, its conflicts and forces. While all agencies report to the same boss, they are obliged by their bureaucratic terms of reference to be concerned mainly with their specialized mission and clientele, not with the performance of government as a whole. In the belief that they are involved in a zero-sum game, wherein if somebody wins somebody else loses, the infrastructure of the executive branch acts in a competitive spirit of wary internal rivalry, especially in capturing budget or policy support by the President and in protecting their mission-defined territory. In the legislative branch, a similar zero-sum game is played, but here the anxieties of congressional committees revolve around jurisdiction.

Policy decisions—in management parlance, nonprogram decisions —are inevitably sharpened by rational analysis of costs, benefits, externalities, and the weighing of the consequences of alternatives. Such cogitation can be an antidote for political passions. Yet the higher that staff papers ascend in the hierarchy, the greater the probability that necessary abridgement squeezes them dry of facts, leaving arguments that deal more with the politics of the issue than with the issue itself. The President looks to his staff to provide objective advice, but such expertise is thinly spread. The President is exposed to ploys and strategies of those who seek to gain his support by exploiting his pride, vanity, and biases. Under pressures of the office, the President inevitably delegates a considerable amount of power to instruments of government not politically accountable—to his White House and Executive Office staff. For example, the Office of Management and Budget acts as a surrogate to set presidential policy because the budget process they manage is the government's primary instrument of priority setting. Their operations are designed to force decisions and thus counteract fence-straddling, watered-down compromise or obfuscation by agencies and political opportunism of White House staff. Actions are often required, however, to outmaneuver some bureaucratic enemy and its outside friends. In dealing with the cost-benefit equation, the Office speaks authoritatively to internal costs and "professes faith in economy and efficiency for fear of excommunication."[1] But OMB is seldom qualified to present equitably the external costs, benefits, or alternatives to guide the presidential decision. Very often information on presidential alternatives considered and rejected reaches Congress and the public only through leaks, in which the press plays a vital role to keep the system honest.

Bureaucrats are generally sensitive to the impact of government policy on private institutions, state and local government, and individuals, and are increasingly inclined to consult those affected in advance of a decision; all agencies thus maintain some liaison with the respective congressional committees, special interests and public constituencies. But in this legitimate communication with their clientele, agencies may open up a gulf of independence in their allegiance to the president's program. Staff tend to believe that their agency's fate depends on symbiosis with vested interests. As Seidman has pointed out, "Sometimes the executive branch takes on the appearance of an arena in which the chiefs of major and petty bureaucratic fiefdoms, supported by their auxiliaries in the Congress and their mercenaries in the outside community, are arrayed against the President in deadly combat."[2] Again, the budget is important as the President's main device to enforce discipline.

More and more problems cross departmental (or congressional committee) lines, and the structure of government is less and less able to cope by shoehorning new functions into existing compartments. If national goals are ever to supersede parochial departmental goals, and if the collection of specialized missions is to be integrated into program systems, powerful incentives and coordination mechanics to orchestrate the bureaucracy are required to induce government-wide cooperation. Yet the rigidity of government structure and its sectorial fragmentation resist forces of change.

During much of the Vietnam conflict, the surplus of public risk capital of productive free enterprise was drained off, and with Presidents often insisting (albeit unsuccessfully) on economic strategies of balanced budgets, resources were poorly matched to legislative assignments. Bureaucrats try valiantly to respond, but the system is winded. This exhaustion coupled with frustrations over a succession of budget squeezes and the fire-alarm reflex to crisis has tended to squeeze out predilections to look ahead and plan ahead. The future is thus too readily abandoned to a linear, mundane extrapolation of the past or to the elbowings of crisis.

Government and industry are too often cast and act as protagonists, polarized at opposite ends regarding the public interest. Actually, public and private interests are increasingly difficult to separate. Moreover, the harnessing of science and technology to deliver desired goods and services to the public will require an imaginative, vigorous partnership with industry. Conversely, industry needs government to share certain risks as new arrangements are sought to steer technology to meet broad public goals that an affluent society can afford and now discovers it wants. Industry, however, will need to develop new sensitivities and understandings of nonmarket needs, and a new sense of social responsibility. And government's arrangements with industry must be more visible.

No theory of public administration or political ideology can replace staff competence as the vehicle of effective government. Not only is the quantity of governmental manpower proving inadequate to meet the new responsibilities, the quality also suffers as fewer motivated and competent professionals are recruited. Waves of talent as attracted by Franklin Delano Roosevelt in the 1930s and during World War II, then by John F. Kennedy, have not recurred. In the Nixon administration, candidates for civil service as well as advisory bodies have been calipered more intensively than by other recent administrations to determine partisan political qualifications and ideological stripes. With personnel ceilings clamped down by budget stringency, new leadership talent dedicated to public service may be so sparse as to imperil the government's ability to function.

Given these constraints that importune the American system of government, some entrepreneurial yeast even in small portions serves as the primary antidote to diseases of massive organization. A few blithe-spirited public servants, both elected and appointed, labor with great dedication toward public goals, often anonymously. There are "movers and shakers" on both sides of Washington at every level, activists who understand the intricacies of political process and who are motivated by a sense of humanity, the grandeur of the democratic process or by dedication to a particular cause; by the desire to leave an anonymous fingerprint on history or for personal identity and power; by the tendency common to vigorous bureaucrats to expand the power base of their host enterprise, or even by an individual's "David and Goliath" complex to get his kicks simply out of goading an unresponsive bureaucuracy. Being involved in the decision-making process provides unique opportunities for service to the nation, not only a grandstand seat close to where the action is and to the people who make things happen, but also opportunities to participate in the political sport of moving the system and influencing the future.

To counteract foot-dragging, fence-straddling, and pettifogging, to leapfrog the many hurdles in the vertical hierarchy, and to pacify or neutralize rivals in the horizontal structure requires copious physical stamina, tolerance for aggravation, willingness to invest sizable energies in persuasion and to take risks. Many public servants, facing the hard challenge to integrity, choose to be right rather than be loved. Most of these sparkplugs dispersed in the bureaucratic engines are more than leaders. Intellectually, they demonstrate the horizontal mind of a generalist and differ from the specialist not only in what they think about, but how they think.

Thus, no matter how imposing the architectonics of federal organization, it is the inhabitants, from Presidents to personal secretaries, that animate the public process. The outstanding lesson extracted from the marine affairs experience was the simple fact that most of the events could be associated with live personalities. The characterization of government as a faceless bureaucracy is a fallacy.

Given, then, that a necessary condition for effective government is "more good men," this is yet not sufficient. Of almost equal importance is the policy framework and associated organizational structure in which they persevere. Marine affairs furnish ample evidence that changes in the system were wrought by processes of public policy. For while the kinetics of policy development suggest a maelstrom of restless conflict, irrationalities and anachronisms, underlying that surface turmoil was a deeper stream of slow but trenchant progress toward the purposes of the legislative mandate. At a strategic level, neither crisis nor pressure was severe enough to have induced the resulting action. Two forces proved stronger than the divisive stresses and strains. The first was the focus of a common cause to use the sea more effectively: a commitment to an objective that transcended splinters of scientific disciplines, marine activities, and institutions, differences in ideology, competition for power and prestige, and personal petty rivalries. The second factor was a single national mandate that, since 1966, underpinned all marine activities and called for implementation through a concentrated presidential authority. From 1966 to 1971, they were steered by a common policy-planning apparatus. In one sense the Marine Sciences Council was an experiment in governmental machinery, a device to bring coherence to the granulated bureaucracy in the face of excruciatingly taut budgets that engendered an abnormally tense, potentially nihilistic attitude.

The experience with marine affairs suggests that the legislative mandate of the Marine Resources and Engineering Development Act of 1966 and implementation by the President from 1966 to 1971 with the Council machinery may be a prototype for other fields. Certainly some such machinery is needed in the development of public policy for science and technology, and of procedural innovation when accelerating new technical programs that cross departmental lines so conspicuously as to stump the traditional departmentalism.

THE SCIENCE POLICY PERSPECTIVE

Marine science affairs have shared in the long history of public policy dealing with science and technology that dates back to the patent clause of the Constitution. Benjamin Franklin and Thomas Jefferson were scientists as well as statesmen. They foresaw a congenial blend of democracy and science as common harbingers for human progress. Both activities reflected a striving for progress; both tested emerging truths through critical inquiry; both were subject to constant revision and revitalization, not by authoritarian edict but by public democratic examination. Progress of science and progress of society were thus expected to have a mutually reinforcing, synergistic effect.

After Franklin, Jefferson, and John Quincy Adams, the philosophical connection was lost; and when it was recovered during the industrial revolution, science, and especially technology, had lost its standing as an affair of state and had become primarily a concern of the private sector. Urbanized populations had manifested an appetite for increased standards of living that could be satisfied only by mass production of goods. And the technologies of energy conversion through the steam engine and metallurgy were ripe. Industrial enterprises evolved in free market processes of the Western world to meet this challenge. From this combination of capital, ingenuity, and entrepreneurship came cheap abundant energy, railroads, iron ships, telephones, and automobiles. From land grants to subsidies, government was a silent partner of capitalism. One important exception was in agriculture, where the government rather than industry had publicly funded research, education, and the diffusion of knowledge by extension services. Otherwise, science and government were almost totally independent until World War II.

A second era opened in the 1930s when technology was employed by German, Italian, and Japanese leaders to propel expansionist ambitions. The free world countered on the same terms, basing its strategy on technological skills rather than on a quantitative superiority of men and arms alone. Ever since, science and technology have been conspicuously employed as instruments of foreign policy in working toward societal goals. During this interval, new forms of technological enterprise have spawned such federal agencies as AEC and NASA. Development of thermonuclear weapons and landing a man on the moon probably represented the acme of success in this new style of endeavor, incidentally involving a new federalism with the critical (if controversial) creation of the military-industrial-scientific complex whose power led to admonitions by President Eisenhower.[3]

The nation concluded World War II with a warm glow of gratitude to science and engineering for revolutionary armaments that contributed to a stunning victory with relatively low loss in manpower. Atomic weapons elicited new strategies, and scientists were called upon to join the military professionals in formulating them. In view of the potency of new armaments and the possibility of surprise delivery, our national security could no longer afford the old luxury of peacetime decline of its innovative technical core, with frantic mobilization in the event of a new conflict. The cutting edge of scientific research had to be kept sharp, in a state of constant technological readiness to guard against the unpredictable. Government became the patron of American science, determining the scale and mode of research and influencing applications. The locus of scientific creativity that before the war resided in Europe was translated to America. Paradoxically, military applications of research induced such rapid obsolescence that a continuous refreshment of armaments was required. We had to run faster just to stand still; and at one scale or another, our rivals for world power did the same. As costs skyrocketed in the 1960s, only the Soviet Union remained with the United States in the competition, and both were obliged to inquire at disarmament conferences whether they could afford this technological addiction.

In the 1950s, the attitude began to develop that some form of magic was wrought by the atom and the rocket, so that science became "good for you." Unquestioning faith along with optimistic flights of fancy hinted that science might solve all our social problems. In 1957, Sputnik further precipitated an urgent groping for science nostrums, and governmental support sharply increased for both research in universities and engineering development in the aerospace industry. Partly as a political antidote to anxieties over Soviet space initiatives, partly to strengthen the direct availability of scientific discoveries to the President and to provide advice on security-related affairs, science was moved into the White House with the appointment of a presidential science advisor. This office was soon expanded to a troika with the appointment of the President's Science Advisory Committee and the Federal Council for Science and Technology.

The scientific community applauded this move because it opened a hot line of communication between them and the President. Scientists had created nuclear weapons; now they were concerned with ameliorating the threat to humanity. The prestige and visibility offered science by this proximity to the President were also welcomed. If this advisory apparatus was seen as a promoter of federal funding for new programs, it also could serve to protect science against excessive federal control. The scientific community had historically sought to keep governmental politics out of science. But as budgets grew and as public policy issues were raised of "big" versus "little" science or of geographical distribution of funds for research, the isolation of science from politics tended to fade. At this stage in history, accelerated support for science at the presidential level incidentally provided a backdoor approach for federal aid to higher education.

The first few years of the Kennedy administration saw the sharpest growth of support for research and development in the nation's history.

In 1963 funds for R&D accounted for $14.7 billion—over 12 percent of the total federal budget, and actually about 30 percent of its controllable part. The federal government had proudly become the sponsor of 70 percent of the entire nation's R&D enterprise, thus controlling a major fraction of talented scientific and technical manpower. With funding levels doubling every three years, the Budget Bureau was asking questions privately and Congress was asking questions publicly about an activity so large that it was conspicuous, so exotic and complex that it was hard to understand, and so diffused throughout the government that it was difficult to monitor, yet so close to the President as to circumvent usual strictures on power. What had matured quickly to an influential "science establishment" began to encounter friction with its patron, especially on the issue of accountability. There was little doubt that science and technology had radically altered the relationships of political and economic processes, traditionally based on property rather than on knowledge.

Between 1959 and 1964, three incongruous roots of science policy were emerging. The insiders in the executive branch were practicing a de facto policy of support for all fields of science to maintain U.S. supremacy, for reasons of national prestige and security. The insiders in Congress were arguing for greater control, largely by new apparatus; suggestions ranged from a Department of Science to such coordinating machinery as an Office of Science and Technology, which was finally created in 1962[4] (to serve as a Secretariat of the three precursor units). When research and development funding for security reasons began to level off, the affected outsiders among the scientific elite advocated a different policy—a carte blanche increase for science of 15 percent annually and freedom from governmental interference.[5]

There was no single science policy, however. In the American political system, policy including that for science and technology develops inductively from pragmatic programs rather than deductively from doctrine or party ideology. Nevertheless, in retrospect, relatively little policy analysis was focused on questions regarding the social ends of research, only the means. In the mid-1960s the Science, Research and Development Subcommittee of the House Science and Astronautics Committee was almost the lone congressional unit probing science policy. Their determined inquiry brought forth a new concept of technology assessment and an added emphasis to applied research, but they never fully exposed the roots of the policy problem. Their reluctance to speak candidly about their scientific constituency, their constricted jurisdiction that limited examination to means rather than ends, and their relatively modest position to influence the entire Congress blunted accomplishment by an unusually creative chairman, Emilio Q. Daddario, and his subcommittee.[6]

President Kennedy early in his term called on his Science Advisor's Office and the National Academy of Sciences to study the potential of science and technology to reverse a decline in economic growth and to meet social needs, to probe the adequacy of natural resources, adverse effect of pesticides, and expansibility of the nation's water resources. President Johnson inherited many of these unresolved issues, but the initial quest for social applications of science and technology was aborted. The escalation of the Vietnam conflict dominated both the political and the budget scene. Initially stimulated by a tax cut and expanding so fast as to threaten inflation, the heated economy no longer required technological succor. With ardor for defense spending also dampened by public antipathy toward Vietnam involvement, the clamor by scientists for funding could not alone sustain the momentum of earlier efforts. By the late 1960s and early 1970s universities and the aerospace industry were forced to cut back. That trend continued into the Nixon administration and the plateau of funding tilted downward in real money terms because of inflation. Then suddenly in 1971, there was a new and urgent inquiry by Nixon, with a study led by staff aide William M. Magruder, on how to utilize technology to aid the stuttering economy, counter unemployment and reverse the unfavorable balance of payments.[7] The 1963 quest was repeating itself, but now it was technology rather than science that was to provide the miracle cure. There was, however, a new element, a recognition that defects in the technological enterprise were not exclusively and perhaps not primarily deficiencies in the R&D, knowledge production end of the process, but in the transfer mechanism. Thus, when the product of the Magruder exercise was released, by then made a collaborative project with OST, the remedies were a limited injection of funds into new research plus an appeal for technological experimentation through incentives to industry. Caught in a budget bind, President Nixon in his March 1972 message on science and technology found the concentration on software far less costly than that on hardware, and potentially more potent in meeting the problems. Details of this new thrust, however, were yet to be developed by the National Science Foundation and the National Bureau of Standards.

These oscillations were occurring so fast and so without benefit of policy stabilization that few either in politics or in science and technology realized that the technical revolution was having as swift and subtle an influence on our public policy and the values of society as two or three generations before it had exerted on business and industry. Unfortunately, in both cases it was treated only as another tool. Moreover, technologies conceived with limited purpose and implemented in isolation were found not to operate in isolation. Often they produce inadvertent, unexpected, second- or third-order consequences that impinge on the natural and social environments, effects which may be either harmful or beneficial. In the civilian sector, historically keyed to a venturesome spirit of rugged individualism, technological strategies and tactics, policies and programs, were directed toward exploitation and growth. But the theater for innovation had become so crowded that narrowly conceived developments collided with other goals. Regrettably, the consequences could often not be detected until long after the enterprise had developed high momentum, so that remedies were economically expensive, politically difficult, or both.

Belatedly this oversight was discovered in the voices of protest late in the decade of the 1960s. Concerns were expressed in two contrasting themes. Science and technology had been highly successful in meeting consumer needs and powering the military engine; they had contributed to man's freedoms from hunger, from disease and disability, from arduous physical labor, from ignorance; indirectly, they had produced a surplus permitting higher education for more citizens, opportunities for career choice, and time for selective avocations. Now there were still more neglected or dormant social and economic needs; problems of inadequate delivery of health services, of traffic congestion, crime, urban decline, and environmental decay.

The second issue, emerging as the theme of at least three "best sellers,"[8] recited the same compelling litany of painfully familiar grievances, often cast as an appetite for energy and raw materials fired by an uncritical quest for economic growth. But in underscoring the neglect of long-term consequences of degrading a perishable habitat and the severity and irreversibility of the unwanted consequences, these interpreters of the contemporary scene warned that the time scale of their occurrence was being so compressed that their impact would threaten not some vague future generation, but today's children. Eliciting as much heat as light, some voices argued that technological determinism was a new social disease so morbid that our only salvation was to "turn it off." Proponents of economic growth and environmental conservancy crashed head on. While the smoke has not yet cleared there was evidence that neither side was prepared for the encounter. Some environmentalists seasoned their arguments as much with passion as with a meaningful array of verifiable fact, while some troglodyte industrialists endeavored to brush the evidence of harmful effects aside with steam-roller tactics to batter down all dissent. Sometimes they were aided by alliances with government agencies who tended to become promoters. The SST debate was a beautiful example of the enigma.[9]

Assuming even the best motivation in applications of science, questions of technological determinism were raised. Had science-based technology acquired a life of its own that mindlessly absorbed people, policies, and institutions, modified them, and fitted them into a monstrous process over which man had manifestly lost control? At what critical size does a technological behemoth become immune to social control, its rampant energies unleashed from any purposeful goals and beyond the reach of any destruct button? The gathering storm of protest precipitated a direct confrontation between technology and society. What makes this trend so ironic is that in 1963 science and technology were riding high as prestige activities and welcome guests at the public budget; by 1970, they were on the defensive.

To be sure, federal research and development budgets had leveled off, and while removing some arguments of opponents, the spasmodic, aimless policy of turning the budget faucet on and off has been to the detriment of both the high technology industries and universities. The underemployment of our technological manpower built with public funds was not only a personal tragedy for the individuals affected, but a national tragedy in failing to match these eroding skills to the solution of civilian problems.

Because of the high public investments involved, the guidance of science and technology has become increasingly a matter of public responsibility. Today there is no technology completely independent of governmental stimulation or regulation. With federal involvement so increased in range and degree, the President and Congress had to assume a prime responsibility and make many key decisions for nourishing the nation's science resources of research and manpower, for deploying these resources toward priority objectives, and for exploiting the technological gains to meet public purposes. The technological age had injected a new order of complexity to public administration as well as increased pressure on the budget. The problem, however, is not whether science and technology can be servants of society, but toward what goals and how.

Relationships of science and technology to public policy, that is, science policy, thus involved two sets of issues: one related to maintenance of a first-rate scientific enterprise in the national interest, and another set involving the purposeful contribution of science to social and economic goals. The former might be termed "science affairs" and the latter "technology affairs." To deal with either science or technology affairs, technical expertise is essential to aid the President and Congress. Historically, the President's advisory apparatus focused more on science affairs than on technology affairs, despite the fact that most policy decisions and fund allocations of the President cluster around such explicit objectives as national security, public health, space exploration, agricultural productivity, enhanced economic growth, public transportation, marine resources and so forth. With the international environment changed in that the military confrontation with the Sino-Soviet block has begun to stabilize, the public targets for application of science and engineering have broadened from primarily military concerns to serving urgent economic and social needs. But the processes of bringing science and technology to the civilian area seem to be far less successful than with national security and have bogged down in futility.

A vast difference has been recognized between the military and the civilian customer. Unlike the military, civilian requirements for technical prescriptions are less quantitative, the play of the marketplace unfamiliar, customer consensus less assured, and the locus of decision-making less concentrated. The transfer route for information between knowledge producer and knowledge consumer (unlike that in NASA, which is both supporter and user of research) is confused, ambiguous, undefined in organizational hierarchy and in time. Except for agriculture, and to some extent medicine, the information transfer process has been uneven.

It is small wonder that the influence of the technical advisory apparatus to the President has diminished since 1965.[10] As nuclear weapons and delivery systems reached points of diminishing returns as to speed, precision and explosive power, the requests for advice in relation to military security declined. Foreign policy questions have been routinely answered by prescriptions of more international cooperation and aid to developing nations. And static budgets have permitted few new starts. In some ways, the President's science advisory apparatus has failed to update itself, to deal with technological priorities involving public goals rather than science priorities, to provide insightful objective advice in the civilian sector as well as it once did in the national security sector, to seek ways to couple scientific discovery with unmet social needs, to anticipate unwanted effects of technology in the new era of environmental concern and consumer protection, to blend social values and techniques, and to broaden the mix of disciplines necessary for that step.

Technical expertise in the President's office can no longer ignore the social fabric of the nation. While maintaining a stern ethic of internal objectivity, science has no predilection toward external evil or good. Whether it is utilized to benefit or to sabotage society depends on man's value system and moral decisions, collectively expressed through his social institutions—especially government. Additionally, technologists who assume they control the engine of social change overlook the fact that the general state of society and the value system critically influence goals of technology and strategies. As to technological determinism, it would be a travesty if American society proved so spineless as to have no inner defenses against the impact, even domination, of technology that itself is devoid of spiritual values and cultural ideals. Treatises of social protest over technology, while lacking in balance and accuracy, have joined the new consumerism in demanding improvement in integrity of our products. But the movement also reflects a new view of politics: the citizen wants "a piece of the action."

The very essence of Western thought demands that man *not* lose control over his own destiny. Society should be the keeper of its technology. While such guidance has been thought by practitioners of science policy in various advisory roles to be an implicit purpose of their craft, there is a growing evidence that the formulation of science policy has been neglected,[11] and an admission of error in attacking science rather than technology affairs.[12] It is not simply that practitioners selected the wrong target. It is also possible that the advisory machinery has lacked two characteristics vital in assuring management of such a complex system as technology in society: social awareness to set goals and feedback to correct course.

The viability of technological enterprises requires sensitive integration of broad human values with technical competences. Otherwise, as exemplified by the SST debate, what is feasible may be confused with what is desirable. Management of technology depends increasingly upon a wide range of social, economic, legal, and political considerations that extend well beyond individual goal-oriented enterprises, including a capacity for error detection and self-correction. Unwanted effects must be forecast, and the public interest served in their prevention, a process of "technology assessment."[13] When the government directly promotes certain technologies or finds a comfortable and stable relationship with private interests, bureaucratic inertia and loss of objectivity tend to produce a built-in resistance to the public interest and a deafness to protest and to feelings of alienation.

The country has lacked a national policy that would provide a deliberate, coherent basis for national decisions of investment and institutional arrangements that take into account future effects of existing and new technologies on our nation's social goals, our quality of environment, the health and safety of its citizens. We have lacked institutional means either to make citizens aware of questions and facts that affect their welfare or to encourage participation. The consequences of alternative policies have not been made clear to the electorate: who wins and who loses in each case, and how much. Facts, value judgments, popular preferences, and capabilities of resources, have not been interchanged and mixed.[14]

Strategies for smoothing the application of scientific discovery within the framework of political process are viscous at best; they become congealed when so many different considerations must be taken into account to match technological prowess with social wisdom, unless strong leadership and clearly outlined policy objectives give needed orientation. The federal government occupies the principal role in this enterprise, and the pre-eminent single manager of our society's technology is the President of the United States. In a world vulnerable to extinction by the power of nuclear warfare, decisions at the White House now become a concern of all humanity. Facing ever more serious consequences of error, the President clearly needs to have available to him a source of authoritative, prompt, objective analysis, able to respond to pressing technological questions. The need is exactly parallel in both urgency and importance to the long-standing requirement that the President have at hand capable advisors to deal with unexpected events of key military significance. Several studies have proposed that units in the President's Office be given this explicit task of technological assessment,[15] helping technological enterprise to foster an effective, graceful and humane orientation of technical means to serve social ends; and concurrently protecting the people from unbridled promotion and unwitting handicaps. The author believes that some of these functions must be cast entirely outside of presidential prerogative,[16] in a role for a technological ombudsman sheltered from waves of parochial interests and storms of political emotion, to assure candor in acquainting those affected by technology with the implications, hopefully completed before the political contest itself was incandescent.

Congress also needs its own capability for technology assessment, a matter that Congressman Daddario proposed to the 91st Congress in H.R. 17046. His successor, John W. Davis picked up the cudgel and in partnership with Charles A. Mosher saw H.R. 10243 accepted in amended form by the House of Representatives on February 8, 1972. The Senate version, S. 2302, introduced by Everett Jordan, was strongly supported in hearings,[17] thus raising hopes for the first such statutory capability created in response to the growing interest in sociotechnological goal setting and public management.

The corresponding needs of the President were, in 1972, still unmet.

NEW AIDS TO PRESIDENTIAL DECISION-MAKING FOR TECHNOLOGICAL POLICY

The President of the United States is advised on matters of science and technology by four bodies headed by the same individual. Three are subject to presidential preferences and a fourth was created by legislation, the Office of Science and Technology. OST was formed at a time when federal research budgets were growing rapidly and new, disparate agencies were proliferating. Subsequently, the expansion of science support has slowed, the National Science Foundation has been strengthened, and additional instrumentalities have been created to assist the President generally in coordination, analysis, and decision-making. But the original substance and style of the science advisory apparatus remains.

In 1963, several congressional committees were hinting that their expectations of OST accomplishments had not been met; others were proposing it manage such new functions as oceanography. In 1964 the House Select Committee on Government Research under Carl Elliott recommended that the Executive present an analysis of government-wide research and development impacts,[18] and a subcommittee under Congressman Henry Reuss proposed that the Executive Office publish an annual science and technology report.[19] To carry out either recommendation required strengthening OST, yet the appropriations committees regularly turned down requests for more staff and funds. Congress showed an uneasy desire for change in the Office but with no clear diagnosis of the ailment. In 1966, the House Committee on Government Operations requested the author to prepare a comprehensive analysis on the mission, structure, and performance of the Office of Science and Technology.[20] Since the issues then illuminated sustain a nagging durability, and since in 1972, ten years after OST's birth, no modifications in directions or emphasis have been proposed either by Congress or by any President, key points of the report are reviewed here.[21]

The analysis did not endeavor to test OST against some theoretical model, but rather to study its role in the dual perspective of science and technology in national affairs and of decision-making by the President. Major OST functions in advising and assisting the President were reported to include:

1. development of national policies critically dependent on science,

2. allocation of science resources,

3. organization of federal government to undertake research and development.

4. evaluation of research and development in individual agencies to assure quality in both content and management,

5. coordination of programs that cross agency lines to detect and eliminate duplication or gaps,

6. initiation of new policies and programs that bring discoveries or new approaches to the solution of problems that fall within the federal domain,

7. support of international science affairs,

8. the introduction of scientific and technical leadership to stimulate both the scientific community and the federal government to maintain the . . . implicit policy of world leadership in science, and

9. response to the Congress by providing information and analysis.[22]

While these functions had continued through 1971, two aspects of the political climate that influenced OST's choice of agenda were in transition at least as early as 1965. The first, already discussed, is the broadened emphasis to cover civilian as well as national security issues. The second concerns a shift in emphasis from intramural coordination to long-range planning and goal identification. On these matters the congressional report had already inquired "whether OST provides the necessary assistance to the President—to identify the role of science and technology in meeting different national goals; to illuminate alternatives for investment and the criteria for choice; to facilitate allocation of resources as between these alternatives; to clarify the role of the Federal Government in relation to States and to private institutions." The 1966 diagnosis concluded that "Congress thus seems to be asking more for long range targets than for directives. And it is asking for a delineation of opportunities in the context of broad social goals that comprise the framework for political action." The Congress not only asked, "What have you done?" and, "What are you doing?" but also, "Where are you going?" The study observed that if OST were more technology oriented, "such policy planning might require diversification of staff specialties to embrace economics, foreign affairs, public administration, law, and other social science fields not now represented on the OST staff or on the President's Science Advisory Committee."

In short, structural reform, new planning functions, and breadth of perspective were deemed needed to guide the creative, innovative spirit of the Office and PSAC so as to help the President in a new arena of civilian needs.

By way of a prescription for dealing with special programs, the report noted that:

Several proposals would assign OST a leadership role in fields warranting special attention. . . . Any such role invites the dilemma of OST serving concurrently as advocate and as adviser. If protected from continuing responsibility in a field, the problem of OST becoming parochial about an area and building up a constituency would be mitigated. On the other hand, it offers an organizational solution to the problem of getting something new started in Government without initially having a new agency. Some experimentation in governmental machinery seems essential, first to identify such areas as in the past earned status as "national programs," and second in cases where the field crosses jurisdiction of several agencies and circumstances mitigate against one agency's serving as delegated agent.[23]

The Marine Sciences Council provided an opportunity on a smaller scale to experiment with the very prescriptions for improved assistance to the President advanced for OST.[24] Yet there were major differences in substance and in style between the two bodies. As an advisory mechanism, the Council functioned in an explicit area where the "national program" to be accelerated was set forth in legislative terms, a key impetus to shifting the focus of marine activity to social purposes. Second, the President was given unequivocal responsibility for leadership, both to achieve selected goals and to mobilize all of the separate agency machines to a common purpose; and he was held accountable, to report annually to the Congress on progress. Third, the

Council was accorded political muscle since it was chaired by the Vice President, the only individual except the President who could facilitate voluntary integration by virtue of his stature, elective rather than appointive origins, and his departmental neutrality. Fourth, with separate appropriations and its own staff, the Council Secretariat was really an independent agency headed by an interdepartmental board that represented at the highest level all the implementation arms of the government that would have to be mobilized and harmonized to satisfy the mandate. In its resource allocation cum policy-planning function, the Council operated in the shadow of the presidency with White House clout; with presidential assent, especially regarding budget priorities, it could be influential as well as substantively creative. With access to the top level policy processes, the Council staff could be a legitimate "maritime presence" in the White House, and throughout the entire bureaucracy. At the same time, located in the Executive Office of the President and serving as a shock absorber to protect the President from interagency conflict as issues were sorted out, the Council left the President free to serve as a final court of appeals.

The Marine Sciences Council endeavored to contribute to presidential choices of priorities by counteracting deficiencies in advocacy through the normal budget processes, and in balancing parochial departmentalism with a systems approach. The Council nominated maritime solutions to broad social goals and tried to anticipate impediments through longer range planning, to assist all participants in looking ahead through an annual report as a planning document, and to develop viable initiatives by drawing a bead on a relatively few targets rather than the easier tactic of across-the-board increases that could have by-passed the agony of priority-setting and interagency bickering. It did not hesitate to rock the bureaucratic boat. It established decorous communications with all constituencies and with Congress and the press. The Council was also not short on idealism. Recognizing that no social forces were greater than self-interest, it found short-range altruism often conferred long-term benefits, although this approach encountered obstacles because rewards and penalties were usually short-range and politically potent.

After the President's response to its recommendations, the Council endeavored to follow through within the government in dealing with both the bureaucracy and Congress. Its topical agenda, though narrow in dealing only with the oceans, did not cater to a narrow clientele.

The Council placed strong emphasis on technology as well as science, on software as well as hardware, blending considerations of economics, law, public administration, foreign affairs, resources, and environmental management. It focused far more on "technology affairs" than on "science affairs," but in so doing, it probed for far more in technology than simply technique.

Apart from devotees to oceanography, other outside observers have deemed the Council something of a success. Said William D. Carey, former Assistant Director of the Budget Bureau:

To balance the books, let me say that I have a very different estimate of the Council on Marine Resources and Engineering Development. This body, too, is an interagency committee which advises the President. What is different is that the Council was created by an act of Congress for a fixed time period, its chairman is the Vice President of the United States, its executive secretary is a Presidential appointee, and its money comes through a direct appropriation. While, even so, this Council might have been mere window-dressing, in fact it has been a very lively body. When Vice President Humphrey had the chairmanship, he never failed to show up. He ran the meetings as though he were presiding over a working committee of the Congress. Before each meeting, the individual members of the Council received pointed notes from the Vice President suggesting that he would be disappointed not to see them, and there was a remarkable attendance record. The Marine Council did not hesitate to advise the President; in fact, I have reason to believe that he received somewhat more advice than he cared for. It comes down to saying that this interagency body was an outstanding success story. Was it because of Mr. Humphrey, or because its executive secretary held a Presidential commission, or because it had a time-limited lease on life, or because oceanography is an important subject, or because an appropriations hearing lay around the corner—or did all these factors have a bearing on the results?[25]

Marine science as a subset of science-related activities would be expected to follow in the wake of their fortunes. Considering the lethargic state of marine activities as a whole, it was to be expected that marine science would not even keep pace with other fields during the upsurge of support for science from 1956 to 1963. But, paradoxically, since 1966 when other fields of science and technology, except health, have languished, federal support of marine sciences has increased. Before 1966, one of the endemic problems of marine affairs was its inability to engage presidential support, then to secure funds. The 1966 policy renascence marked a difference. The evidence lay in funding for research and development for marine affairs, remaining a constant fraction of the federal budget during the interval when R&D across the board dropped correspondingly by 50 percent.

Some of that result can be attributed to the multidisciplinary, problem-rather than science-oriented, staff of the Council Secretariat. While not composed primarily of oceanographers as many proposed, the Secretariat did not neglect science for application and equally emphasized both basic and applied research. In fact, application lent dramatic point to basic research.

Despite the Budget Bureau's anxieties, the marine program did not get out of hand. To be sure, the monolithic qualities of the enterprise linking bureaucracy with a powerful constituency were absent. Inside the establishment, specialists who entered the oceanographic enterprise and were promoted in it were scientists by training, temperament, orientation, and life style; they looked on technology largely in terms of producing new and better tools for environmental observation, thus more concerned with "how" rather than "what." Outside, the private sector proved to be either politically unsophisticated or inhibited by a deep prejudice against building up governmental power that could ricochet into regulatory authority. Marine industries were fragmented and not given to effective organization.

With the Council's retirement in 1971, its responsibilities to aid the President to implement P.L. 89–454 across the entire government were implicitly lodged in OST. Considering its past style, it is doubtful if the economic, legal, institutional issues in marine affairs can be resolved if OST follows a science rather than technology orientation. The future of marine affairs thus rests on the general treatment of technology affairs by the President, with his advisory apparatus playing a key role.

In 1972, the nation seemed poised on the threshold of a new technological thrust, but susceptible to far more careful consideration than the 1957 response to Sputnik. To help the President in social management of technology, OST's legislative skeleton and implementation dynamics deserve appraisal. Among other changes such as the congressional study suggested, allocation decisions might well be focused on *objects* of research and development as much as it has been on *subjects*.

To draw on the Marine Sciences experience, technology affairs lack (1) a clear legislative mandate for the social management of technology; (2) a mechanism for assisting the President in sorting out priorities, concerned with ends as much as means; (3) machinery to energize a faster start and to integrate separate public service delivery functions from appropriate members of the federal family into a system when special programs so require; and (4) communication with constituencies affected by federal policies, public and private.

The Office of Science and Technology has a powerful and relevant base; with suitable renovation it could fulfill the needs. First, however, statutory emphasis on technology must be increased well beyond that transferred to OST in 1962 of then-dormant functions of the National Science Foundation.[26] These were specified as "federal policies for the promotion of basic research and education in the sciences" and evaluation of "scientific research programs carried out by agencies of the Federal Government."[27] That limitation in statutory authority did not completely pre-empt OST from technologically based issues. On the other hand, three factors in addition to its restricted statutory basis have hampered OST: the functional difficulty in shifting from military to civilian technology described earlier in this chapter, the resistance of entrenched

bureaucracy in the large operating agencies to Executive Office interference, and the Budget Bureau's objection to an advocate in the President's staff. To overcome these handicaps, to break with the past stewardship of science affairs via a new and more catholic orientation, requires
a fresh legislative mandate for OST.

As to structure to deal with technology, the congressional report
proposed that the four-pronged Office should be split,[28] a notion that
was endorsed by the Daddario subcommittee in 1970.[29] The roles of the
Special Assistant for Science and Technology and of the Science Advisory Committee could be oriented toward independent advice on
national security affairs. They would thereby be separated from a
civilian focus of OST and the Federal Council for Science and Technology and headed by a separate man. National security affairs demand
a sharp cutting edge in both weapons development and arms control.
Urgent too, is a direct pipeline of scientific advice to the President, to
counterbalance proposals from the Department of Defense, NASA and
AEC, and to afford swift communication in the event of portentous discoveries such as occurred in 1939, when Albert Einstein wrote President Roosevelt about the military implications of atomic physics. Neither
the House Government Operations Committee after the release of the
report on OST nor the Science and Astronautics Committee after its
diagnosis has picked up the questions of legislative refurbishment for
OST or of its direction some ten years after establishment. Few federal
agencies have enjoyed such indifference.

The civilian component of OST could be strengthened by legislation
into a statutory Council of Technological Advisors, a mission symmetrical with the functions assigned to the Council on Environmental
Quality in P.L. 91–190, to perform more as a long-range policy-
planning body, than as a quick-response team for administrative
emergencies. In this transformation, there would have to be recognition
that research and development on which OST has been focused is only
one sector of the science and technology charge mandated by the nation's
needs. Technology is more than technique; it calls into play the full
range of ingredients to transform scientific discovery to application:
capital, specialized manpower, institutional devices and management.
For the social management of technology to meet a new set of public
goods and services that are not presented in familiar signals from the
market place, that management will have to be all the more sensitive
and skillful in detecting and satisfying needs.

Policy planning by CTA staff would deal with governmental goals
and strategies, priorities and programs; with impediments to progress
and tactics for their circumvention; with unwanted consequences to
the social or natural environment and alternatives to mitigate them; and

with the definition of beneficial new institutions or relationships between the public and private sector.

One fundamental requisite of such policy planning is "precrisis" not "postcrisis" planning. No matter how cruel the discipline, government administration requires clear goals and a map of the pitfalls that lie in the way. The essence of planning is not only the illumination of alternatives in terms of goals, but details as to implementation strategies, the resources needed, and timetable. President Nixon on July 13, 1969 established a White House staff to determine the direction the nation was moving and to offer alternative courses. As the President said: "We can no longer afford to approach the long-range future haphazardly. As the pace of change accelerates, the process of change becomes more complex. Yet, at the same time, an extraordinary array of tools and techniques has been developed by which it becomes increasingly possible to project future trends and thus to make the kind of informed choices which are necessary if we are to establish mastery over the process of change."[30] While that operation was discontinued soon after it rendered its first report, *Toward Balanced Growth: Quantity with Quality*,[31] the theme should be a *sine qua non* as it applies to change-inducing technology.

This theme, incidentally, was picked up in the March 1972 annual report of the National Science Board. Entitled *The Role of Engineers and Scientists in a National Policy for Technology*, the report made five recommendations—toward government aid in support of industrial technology, technological support for public goods and services, exploration of future alternatives, public understanding of technology, and establishment of institutions for technology assessment. The Board's emphasis on policy planning and on technology rather than on science reflected a 90-degree turn from the past emphasis on science, along arguments presented here previously.

One further required step was exemplified by the very publication of the Board report—the need for an annual review of the public purposes, resources, problems, opportunities, external costs and alternative courses of technological action that are selected by the President to improve social progress and industrial productivity. These priorities should then be subject to congressional review and concurrence. Responsibility for implementation should be assigned to operating agencies, but where programs cross agency lines and suffer dilution in leadership, CTA should follow through as did the Marine Council. Parenthetically, President Nixon asked OST for such a report in the fall of 1970, but by 1972 none had been released.[32]

With the strengthening of the National Science Foundation, its Director and the National Science Board should assume the specific responsibility for assuring the health of the scientific establishment, government-wide. The annual reports of the Board might thus complement the technologically-oriented report of CTA.

While the management of technology to meet the needs of man is a matter of presidential concern, its essence cannot devolve only from the top down. The interaction of technology and society must be understood at every step of technological development and at the interfaces between our fragmented social institutions. In their day-to-day business, every professional man, every industrial enterprise, every university, every state and, on one planet, every nation, is challenged to manage technology, with a far greater sensitivity to the backdrop of values and the changing preoccupation with economic growth, to a sensitive concern for social needs and new technological investments to meet social purposes. In dealing with these problems, the federal government can provide only the resources and the climate. A legislative mandate is necessary but not sufficient. Renewing the constitutional principle that our government derives its authority—and direction—from the governed, participation must be expected from technical and professional communities, from involved business and academic communities, and from the concerned citizen. Such involvement will require better understanding of the role of science and engineering in our society.

While the scientist clings to his preference for undirected basic research, most will admit a private hope that their discoveries can be of utility to mankind. Obviously, the scientific community must have a hand in technology management. But so must engineering, which is numerically larger, institutionally wider spread, and more conspicuously involved in management. By training, by orientation toward purpose, and by career inclination, the engineer is professionally suited to deal with technology. In his employment, he performs the role of technologist— assembling the components of the technological enterprise, translating science into practical terms, and steering the purpose to successful accomplishment. But, as a group, engineers have generally kept silent on major issues of public policy that define technological goals.

C. P. Snow has written of the engineer, "The people who made the hardware, who used existing knowledge to make things go, were . . . interested in making their machine work but indifferent to long-term social consequences." James Killian, President Eisenhower's science advisor, has written, "The engineer's concern for social problems must grow steadily as his work affects society more profoundly." And Whitlock and Edington before the American Society for Engineering Education have said that the problems are those whose answers are not found in engineering courses but which concern people, sociological needs, aesthetic judgments, political decisions.

The engineer is more likely to play the role of synthesizer, but he has lacked the professional training, especially in comprehending the value system, processes, and strategies of the society he wishes to serve. During the 1960s, the profession saw its future as partaking more of applied science, but this did not remedy its narrow disposition to deal with things people want, rather than with people.

Today, scientists are looking for other fields to conquer as their governmental support dries up. It would be a mistake if they abdicated their important role as knowledge producers to move too far toward application. The engineer has an opportunity to bridge the two cultures. But his training and orientation require a far greater breadth. There is a fundamental difference between scientist and engineer. Both have complementary roles to play—an impossible task if either tries too hard to be like the other. In different ways, both can contribute to the rationality of the political decision-making process.

A Systems Approach to Public Administration

Mobilizing all of the technological engines of federal agencies to focus on explicit social goals is one of the most bewildering enigmas of modern democratic government. The problem is how to gain a sense of unity and direction when the compartmented bureaucracy, created one step at a time, is constantly stressed by pluralistic goals of our society and outside clientele. Under such battlefield conditions, the fragmentation is a price for democracy; at the least, it leads to ineffective management in achievement of goals; at most, it can generate a stalemate. The fundamental process for gaining coherence is coordination. Yet many who realize how important is this management device delight in deriding it.

According to Seidman "interagency committees as a general institutional class have no admirers and few defenders."[33] To make his point, he quotes former Secretary of Defense Robert Lovett as saying that committees have blanketed the whole executive branch so as to give it "an embalmed atmosphere," composed of "some rather lonely, melancholy men who have been assigned a responsibility but have not the authority to make decisions at their levels, and so they tend to seek their own kind. They thereupon coagulate into a sort of glutinous mass. . . ." And he quotes W. Averell Harriman as condemning committees "as organs of 'bureaucratic espionage' employed by agencies to obtain information about the plans of other departments which could be used to 'obstruct programs which did not meet with their own departmental bureaucratic objectives.'" But after poking fun, Seidman poses the importance of such functions and admits the Budget Bureau "performed least well in the area of operational coordination," because "they were not and never intended to be a systems manager."[34]

The alternative to decentralized authority with presidential coordination is centralization. Yet there is evidence that the concentration of programs and power in a single agency carries a built-in hazard of loss of control by the President as well as the Congress. And the absence of competition diminishes quality; mistakes by an agency having an exclusive franchise could result in serious undetected consequences. There is strong evidence that the Congress is ambivalent about coordination. Despite their compulsive rhetoric to exorcise the devils of waste, committees do not want presidential interference by doctrines that might interrupt agency collusion with the Congress. So the attitude toward centralization versus decentralization has been ambiguous. Like so many management questions in public administration, *it all depends*.[35]

One great puzzle to the outsider is how government agencies can be so independent. In fact, no one really orders them around. The President is, of course, the chief executive of all, and all are responsible to the Congress. But on a day-to-day basis the division chiefs and their subordinates have enormous freedom to exercise initiative, in achieving the broad goals set out by the basic legislation or presidential policy. Especially to those in the business world who consider organizational structure as a management technique of specialization for economy and efficiency, this is unthinkable. But then, government is not, and cannot be, structured like a business, with effectiveness measured by a profit yardstick.

Few carrots and sticks are available to the President and Congress to foster coordination. Each advocate is expected to advance his function in the face of impediments and to demonstrate his machismo in terms of capturing budgets or enlarging his power base. In the reward structure for managers, promotions are usually appoved by senior officials in their own chains of command, and few credits are awarded for advancing toward common goals.

A scale in five stages to measure effectiveness of coordination was offered to the Elliott Select Committee on Government Research when discussing the coordination role of the Federal Council:

1. interagency exchange of information on current plans . . . an inventory of on-going work, followed by an analysis of possible duplication or gaps;
2. the development and understanding of common goals toward which Government-wide policy and program planning should be directed . . . preparation of staff studies setting forth policy alternatives together with an evaluation of their consequences;
3. communication among agencies of their future plans . . . ;
4. mutual planning ahead and planning together—the development of Government-wide objectives; the comparison of these targets with the aggregate of individual agency plans; collective agreements to fill gaps or eliminate duplication; and finally, the joint use of specialized facilities . . . ;

5. assignment or reassignment of programs to optimize effectiveness of the total effort. This may involve transfer of functions or of funds, or generation of proposals to modify legislative authority.[36]

The collection of federal agencies may thus be thought of as multi-program instruments, sectors of which can be aggregated as needed to serve new clientele. Such a systems approach to government was employed by the Marine Sciences Council. At one time or another, it embraced all five levels of coordination.

The Marine Sciences Council was faced with gaining the best product from an existing array of agencies. In attempting to gain coherence in goals and cooperation in programs, some of the most creative moves of the Council arose from recognition of new cross-connections rather than new elements.

In an environment of departmentalism, coordinating bodies must operate to gain consensus. By and large their techniques, applied in the context of commonly developed fact, reflect the role of mediation and persuasion rather than executive fiat. Coordination may result in a coherent program but it makes few friends. Flawed and imperfect as they are, these coordinating mechanisms become the proving ground for public administration. In the end, it is the sense of community, the suppression of parochial interests to the common weal, the systemic rather than sectorial approach that ultimately tests the degree to which a public enterprise can fulfill its purpose to serve a people that had a noble ideal of individual freedom and sought instruments to enhance and extend it.

What helped provide the coherence in the Council's approach is what Luther Gulick has referred to as the "dominance of an idea."[37] The Council had no additional legal authority beyond what the member agencies contributed, only a license to harmonize the sectors. Yet it saw its role as more than an umpire in adjudicating difficult disputes.

Among the various modes of government operation, the laissez faire, countervailing pressure and federal/state compacts are well known models. So is the alternative of federal leadership. The Marine Sciences Council, its chairman and staff, followed that strategy. Yet, clearly, the Council derived some measure of power from its position in the establishment, regarding both authority relationships among participants and mutual expectations regarding each other. Last, but by no means least, it was chaired by the Vice President of the United States.

New Management Roles for the Vice President

Given that the President must cope with this complex mixture of subject matter and manipulation within a fuzzy mosaic of governmental bureaucracy, under a barrage of pressures from external interests, and against the broad philosophical backdrop of the American dream, a second lesson in the quest for coherence in public administration may be extracted from the Marine Council experience. It demonstrated that more powerful assistance can be afforded the President by way of new management roles for the Vice President, set up either by presidential direction or by constitutional reform.

As for the office itself, John Adams, the first Vice President of the United States (1789–1797) asserted that his country had "in its wisdom contrived for him the most insignificant office that ever the invention of man contrived or his imagination conceived." No wonder that in the stage directions for a parody of American politics, *Of Thee I Sing*, "the name of the vice-presidential candidate, however, is lost in shadow."

The Vice President's key importance lies, of course, in his readiness to ascend to the presidency in the unhappy eventuality cited by the Constitution, "of the removal of the President from office, or of his death, resignation, or inability to discharge the power and duties of his office." Otherwise the Constitution assigns to the Vice President such tasks as opening the certificate listing the votes of presidential electors, presiding over the Senate, and voting in the Senate only in case of a tie. Despite these latter duties and the decree of custom that he officiate at numerous Congressional ceremonial occasions, and despite his official geographical address in the U.S. Senate (not in the White House), the Vice President is not genuinely a member of the legislative branch.

As to the executive branch, both presidents and Congress have toyed with the idea of developing a more meaningful role for the Vice President. According to a brief analysis by Dorothy D. Tompkins,[38] it was President Franklin D. Roosevelt who reinstated the cabinet-seat idea for the V.P. and instituted counseling and liaison activities. Toward the middle of his first term, Roosevelt sent John Nance Garner to the Philippines, Japan, and Mexico—the first time a Vice President was out of the country as an official representative. During World War II, Roosevelt added duties as an administrator, naming the Vice President head of the Board of Economic Warfare and of the Supply Priorities and Allocation Board, and as a member of the Advisory Committee on Atomic Energy. During President Harry S. Truman's administration and at his request, Congress amended the organic legislation of the National Security Council passed two years earlier to make the Vice President a statutory member.[39] Thus, for the first time since writing of the Constitution, a role was assigned by law to the Vice President enabling the standby to keep informed about high level affairs and to participate in policy formulation. The second statutory assignment in 1961 was to chair the National Aeronautics and Space Council. Chairmanship of the interim Marine Sciences Council was the third.

President Eisenhower in a May 31, 1955 news conference said of Richard M. Nixon: "I personally believe the Vice President . . . should never be a nonentity. I believe he should be used. I believe he should have a very useful job." But then he could not answer a reporter's question on what his Vice President had ever done.

Presidents Kennedy, Johnson, and Nixon all used their Vice Presidents in a variety of administrative roles. Johnson was chairman of Kennedy's Committee on Equal Employment Opportunities and of the Peace Corps National Advisory Council. Humphrey served as chairman of numerous councils, and was asked to oversee enforcement of the new Civil Rights Act and to coordinate the antipoverty campaign. Agnew was also assigned the chairmanship of many committees. Vice President Humphrey used to joke about his role as fifth wheel except for these assignments with the Marine and Space Councils as exotic domains "out of this world or under the sea," but he appreciated the two roles cast for the lonely incumbent of the stand-in post.

That a Vice President should be totally knowledgeable about governmental affairs is of course unarguable. He is in direct line to occupy the presidency, and this emergency stepping stone has been utilized eight times in our history. During Eisenhower's sudden illness in September 1955, the issue of presidential succession flashed into headlines and reawakened concerns engendered by the lingering death of President Garfield and by President Wilson's disability before his death. The confusion and uncertainty attending presidential inability were highlighted by the death of President Kennedy and the makeshift quality of Eisenhower-Kennedy-Johnson letters of understanding prepared as stop-gap measures. Early in 1964, the Senate Subcommittee on Constitutional Amendments, chaired by Birch Bayh of Indiana, took a searching look at the succession procedure. Their findings were intended

to assure a smooth transition from one Executive to the next in an emergency and protect against mid-term reverses in government policy; to set up the machinery and the procedure for declaring a presidential inability and to make it absolutely clear that while a Vice President became President when his Chief died, or resigned or was removed from office, he became only acting President when the President was too sick to serve, and gave up the powers and duties when the President was again able to assume them.[40]

Bayh's constitutional amendment passed the Congress July 6, 1965, and was ratified by the necessary 38 states by February 10, 1967.

The most compelling new issue regarding the vice-presidency centers on the obvious fact that the President is grossly overworked and that modern telecommunications increase the pace and variety of workload. While the Vice President is available as ad hoc advisor and confidant, his energies and talents might well be combined with prestige of office to realize a new supportive role.

Legislation provides explicit assistance for the President in functions of the Executive Office, but all of the aides are in categorical fields so their capability to relieve the President is restricted. Furthermore, the Constitution is silent on this matter. Its framers clearly wanted to avoid ambiguities of divided responsibility, and the President shares his power with no one. Rivalry for prestige has been avoided by custom. But the Chief Executive is a lonely man, liable to exhaustion less from sheer volume of work than from the crushing responsibility for decisions which have such far-reaching impact on mankind. Scholars have repeatedly noted that the President "does need someone with Constitutional standing similar to his own and stature and prestige not too far beneath his. He must have someone to help him to oversee national affairs and to confer with him constantly on fundamental decisions."[41] Said Irving Williams:

The assistance of the second officer of the land in smoothing the President's path with Congress, within the executive branch, or at large among the people, is not only called for as an element of a proper understudy role for the Vice President, but is required as an integral part of a presidential administration that will operate at maximum efficiency. . . . The running mate requires a fixed position in the bureaucratic hierarchy so that he may be both well-informed on presidential staff problems and also be the buffer to relieve the President of numerous duties auxiliary to his main job of administering large policies as well. . . . The Vice President must become a lieutenant to the President as well as his heir apparent.[42]

Gaps in the offices of both the President and the Vice President conceivably could be filled by a new, permanent executive role for the Vice President.

In regard to statutory changes in the office itself, numerous proposals have been advanced—to increase the Vice President's legislative powers, to make him a business or operating manager of some explicit functions in the executive branch, to make him an agency head, to make him a presidential advisor or to give him temporary assignments. Some scholars have proposed the Vice President be given a stronger role in foreign negotiations, others in internal management.

Both Hoover Commissions recommended easing the President's administrative workload. The second explicitly proposed that Congress create a post of Administrative Vice President, a matter examined in January 1956 hearings by the Senate Subcommittee on Reorganization chaired by John F. Kennedy.[43] Former President Hoover suggested that an Administrative Vice President (possibly but not necessarily the Constitutional Vice President) might discharge such duties as (1) supervision of independent agencies; (2) performance of certain statutory functions which have no appropriate home in the executive branch; (3) appointment of certain officials such as postmasters; (4) coordination of interdepartmental committees; (5) supervision of reorganization within the executive branch; (6) resolution of conflicts among various executive agencies; (7) supervision of liquidations of executive agencies.

Truman felt any such changes would require Constitutional amendment. Sherman Adams, then an assistant to President Eisenhower, indicated, "I know of no objection which we would have if the Congress should decide to make such an office available to the President for his use," later writing that Hoover's proposal "might prove upon thorough analysis and examination to have advantage."

During these hearings, two veterans of the Senate from opposite sides of the aisle—Maine Republican Margaret Chase Smith and Arkansas Democrat John L. McClellan—pointed out the peculiarly undemanding role in the Senate of the elected Vice President while urgent needs in the executive branch, including better training as a standby, go unfulfilled. Speaking from his experience as Truman's special assistant, Clark Clifford raised the provocative question of a Constitutional amendment to divorce the Vice President from the Senate chair and to strengthen his role in the executive branch. Nevertheless, like most other witnesses, he noted the inherent problem if the Vice President were given a statutory role as administrative assistant: if he and the President disagreed, the President had no option of removing him from office. All agreed that the Vice President could discharge any new responsibilities delegated by the President. Hoover, however, strongly objected to any assignment of administrative functions to the elected Vice President, preferring a new legislatively based post.

From these 1956 hearings the subcommittee concluded the President already has adequate authority to delegate performance of administrative functions to other subordinate officials. The Executive Office of the President had been created in 1939. This, together with enlarged White House staff and new Executive Office arms, gave the President substantial administrative assistance. The McCormack Act of 1950 (P.L. 81–673) had authorized the President to delegate functions vested in him by statute to officers of the executive branch whose appointments require Senate confirmation. And by the Reorganization Act of 1949 (renewed most recently in 1971) the President has ample precedent to abolish, transfer, consolidate, and relocate his agencies for better management, subject to congressional approval. Thus, the subcommittee asserted that "Congress should not take the lead in diluting the President's responsibilities or functions in order to lessen his burden, unless such legislative authority is actively sought from Congress by the President. Authority for the establishment of the Office of Administrative Vice President is not now being so sought."

Subsequently, John Gardner has proposed through Common Cause that there be an executive Vice President to deal with overlapping and conflicting agencies on domestic problems, arguing that the Budget Bureau and White House staff have proven wholly inadequate to the task of coordinator.[44]

The Marine Sciences Council under Humphrey demonstrated what can be done, given the statutory authority, the willingness of the President to utilize it, and the personal inclinations of the Vice President. By virtue of these prerequisites, the Vice President assisted the President in identifying unmet needs and in developing programs and policies to serve them; in recommending priorities and matching resources to goals; in clarifying and coordinating responsibilities of various participating agencies where the field crossed departmental lines, coordinating their activities, and resolving differences; in developing long-range evaluation of future developments and conflicts; in assessing the quality of on-going programs to eliminate the marginal; and in integrating diverse technical, economic, and political considerations. The Council had its share of failures as well as successes, but the bureaucracy, the Congress, and the citizens knew it was there and what it was doing. Humphrey contributed to success of the Council largely from his own qualities of intellect, style, enthusiasm, and leadership. Indeed, the Council revealed qualities of Humphrey unknown to the general public: conciseness in addressing issues, sharpness in phrasing alternatives, impatience with bureaucratic red tape, and breadth of vision in relating government to needs of future generations. In the theater of action associated with marine affairs, Humphrey *was* an Administration Vice President, and an effective one.

Dilution of authority feared by the Senate subcommittee did not occur in the vice-presidential chairing of either the Space or the Marine Council. Because the functional responsibilities for both activities were lodged in the presidency, the Councils clearly served ony to advise and assist the President with implementation.

Nevertheless, some Budget Bureau and especially White House staff were apprehensive over elevation of the office of Vice President because a competitive center of power might sprout close to the presidency and indirectly subvert its function. At a tactical level, the President's closest advisors seem to prefer leaving the Vice President out of highly sensitive matters, a practice having some justification in logistics, because the Vice President is often in transit as the President's representative and simply is not available on short notice. Such assignments inevitably help the President, either in easing his ceremonial burdens or occasionally, as with Humphrey's trip to Europe in 1967, in providing a diplomatic spokesman.

There is ample reason to believe that Nixon had an incipient vision of this role in mind early in his administration when for the first time in history, with much fanfare, he gave his Vice President office space in the White House. But Nixon's early flush of enthusiasm, probably based on his own favorable experience under a staff-oriented (sometimes acting) President, dwindled, perhaps eroded by traditional antipathy on the part of the Budget Bureau and White House staff, possibly because the incumbent was more clearly skilled as political spokesman than as manager. Agnew had no desire, much less experience, to rein the wild bureaucratic horses.

A number of problems attend the vice presidency, inherent in the need to keep politically under wraps, to be subordinate always to the President. Incumbents having to curb all the animal energies that earned stature for nomination can easily get cabin fever, disillusioned by the feeble requirements for public service. While the President needs the Vice President, the inverse is also clearly true, because at present only the President can assign him new tasks. An interdependence, albeit asymmetrical, attends their relationship.

Should such a role be regularized in other areas of national importance? At times it could make real sense: when governmental response fails to match a pressing requirement; when issues, policies, and programs conspicuously cross departmental lines; when a temporary assignment could get swift action on an urgent project which could subsequently be either returned to a normal level or promoted by legislation to a permanently higher status.

The flexibility of such apparatus could give it the advantages of an Ad Hocracy—a mode of administration increasingly employed in business enterprise where rigidity in structure impedes adaptability to changing situations—an adaptive, rapidly changing, temporary system, organized around the problem to be solved.[45] Governmental structure cannot be modified too rapidly or all those elements of society related to it would plunge into chaos; but too slow a response can be critically debilitating. Situations deserving of high presidential priority often suffer from neglect in the separate agencies because they bear lower priority in the range of missions of those particular bodies. The future of the social management of technology, for example, may depend a great deal on strengthening capabilities around the President.

On the basis of the Marine Council experiment, it might be instructive to re-examine the over-all role of the Vice President. One possibility would be to amend the Constitution with a general provision that "the Vice President shall serve the President in such administrative capacity as the latter shall deem appropriate"[46]—a prod to the presidential candidate to choose the best available running mate and an inducement for qualified men not just to accept but actively to seek the nomination. The essential qualities that would make for a competent Assistant President seem often to be overlooked in the heat of compromise at party conventions. If the office were elevated by Constitutional amendment (even at the risk of potential internal conflicts between its incumbent and his Chief) more aspirants might be attracted and the public might demand that the role be respected in the selection of candidates. It may just prove that management ability and leadership of the vice-presidential nominee are as vital in strengthening the ticket as his home state, his religious affiliation, or his ideology. It would certainly be in the national interest.

CHAPTER 10

Alternative Marine Futures

THE PAST AS PROLOGUE

While marine science affairs in the United States from 1959 to 1971 may contribute indirectly to advances in science policy and to a systems approach in public management, they directly set the future trajectory of man's involvement with the sea. Earlier chapters portrayed that development in three distinct stages: (1) 1959–65, intensified preparation by building research tools, (2) 1965–70, policy formulation to turn the seas toward practical benefit, and (3) in 1970, the creation of new marine-related institutions.

The early 1960s saw a reawakening of national interest in the seas, a groping for delineation of benefits and for a strategy to evoke official attention. Even without explicit goals, it was apparent that knowledge about the remote behavior and hidden contents of the sea was so deficient as to merit governmental intervention. In response, President Kennedy took steps to accelerate research by providing ships, laboratories, and trained manpower.

The second stage opened about 1965 with a determined inquiry by the Congress as to structural and conceptual weaknesses in the executive branch, followed by resolute legislation to study and utilize the seas more continuously and effectively. The President of the United States was assigned the key role in deciding marine priorities, assisted by the interim National Council on Marine Resources and Engineering Development, chaired by the Vice President, and by an advisory Commission on Marine Science, Engineering, and Resources. Soon after, the oceans gained more alert attention beyond its role in national security as a source of fuel, minerals, and protein. The beauty and ecological significance of the littoral were rediscovered, but so was its mundane function as a sink for the wastes of an industrialized society. And finally, the oceans entered into international relations as a subject for comity and for conflict. On the basis of Council recommendations, both President Johnson and President Nixon acted to define goals and to provide funds by which the reinforced marine research and engineering capabilities could be channeled to meet needs of economic well-being, a wholesome environment, and human satisfaction. In just one decade, some twenty-five ocean-related initiatives had been advanced by three Presidents. As one barometer of their attention, federal funding for marine science affairs had increased by roughly a factor of five, proportionately more in the civilian as compared to the military sector.

The early 1970s opened a third era with steps to strengthen state and federal institutions, both to engage the sea more vigorously and to protect the quality of the marine environment against human abuse. The establishment of the National Oceanic and Atmospheric Administration, similar to the centralized civilian agency proposed by the Stratton Commission, and the National Advisory Committee on Oceans and Atmosphere were intended to bestow higher national priorities on marine affairs: the medium was the message.

Given the conservative tradition of ocean activities, their periodic neglect and twentieth-century isolation from the main stream of human concerns, the fragmentation of participating interests and their low-keyed advocacy, and given the absence of palpable crisis to excite public attention, the advances in the dozen years from 1959 are remarkable. That progress, so painfully achieved, is often taken for granted.

By no means, however, has the transformation from awareness to action been completed. Certainly interest in the seas is here to stay. Further developments of marine affairs are inevitable, but their direction and pace are ill defined. Pronounced internal trends as well as external forces of change and broad social strategy will determine that course.

Technology, which has already opened the way to exploitation of the sea, can be expected to go into higher gear as a driving force, although now steered to protect as well as consume marine resources. Human needs may increasingly find satisfaction from the sea. Opportunities for the remainder of the 1970s are well enough documented that marine-related activities could well increase at a rate exceeding ten percent annually, faster than either the population or the over-all economic development. Extraction of offshore oil and gas, and the bulk shipment of fuels, ore, and minerals are expected to mount swiftly. Contemporary, random development of the coastal zone may well increase with the encroachment by urban sprawl. Consumption of fishery products seems destined to grow, whether or not resources are harvested by American fishermen. Scientific exploration of the seas and the conveyance of environmental services and other measures to protect life and property will certainly expand.

These economic and demographic indicators, however, are surface manifestation of a new and complex phenomena. When oceanography came of age, the marine enterprise moved from specialized, low-visibility, military applications and from the quiet of the laboratory to the brightly lighted and noisy stage of the political theater where scores of public and private actors promptly took their cues. Marine affairs thus became subject to the same mix of values, technical, economic, legal, and institutional factors and conflicts between special interests that determine public policy in a democratic, pluralistic society. The marine component of public policy will thus be put to new tests of its substantive viability and its political energies.

Internationally, attitudes toward marine affairs have been shaping up along similar lines. Casual appreciation has changed to evaluation and movement based on national self-interest, sometimes enlightened, sometimes not. With a few conspicuous exceptions (particularly the U.S.S.R.) developments lag well behind the United States, in most cases by some five years but in others perhaps by fifty. The United Nations and its many specialized agencies have stepped up their involvement, putting one sector of the marine environment off limits to weapons of mass destruction, and after lengthy debate, declaring the resources of the seabed beyond limits of national jurisdiction to be a common heritage of mankind. Widespread pessimism, however, greets the question whether the United Nations can succeed in extending marine cooperation. Nationalism has continued to dominate action, whether by large or small countries. Many nations have chosen to assert unilateral claims to marine resources and have even sought to control oceanographic research off their shores. The United Nations' Conference on the Human Environment in 1972 and Law of the Sea Conference in 1973 provide opportunities to deal with such new problems as pollution control and with old ones that have been amplified by technology.

As individual nations and blocs prepare positions in relation to these potentially seminal convocations, the outcome is doubtful. Certainly the status of international cooperation will have repercussions on domestic policies and programs of every nation, and vice versa. And to complicate the problem further, the functional uses of the sea, the ecological implications, and the policies have become so entangled as to render obsolete historical sector-by-sector treatment. Internationally as well as domestically, a wide range of alternatives lies ahead.

THE DOMESTIC SETTING FOR MARINE DEVELOPMENT

In marine affairs, the federal government will continue to play the major role in leadership. With the creation of so many new components of machinery, however, bureaucratic power relationships could well be upset. NOAA added an operating-level advocate to the enterprise, but it will have strenuous competition and its ability to command funding priorities and to develop an energetic constituency remain unconvincing. The loss of Secretary of Commerce Maurice Stans to help in the 1972 presidential campaign of his close friend Richard M. Nixon introduced further uncertainties as to presidential support of NOAA; past budgetary battles have been stormy, and demands on the public purse to meet mounting social needs can only aggravate the competition for funds.

The dismantling of the Marine Sciences Council may have two effects. First, opportunities to present government-wide policy options in marine affairs directly to the President will be lessened. Second, lacking this central, politically powerful, and potentially creative leadership, the marine programs of various agencies that respond to the diverse clientele outside of government may be shattered by family quarrels among rival agencies. Whether NOAA and the Office of Science and Technology can assume the functions of coherent, government-wide policy planning and coordination, and whether the statutory National Advisory Committee on Oceans and Atmosphere can serve (or is desired by the President) as a new source of advice on policy and priorities, remains to be seen.

Two other new agencies have marine cognizances and confuse the picture as to NOAA's ultimate role and influence: the Council on Environmental Quality created in 1970 to advise and assist the President, and the Environmental Protection Agency created in 1971 to set and enforce standards of environmental quality.

As to politics, marine affairs have been historically nonpartisan; since they have little prospective impact as election issues, they are likely to remain that way. On the other hand, tension between Congress and the Executive about marine policy is certain to continue. If past experience is a reliable clue, such conflict would be to the benefit rather than detriment of the field. Much depends on decisions of key members of Congress as to the time and energy they are willing to continue to invest in a cause that has brought limited political rewards to its devotees. The voluntary retirement of Congressman Alton Lennon from the House of Representatives in 1972 leaves a critical role for his successor.

Another determining aspect lies in the general social setting. By and large, marine matters have been ignored in the demands of the nation's more turbulent youth and others who feel unrepresented in the power elite. The oceans are not on their social agenda. True, a new entourage of concerned citizens and of watchdog environmental organizations has gained a place at the political table, and the National Environmental Policy Act in the late 1960s turned a spotlight on the health of the oceans and especially of the coastal zone. By 1971 the courts had been drawn into the ferment of environmental protection, and their decisions had deep-reaching effects. Those concerned with the sea were also challenged to identify potentially harmful effects of pollutants, to specify safe limits, to evaluate other forms of man's intervention in the natural environment, and to develop prophylactic measures. The ideological collision between economic growth and environmental protection implicit in this issue immediately set up skirmishes likely to continue for many years.

Related to this development was the recognition by President Nixon and a number of members of Congress of the need for new national policy governing land use. Where the traditional, laissez-faire attitude had once permitted short-term market economics to preempt future uses in the pattern of urban development, it was now deemed essential to determine the use of natural resources by more rational processes. Conflicts arising from unconstrained exploitation of the coastal zone for quick profit epitomize this need to plan for multipurpose use and protection, but steps to serve the public interest are strenuously resisted.

Two other issues that erupted into public notice in 1971 may influence marine futures. The first concerns the unfavorable balance of payments. Since the United States imports 70 percent of its fishery products, ships its goods largely in foreign bottoms, and imports both fuel and minerals that could be extracted from the seabed by U.S. industries, questions are naturally raised about the potential of U.S. maritime policy to redress that balance. The government can help domestic industries revitalize their operations and strengthen their competitive position by a variety of instruments including, but by no means limited to, the acceleration of technological development. The second issue concerns more effective utilization of the nation's scientific and engineering talents. Assembled as a public investment over a period of 25 years to protect our national security, but then underemployed because of military cutbacks, these skills and their virile systems-management organizations could be transferred from the aerospace field to maritime enterprises and other new areas that offer payoff in social and economic goals.

All of these assumptions tend to be accommodations within the present value system. Other assumptions might envision changes in our perspective and institutional settings wrought by countercultures. The quality of human life may become paramount, and acceptance of dissimilar life styles may reduce the alienation that has beset our society.

While it seems unlikely that even such radical changes as these could significantly alter marine affairs in the near term, they could spell a vastly different future in the long term. For if society assumes responsibility for a stewardship of the global environment, as discussed at the end of this chapter, this may be an indication of a broader movement toward a world free of violent strife and disorder.

MARINE STIMULI AND IMPEDIMENTS

Advances in marine affairs, as in other areas of public policy, generally await crisis or legitimate outside pressure to force action. Neither stimulus seems potent enough at present to alter maritime priorities appreciably. In the near term, except for advocates of environmental protection and coastal management, the maritime constituency is unlikely to broaden or to grow more articulate. The National Academies of Sciences and of Engineering have upgraded their ocean committees to board status, to represent a new breadth of approach, but their 1959 influence on governmental priorities has not recurred. The most powerful private interests have been in the petroleum business, but since they are obliged to reconcile many different issues when optimizing their economic stance, offshore aspects of petroleum development cannot claim a special status in their strategy. The large number of other maritime constituencies has yet to develop a workable coalition. Moreover, some political forces may inhibit marine initiatives; local political subdivisions and entrenched coastal industries, for example, have opposed any policy for environmental management. Fishermen have resisted federal rather than state resource management, and have fought concepts of limited entry that would give the industry some modicum of competitive edge.

As to private venture enterprise, there are few indications that incentives for investment are great enough to conspicuously alter rates of growth or modes of operations in the fishing, shipping, or other marine industries. At least the prospects seem dim in the absence of new national policies that would include governmental assistance.

This account is not likely to kindle unabashed optimism. On the other hand, the two economic motivations cited earlier could conceivably impel accelerated marine activity. The United States is a heavy consumer of all the resources that the oceans at this time can potentially deliver: fish, petroleum, manganese, copper, nickel, tin, magnesium, phosphate rock, and sulfur. Except for magnesium and sulfur, we import far more than we export, resulting over the period of 1950 to 1970 in a cumulative balance-of-payments deficit of almost $40 billion. The annual demand for these industrial materials is increasing, and there is no reason to expect a major shift in the trend. Spangler[1] estimated that from 1960 to 2000 about $67 billion will go toward imports of cobalt, nickel, copper, and manganese alone. Economists may debate how serious such a deficit is to the economic position of the United States, although none would assert that it is healthy especially when, as occurred in 1971, total imports exceeded exports. Nevertheless, there are other subtle considerations. Some of the imported metals are produced by American-owned companies. More importantly, the sale by less developed countries of raw materials constitutes their main source of hard currency available, among other things to buy needed manufactured goods, thus establishing important indirect assistance while creating potential markets for the products we can produce efficiently. The private American companies involved also argue that lower labor costs abroad create a price differential that makes foreign sources attractive, even necessary, when competing with their manufactured goods in a world market. The balance-of-payments argument may also lose some potency with the December 1971 devaluation of the dollar and with troop withdrawals from Southeast Asia. But the fact remains that because the United States has access to new technologies and the capability to form the heavy capital investment required for offshore mineral extraction, it is likely to be in a competitive position to operate world-wide, in partnership with any number of countries or even with international bodies, in helping meet the world-wide demand for energy and ore.

Apart from economic considerations, a number of current environmental trends approach the level of a slow crisis that forces attention to the sea. The depletion of certain fishery stocks on both coasts of the United States by foreign fishing fleets, the unconstrained injection of pollutants into ocean waters, the growing ocean traffic in oil, and the destruction of natural ecological habitats along the shoreline all have potentially irreversible consequences and urgently demand solution.

National security in its broadest sense should provide another powerful stimulus for marine development. Both the Marine Sciences Council and the Commission have sought to evolve a new strategy of the oceans based on foreign policy, but their efforts have not produced any over-all demand for a more visible U.S. presence at sea. Notwithstanding trends toward isolationism, the treaty obligations of the United States and the staccato of threats to world order amid withdrawal of U.S. troops from Southeast Asia demand that we maintain a strong mobile naval force. Near East and Indian Ocean uncertainties reinforce that need. Yet seapower has many dimensions that go well beyond gunboat diplomacy. Measures toward peaceful use of the sea could well earn us respect and support in the international arena.

National prestige as a reason for strengthening our maritime activities has not been forcefully argued in the past few years. To some extent the coolness to this rationale is a backlash following the U.S. manned lunar landing. But as a desirable alternative to force, prestige remains a critically important diplomatic tool in a world still jarred by militant nationalism. Different aspects of our national prestige have earned lasting admiration in the world community—our spirit of independence and love for freedom, our open society, and our willingness to accept, even promote, progress. Prestige associated with ocean endeavors could be equally admirable if the goals were clearly aimed away from short-term self-interest, and toward the sharing of benefits with all nations. The United States was the first nation in history to adopt such a policy as a basis for exploration and use of the sea, in line with a national tradition that emphasizes the spirit of discovery, an affection for creative innovation, and a willingness to adapt to technologically induced change. The challenge of an ocean frontier seems to "turn on" youth everywhere.

Both President Johnson and President Nixon have renounced territorial aggrandizement and proclaimed principles of freedom of the seas that are compatible with aspirations of other nations. Ironically, these initiatives by the United States have been regarded with almost as much suspicion by the less developed countries as would have been attempts to dominate the sea by a policy of unilateral extension of seabed boundaries, as some U.S. interests proposed. To be consistent, however, the United States may be obliged to defend the notion of freedom of the seas to include not only navigation for transit of naval vessels but fish harvesting, mineral extraction, and exploration. Arguments are no longer convincing that since different modes of use are functionally separable, so are different legal regimes. Creeping jurisdiction has been taking over as different international legal regimes have become eligible for bargaining. If the 1973 Law of the Sea Conference produces agreements on the basis of exigencies of the moment, however, the international rules of the game may be set arbitrarily in patterns that in the long run could paradoxically defeat the self-interest of all participants. Yet another set of problems will be generated domestically if the long-term ramifications of international sea law are bargained away from myopic expediency. The United States is thus challenged to implement its portentous 1966 commitment that the seas shall be used to benefit all men, despite pressures for interim unilateral steps.

UNDEVELOPED INDUSTRY-GOVERNMENT PARTNERSHIP

In the United States, marine activities involve both the government and industry. Their most familiar partnership has long concerned research and development. A customer-vendor relationship has prevailed, where the federal government has provided the funds and industry or the universities have done the research. But in shipbuilding, ship operations, fishing, offshore oil development, port and harbor development, coastal modification, and other activities, such indirect instruments as federal subsidies, tax incentives, environmental services, and other devices also have been widely used to encourage industrial participation.

These follow a pattern of informal alliances to achieve an overlap of public and private goals where economic processes of the market place have proved inadequate to produce action. The government has thus intervened to satisfy needs articulated by our society through induce-ment to the private sector rather than by direct production of the requisite goods and services. Price characterizes this strategy as a form of socialization "less troublesome than assuming ownership."[2] By and large, these measures compromise a patchwork where neither side has been willing to admit freely to the arrangements; on one hand, industry likes to regard itself as independently able to perform as a textbook example of free private enterprise, and on the other, government tends to suppress visibility of actions that may be criticized as meddling. The other side of this relationship appears in federal regulation of industry for safety, environmental protection, or antitrust cases, where it acts to meet other expressions of the "public interest."

In the realm of marine industry, three intrinsic characteristics are so fundamental and pervasive as to warrant examination of special require-ments for government-industry partnerships. First, virtually all living and nonliving marine resources are common resources held in a public trust, but their exploitation is expected to be undertaken by private initiative. Second, in many cases where U.S. industry must compete in the international market or in harvesting of common resources beyond national sovereignty, foreign commercial interests are heavily and sys-tematically subsidized. Third, marine activities frequently entail high cost, high risk, or uncertain and distant payoffs compared to other al-ternatives for investment, with generally low profit margins and diffi-culties in preserving proprietary positions. All three factors have resulted in very modest investments by maritime industries in engineering re-search and development, not compensated by the government's research and development initiatives because most of this support has been oriented toward military, not commercial, technology.

Further disenchanting to private enterprise are institutional and legal constraints, such as counterproductive state-by-state regulation in the fishing industry, labor-management disputes in shipping, and other circumstances that historically have retarded progress and reduced productivity. Free enterprise traditionally responds to the call of the market place with goods and services focused on a consumer body of some predictable strength. The marine theater certainly offers enticing opportunities for ultimate profit, but its prospects are apparently too remote to counter the obstacles. Additionally, marine enterprises entail a complicated mix of public goals (such as national security, full em-ployment, and environmental protection) with private goals of profit. Neither government nor industry can handle the combination independ-ently. Old models of government-industry relationships give no promise of performing well here, and further small-scale and short-term tinkering seems unlikely to produce remedies in proportion to the disabilities. It is also possible that traditional marine industries have reached a stage of maturity where entrepreneurship and temperament for risk taking have evaporated, perhaps too long used to a stable protectionist environment; far more radical steps may be needed to infuse the leadership and creativity that characterize other more effervescent sectors of American industry. A conspicuous exception lies in offshore oil development, where scintillating technical activity and investments have grown sharply in the last few years, and are accelerating. Anticipated profits have driven the enterprise, but its prospects have been greatly facilitated by governmental protection.

Some new form of collaboration seems incumbent, most urgently of all, perhaps, in the domestic fishing industry. In the past, no suitable focus had existed in the federal government for setting up such arrangements. The emergence of NOAA furnished that vehicle. Yet, except for a small component of the National Marine Fisheries Service, virtually all of NOAA's energies have been focused on the generation and dissemination of knowledge. It has seldom worked with industry toward joint goals, although its location in the Department of Commerce opens to it all manner of conveniently available expertise to study the full range of governmental devices that could be invoked in such a partnership.

Although the birth of NOAA signaled the start of a systemic federal approach toward strengthening domestic institutions to deal with the sea, that evolution is incomplete and will require sensitive attention. The next big step lies in an entirely new examination of the essential synergism between government and industry to undertake tasks traditionally reserved to the private sector.

ISSUES TO SHAPE THE FUTURE

Within this framework of general factors that condition marine futures, a wide range of policy initiatives and responses remains open, domestically and internationally. To deal with marine affairs through some over-all doctrine has been rhetorically and ideologically attractive, especially to advocates who uncritically press for a larger, more powerful enterprise across the board. A similar breadth of approach underlies the Utopian assertion that the oceans are capable of satisfying man's material needs and his desire for a more peaceful world. Ultimately, a grand strategy for the oceans may emerge; but on the basis of experiences during the 1966–1970 interval of energetic policy development, it would appear that only pragmatic, rather than general, propositions can be counted on to engage policy-level interest. Even so, it is possible to identify certain immediate issues that embody many of the problems and opportunities just enumerated, and that could be resolved near term within the present setting of political, social and institutional considerations.

Some fourteen key initiatives could shape the future of marine science affairs domestically, each offering its own range of possible actions.

1. In domestic policy, the Marine Resources and Development Act of 1966 that emphasized more deliberate study and effective use of the sea and its resources should be amended to provide statutory balance of environmental protection and resource conservation.

2. Within that policy framework, explicit legislation should be sought at both federal and state levels to meet conflicts in use and degradation of the coastal zone through rational long-range planning, opportunities for citizen expression of preferences, and creation of administrative machinery by individual states. And more precious coastal land should be acquired for public use.

3. Research on harmful effects of pollution should be sharply accelerated, to determine levels detrimental to humans and to wildlife, and to determine the natural capability of different homogeneous ecological areas to dilute or disperse pollutants. Such new understanding should underlie all decisions posing a risk to the environment, ranging from setting standards for effluents to approval of power-plant siting and supertanker traffic.

4. In foreign policy, the 1966 Act that expressed U.S. intent to regard the seas for the benefit of all peoples was initially implemented so as to establish the United States as a world leader. That status has subsequently diminished. It should be restored to enhance world order through arms control and restatement of freedoms of the sea; extraction of marine protein to meet world hunger; and stronger institutional and leadership roles of the United Nations. U.S. policy should be redefined and given operational credibility at the highest levels of government.

5. Because oceanographic phenomena are not delimited by political boundaries, the influences of marine ecology on migrating fish, the circulation of pollutants, and the prediction of and possible modification of storms originating over ocean areas can be understood only through widespread collection and pooling of data. Unilateral restrictions by some nations on freedom of research could well impede understanding to the detriment of all. By adoption of a strong position on the Law of the Sea to preserve freedom of research, as well as by practical steps to share its own scientific results, the United States could arrest that political fragmentation that is disrupting the collection of essential data. Toward this end, the United States should designate knowledge of the marine environment as a common heritage of mankind.

6. The U.S. proposal in 1968 for an International Decade of Ocean Exploration, which was intended to expand scientific comprehension of

the oceans, should be given greater U.S. budgetary support and stronger policy endorsement as an international cooperative venture from which results would be made available to all nations regardless of their degree of participation. The original concept of the Decade would have provided technical assistance to those nations that front on the sea but lack wherewithal to study and assess even their own coastal resources. That pledge should be renewed, and appropriately funded.

7. Steps should also be taken by the United States to help strengthen international organizations concerned with the sea, to collect, store, and disseminate data, to meet threats of pollution and overfishing, and to improve planning and coordination of the existing array of international bodies. Such mechanisms could identify the importance to each nation both of its own coastal waters and of the seas and seabed beyond national jurisdiction.

8. The United States could forward humanitarian concerns by a renewed program and demonstration projects for production of fish protein concentrate at lower cost to meet domestic and overseas malnutrition.

9. Restrictions on the use of the sea and seabed as a theater for combat using weapons of mass destruction should be extended.

10. In domestic efforts toward both economic and social goals, new measures are needed to improve articulation between the public and private sectors for marine resource development and protection, especially to emphasize their necessary partnership in satisfying joint public and private goals, but with sensitive concern for public interest.

11. Among the most urgent of the areas requiring mutual participation is rehabilitation of domestic fisheries. Progress, however, requires reformulation of a national fisheries policy.

12. An expanded program of industrial marine technology should also be generated by the federal government to meet public goals, to update the technological status of selected marine industries so as to make them more competitive in the international market, and to utilize some talent displaced from aerospace. But such technology should be regarded as more than technique, especially to couple knowledge production better with knowledge consumption.

13. As to federal management practices, most of these objectives involve participation by numerous federal agencies, and their success requires a system of mutually reinforcing elements, not merely a collection of specialized parts. Such integration is more than "coordination." It demands application of policy planning and management skills backed up by an administrative framework of sufficient influence within the federal establishment to foster a coherent systems approach to public administration. Arrangements such as through the Council's Committee on Multiple Uses of the Coastal Zone, discussed in chapter 4, need to be reinstituted at a policy level, especially recognizing that, despite con-

solidation of agencies in NOAA, with creation of CEQ and EPA more different agencies are involved than previously.

14. The role of the National Oceanic and Atmospheric Administration should be clarified. It began as a linear projection of traditional functions of its parent agencies. Will it continue mainly to provide environmental services? Or, with extension of authority, will it serve as a focal point for grants-in-aid to states for coastal management, as sponsor of related coastal research, and as protector of environmental quality? Is it perhaps destined to deal more broadly with technological ends as well as means, to build new institutional and policy bridges across the industry-government gap and so create a powerful marine resource development capability? Or will it try to accomplish all three? Finally, there is the question of leadership, the same question that nagged Congress in 1959. Who is there to recognize that national policy is more than the sum of the components still practiced by eleven agencies? And where is the leadership sited?

Some of these issues are new. It is a melancholy fact, however, that most had their roots exposed years ago, in many cases with a definition of the problem and even proposals for solution that earned broad consensus. Yet, most of the original issues remain, sometimes because resources for implementation were inadequately matched to the problem, sometimes because there was no follow-through. As unfinished business, these questions require a new look at needed legislative, budgetary, management, or organizational moves. The background to all of these issues lies in the preceding chapters, but certain implications for the future deserve emphasis.

As to the coastal zone, almost all nations that front on the sea have begun to realize that the ecologically linked ribbon of land and water along the coastline is of special importance, both in industrial development and in nature conservancy. Paradoxically, the heightened desire to employ this resource has in itself degraded the unique qualities that make it valuable. And we have come to realize that the littoral cannot indefinitely absorb insults of man-induced change. The coastal margin and the oceans must be treated as a complete ecological system rather than through scattered stop-gap measures. In the United States, development has taken place largely in response to the pressure of individuals, industry, and local units of government seeking short-term benefits through real estate speculation and industrial development. Little recognition has been accorded their environmental and aesthetic values, scarcity, and proximity to common resources of coastal waters, fish, and seabed, which confer on the contiguous land a separate unique quality. Citizens should be afforded an opportunity to enjoy those resources that belong to them collectively as a public trust, and the coastal zone should be controlled by rational public management. While some states that have legal authority over coastal resources have acted with courage and foresight, most have been awaiting federal leadership. Federal and state governments should be more than umpires in deciding on uses among competitors; they should serve as stewards of the public trust, for unborn generations who have no voice in political bargaining. An explicit national policy is required to balance protection and development of coastal resources. Measures such as S. 3507 and H.R. 14146 in the 92d Congress would amend the 1966 Marine Resources and Engineering Development Act with a new Title III on coastal and estuarine management, sufficient to animate states to take complementary initiatives. Such legislation should have sufficient teeth to meet the issue if states choose to ignore their responsibilities. Although compatible with proposals for national land policy, steps for rational and balanced use of the coastal zone need not indefinitely await the broader, more controversial measure. In fact, successful experience in this coastal sector, for which techniques of management have been carefully studied since 1967 by governmental and private specialists alike, could pave the way for the more extensive policy.

One aspect of coastal management, that of control of coastal pollution, has been pressed by a zealous Environmental Protection Agency. Its efforts are hampered, however, by lack of firm data on acceptable levels of pollution. Zero pollution seldom exists in nature, is impracticable to enforce, and is probably an overly demanding goal. Since different bodies of water have varying capacities to dilute and disperse pollutants, uniform standards may impose unnecessary economic burdens on some areas. Until more knowledge is available, it makes sense to take a conservative stance, but a sharp acceleration of research is clearly required. New oceanographic data can also facilitate decisions about alternate uses of coastal resources.

Such information is vital to the sequence of understanding, predicting, and controlling all sectors of the marine environment. Yet the conduct of basic research is threatened by some nation-states that are imposing unilateral prohibition or stringent regulation at distances off their shores that were historically regarded as the high seas. What is so paradoxical about such moves is that every nation benefits from access to knowledge about the ocean phenomena that are never constrained by political boundaries, and all stand to lose by this philosophy.

Those nations seeking to restrict research cite a number of real or imagined reasons—national security, protection against foreign exploitation of resources and against pollution, preservation of a bargaining card in international negotiations. To a great extent, this rationale is another manifestation of the polarization of the more numerous less developed countries against the technologically advanced nations that seasons every UN debate in marine affairs. The less developed countries regard science and technology as one route to power by the more advanced, and view the gathering of marine information as gaining yet further advantage because only the wealthy have the wherewithal to utilize that knowledge. They trace a quick progression from exploration to exploitation, and on the basis of a few unhappy experiences with expeditions, they distrust the innocent uses of data developed for scientific purposes. Laboring under anxieties that advanced countries doing research off their shores will learn more about the coastal nation than less developed countries know about themselves, they see territorial claims as their only protection; with understandable impatience, they discount the future.

Historically, ocean exploration was motivated by expectations of territorial claims, but first with regard to pure science, then to military implications, and most recently to resource delineation and environmental protection, maritime research and exploration have taken on new meanings. In the 1958 Law of the Sea Conference, it was not even considered necessary to include on the agenda the explicit question of freedom of research. The topic has suddenly gained prominence because of the role of knowledge in use of the sea, and paradoxically because individual nations have overtly restricted such freedom, holding it hostage in the conflict over marine resource control and legal rights to the sea. The primary defenders of freedom in research are ocean scientists and a few maritime lawyers, acting through the International Marine Science Affairs Panel of the National Academy of Sciences. Their February 1972 report contained a strong recommendation on securing maximum freedom of access for scientific exploration.

The issue as a fundamental precept, however, transcends these professional interests. Granted that coastal states would be expected to exert various rights close to their shores, the importance of scientific study is not a matter for scientists to defend alone. It is a moral principle in foreign policy.

Scientists operating under such a framework, however, will have to recognize that they are no longer outside of the political arena and must accept the need of advance notification in conducting research and subsequent full data sharing. They have been missionaries of good faith and good will on other occasions and have the opportunity again to provide that bridge. But first and foremost, the United States must prove its own credibility with regard to the sea.

In 1971, preparations by the United States for the 1973 Law of the Sea Conference were reported as trading off some freedom of research in exchange for freedom of navigation for naval vessels. Important as the military consideration may be, for the United States to accept such a proposition nullifies its role as a leader in scientific study of the seas.

The United Nations in December 1970 resolved that the seabed and its resources beyond limits of national jurisdiction were the common heritage of mankind. A similar appellation could be logically applied to fishery resources and to the entire marine environment beyond national jurisdictions; ocean space is increasingly recognized as a coherent functional regime rather than an artificially fragmented geographical territory. That trend is implicit if not explicit. If the marine environment were to be considered the common heritage of mankind, so would information about it. As will be noted later, the common heritage in its shared management approach could well transcend freedom of the seas as a basic concept for the future.

The United States started down the road to dissolving anachronistic territorial imperatives with President Johnson's July 1966 declaration, later with proposals for expanded cooperation to comprehend the sea through an International Decade of Ocean Exploration, then with President Nixon's draft treaty on seabed resources that was based on relatively narrow territorial limits on seabed sovereignty. The rationale was developed largely on premises of enhancing world order; the U.S. Navy's requirements for freedom of transit through narrow straits was accompanied by freedom for commercial shipping of all nations so it would not be subject to the uncertainties of goodwill of nations bordering such straits. There was a firm belief that all nations would benefit. Similarly, as discusssed on page 261, on the average all less developed countries would benefit from narrow territorial seas and seabed sovereignty, leaving the largest possible area under universal jurisdiction. The activist U.S. policy in international cooperation began to gain credibility. Then, late in 1970, the situation changed sharply, with decline in leadership and insufficient funds for the Decade. With U.S. motives suspect by the all-too-apparent anxiety of the U.S. Navy over restriction of navigation, the sprouts of comity over peaceful uses of the sea began to wither. The enlightened Nixon policy on the seabed regime became another trading stamp. And some oil and mining interests sought to drown the policy.

The rationale for a cooperative effort was and still is compelling: the seas are too extensive for any single nation to mount the necessary efforts; marine phenomena do not respect territorial boundaries; research is almost exclusively conducted by individual nations, not by regional or international, multinational organs. Moreover, the urgent targets for study require integration and renewed stimulation, and the United States could well regain the leadership it manifested in the 1966–69 interval by substantiating its long-standing support for scientific research and its more recent policies to relate the seas to the needs and aspirations of all mankind. This was the lofty purpose of the International Decade of Ocean Exploration. But as its funding for FY 1973 sank below the prior year's level, the commitment by the United States seemed shaky. The absence of foreign technical assistance to buttress the capabilities of other nations to undertake ocean studies, at least to interpret data furnished them by the advanced nations participating in ocean exploration, further diminishes the potential of the Decade to neutralize the anxieties of LDC's over freedom of research. Finally, given the threats to world peace and to health of the oceans, and the needs for management of living and nonliving resources, freedom of research should not be regarded by the U.S. as negotiable.

In no area would that articulation assume greater humanitarian benefit than in helping nations to check and even to eliminate malnutrition by extracting protein from the sea. Though representing only 15 percent of all protein foods, marine sources are a major ingredient of diets of over one billion people. The development of fish protein concentrate was acclaimed by the United States in 1966 as a potent weapon in the war on hunger. By 1971 it was in the doldrums. For one thing, successful extraction of protein at lower cost depends on economies of scale, and so far demonstrations had been conducted at too low a level of activity, or with clumsy management. The FDA restriction to one-pound packaging also represents a serious obstacle. The single pilot plant at Aberdeen, Washington should be extended into a full system that carries from fish in the sea to delivery of enriched products to hungry people, and the original Marine Council goal of meeting needs of 200 million people by 1980 without depleting fish stocks should be reexamined. Not to be overlooked as a source of raw material is the productive artificial culture of such marine species as mussels in coastal waters.

Quite apart from these opportunities for American leadership to foster world order indirectly, the potential exists for direct steps. One lies in reaffirmation of freedom of the seas, not simply for naval vessels through straits, but for all shipping which constitutes the medium of trade essential for world-wide economic development and for peace. If strictured by territorial claims and subject to uncertainties, world order is seriously threatened. This is a second principle that the United States should defend as nonnegotiable.

Another step lies in arms control. Since the mid 1950s the Soviet Union and the United States have increasingly considered the ocean locale vital to their national security, especially to base second-strike nuclear capabilities. In the U.S. arsenal, the submarine-mounted ballistic missile, its hydra-headed successor, and emerging generations of longer range weapons have been key elements. To catch up, the Soviets have fashioned analogous systems. The end of such evolution is far from clear, and in the struggle for offense superiority and defensive counter systems, the number, potency, and costs of weapons steadily mount. That such escalation does not necessarily purchase greater security is at the heart of the bilateral strategic arms limitation talks.

Placing the seabed off limits to fixed emplacement of weapons of mass destruction was an easy step for the United States and U.S.S.R. to take because neither abandoned any near-term advantages for such weapon systems. This leaves, however, an enormous volume of ocean for potential excursions of missile launching submarines. With the exception of experimental, research, or special craft, these vehicles were restricted in the early 1970s by design limitations to depths no greater than 1,000 to 2,000 feet, in contrast to the 12,500-foot average depth of the sea floor, with holes that plunge to 35,800 feet. New technologies could alter that situation, and a fresh cycle of weapons competition with super-deep-diving vehicles and counter-threat antisubmarine technology could result. Steps to arrest that expensive and probably fruitless rivalry could begin by setting progressive layers of the sea off limits to missile-carrying subs, beginning at the bottom and working upward toward existing operating depths. Many details would require solution, especially on inspection, but this mode of arms limitation could lead to others by which new areas of the ocean could be progressively sanitized. Such are the issues of international import where the United States could exert leadership.

Turning to domestic issues, the state of fisheries deserves priority treatment. The vacuum of national policy reflects both the low priority accorded to fisheries in our national economic schema and the bewildering array of problems the industry confronts. But the issue can no longer be dealt with by Band-Aids. To supplant fragmented, archaic law and pseudo-conservation measures formulated state by state and species by species, bold policy steps must be taken on balanced utilization and protection. That policy must include teeth to enforce such controversial issues as limited entry to improve the productivity of the industry as a whole. Subsidies and technological innovation can help, but not unless basic changes are made in international agreements on fishery management, the institutional and economic structure of the industry, and in the attitude of the fishermen themselves. Sensitive to their plight and to the way fishermen feel about their own dilemma, the government should explore ways to ease the stresses of a transition to a modern industry, but should not defer rational approaches indefinitely.

In considering public as well as private objectives for rehabilitating the fishing enterprise, a full range of policy alternatives deserves to be examined, including nationalization. While that extreme remedy is often rejected as inimical to the American image of free enterprise, other sick industries of crucial import have required such steps, whatever euphemism is employed. Passenger rail service is one such example. The AMTRAK construct was a last desperate move to salvage this capability. It could well set an example when the national interest is involved and not satisfied by present arrangements of private enterprise, in this case regarding foreign policy, a voice in ocean management, domestic employment, and balance of payments. Even federal rather than state regulation of fisheries, using a resource management approach, would be a major step.

Finally, aquaculture to farm the inshore waters promises to be more productive in pounds of protein per acre than any agriculture on land. Many other countries find this technology profitable, and the United States may yet discover this new industry.

Without a reversal in fortunes of the fishing industry, relative to other fishing nations, the U.S. voice at the international conference table is less and less important. Such intervention is essential to meet the overfishing in open seas that has seriously depleted stocks everywhere and necessitated more stringent international conservation measures. While inshore and deep-sea fishermen disagree on the appropriate posture in such matters, the integrity of the United States in negotiations will suffer if efforts are compromised to please all domestic interests simultaneously. The nation must go to the conference table armed with consistency as much as with virtue.

Another domestic problem with international ramifications concerns oil spills. While monitoring networks and quick reaction teams are necessary to deal with spills where they occur, the technology of containment and cleanup is still primitive. Under any circumstances there should be greater emphasis on prevention. There are now 55 collisions per year involving tankers. But other ships spill oil too, and the insurance companies report a staggering 8,600 strandings, collisions, or other serious damage a year in 30,000 vessels of the world fleet. As surveillance has improved the safety of aviation, so harbor radar and disciplined ship control in and around ports could reduce ship casualties and the hazards of oil spills. Advances in the technology of offshore oil loading and unloading may preclude the need for supertankers to enter valuable estuaries, heading off dredging and port construction costs and environmental risks. The very use of supertankers needs re-evaluation.

This technology, as with so many involving the sea, has yet to flower. In our constellation of federal agencies dealing with the sea, only the Navy has built sophisticated engineering, research, and development capabilities. The civilian counterpart did not previously exist in the agencies that were consolidated into NOAA; it is lacking today.

Examples of retarded technological potential abound. Beginning in 1959 with *Aluminaut*, manned submersibles were attractive as research tools because they afforded subsurface observation, sampling, and manipulation, less feasible from unmanned vehicles. American industry invested capital to build a fleet of submersibles that by 1971 exceeded all other nations combined. But these submarines were slowly deteriorating, underutilized, or mothballed because funds were insufficient to sponsor their use. The Marine Council attempted as early as 1967 to spur government leasing of these vehicles for use by oceanographers who yearned for such capabilities but whose meager project funds would not support them. (Funding planned in 1968 was subsequently withdrawn.) In the meanwhile, projects of major scientific interest await that step. For example, in January 1970 the Council Secretariat negotiated a bilateral agreement with France for an exciting scientific project of as great significance as lunar studies. It involved a research dive of 8,000 feet to the mid-Atlantic Ridge where new material from inside the earth is believed to exude as the continents move apart and pull open the seabed. Under the International Decade of Ocean Exploration, this was to be an important bilateral enterprise. In 1972, plans were still on the drawing board.

An alert family of technological projects was expected by the Stratton Commission to be bred and reared by NOAA. By 1972 those expectations were beginning to wilt. There is no magic in tour de force projects conducted simply because they are feasible. Floating cities, for example, are extremely expensive, do not solve problems of urban crowding, and do not automatically promise the quality of human life now so earnestly sought by those most needing housing in the central city. A full range of targets was outlined in the March 1972 report by the Marine Board of the National Academy of Engineering, entitled *Toward Fulfillment of a National Ocean Commitment*, concluding the study it began in 1968. Thirteen major and twelve detailed recommendations were offered. While their technical propositions are compelling, they are not sufficient. What is needed is a deeper study and action on how the government and industry can assess and equitably share risks. This requires selection of goals, identifying impediments to their achievement, and determining appropriate public and private roles and assistance to industry that may be necessary where market forces are inoperative because the market is less aggregated, customer consensus is less assured, technical prescriptions are vague, and the transfer route for information from knowledge producer to knowledge consumer is ambiguous or institutionally mismatched. In the government, this role can be filled primarily by NOAA. NOAA will thus have to consider, in addition to hardware, the economic, institutional, and cultural factors whereby a twenty-first century society can assimilate technology compatible with its needs and wants.

FEDERAL STRUCTURE—THE MEDIUM AS THE MESSAGE

Nine months after its creation, the National Oceanic and Atmospheric Administration that was expected to symbolize higher marine priorities was encountering an identity crisis. The hopes of its supporters were shaved by gloom over constraints on its leadership. None of the five marine areas stamped "priority" by the Nixon administration in the fall of 1969 was assigned NOAA; the designation of an agency to lead coastal management was unresolved; funds to energize civilian marine technology were minuscule; much of the agency's budget and focus of leadership was directed to important past commitments in atmospheric rather than marine affairs.

Since NOAA was created under the Reorganization Act and consolidated only existing agencies with well-defined prior missions, no new missions were added. The Stratton Commission, however, foresaw a number of related new functions that would not only broaden NOAA's traditional dedication to environmental services, but could be the key to revitalization of the entire enterprise. These included responsibility for establishing: (1) concepts and standards for multiple uses of the coastal zone and for grant-in-aid programs to states; (2) associated coastal research capabilities; (3) coordination of weather modification research and services, so as to provide an administrative home for this function that was assigned to, then removed from, the National Science Foundation,[3] (4) a program of engineering research and development, including the creation of a new laboratory and contract and grant program. More broadly conceived than as stated by the Commission, such a function could go beyond sterile hardware development, to consider economic and institutional aspects of marine technology and set the stage for improved understanding of industry's needs and opportunities, and the role of government in providing incentives; (5) exploration and environmental research in the Arctic and Antarctic.

If NOAA had been created as an independent agency through new organic legislation, these functions could have been incorporated at its birth. Instead, hints of a possible new maritime malaise were reinforced when the administration instructed NOAA chief Robert M. White, its only spokesman in marine affairs, to oppose any new legislation that would strengthen NOAA's authority and influence in coastal management. Concurrently, White's boss, Secretary of Commerce Maurice H. Stans, was addressing industrial audiences sympathetic to their discomfort under the impact of new environmental standards and aggressive enforcement.[4] Attempts by Stans to persuade Delaware's Governor Russell Peterson to back away from tough coastal protection in his state did not earn admiration from the conservation-minded, and put NOAA in an awkward position as a potential champion of environmental conservancy.[5]

By the nature of its precursors and origins, NOAA inherited an historical orientation to research and services by all of its components, a focus on means rather than on ends associated with utilization and protection of the marine environment. Second, as an amalgam of specialized bodies, NOAA partakes of competences and relationships to the corresponding outside sectors of interest. With disparate elements assembled in a new administrative tent, without a concept integrating ecological disciplines and sectorial interests, the result may only add a layer of administrative structure and its penalties without meeting the objectives of consolidation.

Nevertheless, NOAA has all the organizational horsepower to realize a destiny projected by the Stratton Commission.

Whatever strengths of NOAA may emerge, it is a fact of life that numerous other federal agencies continue to bear responsibilities for dealing with the sea—the State Department, Navy, National Science Foundation, Coast Guard, Department of Interior, Maritime Administration, to mention but a few. Some integration at a high level is essential, to regain the systems approach of the Council. Amidst the blunt realities of interdepartmental rivalries, some incentive-penalty technique is essential for mustering the best from all the pieces. If the Interagency Committee on Marine Science and Engineering recreated in 1971 under the Federal Council for Science and Technology is no stronger than the Interagency Committee on Oceanography of the 1959–66 era, government-wide management will be pallid indeed. As discussed in the preceding chapter, much depends on the stature and leadership of the FCST and its OST host.

The 1971 plateau of marine activity brings to mind a *bon mot* by Oliver Wendell Holmes. Speaking in another context and taking aim at our ship of state, he said, "We must sail, sometimes with the wind and sometimes against it—but we must sail and not drift nor lie at anchor." The obstacles to progress in 1971 are no more severe than in the 1965–69 phase of policy development when substantial progress was made. But it was support by the President of the United States that advanced the momentum and prestige of the field even when all it had in the way of impetus was pure logic. With the public constituency so shaky, the Marine Council catered to a constituency of one. Moreover, there was both legislative assignment to and active intervention by the Vice President of the United States and the Secretariat. To be sure, the robust peppery leadership of a chairman cannot be generated by legislation, but the political wheel base of the agency itself was a significant factor in advancement of marine affairs.

With the Council dismantled, the key question is how to make NOAA, the President's National Advisory Committee on Oceans and Atmosphere, and the FCST coordinating committee fulfill the roles the Council played in advising the President. That this question is of critical importance was reflected at the April 26, 1972, banquet by the American Oceanic Organization honoring Alton Lennon upon his self-imposed retirement from Congress. Responding to accolades on his contributions, Lennon observed that progress did not seem as assured as he had hoped with creation of NOAA; perhaps, he said, we need the return of the Marine Sciences Council. This dilemma places a premium directly on leadership, allies in the White House, the Congress, the professions, and interest groups. But marine affairs activities of the federal government will fail to thrive without an attractive program oriented toward the goals of the nation. In the words of Socrates, "There is no favorable wind unless the helmsman knows his destination."

The future of marine affairs in the United States is neither riotously exhilarating nor mundane; neither wildly expanding nor static; neither at the top of our nation's agenda of unfinished business nor at the bottom. Marine sciences have many promises to keep. They will be realized during the next decade only with the same investment of labor and dedication that attended progress in the past. The future is in the hands not only of our competent scientists, creative engineers, commercial entrepreneurs, and dedicated bureaucrats, but also those leaders thrust by position and circumstances into the politics of the oceans.

The fate of marine affairs world wide also awaits this outcome.

THE INTERNATIONAL SETTING

While marine affairs were evolving domestically, a three-step progression was taking place internationally. After World War II, a number of individual nations returned to fishing and shipping, and the Soviet Union entered the maritime arena as a new and vigorous contestant. But in general, the seas did not command serious attention. Except for regionally oriented fisheries research and sporadic scientific expeditions, multinational collective efforts were also rare.

Within the United Nations in 1966, a second phase of interest was aroused with adoption by the General Assembly of its first resolution to examine the resource potential of the sea. The new national policy of the United States, the invitation by President Johnson that others join in cooperative study, the proposition of Malta Ambassador Arvid Pardo at the 22d session that the UN should declare the seabed beyond the narrow limits of national sovereignty under its jurisdiction, aroused great interest. The community of nations began to recognize that all peoples cluster on continental islands embedded in the vast ocean, that man's activities are intimately linked to the sea, and that the planet represents a closed ecological system wherein ostensibly local events may have widely distributed effects. After a quick surge of enthusiasm following Ambassador Pardo's overly optimistic assertions to the UN about the immediacy and level of profit from exploitation of marine resources, most nation-states began a more thoughtful appraisal of potential benefits and responses consistent with their self interests.

Additional resolutions were soon adopted by the United Nations General Assembly on the role of the oceans in extending and preserving world order; on the harvesting of marine protein to meet malnutrition; on the extraction of seabed energy and minerals to meet industrial requirements and on narrowing the economic disparity between newly developing nations with the technologically more advanced. Resolutions also treated pollution of the ocean and the importance of scientific research to underpin rational development along all the preceding lines.[6]

This menu of possibilities became even more a tantalizing challenge with appreciation that some 85 percent of the seabed and ocean lies beyond boundaries of national jurisdiction, so that its description, disposition, protection, and employment becomes a collective responsibility. In 1967, the UN created an Ad Hoc Seabeds Committee that was upgraded almost immediately to a Standing Committee on the Peaceful Uses of the Seabed and Ocean Floor beyond Limits of National Jurisdiction to examine legal questions regarding boundaries and requisite policies and machinery to deal with international questions. In the short span of four years, the ocean floor beyond national jurisdiction was deemed by the United Nations in 1970 to be the common heritage of mankind.[7] For such a far-reaching precept to have matured so swiftly revealed that many aspects of maritime law were less constrained by tradition and precedent than would have been believed possible. This left open a wide range of options as to how the community of nations might act collectively to serve their own interests.

After adoption of principles and consideration of draft treaties, the General Assembly in 1970 took steps to prepare for a new Conference on Law of the Sea and update the international legal framework concerning not only the seabed, but also the regime of the high seas, the continental shelf, the territorial sea, and the contiguous zone, the preservation of the marine environment, and scientific research. It took seriously the question of stronger international apparatus to serve community purposes.

A wide range of prescriptions for necessary legal arrangements and machinery resulted. Before long, as the functional, ecological, and legal interconnections became visible, the study, protection, and disposition of resources partaking of such a common heritage became such a serious collective task as to call into question the scale, competence, experience, vitality, and influence of international structure to meet the challenge at hand.

About 1970 the third stage of international development opened even before the second had matured. A quest for institutional solutions ushered in an era of conflict, with debates in various international forums from which two separate, contrasting strategies emerged. The first was an extrapolation of traditional territorial boundaries from landward activity into the marine theater. Such volatile questions as narrow or wide extension of national sovereignty so preoccupied opponents that they lost sight of their rhetorical dedication to global comity. This seaward extension of historical property concepts led to further unilateral claims of jurisdiction over living and seabed resources and scientific research by a growing number of individual nation-states. Proprietorship became the theme, rather than rational management of individual functional uses (for fishing, petroleum extraction, navigation, or waste disposal) that resulted in unwanted interactions with other uses or with conservation and environmental quality having equally legitimate status. As a consequence, there was encroachment on common resources, instability in relationships among nation-states, inequitable distribution of benefits, the hazard of depletion of living resources to the detriment of all by the initiatives of a few on a "first come, first served" basis, and jeopardy to the health of the environment. Ironically, the contributions of science and technology that could have fostered a fresh and hopeful outlook for international collaboration had instead ignited conflict and debilitating competition.

The second, conflicting strategy was based on the interconnectedness of marine activities, increasingly denoted by the term "ocean space." There unfolded, for example, an awareness that waste of national origin dumped at sea may be distributed globally. While such threats were not regarded as immediate or of crisis proportions, the pervasiveness of the fluid media potentially exposed all nations to the same risk and uncertainty. So whatever the geopolitical and geoeconomic considerations in debate, no matter how parochial the arguments, participants came to recognize that all questions shared a central core of scientific, technological, and economic facts not constrained by political or institutional boundaries or ideology. A global information base had become indispensable for rational decision-making in any and all marine areas. Requirements also merged for an "early warning system" that could give advance notice of harmful effects beyond the functional (as well as geographical) boundary of any given initiative.

Behind this groping for international solutions lay several political realities. International organizations did not have a convincing track record of dealing with highly complex or controversial affairs, much less of performing as management agents. Membership in the UN was growing rapidly, and the one nation–one vote process weakened the UN generally. The major powers were continuing their stingy support for UN peacekeeping. As the mirage of deep-sea treasure evaporated, many less developed countries viewed environmental protection as a barrier to their economic self-sufficiency and preferred to exert jurisdiction over nearby, albeit hypothetical, marine resources rather than to take a chance on sharing in future common benefits from a larger area under universal sovereignty. Some developed countries held to grandfather clauses in protecting past practices in fisheries, sheltering their domestic industries in partnership with developing countries rather than chancing involvement with an international organ. Finally, knowledge about resource richness and distribution was so incomplete as to preclude a truly rational assessment of alternative policies, and most nation-states were unprepared scientifically or politically to deal with this rash of ocean-related issues. In the turmoil of preparation for the 1973 Law of the Sea Conference, the nationalism associated with ownership was transcending present realities and future certainties of internationalism of commerce.

UNSOLVED INSTITUTIONAL PROBLEMS

As ocean affairs attracted international attention, bold and innovative, even Utopian, concepts for ocean space were advanced, along with a new philosophy based on a systemic rather than the traditional sectorial approach. The call for reform acknowledged the ecological integrity of ocean space and the power of human intervention in the environment to induce efforts geographically remote and functionally unconnected. It demanded a wide perspective in dealing with resources increasingly regarded as the common heritage of mankind, lying outside the sovereignty of individual nations and thus immune to traditional ways of dealing with formation of capital, application of technology to exploitation and problems of conservation. Proposals for institutional reform introduced in numerous international, national, and professional bodies ranged over a wide spectrum in their terms of reference and potency of transnational governance.[8]

The thrust for institutional reform partly obscured the fact that within the past twenty-five years the United Nations General Assembly, the Economic and Social Council (ECOSOC) and many of its specialized agencies held cognizance over particular sectors of marine interests within the ambit of their treaty-defined functions. The UN Educational, Scientific and Cultural Organization (UNESCO) and its Intergovernmental Oceanographic Commission (IOC), the Food and Agriculture Organization (FAO), the World Meteorological Organization (WMO), the Intergovernmental Maritime Consultative Organization (IMCO), the International Atomic Energy Agency (IAEA), and the World Health Organization (WHO) shared in such responsibilities.

This inventory of vehicles for multinational cooperation reflects the diversity and fragmentation of marine-related functions that were structured in categories of comprehensible and reasonable scope, to be articulated with like interests in individual countries.

Established under the UN charter as the primary organ dealing with economic and social progress and development, and serving as informational switchboard and coordinating agent for affiliated specialized agencies, ECOSOC in the late 1960s attempted to unify approaches to ocean problems. It called on the Secretary General to accelerate studies and to integrate ocean-space functions, and endeavored to foster consistency among the specialized program elements through its Committee for Programme and Coordination, its Administrative Committee on Coordination (ACC), and its Subcommittee on Marine Science and its Applications.

Coordination was also called for regarding scientific information, services, and amenities; marine resources both living and nonliving; effects on the human environment; the law of the sea; and administration. An Intersecretariat Committee on Scientific Programs Related to Oceanography (ICSPRO) was created with staff representatives assigned from the UN, FAO, WMO and IMCO to the IOC secretariat in Paris, thus removing some obstacles that had resulted from geographical dispersion of their headquarters in New York, Rome, Geneva, London, and Paris. All of the specialized agencies as well as individual member states looked to the IOC to deal broadly with marine sciences; yet it was inadequate in fulfilling its own mandate. Two preparatory bodies for the UN Conference on the Human Environment—on marine pollution and global monitoring—desired to add further responsibilities to IOC in 1971 but expressed serious reservations as to its capabilities. Its weaknesses in responding to existing, much less enriched missions, were openly discussed by the IOC itself at its 7th Session, October 1971, recognizing structural, staffing and legislative as well as funding limitations.

The UN Department of Economic and Social Affairs through its Ocean Economics and Technology Branch, the Seabed Section in the Department of Political and Security Council Affairs, and the Legal Office have all cooperated to service the Seabed Committee and thus enhance coordination. While this array of separate functions, institutions, and jurisdictions dictated by the UN organizational framework is impressive, debates within the UN repeatedly called attention to the haphazard, random, and piecemeal attack being made on problems of ocean space.[9] Attempts at coordination were deemed makeshift improvisations. The pieces did not add up to provide the comprehensive, systematic, methodical approach required, and no basic plan or policy had welded the separate elements into an effective instrument to deal with emerging issues of the sea. Jurisdictional rivalry among the international organizations augmented the division, with the obverse hazard of unwitting duplication and overlap. Moreover, none of the specialized agencies dealt comprehensively or methodically with technical questions of seabed resources or of marine pollution.

That some organizational step was necessary was illuminated in a number of UNGA and ECOSOC resolutions, but the questions of institutional adequacy were inevitably interwoven with questions of goals and values, concepts and legal arrangements. A wide range of positions divided nation-states on almost all these questions.

One consensus was reached. On the organizational issue no nation favored immediate establishment of a new superagency to deal with ocean space, either within or outside the United Nations. The LDC's were traditionally apprehensive that such a creation could become another "rich man's club." The larger nation-states, on the other hand, expressed reluctance to provide what they considered a disproportionate share of additional financial support.

Yet some institutional innovation was required.

Toward Rational Management of Ocean Space— Innovation at the United Nations

Given the growing importance of ocean space to the world community, the elaborate cast of organizations involved, and the potential for conflict, and given the responsibility that the community of nations implicitly assumes in declaring a sizable fraction of the planet a common heritage of mankind, a single major question arises: *What steps are necessary for the rational management of ocean space—to maximize the effective use of the limited resources, to distribute benefits equitably and rationally, and to consider the long-term dilemmas as well as short-term needs?* Subsumed within this broad question is recognition that individual countries have problems entirely within limits of national jursidiction, such as of coastal pollution or conflicts in use, that must be solved locally but toward whose solution experience shared by others would be of value.

This problem of management at both a local and a global level dwells upon a relatively new and abstract concept whose definition can best be sharpened by an inquiry into its operational mechanics. Management here is defined as a capability for decision-making and implementation that includes the administrative, technical, and coordination machinery and infrastructure necessary to:

1. Identify goals and commonality of purpose, unmet needs and opportunities to which international programs could be directed, and especially those gaps in programs that cross lines of existing international bodies;

2. Develop and recommend priorities to these international operating bodies and to their constituent nation-states that are the primary source of funds and vehicles of implementation;

3. Develop policies by which objectives and programs of individual interests may be harmonized to avoid inadvertent conflict or the preempting of options for future developments;

4. Integrate programs that involve two or more international bodies and delegate responsibilities for initiatives where appropriate;

5. Identify international resources needed to achieve goals;

6. Foster exchange of new information and techniques for optimal

use of the sea, as well as analysis requisite to policy-making by all national, international, and private bodies whose activities contribute to or respond to international policy;

7. Develop necessary legal, economic, and technological background studies for identifying alternative policies and criteria for choice, with options outlined for decision-making at all levels, and with recognition of the changing framework of international law;

8. Inventory relevant programs of national governments and of international and private bodies;

9. Engage in future-oriented planning;

10. Identify future institutional needs so that international machinery may be evolved to meet additional responsibilities as they arise;

11. Identify gaps in information-gathering and monitoring capabilities and make recommendations on ways and means to supplement national capabilities to assure a necessary base of data;

12. Develop an information system and instruments of technical cooperation by which experiences of individual nations in problems of coastal environment could be pooled and made available to all nations.

To satisfy these management functions, a new mechanism would have to be created. It would be based (but not exclusively dependent) upon facts developed by members. It would undertake methodical and rational planning, a "precrisis," early warning system, and a capability to respond to crisis should one arise. It should be designed to provide facts and analysis as much to member states as to international bodies.

All of these technical, economic, and policy-planning functions must be erected on a base of scientific knowledge. For it is this neutral core of physical reality that sharpens the value choices to be made, widens the awareness of technical options, permits an evaluation of unwanted consequences, and illuminates benefits, burdens, and the potential recipients of each. Because phenomena on the entire planet are involved, because data are largely collected by individual nation-states and because requisite capabilities are beyond the resources of many, a *sine qua non* of operational performance is a multifaceted cooperative effort.

Four key functions are thus involved: (1) information processing, (2) analysis, (3) planning, and (4) coordination. As to data collection and analysis, steps are required to develop information and monitoring systems for collection, storage, retrieval, and analysis of essential data. But raw information gains value from validation, identification of gaps and ranges of uncertainty, and from structuring to reveal relationships between parameters of the natural environment and their sensitivity to human intervention, to illuminate alternatives of projected actions, and the consequences of each. Scientific facts are often important only in relation to other considerations derived from engineering, economics, and law. Here, objectivity is often threatened if policy research is too closely linked with value judgments and advocacy; at the same time, it loses its relevance if it is separated too far from the decision-making locus by life styles, jargon, or unwillingness by analysts to express conclusions, no matter how qualified, until *all* the data are in.

Planning, the third function, is not a blueprint for explicit initiatives but rather a road map of alternative routes to jointly established goals, and the consequences of each, leaving the decision-maker free to make his choice. Planning cannot be done in the abstract but must be oriented to goals established at the highest practicable legislative body expressing a consensus as to global purposes, based on the choices it has and trade-offs involved.

Coordination animates the effective management of such a complex enterprise. Far from being merely a passive data bank and source of analysis, coordination must include the active role of integration to provide a total enterprise that is more than a sum of its parts. Within its range of functional responsibilities, a coordinating agent must influence its constituent members so as to synchronize initiatives, reduce duplication, fill gaps, and inject quality standards. In the absence of authority, it must operate by persuasion and incentives to achieve consensus by which all participants agree to future plans, including their share of the total responsibility. In the first instance, this would anticipate agreement by the various international bodies as to their respective roles, leaving to each the extraction of commitments from component national entities. Such persuasion might be facilitated by the very facts and analysis collected by the secretariat.

The postulate of an enhanced coordinating mechanism assumes that participating international bodies are now adequate for implementing agreed upon tasks. That does not seem to be the case. Thus one of the first steps would be to identify the functional areas that require critical strengthening and the organizational routes for that attainment.

Sooner or later, the coordinating body must be able to reinforce the capabilities of its operating arms or of member states in order to round out needed information. Ultimately, it could also serve as an entrepreneurial agent to set in motion the operational initiatives for meeting internationally agreed-upon goals. Through such an international instrument for research, monitoring, analysis, planning, and coordination, all nations, including those that may presently lack technological capabilities, would acquire equal access to knowledge about ocean space.

Given the existing tapestry of nation-states and international organs involved, integration seems possible only at the highest level, and involves a mandate from the most transcendent organ—the UN itself. It would also require institutional reform. A number of organizational remedies have been proposed, but convincing as they are in terms of ultimate development, they contrast so sharply with existing international arrangements as to present serious doubt that they could evolve without causing sharp institutional discontinuity. And most nations oppose immediate creation of a new superagency. There are several alternatives. All of these are postulated within the UN system, because they deal in one way or another with elements of the UN family and it makes no sense to consider excising programs from existing arrangements.

As to alternatives: (1) All of the existing agencies engaged in ocean space could be individually strengthened, with their jurisdictions extended to meet gaps; something would also have to be done about coordination. (2) Properly articulated with its sister components, a lead agency could be selected from one of the specialized agencies to steer the collective enterprise. However, this solution is complicated by existing rivalries and the organic and organizational constraints and traditions that make it awkward for one agency to be given a lead role involving the others. This became evident when IOC was tacitly given such a role. Considering factors of size, prestige, and funding, the IOC was handicapped as a subordinate component of a treaty organization in providing leadership to sister institutions themselves having treaty stature. It could perform only if its status were sharply upgraded. (3) New policy-planning and coordinating machinery could be established to confer unity and direction to meet the operational criticisms leveled at the existing structure. This latter possibility, strongly suggested by ECOSOC Resolution 1537(XLIX) of June 30, 1970, appears to meet criteria for rational management of ocean space, and should permit a graceful transition from present arrangements to meet long-term as well as more immediate requirements, without pre-empting even the boldest future options.

To fulfill the functions just outlined for policy planning and coordination, five complementary organs may be required:

1. A *Policy Commission* to extract a coherent set of collective global goals and priorities for ocean space, with a high-level Director General independent of any specialized agencies or functions they represent, and reporting directly to the Secretary General of the UN;

2. A *Coordinating Board* to represent existing intergovernmental bodies that would deal with functional sectors of ocean space and be responsible for implementation of the objectives set forth by the policy body; the Board must be represented at a level senior enough so it can make commitments for follow thru on any coherent over-all strategy developed;

3. A *Secretariat* of both administrative and technical staff to collect, structure and analyze relevant data, to act as an information switchboard, to generate policy papers, and to provide logistics support to the previous two collegiate activities;

4. An *Information Center;* and

5. An *Advisory Network*, calling as necessary on individual specialists and nongovernmental organizations.

Roots for all five of the components already exist. The Policy Commission could be incorporated in either the UNGA or ECOSOC, but in either case existing subsidary elements having jurisdiction would have to be substantially revamped. Jurisdiction of the General Assembly's Seabed Committee could be extended in scope, or because of the Committee's present orientation to legal questions, a new committee could be created to deal with ocean space management. ECOSOC's original charter also makes it a prime candidate for these planning and coordinating functions, but there are widespread criticisms of its past feeble performance. Even revitalized, ECOSOC would have to be equipped with a new infrastructure articulated with existing committees on natural resources and science and technology, and possibly new bodies dealing with the human environment.

The Coordinating Board could be developed from the present Administrative Committee on Coordination (ACC) or, alternatively, from the Intersecretariat Committee on Scientific Programs Related to Oceanography (ICSPRO). However, its membership would have to function at the intended, uppermost policy level and its scope of responsibility be expanded beyond its focus on science. Communication would have to be substantially strengthened and made more frequent and continuous, if criticisms of its flaws are to be satisfied.

The needed Secretariat also exists in rudimentary form in the UN Secretariat itself and in the secretariats of various agencies. Additional staff for this Secretariat could be obtained through consolidation and through outposting of specialists from various international bodies to bring in specialized expertise. The demands for staff should not be underestimated, and additions would be required, keeping in mind a necessary balance of disciplines, policy experience, and coordination skills, as well as distribution by national origins that by itself could unfortunately subordinate qualifications of competence.

Arrangements for extracting advice from the community of experts are already widely exercised and developed through the specialized agencies, ICSU committees, and through the new Engineering Committee on Ocean Resources. These could be structured and regularized. Additionally, expertise should be sought in economics and technological areas.

As to the information center, a wide variety of data banks already exists in various fields. Rather than centralize storage of all ocean-related data, an inventory of existing banks should be developed with means for rapid access to and exchange of contents. Ultimately, this may require a network of regional data banks available to all interests, national and international. It may well be that one of the crucial tasks of any management apparatus developed along the lines suggested would be to organize the necessary informational network.

To fulfill any of these functions, however, will require availability of information on how things stand today. The first of a series of comprehensive reports on ocean space and its relationship to the affairs of man would meet that need.

This predicament of international mechanisms in dealing with ocean space finds an interesting analogue in problems, diagnoses, and remedies that evolved in the United States. The specialized agencies of the UN greatly resemble the U.S. government agencies in jurisdiction as well as in bureaucratic and interagency behavior. The coordinating agencies of the UN have many of the same objectives as corresponding U.S. instruments, with ICSPRO and ACC resembling the 1956–60 Coordinating Committee on Oceanography in performance.

Strengthening UN policy planning and coordination along lines proposed corresponds to policy apparatus provided by the National Council on Marine Resources and Engineering Development.

To be sure, there are vast differences between international and national structures. The authority and responsibility of a chief executive in each nation-state are unparalleled in the international forum as to setting and implementing goals, but the proposed policy body representing international consensus would serve that function.

The incentives and penalties that accompany the kinetics of international and national coordination will also differ. In national governments as in the international area, severe problems inhibit the development of unified goals and policies among various components. Consensus cannot ordinarily be induced by executive fiat. Persuasion, mediation, and accommodation are essential to gain collaboration. In the case of the Marine Council, a promise of high level support, the desire for esteem, and fear of disapproval provided powerful, albeit invisible, incentives. The same factors arise when considering an analogous international schema.

Given the normal centrifugal forces of bureaucratic organization at any level, the achievement of coordination ultimately lies in an ability to control funds, either indirectly by recommendations to the body setting priorities or directly in reallocation by the coordinating authority to the operating agencies. Although the latter control will ultimately be required if the proposed apparatus is to be fully effective, the power of a marine policy body initially to recommend funding allocations to the UN Development Fund or World Bank on the basis of a comprehensive and rational plan may be significant for improved coordination.

While this pattern is readily committed to a neat organizational chart, the wiring diagram is not the solution. Rather the answer lies in the legislative mandate, leadership, and funds necessary to animate this complex array of interests. International bodies are voluntary associations. At the level of the General Assembly and even the Seabed Committee, debates tend to become highly political. Short-term individual self-interest and traditions of sovereignty are unlikely to be replaced unless the benefits of cooperation exceed the penalties. Consensus in goals among constituents—and especially operational coherence—will depend on incentives and penalties, and some degree of insulation from the highly visible and heated stage of international politics. The most immediate incentive lies in prompt, effective access, through collective arrangements, to knowledge about the sea and to analytical capabilities to deal with economic and political questions as well as scientific ones in phase with the evolving legal framework. Facts often serve to soften positions that may have been adopted prematurely and polarized into conflict, when data disclose that the realities of the situation do not support one or another viewpoint. Information is the basic commodity of collaboration. The penalty for failing to work toward cooperative use of the oceans will be protraction or even extension of conflicts associated with the status quo, deferment of benefits from ocean resources, or conceivably an irreversible ocean contamination harmful to all mankind.

Goals and policies established by the proposed Policy Commission, directed downward, could be harmonized with substantive and assessment considerations from the Coordinating Board and Secretariat, directed upward.

A common purpose around which efforts could be mobilized to mutual benefit would be international stewardship of the blue planet. At a time when the UN is facing so many difficult problems, all members might well consider an issue around which a broad consensus has a reasonable possibility of success. Determination, however, must be accompanied by the institutional engines by which the desires of the world community can be geared into reality. The present fragmented approach must be so strengthened with cross connections that the institutional framework can meet the stresses and strains of the modern world, and reflect the environmental integrity of ocean space itself.

INTERNATIONAL STEWARDSHIP OF THE BLUE PLANET

During the millennia that life has blessed this blue planet since its origin in the oceans, the same oceans have patiently received detritus eroded from the land, deposited as sediments, and later uplifted into new mountains. These actions could be measured only in geologic time. For the past hundred years, man's technological feats to protect himself from a hostile environment and to enjoy the fruits of affluence have resulted in the production of waste that now in a flick of historical time has not only increased the volume of materials directly cast into the sea or transported and then deposited by the winds, but has altered the natural mixture with new chemical ingredients. The passive, aquatic receiver collects and circulates these products without complaint and without respect for national boundaries. But side effects are now becoming obvious. Mercury expelled by one nation may show up in tuna, halibut, or swordfish consumed by others. Biologically nondegradable DDT, effectively utilized to suppress malaria in the tropics, spreads to colder waters, there to accumulate and interfere with natural production of phyto- and zooplankton and alter the dynamics of the marine food chain, perhaps threaten the health of the ocean itself. The construction of the Aswan Dam is known to have seriously affected the ecology of the eastern Mediterranean, including the international fisheries. Tanker traffic and offshore oil extraction projected to quadruple over the next thirty years are almost certain to be accompanied by an increase in spills. A single runaway oil well in 1971 in the Persian Gulf spread slicks for hundreds of miles. No coastline will remain immune to visible stains and invisible biological effects.

We are faced with a task of separating fact from myth in assessing contemporary threats to the global environment; we are also challenged to put such knowledge to use through national and international institutions and to inform citizens of all nations of threats to our habitat and of their alternative futures. Ecological neglect, amid the pressure of commercial utilization, has only scanty protection by a myriad of legal jurisdictions. Only recently have these been balanced by a new perception of responsibility by the human inhabitants of the planet for its stewardship. Perception must beget principles as a basis for action.[10]

All nations of the world share responsibility for stewardship of the global environment for future generations: indeed, they have a fiduciary duty. All nations of the world are likely to agree on the need for collective action to manage human affairs so as to protect the global environment from serious or irreversible modifications. Detection and diagnosis of environmental damage has thus far been largely random, haphazard, and often fortuitous. Lacking are any international efforts at establishing uniform standards, much less coordinated remedial actions. Facts as to the harmful effects of pollutants are generally incomplete and their interpretation open to debate. Definitive data are needed on the magnitude and distribution of sources, levels, retention, and sinks, and as to the adverse effects of foreign substances unwittingly introduced into the natural environment by technological societies.

In operational terms, management of the quality of the environment requires management of potentially harmful technology. Control of pollution should be focused more directly on the sources of pollution than on alleviating symptoms or generating countermeasures. The greatest threats in terms of pollution loads come from industrial nations, with but a few important exceptions such as DDT from less developed countries. The vanguard of the attack on management of the global environment should thus be assumed by these same nations who, coincidentally, can best afford the costs of pollution control at their source. These same nations can also best afford the costs of research and monitoring to develop information required in the feedback processes for rational management of their own technologies.

While information is necessary, it is not sufficient. New apparatus must evaluate implications as to environmental threats; to establish criteria as to uncertainty and risk; to publicize results through environmental-impact analyses, counting on individual governments and world opinion to force remedial action; to encourage global education, beginning at elementary school levels, on fundamental attitudes needed to protect the environment; to encourage nations to develop their own environmental policies.

Environmental impact statements prepared by an authoritative international body would parallel U.S. initiatives on a domestic scale. Publication should not be subject to censorship by individual governments but should accept and distribute official dissenting opinion.

In the absence of international police forces, remedial actions must be taken at the source by national regulation and enforcement. Resistance to such regulation by the industries of any one nation would be eased if standards were adopted and applied internationally to avoid unfair competitive advantage by nations unwilling to assume the costs of pollution abatement. Ample economic sanctions exist to encourage this development. International arbitration should be compulsory.

The success of such regulation depends heavily on political leadership supported by an interested, informed public. Citizens in most advanced nations seem to be adopting an ethic of environmental conservancy that may be expected to demand corrective action in the presence of known dangers.

The less developed countries have widely opposed such concepts, on grounds that their aspirations for economic development will be blunted. They assume that their route to swift technological maturity must employ available techniques that indeed do now pollute. That assumption can be demolished by the determined application of science and engineering ingenuity to invest nonpolluting technology. That goal should be the prime target of American technical assistance for the next decade: an unequivocal subsidy to underemployed industry, with a challenge to effect a quantum jump in sophistication of manufacturing processes that could then foster economic growth to meet the disparity between the two hemispheres while reversing the present decline in environmental quality.

The projection of marine technology to the twenty-first century requires some assumption as to what the world will be, or could be, like at that time. One could assume a static condition of national rivalries, with group loyalties still internalized. Further, one could assume that profit-motivated private enterprise and public-purpose-oriented government will maintain a constant dichotomy, with unremitting strife between economic growth and environmental conservancy. With social values also assumed to be static, national boundaries of sovereignty would be extended to slice up the seabed; deals would continue between narrow vested interests whether public or private to maintain the status quo. But in sketching alternative futures, it is clear that the market mechanism cannot provide the needed protection to the planet. When realization of contemporary limits to mankind's destiny is widely understood, it is possible to assume changes in values that place more emphasis on improving the human condition, weakening of national loyalties in favor of multinational considerations, even a sense of humanity. Political subdivisions could then assume a clearer role in terms of necessary local capabilities to carry out broad social policy, rather than units competing for position and power. Such global objectives as peace, conservation of the environment and of resources for future generations, reduced disparity in eonomic well being among different peoples, greater individual freedom and equality of opportunity, health, safety, and nutrition—all would be enhanced.

A concept of management advances the theme of the global environment becoming a common heritage of mankind. For the oceans, it could be as portentous as the traditional notion of freedom of the seas. With that shift in emphasis, the gnawing uncertainties as to richness and distribution of living and nonliving resources that attends the current nervous twitching of contending nationalistic claims will become far less important than the resolution of more fundamental questions of collectively determining what we want these resources for.

The oceans offer unique opportunity for a great social experiment that could be started within the context of present institutions. The United States' support of such a precept would not be suspect in the eyes of our own or other people if motivation to steer technology for world service were clear.

Many acts of leadership are needed to bring rational management of the oceans to the world community. Indeed, the sense of community is a hallmark of civilization. While such management is itself a new abstraction, collaboration on a global scale is an even more unprecedented and challenging concept. A strategy for the oceans cannot be thought of as belonging to the U.S. or to the U.S.S.R., to the northern hemisphere or to the southern.

It must be a strategy for mankind.

APPENDIX 1

Countries with Extensive Adjacent Shallow Ocean Areas

	Approx. coastline (miles)	Approx. area (square miles)	
		(Less than 100 fathoms)	(Between 100–1000 fathoms)
Argentina..............................	2, 415	331, 000	52, 960
Australia (including New Guinea)..........	20, 125	827, 500	556, 080
Indian Ocean Islands................	N.A.	2, 648	112, 540
Brazil.................................	4, 255	264, 800	90, 032
Burma................................	1, 414	83, 412	26, 480
Canada................................	12, 650	926, 800	397, 200
China (Mainland)......................	4, 025	264, 800	39, 720
Denmark..............................	N.A.	N.A.	N.A.
Faeroe Islands......................	178	7, 944	52, 960
Greenland.........................	5, 750	79, 440	331, 000
France................................	1, 591	54, 284	12, 578
Indian Ocean Islands...............	N.A.	23, 832	140, 344
Pacific Ocean Islands................	N.A.	38, 396	152, 260
Iceland...............................	1, 242	29, 128	152, 260
India.................................	3, 162	105, 920	66, 200
Indonesia.............................	23, 000	503, 120	N.A.
Ireland...............................	759	47, 664	29, 128
Malaysia..............................	2, 127	165, 500	N.A.
Mexico...............................	5, 750	132, 400	66, 200
New Zealand..........................	3, 185	79, 440	529, 600
Norway...............................	1, 897	39, 720	152, 260
Portugal (including dependencies)..........	N.A.	79, 440	185, 360
South Africa...........................	1, 644	60, 904	101, 948
South West Africa......................	897	26, 480	59, 580
Spain.................................	1, 725	26, 480	60, 904
Atlantic dependencies...............	N.A.	31, 776	34, 424
Thailand..............................	1, 495	99, 300	19, 860
U.S.S.R...............................	26, 450	1, 324, 000	926, 800
U.K..................................	3, 220	52, 960	99, 300
Bahamas...........................	1, 610	48, 988	25, 156
Falkland Islands....................	N.A.	39, 720	139, 020
Indian Ocean Islands...............	N.A.	63, 552	72, 820
Pacific Ocean Islands...............	N.A.	22, 508	68, 848
U.S...................................	13, 112	860, 600	440, 000
Venezuela.............................	1, 150	35, 748	39, 720
Vietnam..............................	995	111, 216	55, 608

Source: *MSA*, 3d report (1969), p. 19.

APPENDIX 2

Length of Coastline of the United States, by Coastal Reach

[In statute miles]

Coastal reach	General [1] coastline	Tidal [2] shoreline	Tidal [3] shoreline detailed
Atlantic coast	2,069	6,370	28,673
New England	(473)	(1,395)	(6,130)
Middle Atlantic	(285)	(947)	(4,112)
Chesapeake	(143)	(1,019)	(6,505)
South Atlantic	(1,168)	(3,009)	(11,926)
Gulf coast	1,631	4,097	17,141
Pacific coast	7,933	17,542	41,767
Hawaii	750	900	1,052
U.S. territories and possessions	729	820	1,487
Total U.S. seacoast	13,112	29,729	90,120
Great Lakes [4]	3,533	3,533	3,533
Exterior and interior coastlines	16,645	33,262	93,653

[1] Measurements were made with a unit measure of 30 minutes latitude. The corresponding mileage varies slightly, but at the latitude of San Francisco, 30″ is about 34.5 miles. Shoreline of bays and sounds is included to a point where such waters narrow to the width of the unit measure, and the distance across at such point is included.

[2] As above, except that a unit measure of 3 statute miles was used.

[3] As above, except that a unit measure of 100 feet was used.

[4] Measurements on detailed basis not available; approximated here by using "general coastline" miles.

Source: *MSA*, 1st report (1967), p. 140.

APPENDIX 3

Area of the United States Continental Shelf, by Coastal Regions

[Thousands of square statute miles]

	Area [1] measured from coastline bounded by—		
	3-nautical-mile band	100-fathom [2] contour	1,000-fathom [2] contour
Atlantic coast	6	140	240
Gulf coast	5	135	210
Pacific coast	4	25	60
Alaska coast	20	550	755
Hawaii	2	10	30
Puerto Rico and Virgin Islands	2	2	7
Total	39	862	1,302

[1] That part of the sea floor extending from the low water line at the coast seaward to the indicated distance or depth.

[2] Fathom is a unit of length equal to 6 feet.

Source: *MSA*, 1st report (1967), p. 141.

APPENDIX 4

Petroleum Potential of Continental Shelves of the World[1]

(Areas in thousands of square miles)

Region	Shelf Area [2]	Excellent potential [3]			Fair potential [4]		
		Area	Percent of total Shelf of region	Percent of world excellent Shelf area	Area	Percent of total Shelf of region	Percent of World fair Shelf area
Total World	10,763	188	1.8	100.0	1,657	15.3	100.0
North America	2,140	40	1.9	21.3	315	14.7	19.0
South America	910	20	2.2	10.6	150	16.5	9.1
Middle East, Asia	200	40	20.2	21.3	65	32.5	3.9
East Indies Islands (Incl. Philippines)	1,350	35	2.6	18.6	305	22.6	18.4
Iron Curtain Countries	2,718	35	1.3	18.6	385	14.2	23.2
Other Areas [5]	3,445	18	5.2	9.6	437	12.7	26.4

[1] Adapted from "Offshore Operations Around the World"; Weeks, Lewis G., *Offshore*, June 20, 1967, Vol. 27, No. 7.

[2] Areas to depth of 1000 feet of water.

[3] "Excellent Potential" rating is given to areas containing or in continuity with excellent producing areas and with like geology.

[4] "Fair Potential" rating is given an area containing or in continuity with a fair producing area, or when geology is similarly favorable for commercial production.

[5] Includes Europe, Africa, Far East, Oceania, and Antarctica.

Source: *MSA*, 2d report (1968), p. 218.

APPENDIX 5

Executive Order No. 10807 Establishing the Federal Council for Science and Technology

V–E. FEDERAL COUNCIL FOR SCIENCE AND TECHNOLOGY (1959–66)

Text of Executive Order No. 10807, March 13, 1959, 24 F.R. 1897, establishing the Federal Council for Science and Technology, amending Executive Order 10521 of March 17, 1954, and revoking Executive Order 9912 of December 24, 1947

EX. ORD. 10807. FEDERAL COUNCIL FOR SCIENCE AND TECHNOLOGY

Ex. Ord. No. 10807, Mar. 13, 1959, 24 F.R. 1897, provided:

SECTION 1. *Establishment of Council.* (a) There is hereby established the Federal Council for Science and Technology (hereinafter referred to as the Council).

(b) The Council shall be composed of the following-designated members: (1) the Special Assistant to the President for Science and Technology, (2) one representative of each of the following-named departments, who shall be designated by the Secretary of the Department concerned and shall be an official of the Department of policy rank: the Departments of Defense, the Interior, Agriculture, Commerce, and Health, Education, and Welfare, (3) the Director of the National Science Foundation, (4) the Administrator of the National Aeronautics and Space Administration, and (5) a representative of the Atomic Energy Commission, who shall be the Chairman of the Commission or another member of the Commission designated by the Chairman. A representative of the Secretary of State designated by the Secretary and a representative of the Director of the Bureau of the Budget designated by the Director may attend meetings of the Council as observers.

(c) The Chairman of the Council (hereinafter referred to as the Chairman) shall be designated by the President from time to time from among the members thereof. The Chairman may make provision for another member of the Council, with the consent of such member, to act temporarily as Chairman.

(d) The Chairman (1) may request the head of any Federal agency not named in section 2(b) of this order to designate a representative to participate in meetings or parts of meetings of the Council concerned with matters of substantial interest to the agency, and (2) may invite other persons to attend meetings of the Council.

(e) The Council shall meet at the call of the Chairman.

SEC. 2. *Functions of Council.* (a) The Council shall consider problems and developments in the fields of science and technology and related activities affecting more than one Federal agency or concerning the over-all advancement of the Nation's science and technology, and shall recommend policies and other measures (1) to provide more effective planning and administration of Federal scientific and technological programs, (2) to identify research needs including areas of research requiring additional emphasis, (3) to achieve more effective utilization of the scientific and technological resources and facilities of Federal agencies, including the elimination of unnecessary duplication, and (4) to further international cooperation in science and technology. In developing such policies and measures the Council, after consulting, when considered appropriate by the Chairman, the National Academy of Sciences, the President's Science Advisory Committee, and other organizations, shall consider (i) the effects of Federal research and development policies and programs on non-Federal programs and institutions, (ii) long-range program plans designed to meet the scientific and technological needs of the Federal Government, including manpower and capital requirements, and (iii) the effects of non-Federal programs in science and technology upon Federal research and development policies and programs.

(b) The Council shall consider and recommend measures for the effective implementation of Federal policies concerning the administration and conduct of Federal programs in science and technology.

(c) The Council shall perform such other related duties as shall be assigned, consonant with law, by the President or by the Chairman.

(d) The Chairman shall, from time to time, submit to the President such of the Council's recommendations or reports as require the attention of the President by reason of their importance or character.

SEC. 3. *Agency assistance to Council.* (a) For the purpose of effectuating this order, each Federal agency represented on the Council shall furnish necessary assistance to the Council in consonance with section 214 of the act of May 3, 1945, 59 Stat. 134 (31 U.S.C. § 691). Such assistance may include (1) detailing employees to the Council to perform such functions, consistent with the purposes of this order, as the Chairman may assign to them, and (2) undertaking, upon request of the Chairman, such special studies for the Council as come within the functions herein assigned to the Council.

(b) Upon request of the Chairman, the heads of Federal agencies shall, so far as practicable, provide the Council with information and reports relating to the scientific and technological activities of the respective agencies.

SEC. 4. *Standing committees and panels.* For the purpose of conducting studies and making reports as directed by the Chairman, standing committees and panels of the Council may be established in consonance with the provisions of section 214 of the act of May 3, 1945, 59 Stat. 134 (31 U.S.C. § 691). At least one such standing committee shall be composed of scientist-administrators representing Federal agencies, shall provide a forum for consideration of common administrative policies and procedures relating to Federal research and development activities and for formulation of recommendations thereon, and shall perform such other related functions as may be assigned to it by the Chairman of the Council.

SEC. 5. *Security procedures.* The Chairman shall establish procedures to insure the security of classified information used by or in the custody of the Council or employees under its jurisdiction.

SEC. 6. *Other orders; construction of orders.* (a) Executive Order No. 9912 of December 24, 1947, entitled "Establishing the Interdepartmental Committee on Scientific Research and Development," is hereby revoked.

(b) Executive Order No. 10521 of March 17, 1957 [set out as a note under this section], entitled "Administration of Scientific Research by Agencies of the Federal Government," is hereby amended:

(1) By substituting for section 1 thereof the following:

"SECTION 1. The National Science Foundation (hereinafter referred to as the Foundation) shall from time to time recommend to the President policies for the promotion and support of basic research and education in the sciences, including policies with respect to furnishing guidance toward defining the responsibilities of the Federal Government in the conduct and support of basic scientific research."

(2) By inserting before the words "scientific research programs and activities" in section 3 thereof the word "basic".

(3) (i) By adding the word "and" at the end of paragraph (a) of section 8 thereof, (ii) by deleting the semicolon and the word "and" at the end of paragraph (b) of section 8 and inserting in lieu thereof a period, and (iii) by revoking paragraph (c) of section 8.

(4) By adding at the end of the order a new section 10 reading as follows:

"SEC. 10. The National Science Foundation shall provide leadership in the effective coordination of the scientific information activities of the Federal Government with a view to improving the availability and dissemination of scientific information. Federal agencies shall cooperate with and assist the National Science Foundation in the performance of this function, to the extent permitted by law."

(c) The provisions of Executive Order No. 10521, as hereby amended, shall not limit the functions of the Council under this order. The provisions of this order shall not limit the functions of any Federal agency or officer under Executive Order No. 10521, as hereby amended.

(d) The Council shall be advisory to the President and to the heads of Federal agencies represented on the Council; accordingly, this order shall not be construed as subjecting any agency, officer, or function to control by the Council.

DWIGHT D. EISENHOWER.

APPENDIX 6

Message to the Congress of the United States by President John F. Kennedy Transmitting Reorganization Plan No. 2 of 1962, Providing for Certain Reorganizations in the Field of Science and Technology

To the Congress of the United States:

I transmit herewith Reorganization Plan No. 2 of 1962, prepared in accordance with the provisions of the Reorganization Act of 1949, as amended, and providing for certain reorganizations in the field of science and technology.

Part I of the reorganization plan establishes the Office of Science and Technology as a new unit within the Executive Office of the President; places at the head thereof a Director appointed by the President by and with the advice and consent of the Senate and makes provision for a Deputy Director similarly appointed; and transfers to the Director certain functions of the National Science Foundation under sections 3(a)(1) and 3(a)(6) of the National Science Foundation Act of 1950.

The new arrangements incorporated in Part I of the reorganization plan will constitute an important development in executive branch organization for science and technology. Under those arrangements the President will have permanent staff resources capable of advising and assisting him on matters of national policy affected by or pertaining to science and technology. Considering the rapid growth and far-reaching scope of Federal activities in science and technology, it is imperative that the President have adequate staff support in developing policies and evaluating programs in order to assure that science and technology are used most effectively in the interests of national security and general welfare.

To this end it is contemplated that the Director will assist the President in discharging the responsibility of the President for the proper coordination of Federal science and technology functions. More particularly, it is expected that he will advise and assist the President as the President may request with respect to:

(1) Major policies, plans and programs of science and technology of the various agencies of the Federal Government, giving appropriate emphasis to the relationship of science and technology to national security and foreign policy, and measures for furthering science and technology in the Nation.

(2) Assessment of selected scientific and technical developments and programs in relation to their impact on national policies.

(3) Review, integration, and coordination of major Federal activities in science and technology, giving due consideration to the effects of such activities on non-Federal resources and institutions.

(4) Assuring that good and close relations exist with the Nation's scientific and engineering communities so as to further in every appropriate way their participation in strengthening science and technology in the United States and the Free World.

(5) Such other matters consonant with law as may be assigned by the President to the Office.

The ever-growing significance and complexity of Federal programs in science and technology have in recent years necessitated the taking of several steps for improving the

445

organizational arrangements of the executive branch in relation to science and technology:

(1) The National Science Foundation was established in 1950. The Foundation was created to meet a widely recognized need for an organization to develop and encourage a national policy for the promotion of basic research and education in the sciences, to support basic research, to evaluate research programs undertaken by Federal agencies, and to perform related functions.

(2) The Office of the Special Assistant to the President for Science and Technology was established in 1957. The Special Assistant serves as Chairman of both the President's Science Advisory Committee and the Federal Council for Science and Technology, mentioned below.

(3) At the same time, the Science Advisory Committee, composed of eminent non-Government scientists and engineers, and located within the Office of Defense Mobilization, was reconstituted in the White House Office as the President's Science Advisory Committee.

(4) The Federal Council for Science and Technology, composed of policy officials of the principal agencies engaged in scientific and technical activities, was established in 1959.

The National Science Foundation has proved to be an effective instrument for administering sizable programs in support of basic research and education in the sciences and has set an example for other agencies through the administration of its own programs. However, the Foundation, being at the same organizational level as other agencies, cannot satisfactorily coordinate Federal science policies or evaluate programs of other agencies. Science policies, transcending agency lines, need to be coordinated and shaped at the level of the Executive Office of the President drawing upon many resources both within and outside of Government. Similarly, staff efforts at that higher level are required for the evaluation of Government programs in science and technology.

Thus, the further steps contained in Part I of the reorganization plan are now needed in order to meet most effectively new and expanding requirements brought about by the rapid and far-reaching growth of the Government's research and development programs. These requirements call for the further strengthening of science organization at the Presidential level and for the adjustment of the Foundation's role to reflect changed conditions. The Foundation will continue to originate policy proposals and recommendations concerning the support of basic research and education in the sciences, and the new Office will look to the Foundation to provide studies and information on which sound national policies in science and technology can be based.

Part I of the reorganization plan will permit some strengthening of the staff and consultant resources now available to the President in respect of scientific and technical factors affecting executive branch policies and will also facilitate communication with the Congress.

Part II of the reorganization plan provides for certain reorganizations within the National Science Foundation which will strengthen the capability of the Director of the Foundation to exert leadership and otherwise further the effectiveness of administration of the Foundation. Specifically:

(1) There is established a new office of Director of the National Science Foundation and that Director, ex officio, is made a member of the National Science Board on a basis coordinate with that of other Board members.

(2) There is substituted for the now-existing Executive Committee of the National Science Board a new Executive Committee composed of the Director of the National Science Foundation, ex officio, as a voting member and chairman of the Committee, and of four other members elected by the National Science Board from among its appointive members.

(3) Committees advisory to each of the divisions of the Foundation will make their recommendations to the Director only rather than to both the Director and the National Science Board.

After investigation I have found and hereby declare that each reorganization included in Reorganization Plan No. 2 of 1962 is necessary to accomplish one or more of the purposes set forth in section 2(a) of the Reorganization Act of 1949, as amended.

I have found and hereby declare that it is necessary to include in the reorganization plan, by reason of reorganizations made thereby, provisions for the appointment and compensation of the Director and Deputy Director of the Office of Science and Technology and of the Director of the National Science Foundation. The rate of compensation fixed for each of these officers is that which I have found to prevail in respect of comparable officers in the executive branch of the Government.

The functions abolished by the provisions of section 23(b) of the reorganization plan are provided for in sections 4(a), 5(a), 6(a), 6(b), and 8(d) of the National Science Foundation Act of 1950.

The taking effect of the reorganizations included in the reorganization plan will provide sound organizational arrangements and will make possible more effective and efficient administration of Government programs in science and technology. It is, however, impracticable to itemize at his time the reductions in expenditures which it is probable will be brought about by such taking effect.

I recommend that the Congress allow the reorganization plan to become effective.

JOHN F. KENNEDY

APPENDIX 7

Marine Resources and Engineering Development Act of 1966 (P.L. 89–454)

AN ACT To provide for a comprehensive, long-range, and coordinated national program in marine science, to establish a National Council on Marine Resources and Engineering Development, and a Commission on Marine Science, Engineering and Resources, and for other purposes.

Be it enacted by the Senate and House of Representatives of the United States of America in Congress assembled, That this Act may be cited as the "Marine Resources and Engineering Development Act of 1966".

DECLARATION OF POLICY AND OBJECTIVES

SEC. 2. (a) It is hereby declared to be the policy of the United States to develop, encourage, and maintain a coordinated, comprehensive, and long-range national program in marine science for the benefit of mankind to assist in protection of health and property, enhancement of commerce, transportation, and national security, rehabilitation of our commercial fisheries, and increased utilization of these and other resources.

(b) The marine science activities of the United States should be conducted so as to contribute to the following objectives:

(1) The accelerated development of the resources of the marine environment.

(2) The expansion of human knowledge of the marine environment.

(3) The encouragement of private investment enterprise in exploration, technological development, marine commerce, and economic utilization of the resources of the marine environment.

(4) The preservation of the role of the United States as a leader in marine science and resource development.

(5) The advancement of education and training in marine science.

(6) The development and improvement of the capabilities, performance, use, and efficiency of vehicles, equipment, and instruments for use in exploration, research, surveys, the recovery of resources, and the transmission of energy in the marine environment.

(7) The effective utilization of the scientific and engineering resources of the Nation, with close cooperation among all interested agencies, public and private, in order to avoid unnecessary duplication of effort, facilities, and equipment, or waste.

(8) The cooperation by the United States with other nations and groups of nations and international organizations in marine science activities when such cooperation is in the national interest.

THE NATIONAL COUNCIL ON MARINE RESOURCES AND ENGINEERING DEVELOPMENT

SEC. 3. (a) There is hereby established, in the Executive Office of the President, the National Council on Marine Resources and Engineering Development (hereinafter called the "Council") which shall be composed of—

(1) The Vice President, who shall be Chairman of the Council.

(2) The Secretary of State.

(3) The Secretary of the Navy.

(4) The Secretary of the Interior.

(5) The Secretary of Commerce.

(6) The Chairman of the Atomic Energy Commission.

(7) The Director of the National Science Foundation.

(8) The Secretary of Health, Education, and Welfare.

(9) The Secretary of the Treasury.

(b) The President may name to the Council such other officers and officials as he deems advisable.

(c) The President shall from time to time designate one of the members of the Council to preside over meetings of the Council during the absence, disability, or unavailability of the Chairman.

(d) Each member of the Council, except those designated pursuant to subsection (b), may designate any officer of his department or agency appointed with the advice and consent of the Senate to serve on the Council as his alternate in his unavoidable absence.

(e) The Council may employ a staff to be headed by a civilian executive secretary who shall be appointed by the President and shall receive compensation at a rate established by the President at not to exceed that of level II of the Federal Executive Salary Schedule. The executive secretary, subject to the direction of the Council, is authorized to appoint and fix the compensation of such personnel, including not more than seven persons who may be appointed without regard to civil service laws or the Classification Act of 1949 and compensated at not to exceed the highest rate of grade 18 of the General Schedule of the Classification Act of 1949, as amended, as may be necessary to perform such duties as may be prescribed by the President.

(f) The provisions of this Act with respect to the Council shall expire one hundred and twenty days after the submission of the final report of the Commission pursuant to section 5(h).

RESPONSIBILITIES

SEC. 4. (a) In conformity with the provisions of section 2 of this Act, it shall be the duty of the President with the advice and assistance of the Council to—

(1) survey all significant marine science activities, including the policies, plans, programs, and accomplishments of all departments and agencies of the United States engaged in such activities;

(2) develop a comprehensive program of marine science activities, including, but not limited to, exploration, description and prediction of the marine environment, exploitation and conservation of the resources of the marine environment, marine engineering, studies of air-sea interaction, transmission of energy, and communications, to be conducted by departments and agencies of the United States, independently or in cooperation with such non-Federal organizations as States, institutions and industry;

(3) designate and fix responsibility for the conduct of the foregoing marine science activities by departments and agencies of the United States;

(4) insure cooperation and resolve differences arising among departments and agencies of the United States with respect to marine science activities under this Act, including differences as to whether a particular project is a marine science activity;

(5) undertake a comprehensive study, by contract or otherwise, of the legal problems arising out of the management, use, development, recovery, and control of the resources of the marine environment;

(6) establish long-range studies of the potential benefits to the United States economy, security, health, and welfare to be gained from marine resources, engineering, and science, and the costs involved in obtaining such benefits; and

(7) review annually all marine science activities conducted by departments and agencies of the United States in light of the policies, plans, programs, and priorities developed pursuant to this Act.

(b) In the planning and conduct of a coordinated Federal program the President and the Council shall utilize such staff, interagency, and non-Government advisory arrangements as they may find necessary and appropriate and shall consult with departments and agencies concerned with marine science activities and solicit the views of non-Federal organizations and individuals with capabilities in marine sciences.

COMMISSION ON MARINE SCIENCE, ENGINEERING, AND RESOURCES

SEC. 5. (a) The President shall establish a Commission on Marine Science, Engineering, and Resources (in this Act referred to as the "Commission"). The Commission shall be composed of fifteen members appointed by the President, including individuals drawn from Federal and State governments, industry, universities, laboratories and other institutions engaged in marine scientific or technological pursuits, but not more than five members shall be from the Federal Government. In addition the Commission shall have four advisory members appointed by the President from among the Members of the Senate and the House of Representatives. Such advisory members shall not participate, except in an advisory capacity, in the formulation of the findings and recommendations of the Commission. The President shall select a Chairman and Vice Chairman from among such fifteen members. The Vice Chairman shall act as Chairman in the latter's absence.

(b) The Commission shall make a comprehensive investigation and study of all aspects of marine science in order to recommend an overall plan for an adequate national oceanographic program that will meet the present and future national needs. The Commission shall undertake a review of existing and planned marine science activities of the United States in order to assess their adequacy in meeting the objectives set forth under section 2(b), including but not limited to the following:

(1) Review the known and contemplated needs for natural resources from the marine environment to maintain our expanding national economy.

(2) Review the surveys, applied research programs, and ocean engineering projects required to obtain the needed resources from the marine environment.

(3) Review the existing national research programs to insure realistic and adequate support for basic oceanographic research that will enhance human welfare and scientific knowledge.

(4) Review the existing oceanographic and ocean engineering programs, including education and technical training, to determine which programs are required to advance our national oceanographic competence and stature and which are not adequately supported.

(5) Analyze the findings of the above reviews, including the economic factors involved, and recommend an adequate national marine science program that will meet the present and future national needs without unnecessary duplication of effort.

(6) Recommend a Governmental organizational plan with estimated cost.

(c) Members of the Commission appointed from outside the Government shall each receive $100 per diem when engaged in the actual performance of duties of the Commission and reimbursement of travel expenses, including per diem in lieu of subsistence, as authorized in section 5 of the Administrative Expenses Act of 1946, as amended (5 U.S.C. 73b–2), for persons employed intermittently. Members of the Commission appointed from within the Government shall serve without additional compensation to that received for their services to the Government but shall be reimbursed for travel expenses, including per diem in lieu of subsistence, as authorized in the Act of June 9, 1949, as amended (5 U.S.C. 835–842).

(d) The Commission shall appoint and fix the compensation of such personnel as it deems advisable in accordance with the civil service laws and the Classification Act of 1949, as amended. In addition, the Commission may secure temporary and intermittent services to the same extent as is authorized for the departments by section 15 of the Administrative Expenses Act of 1946 (60 Stat. 810) but at rates not to exceed $100 per diem for individuals.

(e) The Chairman of the Commission shall be responsible for (1) the assignment of duties and responsibilities among such personnel and their continuing supervision, and (2) the use and expenditures of funds available to the Commission. In carrying out the provisions of this subsection, the Chairman shall be governed by the general policies of the Commission with respect to the work to be accomplished by it and the timing thereof.

(f) Financial and administrative services (including those related to budgeting, accounting, financial reporting, personnel, and procurement) may be provided the Commission by the General Services Administration, for which payment shall be made in advance, or by reimbursement from funds of the Commission in such amounts as may be agreed upon by the Chairman of the Commission and the Administrator of General Services: *Provided,* That the regulations of the General Services Administration for the collection of indebtedness of personnel resulting from erroneous payments (5 U.S.C. 46d) shall apply to the collection of erroneous payments made to or on behalf of a Commission employee, and regulations of said Administrator for the administrative control of funds (31 U.S.C. 665(g)) shall apply to appropriations of the Commission: *And provided further,* That the Commission shall not be required to prescribe such regulations.

(g) The Commission is authorized to secure directly from any executive department, agency, or independent instrumentality of the Government any information it deems necessary to carry out its functions under this Act; and each such department, agency, and instrumentality is authorized to cooperate with the Commission and, to the extent permitted by law, to furnish such information to the Commission, upon request made by the Chairman.

(h) The Commission shall submit to the President, via the Council, and to the Congress not later than eighteen months after the establishment of the Commission as provided in subsection (a) of this section, a final report of its findings and recommendations. The Commission shall cease to exist thirty days after it has submitted its final report.

INTERNATIONAL COOPERATION

SEC. 6. The Council, under the foreign policy guidance of the President and as he may request, shall coordinate a program of international cooperation in work done pursuant to this Act, pursuant to agreements made by the President with the advice and consent of the Senate.

REPORTS

SEC. 7. (a) The President shall transmit to the Congress in January of each year a report, which shall include (1) a comprehensive description of the activities and the accomplishments of all agencies and departments of the United States in the field of marine science during the preceding fiscal year, and (2) an evaluation of such activities and accomplishments in terms of the objectives set forth pursuant to this Act.

(b) Reports made under this section shall contain such recommendations for legislation as the President may consider necessary or desirable for the attainment of the objectives of this Act, and shall contain an estimate of funding requirements of each agency and department of the United States for marine science activities during the succeeding fiscal year.

DEFINITIONS

SEC. 8. For the purposes of this Act the term "marine science" shall be deemed to apply to oceanographic and scientific endeavors and disciplines, and engineering and technology in and with relation to the marine environment; and the term "marine environment" shall be deemed to include (a) the oceans, (b) the Continental Shelf of the United States, (c) the Great Lakes, (d) seabed and subsoil of the submarine areas adjacent to the coasts of the United States to the depth of two hundred meters, or beyond that limit, to where the depths of the superjacent waters admit of the exploitation of the natural resources of such areas, (e) the seabed and subsoil of similar submarine areas adjacent to the coasts of islands which comprise United States territory, and (f) the resources thereof.

AUTHORIZATION

SEC. 9. There are hereby authorized to be appropriated such sums as may be necessary to carry out this Act, but sums appropriated for any one fiscal year shall not exceed $1,500,000.

(Public Law 90–242, January 2, 1968)

AN ACT To amend the Marine Resources and Engineering Development Act of 1966, as amended, to extend the period of time within which the Commission on Marine Science, Engineering, and Resources is to submit its final report and to provide for a fixed expiration date for the National Council on Marine Resources and Engineering Development.

Be it enacted by the Senate and House of Representatives of the United States of American in Congress assembled, That the Marine Resources and Engineering Development Act of 1966 is amended as follows:

Subparagraph (h) of section 5 is amended by striking out "eighteen" and inserting "twenty-four" in lieu thereof.

SEC. 2. Subparagraph (f) of section 3 is amended by striking out "one hundred and twenty days after the submission of the final report of the Commission pursuant to section 5(h)." and inserting in lieu thereof " 'on June 30, 1969.' ".

(PUBLIC LAW 90–477, AUGUST 11, 1968)

AN ACT To amend title II of the Marine Resources and Engineering Development Act of 1966

Be it enacted by the Senate and House of Representatives of the United States of America in Congress assembled, That title II of the Marine Resources and Engineering Development Act of 1966 is amended as follows:

(1) Section 203(b)(1) of the Marine Resources and Engineering Development Act of 1966 is amended by inserting immediately after "for the fiscal year ending June 30, 1968, not to exceed the sum of $15,000,000," the following: "for the fiscal year ending June 30, 1969, not to exceed the sum of $6,000,000, for the fiscal year ending June 30, 1970, not to exceed the sum of $15,000,000,".

(2) Section 204(d)(1) of the Marine Resources and Engineering Development Act of 1966 is amended by deleting the phrase "in any fiscal year" each time it appears therein.

(Public Law 91–95, May 23, 1969)

AN ACT To amend the Marine Resources and Engineering Development Act of 1966 to continue the National Council on Marine Resources and Engineering Development, and for other purposes.

Be it enacted by the Senate and House of Representatives of the United States of America in Congress assembled, That subsection (f) of section 3 of the Marine Resources and Engineering Development Act of 1966 (33 U.S.C. 1102(f)) is amended by striking out "June 30, 1969" and inserting in lieu thereof "June 30, 1970".

SEC. 2. Section 9 of such Act (33 U.S.C. 1108) is amended by striking out "$1,500,000" and inserting in lieu thereof "$1,200,000".

(Public Law 91–414, September 25, 1970)

To amend the Marine Resources and Engineering Development Act of 1966 to continue the National Council on Marine Resources and Engineering Development.

Be it enacted by the Senate and House of Representatives of the United States of America in Congress assembled, That subsection (f) of section 3 of the Marine Resources and Engineering Development Act of 1966 (33 U.S.C. 1102(f)) is amended by striking out "June 30, 1970" and inserting in lieu thereof "June 30, 1971".

APPENDIX 8

National Sea Grant College and Program Act of 1966 (P.L. 89–688)

AN ACT To amend the Marine Resources and Engineering Development Act of 1966 to authorize the establishment and operation of sea grant colleges and programs by initiating and supporting programs of education and research in the various fields relating to the development of marine resources, and for other purposes.

Be it enacted by the Senate and House of Representatives of the United States of America in Congress assembled, That the Marine Resources and Engineering Development Act of 1966 is amended by adding at the end thereof the following new title:

"TITLE II—SEA GRANT COLLEGES AND PROGRAMS

"SHORT TITLE

"Sec. 201. This title may be cited as the 'National Sea Grant College and Program Act of 1966'.

"DECLARATION OF PURPOSE

"Sec. 202. The Congress hereby finds and declares—

"(a) that marine resources, including animal and vegetable life and mineral wealth, constitute a far-reaching and largely untapped asset of immense potential significance to the United States; and

"(b) that it is in the national interest of the United States to develop the skilled manpower, including scientists, engineers, and technicians, and the facilities and equipment necessary for the exploitation of these resources; and

"(c) that aquaculture, as with agriculture on land, and the gainful use of marine resources can substantially benefit the United States, and ultimately the people of the world, by providing greater economic opportunities, including expanded employment and commerce; the enjoyment and use of our marine resources; new sources of food; and new means for the development of marine resources; and

"(d) that Federal support toward the establishment, development, and operation of programs by sea grant colleges and Federal support of other sea grant programs designed to achieve the gainful use of marine resources, offer the best means of promoting programs toward the goals set forth in clauses (a), (b), and (c), and should be undertaken by the Federal Government; and

"(e) that in view of the importance of achieving the earliest possible institution of significant national activities related to the development of marine resources, it is the purpose of this title to provide for the establishment of a program of sea grant colleges and education, training, and research in the fields of marine science, engineering, and related disciplines.

"GRANTS AND CONTRACTS FOR SEA GRANT COLLEGES AND PROGRAMS

"Sec. 203. (a) The provisions of this title shall be administered by the National Science Foundation (hereafter in this title referred to as the 'Foundation').

"(b)(1) For the purpose of carrying out this title, there is authorized to be appropriated to the Foundation for the fiscal year ending June 30, 1967, not to exceed the sum of $5,000,000, for the fiscal year ending June 30, 1968, not to exceed the sum of $15,000,000, and for each subsequent fiscal year only such sums as the Congress may hereafter specifically authorize by law.

"(2) Amounts appropriated under this title are authorized to remain available until expended.

"MARINE RESOURCES

"SEC. 204. (a) In carrying out the provisions of this title the Foundation shall (1) consult with those experts engaged in pursuits in the various fields related to the development of marine resources and with all departments and agencies of the Federal Government (including the United States Office of Education in all matters relating to education) interested in, or affected by, activities in any such fields, and (2) seek advice and counsel from the National Council on Marine Resources and Engineering Development as provided by section 205 of this title.

"(b) The Foundation shall exercise its authority under this title by—

"(1) initiating and supporting programs at sea grant colleges and other suitable institutes, laboratories, and public or private agencies for the education of participants in the various fields relating to the development of marine resources;

"(2) initiating and supporting necessary research programs in the various fields relating to the development of marine resources, with preference given to research aimed at practices, techniques, and design of equipment applicable to the development of marine resources; and

"(3) encouraging and developing programs consisting of instruction, practical demonstrations, publications, and otherwise, by sea grant colleges and other suitable institutes, laboratories, and public or private agencies through marine advisory programs with the object of imparting useful information to persons currently employed or interested in the various fields related to the development of marine resources, the scientific community, and the general public.

"(c) Programs to carry out the purposes of this title shall be accomplished through contracts with, or grants to, suitable public or private institutions of higher education, institutes, laboratories, and public or private agencies which are engaged in, or concerned with, activities in the various fields related to the development of marine resources, for the establishment and operation by them of such programs.

"(d)(1) The total amount of payments in any fiscal year under any grant to or contract with any participant in any program to be carried out by such participant under this title shall not exceed 66⅔ per centum of the total cost of such program. For purposes of computing the amount of the total cost of any such program furnished by any participant in any fiscal year, the Foundation shall include in such computation an amount equal to the reasonable value of any buildings, facilities, equipment, supplies, or services provided by such participant with respect to such program (but not the cost or value of land or of Federal contributions).

"(2) No portion of any payment by the Foundation to any participant in any program to be carried out under this title shall be applied to the purchase or rental of any land or the rental, purchase, construction, preservation, or repair of any building, dock, or vessel.

"(3) The total amount of payments in any fiscal year by the Foundation to participants within any State shall not exceed 15 per centum of the total amount appropriated to the Foundation for the purposes of this title for such fiscal year.

"(e) In allocating funds appropriated in any fiscal year for the purposes of this title the Foundation shall endeavor to achieve maximum participation by sea grant colleges and other suitable institutes, laboratories, and public or private agencies throughout the United States, consistent with the purposes of this title.

"(f) In carrying out its functions under this title, the Foundation shall attempt to support programs in such a manner as to supplement and not duplicate or overlap any existing and related Government activities.

"(g) Except as otherwise provided in this title, the Foundation shall, in carrying out its functions under this title, have the same powers and authority it has under the National Science Foundation Act of 1950 to carry out its functions under that Act.

"(h) The head of each department, agency, or instrumentality of the Federal Government is authorized, upon request of the Foundation, to make available to the

Foundation, from time to time, on a reimbursable basis, such personnel, services, and facilities as may be necessary to assist the Foundation in carrying out its functions under this title.

"(i) For the purposes of this title—

"(1) the term 'development of marine resources' means scientific endeavors relating to the marine environment, including, but not limited to, the fields oriented toward the development, conservation, or economic utilization of the physical, chemical, geological, and biological resources of the marine environment; the fields of marine commerce and marine engineering; the fields relating to exploration or research in, the recovery of natural resources from, and the transmission of energy in, the marine environment; the fields of oceanography and oceanology; and the fields with respect to the study of the economic, legal, medical, or sociological problems arising out of the management, use, development, recovery, and control of the natural resources of the marine environment;

"(2) the term 'marine environment' means the oceans; the Continental Shelf of the United States; the Great Lakes; the seabed and subsoil of the submarine areas adjacent to the coasts of the United States to the depth of two hundred meters, or beyond that limit, to where the depths of the superjacent waters admit of the exploitation of the natural resources of the area; the seabed and subsoil of similar submarine areas adjacent to the coasts of islands which comprise United States territory; and the natural resources thereof;

"(3) the term 'sea grant college' means any suitable public or private institution of higher education supported pursuant to the purposes of this title which has major programs devoted to increasing our Nation's utilization of the world's marine resources; and

"(4) the term 'sea grant program' means (A) any activities of education or research related to the development of marine resources supported by the Foundation by contracts with or grants to institutions of higher education either initiating, or developing existing programs in fields related to the purposes of this title, (B) any activities of education or research related to the development of marine resources supported by the Foundation by contracts with or grants to suitable institutes, laboratories, and public or private agencies, and (C) any programs of advisory services oriented towards imparting information in fields related to the development of marine resources supported by the Foundation by contracts with or grants to suitable institutes, laboratories, and public or private agencies.

"ADVISORY FUNCTIONS

"SEC. 205. The National Council on Marine Resources and Engineering Development established by section 3 of title I of this Act shall, as the President may request—

"(1) advise the Foundation with respect to the policies, procedures, and operations of the Foundation in carrying out its functions under this title;

"(2) provide policy guidance to the Foundation with respect to contracts or grants in support of programs conducted pursuant to this title, and make such recommendations thereon to the Foundation as may be appropriate; and

"(3) submit an annual report on its activities and its recommendations under this section to the Speaker of the House of Representatives, the Committee on Merchant Marine and Fisheries of the House of Representatives, the President of the Senate, and the Committee on Labor and Public Welfare of the Senate."

SEC. 2. (a) The Marine Resources and Engineering Development Act of 1966 is amended by striking out the first section and inserting in lieu thereof the following:

"TITLE I—MARINE RESOURCES AND ENGINEERING DEVELOPMENT

"SHORT TITLE

"SECTION 1. This title may be cited as the 'Marine Resources and Engineering Development Act of 1966'."

(b) Such Act is further amended by striking out "this Act" the first place it appears in section 4(a), and also each place it appears in sections 5(a), 8, and 9, and inserting in lieu thereof in each such place "this title".

APPENDIX 9

Senate Concurrent Resolution 72—International Decade of Ocean Exploration, 1967

Whereas the Congress finds that an unprecedented scientific and technological readiness now exists for exploration of the oceans and their resources;

Whereas accelerated exploration of the nature, extent, and distribution of ocean resources could significantly increase the food, mineral, and energy resources available for the benefit of mankind;

Whereas improved understanding of ocean processes would enhance the protection of life and property against severe storms and other hazards, would further the safety of maritime commerce, would directly contribute to the development of coastal areas of the Nation, would benefit the Nation's fishing and mineral extractive industries, and would contribute to advancement of a broad range of scientific disciplines;

Whereas realization of the full potential of the oceans will require a long-term program of exploration, observation, and study on a world-wide basis, utilizing ships, buoys, aircraft, satellites, undersea submersibles, and other platforms, advanced navigation systems, and expanded data processing and distribution facilities;

Whereas the inherently international character of ocean phenomena has attracted the interest of many nations;

Whereas excellence, experience, and capabilities in marine science and technology are shared by many nations and a broad program of ocean exploration can most effectively and economically be carried out through a cooperative effort by many nations of the world; and

Whereas the United States has begun to explore through the United Nations and other forums international interest in a long-term program of ocean exploration:
Now, therefore, be it

Resolved by the Senate of the United States (the House of Representatives concurring) [1], That it is the sense of Congress that the United States should participate in and give full support to an International Decade of Ocean Exploration during the 1970's which would include (1) an expanded national program of exploration in waters close to the shores of the United States, (2) intensified exploration activities in waters more distant from the United States, and (3) accelerated development of the capabilities of the United States to explore the oceans and particularly the training and education of needed scientists, engineers, and technicians.

Sec. 2. It is further the sense of Congress that the President should cooperate with other nations in (1) encouraging broad international participation in an International Decade of Ocean Exploration, (2) sharing results and experiences from national ocean exploration programs, (3) planning and coordinating international cooperative projects within the framework of a sustained, long-range international effort to investigate the world's oceans, (4) strengthening and expanding international arrangements for the timely international exchange of oceanographic data, and (5) providing appropriate technical and training assistance and facilities to the developing countries and support to international organizations so they may effectively contribute their share to the International Decade of Ocean Exploration.

Sec. 3. It is further the sense of Congress that the President in his annual report to the Congress on marine science affairs pursuant to Public Law 89–454 should transmit to the Congress a plan setting forth the proposed participation of the United States for the next fiscal year in the International Decade of Ocean Exploration. The plan should contain a statement of the activities to be conducted and specify the department or agency of the Government which would conduct the activity and seek appropriations therefor.

---

[1] Not acted on by the House during 90th Congress.

APPENDIX 10

The President's Message to the Congress Upon Transmitting Reorganization Plans No. 3 and 4 to Establish the Environmental Protection Agency and the National Oceanic and Atmospheric Administration, July 9, 1970

To the Congress of the United States:

As concern with the condition of our physical environment has intensified, it has become increasingly clear that we need to know more about the total environment—land, water and air. It also has become increasingly clear that only by reorganizing our Federal efforts can we develop that knowledge, and effectively ensure the protection, development and enhancement of the total environment itself.

The Government's environmentally-related activities have grown up piecemeal over the years. The time has come to organize them rationally and systematically. As a major step in this direction, I am transmitting today two reorganization plans: one to establish an Environmental Protection Agency, and one to establish, within the Department of Commerce, a National Oceanic and Atmospheric Administration.

ENVIRONMENTAL PROTECTION AGENCY
(EPA)

Our national government today is not structured to make a coordinated attack on the pollutants which debase the air we breathe, the water we drink, and the land that grows our food. Indeed, the present governmental structure for dealing with environmental pollution often defies effective and concerted action.

Despite its complexity, for pollution control purposes the environment must be perceived as a single, interrelated system. Present assignments of departmental responsibilities do not reflect this interrelatedness.

Many agency missions, for example, are designed primarily along media lines—air, water, and land. Yet the sources of air, water, and land pollution are interrelated and often interchangeable. A single source may pollute the air with smoke and chemicals, the land with solid wastes, and a river or lake with chemical and other wastes. Control of the air pollution may produce more solid wastes, which then pollute the land or water. Control of the water-polluting effluent may convert it into solid wastes, which must be disposed of on land.

Similarly, some pollutants—chemicals, radiation, pesticides—appear in all media. Successful control of them at present requires the coordinated efforts of a variety of separate agencies and departments. The results are not always successful.

A far more effective approach to pollution control would:

—Identify pollutants.

—Trace them through the entire ecological chain, observing and recording changes in form as they occur.

—Determine the total exposure of man and his environment.

—Examine interactions among forms of pollution.

—Identify where in the ecological chain interdiction would be most appropriate.

In organizational terms, this requires pulling together into one agency a variety of research, monitoring, standard-setting and enforcement activities now scattered through several departments and agencies. It also requires that the new agency include sufficient support elements—in research and in aids to State and local anti-pollution programs, for example—to give it the needed strength and potential for carrying out its mission. The new agency would also, of course, draw upon the results of research conducted by other agencies.

COMPONENTS OF THE EPA

Under the terms of Reorganization Plan No. 3, the following would be moved to the new Environmental Protection Agency:

—The functions carried out by the Federal Water Quality Administration (from the Department of the Interior).

—Functions with respect to pesticides studies now vested in the Department of the Interior.

—The functions carried out by the National Air Pollution Control Administration (from the Department of Health, Education, and Welfare).

—The functions carried out by the Bureau of Solid Waste Management and the Bureau of Water Hygiene, and portions of the functions carried out by the Bureau of Radiological Health of the Environmental Control Administration (from the Department of Health, Education and Welfare).

—Certain functions with respect to pesticides carried out by the Food and Drug Administration (from the Department of Health, Education and Welfare).

—Authority to perform studies relating to ecological systems now vested in the Council on Environmental Quality.

—Certain functions respecting radiation criteria and standards now vested in the Atomic Energy Commission and the Federal Radiation Council.

—Functions respecting pesticides registration and related activities now carried out by the Agricultural Research Service (from the Department of Agriculture).

With its broad mandate, EPA would also develop competence in areas of environmental protection that have not previously been given enough attention, such, for example, as the problem of noise, and it would provide an organization to which new programs in these areas could be added.

In brief, these are the principal functions to be transferred:

Federal Water Quality Administration. Charged with the control of pollutants which impair water quality, it is broadly concerned with the impact of degraded water quality. It performs a wide variety of functions, including research, standard-setting and enforcement, and provides construction grants and technical assistance.

Certain pesticides research authority from the Department of the Interior. Authority for research on the effects of pesticides on fish and wildlife would be provided to the EPA through transfer of the specialized research authority of the pesticides act enacted in 1958. Interior would retain its responsibility to do research on all factors affecting fish and wildlife. Under this provision, only one laboratory would be transferred to the EPA—the Gulf Breeze Biological Laboratory of the Bureau of Commercial Fisheries. The EPA would work closely with the fish and wildlife laboratories remaining with the Bureau of Sport Fisheries and Wildlife.

National Air Pollution Control Administration. As the principal Federal agency concerned with air pollution, it conducts research on the effects of air pollution, operates a monitoring network, and promulgates criteria which serve as the basis for setting air quality standards. Its regulatory functions are similar to those of the Federal Water Quality Administration. NAPCA is responsible for administering the Clean Air Act, which involves designating air quality regions, approving State standards, and providing financial and technical assistance to State Control agencies to enable them to comply with the Act's provisions. It also sets and enforces Federal automotive emission standards.

Elements of the Environmental Control Administration. ECA is the focal point within HEW for evaluation and control of a broad range of environmental health problems, including water quality, solid wastes, and radiation. Programs in the ECA involve research, development of criteria and standards, and the administration of planning and demonstration grants. From the ECA, the activities of the Bureaus of Water Hygiene and Solid Waste Management and portions of the activities of the Bureau of Radiological Health would be transferred. Other functions of the ECA including those related to the regulation of radiation from consumer products and occupational safety and health would remain in HEW.

Pesticides research and standard-setting programs of the Food and Drug Administration. FDA's pesticides program consists of setting and enforcing standards which limit pesticide residues in food. EPA would have the authority to set pesticide standards and to monitor compliance with them, as well as to conduct related research. However, as an integral part of its food protection activities, FDA would retain its authority to remove from the market food with excess pesticide residues.

General ecological research from the Council on Environmental Quality. This authority to perform studies and research relating to ecological systems would be in addition to EPA's other specific research authorities, and it would help EPA to measure the impact of pollutants. The Council on Environmental Quality would retain its authority to conduct studies and research relating to environmental quality.

Environmental radiation standards programs. The Atomic Energy Commission is now responsible for establishing environmental radiation standards and emission limits for radioactivity. Those standards have been based largely on broad guidelines recommended by the Federal Radiation Council. The Atomic Energy Commission's authority to set standards for the protection of the general environment from radioactive material would be transferred to the Environmental Protection Agency. The functions of the Federal Radiation Council would also be transferred. AEC would retain responsibility for the implementation and enforcement of radiation standards through its licensing authority.

Pesticides registration program of the Agricultural Research Service. The Department of Agriculture is currently responsible for several distinct functions related to pesticides use. It conducts research on the efficacy of various pesticides as related to other pest control methods and on the effects of pesticides on non-target plants, livestock, and poultry. It registers pesticides, monitors their persistence and carries out an educational program on pesticide use through the extension service. It conducts extensive pest control programs which utilize pesticides.

By transferring the Department of Agriculture's pesticides registration and monitoring function to the EPA and merging it with the pesticides programs being transferred from HEW and Interior, the new agency would be given a broad capability for control over the introduction of pesticides into the environment.

The Department of Agriculture would continue to conduct research on the effectiveness of pesticides. The Department would furnish this information to the EPA, which would have the responsibility for actually licensing pesticides for use after considering environmental and health effects. Thus the new agency would be able to make use of the expertise of the Department.

ADVANTAGES OF REORGANIZATION

This reorganization would permit response to environmental problems in a manner beyond the previous capability of our pollution control programs. The EPA would have the capacity to do research on important pollutants irrespective of the media in which they appear, and on the impact of these pollutants on the total environment. Both by itself and together with other agencies, the EPA would monitor the condition of the environment—biological as well as physical. With these data, the EPA would be able to establish quantitative "environmental baselines"—critical if we are to measure adequately the success or failure of our pollution abatement efforts.

As no disjointed array of separate programs can, the EPA would be able—in concert with the States—to set and enforce standards for air and water quality and for individual pollutants. This consolidation of pollution control authorities would help assure that we do not create new environmental problems in the process of controlling existing ones. Industries seeking to minimize the adverse impact of their activities on the environment would be assured of consistent standards covering the full range of their waste disposal problems. As the States develop and expand their own pollution control programs, they would be able to look to one agency to support their efforts with financial and technical assistance and training.

In proposing that the Environmental Protection Agency be set up as a separate new agency, I am making an exception to one of my own principles: that, as a matter of effective and orderly administration, additional new independent agencies normally should not be created. In this case, however, the arguments against placing environmental protection activities under the jurisdiction of one or another of the existing departments and agencies are compelling.

In the first place, almost every part of government is concerned with the environment in some way, and affects it in some way. Yet each department also has its own primary mission—such as resource development, transportation, health, defense, urban growth or agriculture—which necessarily affects its own view of environmental questions.

In the second place, if the critical standard-setting functions were centralized within any one existing department, it would require that department constantly to make decisions affecting other departments—in which, whether fairly or unfairly, its own objectivity as an impartial arbiter could be called into question.

Because environmental protection cuts across so many jurisdictions, and because arresting environmental deterioration is of great importance to the quality of life in our country and the world, I believe that in this case a strong, independent agency is needed. That agency would, of course, work closely with and draw upon the expertise and assistance of other agencies having experience in the environmental area.

ROLES AND FUNCTIONS OF EPA

The principal roles and functions of the EPA would include:

—The establishment and enforcement of environmental protection standards consistent with national environmental goals.

—The conduct of research on the adverse effects of pollution and on methods and equipment for controlling it, the gathering of information on pollution, and the use of this information in strengthening environmental protection programs and recommending policy changes.

—Assisting others, through grants, technical assistance and other means in arresting pollution of the environment.

—Assisting the Council on Environmental Quality in developing and recommending to the President new policies for the protection of the environment.

One natural question concerns the relationship between the EPA and the Council on Environmental Quality, recently established by Act of Congress.

It is my intention and expectation that the two will work in close harmony, reinforcing each other's mission. Essentially, the Council is a top-level advisory group (which might be compared with the Council of Economic Advisers), while the EPA would be an operating, "line" organization. The Council will continue to be a part of the Executive Office of the President and will perform its overall coordinating and advisory roles with respect to all Federal programs related to environmental quality.

The Council, then, is concerned with all aspects of environmental quality—wildlife preservation, parklands, land use, and population growth, as well as pollution. The EPA would be charged with protecting the environment by abating pollution. In short, the Council focuses on what our broad policies in the environmental field should be; the EPA would focus on setting and enforcing pollution control standards. The two are not competing, but complementary—and taken together, they should give us, for the first time, the means to mount an effectively coordinated campaign against environmental degradation in all of its many forms.

NATIONAL OCEANIC AND ATMOSPHERIC ADMINISTRATION

The oceans and the atmosphere are interacting parts of the total environmental system upon which we depend not only for the quality of our lives, but for life itself.

We face immediate and compelling needs for better protection of life and property from natural hazards, and for a better understanding of the total environment—an understanding which will enable us more effectively to monitor and predict its actions, and ultimately, perhaps to exercise some degree of control over them.

We also face a compelling need for exploration and development leading to the intelligent use of our marine resources. The global oceans, which constitute nearly three-fourths of the surface of our planet, are today the least-understood, the least-developed, and the least-protected part of our earth. Food from the oceans will increasingly be a key element in the world's fight against hunger. The mineral resources of the ocean beds and of the oceans themselves, are being increasingly tapped to meet the growing world demand. We must understand the nature of these resources, and assure their development without either contaminating the marine environment or upsetting its balance.

Establishment of the National Oceanic and Atmospheric Administration—NOAA—within the Department of Commerce would enable us to approach these tasks in a coordinated way. By employing a unified approach to the problems of the oceans and atmosphere, we can increase our knowledge and expand our opportunities not only in those areas, but in the third major component of our environment, the solid earth, as well.

Scattered through various Federal departments and agencies, we already have the scientific, technological and administrative resources to make an effective, unified approach possible. What we need is to bring them together. Establishment of NOAA would do so.

By far the largest of the components being merged would be the Commerce Department's Environmental Science Services Administration (ESSA), with some 10,000 employees (70 percent of NOAA's total personnel strength) and estimated Fiscal 1970 expenditures of almost $200 million. Placing NOAA within the Department of Commerce therefore entails the least dislocation, while also placing it within a Department which has traditionally been a center for service activities in the scientific and technological area.

COMPONENTS OF NOAA

Under terms of Reorganization Plan No. 4, the programs of the following organizations would be moved into NOAA:
—The Environmental Science Services Administration (from within the Department of Commerce).

—Elements of the Bureau of Commercial Fisheries (from the Department of the Interior).

—The marine sport fish program of the Bureau of Sport Fisheries and Wildlife (from the Department of the Interior).

—The Marine Minerals Technology Center of the Bureau of Mines (from the Department of the Interior).

—The Office of Sea Grant Programs (from the National Science Foundation).

—Elements of the United States Lake Survey (from the Department of the Army).

In addition, by executive action, the programs of the following organizations would be transferred to NOAA:

—The National Oceanographic Data Center (from the Department of the Navy).

—The National Oceanographic Instrumentation Center (from the Department of the Navy).

—The National Data Buoy Project (from the Department of Transportation).

In brief, these are the principal functions of the programs and agencies to be combined:

The Environmental Science Services Administration (ESSA) comprises the following components:

—The Weather Bureau (weather, marine, river and flood forecasting and warning).

—The Coast and Geodetic Survey (earth and marine description, mapping and charting).

—The Environmental Data Service (storage and retrieval of environmental data).

—The National Environmental Satellite Center (observation of the global environment from earth-orbiting satellites).

—The ESSA Research Laboratories (research on physical environmental problems).

ESSA's activities include observing and predicting the state of the oceans, the state of the lower and upper atmosphere, and the size and shape of the earth. It maintains the nation's warning systems for such natural hazards as hurricanes, tornadoes, floods, earthquakes and seismic sea waves. It provides information for national defense, agriculture, transportation and industry.

ESSA monitors atmospheric, oceanic and geophysical phenomena on a global basis, through an unparalleled complex of air, ocean, earth and space facilities. It also prepares aeronautical and marine maps and charts.

Bureau of Commercial Fisheries and marine sport fish activities. Those fishery activities of the Department of the Interior's U.S. Fish and Wildlife Service which are ocean related and those which are directed toward commercial fishing would be transferred. The Fish and Wildlife Service's Bureau of Commercial Fisheries has the dual function of strengthening the fishing industry and promoting conservation of fishery stocks. It conducts research on important marine species and and on fundamental oceanography, and operates a fleet of oceanographic vessels and a number of laboratories. Most of its activities would be transferred. From the Fish and Wildlife Service's Bureau of Sport Fisheries and Wildlife, the marine sport fishing program would be transferred. This involves five supporting laboratories and three ships engaged in activities to enhance marine sport fishing opportunities.

The Marine Minerals Technology Center is concerned with the development of marine mining technology.

Office of Sea Grant Programs. The Sea Grant Program was authorized in 1966 to permit the Federal Government to assist the academic and industrial communities in developing marine resources and technology. It aims at strengthening education and training of marine specialists, supporting applied research in the recovery and use of marine resources, and developing extension and advisory services. The Office carries out these objectives by making grants to selected academic institutions.

The U.S. Lake Survey has two primary missions. It prepares and publishes navigation charts of the Great Lakes and tributary waters and conducts research on a variety of hydraulic and hydrologic phenomena of the Great Lakes' waters. Its activities are very similiar to those conducted along the Atlantic and Pacific coasts by ESSA's Coast and Geodetic Survey.

The National Oceanographic Data Center is responsible for the collection and dissemination of oceanographic data accumulated by all Federal agencies.

The National Oceanographic Instrumentation Center provides a central Federal service for the calibration and testing of oceanographic instruments.

The National Data Buoy Development Project was established to determine the feasibility of deploying a system of automatic ocean buoys to obtain oceanic and atmospheric data.

ROLE OF NOAA

Drawing these activities together into a single agency would make possible a balanced Federal program to improve our understanding of the resources of the sea, and permit their development and use while guarding against the sort of thoughtless exploitation that in the past laid waste to so many of our precious natural assets. It would make possible a consolidated program for achieving a more comprehensive understanding of oceanic and atmospheric phenomena, which so greatly affect our lives and activities. It would facilitate the cooperation between public and private interests that can best serve the interests of all.

I expect that NOAA would exercise leadership in developing a national oceanic and atmospheric program of research and development. It would coordinate its own scientific and technical resources with the technical and operational capabilities of other government agencies and private institutions. As important, NOAA would continue to provide those services to other agencies of government, industry and private individuals which have become essential to the efficient operation of our transportation systems, our agriculture and our national security. I expect it to maintain continuing and close liaison with the new Environmental Protection Agency and the Council on Environmental Quality as part of an effort to ensure that environmental questions are dealt with in their totality and that they benefit from the full range of the government's technical and human resources.

Authorities who have studied this matter, including the Commission on Marine Science, Engineering and Resources, strongly recommended the creation of a National Advisory Committee for the Oceans. I agree. Consequently, I will request, upon approval of the plan, that the Secretary of Commerce establish a National Advisory Committee for the Oceans and the Atmosphere to advise him on the progress of governmental and private programs in achieving the nation's oceanic and atmospheric objectives.

AN ON-GOING PROCESS

The reorganizations which I am here proposing afford both the Congress and the Executive Branch an opportunity to re-evaluate the adequacy of existing program authorities involved in these consolidations. As these two new organizations come into being, we may well find that supplementary legislation to perfect their authorities will be necessary. I look forward to working with the Congress in this task.

In formulating these reorganization plans, I have been greatly aided by the work of the President's Advisory Council on Executive Organization (the Ash Council), the Commission on Marine Science, Engineering and Resources (the Stratton Commission, appointed by President Johnson), my special task force on oceanography headed by Dr. James Wakelin, and by the information developed during both House and Senate hearings on proposed NOAA legislation.

Many of those who have advised me have proposed additional reorganizations, and it may well be that in the future I shall recommend further changes. For the present, however, I think the two reorganizations transmitted today represent a sound and significant beginning. I also think that in practical terms, in this sensitive and rapidly developing area, it is better to proceed a step at a time—and thus to be sure that we are not caught up in a form of organizational indigestion from trying to rearrange too much at once. As we see how these changes work out, we will gain a better understanding of what further changes—in addition to these—might be desirable.

Ultimately, our objective should be to insure that the nation's environmental and resource protection activities are so organized as to maximize both the effective coordination of all and the effective functioning of each.

The Congress, the Administration and the public all share a profound commitment to the rescue of our natural environment, and the preservation of the Earth as a place both habitable by and hospitable to man. With its acceptance of these reorganization plans, the Congress will help us fulfill that commitment.

RICHARD NIXON

The White House
July 9, 1970

APPENDIX 11

Proclamation 2667 of President Truman Claiming Jurisdiction Over Resources of the Continental Shelf, September 28, 1945

Whereas the Government of the United States of America, aware of the long-range world-wide need for new sources of petroleum and other minerals, holds the view that efforts to discover and make available new supplies of these resources should be encouraged; and

Whereas its competent experts are of the opinion that such resources underlie many parts of the continental shelf off the coasts of the United States of America, and that with modern technological progress their utilization is already practicable or will become so at an early date; and

Whereas recognized jurisdiction over these resources is required in the interest of their conservation and prudent utilization when and as development is undertaken; and

Whereas it is the view of the Government of the United States that the exercise of jurisdiction over the natural resources of the subsoil and sea bed of the continental shelf by the contiguous nation is reasonable and just, since the effectiveness of measures to utilize or conserve these resources would be contingent upon cooperation and protection from the shore, since the continental shelf may be regarded as an extension of the land-mass of the coastal nation and thus naturally appurtenant to it, since these resources frequently form a seaward extension of a pool or deposit lying within the territory, and since self-protection compels the coastal nation to keep close watch over activities off its shores which are of a nature necessary for utilization of these resources;

Now, THEREFORE, I, HARRY S. TRUMAN, President of the United States of America, do hereby proclaim the following policy of the United States of America with respect to the natural resources of the subsoil and sea bed of the continental shelf.

Having concern for the urgency of conserving and prudently utilizing its natural resources, the Government of the United States regards the natural resources of the subsoil and sea bed of the continental shelf beneath the high seas but contiguous to the coasts of the United States as appertaining to the United States, subject to its jurisdiction and control. In cases where the continental shelf extends to the shores of another State, or is shared with an adjacent State, the boundary shall be determined by the United States and the State concerned in accordance with equitable principles. The character as high seas of the waters above the continental shelf and the right to their free and unimpeded navigation are in no way thus affected. . . .

Source: *Federal Register*, 10:12303.

APPENDIX 12

Excerpts from the President's Special Message to the Congress on Natural Resources, February 23, 1961

TO THE CONGRESS OF THE UNITED STATES:

From the beginning of civilization, every nation's basic wealth and progress has stemmed in large measure from its natural resources. This nation has been, and is now, especially fortunate in the blessings we have inherited. Our entire society rests upon—and is dependent upon—our water, our land, our forests, and our minerals. How we use these resources influences our health, security, economy, and well-being.

· · · · · · · · · · · · · · · · · · · ·

This statement is designed to bring together in one message the widely scattered resource policies of the Federal Government. In the past, these policies have overlapped and often conflicted. Funds were wasted on competing efforts. Widely differing standards were applied to measure the Federal contribution to similar projects.

· · · · · · · · · · · · · · · · · · · ·

To coordinate all of these matters among the various agencies, I will shortly issue one or more Executive Orders or directives:

· · · · · · · · · · · · · · · · · · · ·

In addition, to provide a coordinated framework for our research programs in this area, and to chart the course for the wisest and most efficient use of the research talent and facilities we possess, I shall ask the National Academy of Sciences to undertake a thorough and broadly based study and evaluation of the present state of research underlying the conservation, development, and use of natural resources, how they are formed, replenished and may be substituted for, and giving particular attention to needs for basic research and to projects that will provide a better basis for natural resources planning and policy formulation. Pending the recommendations of the Academy, I have directed my Science Advisor and the Federal Council for Science and Technology to review ongoing Federal research activities in the field of natural resources and to determine ways to strengthen the total government research effort relating to natural resources.

· · · · · · · · · · · · · · · · · · · ·

V. OCEAN RESOURCES

The sea around us represents one of our most important but least understood and almost wholly undeveloped areas for extending our resource base. Continental shelves bordering the United States contain roughly 20 percent of our remaining reserves of crude oil and natural gas. The ocean floor contains large and valuable deposits of cobalt, copper, nickel, and manganese. Ocean waters themselves contain a wide variety of dissolved salts and minerals.

Salt (and fresh water) fisheries are among our most important but far from fully developed reservoirs of protein foods. At present levels of use, this country alone will need an additional 3 billion pounds of fish and shellfish annually by 1980, and many other countries with large-scale protein

468

deficiency can be greatly helped by more extensive use of marine foodstuffs. But all this will require increased efforts, under Federal leadership, for rehabilitation of depleted stocks of salmon and sardines in the Pacific, groundfish and oysters in the Atlantic, Lake Trout and other desirable species in the Great Lakes, and many others through biological research, development of methods for passing fish over dams, and control of pollution.

This Administration intends to give concerted attention to our whole national effort in the basic and applied research of oceanography. Construction of ship and shore facilities for ocean research and survey, the development of new instruments for charting the seas and gathering data, and the training of new scientific manpower will require the coordinated efforts of many Federal agencies. It is my intention to send to the Congress for its information and use in considering the 1962 budget, a national program for oceanography, setting forth the responsibilities and requirements of all participating government agencies.

APPENDIX 13

Letter to the President of the Senate on Increasing the National Effort in Oceanography, March 29, 1961

My dear Mr. President:

The seas around us, as I pointed out in my message to the Congress on February 23, represent one of our most important resources. If vigorously developed, this resource can be a source of great benefit to the Nation and to all mankind.

But it will require concerted action, purposefully directed, with vision and ingenuity. It will require the combined efforts of our scientists and institutions, both public and private, and the coordinated efforts of many Federal agencies. It will involve substantial investments in the early years for the construction and operation of ship and shore facilities for research and surveys, the development of new instruments for charting the seas and gathering data, and the training of new scientific manpower.

We are just at the threshold of our knowledge of the oceans. Already their military importance, their potential use for weather predictions, for food and for minerals are evident. Further research will undoubtedly disclose additional uses.

Knowledge of the oceans is more than a matter of curiosity. Our very survival may hinge upon it. Although understanding of our marine environment and maps of the ocean floor would afford to our military forces a demonstrable advantage, we have thus far neglected oceanography. We do not have adequate charts of more than one or two percent of the oceans.

The seas also offer a wealth of nutritional resources. They already are a principal source of protein. They can provide many times the current food supply if we but learn how to garner and husband this self-renewing larder. To meet the vast needs of an expanding population, the bounty of the sea must be made more available. Within two decades, our own nation will require over a million more tons of seafood than we now harvest.

Mineral resources on land will ultimately reach their limits. But the oceans hold untapped sources of such basic minerals as salt, potassium and magnesium in virtually limitless quantities. We will be able to extract additional elements from sea water, such as manganese, nickel, cobalt and other elements known to abound on the ocean floor, as soon as the processes are developed to make it economically feasible.

To predict, and perhaps some day to control, changes in weather and climate is of the utmost importance to man everywhere. These changes are controlled to a large and yet unknown extent by what happens in the ocean. Ocean and atmosphere work together in a still mysterious way to determine our climate. Additional research is necessary to identify the factors in this interplay.

These are some of the reasons which compel us to embark upon a national effort in oceanography. I am therefore requesting funds for 1962 which will nearly double our government's investment over 1961, and which will provide $23 million more for oceanography than what was recommended in the 1962 budget submitted earlier. A summary and comparison of the 1960, 1961 and 1962 budgets is contained in two tables which are enclosed with this letter.

1. *Ship Construction.*

The proposed program for 1962 includes $37 million for ship construction, an increase of $23 million over 1961. This will provide for 10 oceanographic vessels. Only two will replace existing ships. The others will be used to meet needs that have long existed in Federal agencies and other oceanographic institutions conducting research for the Government.

The present United States oceanographic fleet is composed of 27 research ships and 17 survey vessels. All but two were constructed prior to the end of World War II; many are over thirty years old. Only two of the ships were designed specifically for research purposes; the remainder has been converted from a variety of ships designed for other uses. Thus the success of the national oceanographic program will depend heavily on the construction of the new specially designed vessels proposed for 1962.

2. *Shore Facilities and Data Center.*

Shore facilities are urgently required to provide laboratory space for analysis and interpretation of data and to train new oceanographers. In oceanographic research about five scientists and technicians are required ashore for each scientist aboard ship.

For 1962, $10 million is being requested for laboratories and wharfside facilities. This represents a five-fold increase over 1961. It includes, for example, funds for a new Bureau of Commercial Fisheries laboratory to replace a forty-year old structure and additional laboratory space at universities and other oceanographic institutions.

An essential part of the shore establishment is the new National Oceanographic Data Center which will begin its first full year of operation in 1962. This Center will make available to the scientific community oceanographic data collected throughout the world.

3. *Basic and Applied Research.*

The conduct of research is the central purpose of our whole national effort in oceanography. New ships and shore facilities are essential tools of scientific research, but it is the research itself that will yield new knowledge of the earth's "inner space", and new uses of the sea. The proposed program includes $41 million for basic and applied research in oceanography. This is an increase of $9 million over the 1961 level.

Basic research is the cornerstone on which the successful use of the seas must rest. Progress here is largely dependent on the work of scientists at many universities and laboratories throughout the United States and on ships at sea. Their investigations cover all aspects of the marine environment, the motion and composition of ocean waters, the evolution and distribution of marine plants and animals, the shape and composition of the ocean bottom, and many other geophysical and biological problems. Of timely significance is the attempt to penetrate to the earth's mantle to better our understanding of the origin and history of our planet. This undertaking, known as Project Mohole, involves the development of new drilling methods that can be used in the deep seas. This project has recently resulted in a spectacular achievement. Samples from nearly a thousand feet beneath the sea floor were obtained by drilling in three thousand feet of water.

Considerable attention will also be given to applied problems in the marine sciences. Oceanographers will be studying such problems as sound propagation in water, the effects of changes in ocean conditions on the movement of ships, weather forecasting, and fisheries management. Methods of predicting changes in ocean conditions also are being developed. Eventually they may lead to maps of "weather within the sea" much like the atmospheric weather maps of today.

Many advances are being made in methods of exploring the seas. Oceanographers are now able to descend to the great depths in bathyscaphes. New electronic equipment will allow them to probe the ocean and to "see" with sound pulses what before has been opaque. Using these new techniques, our scientists already have discovered vast currents below the ocean surface a thousand times larger than the flow of the Mississippi.

4. Training of Oceanographers.

The most important part of our long-range program in oceanography is the training of young scientists. Scientific manpower of every sort will be needed—technicians, college graduates, and post-graduate researchers—and they must be trained in many scientific disciplines. This training should go hand in hand with the conduct of research at universities and other oceanographic institutions. By their support of these institutions, the programs of the National Science Foundation, the Office of Naval Research, and the Department of Health, Education and Welfare will be of major importance to an expanding program in oceanography; for they can result in the education of new young scientists as well as in the production of new knowledge. In the coming year, these agencies are undertaking to increase the number of fellowship awards and graduate student research contracts, and they also will encourage the development of new university programs in oceanography.

5. Ocean Surveys.

World-wide surveys of the oceans—their properties, their contents and boundaries—are needed to make charts and maps for use of scientists in their research programs and for a variety of commercial and defense applications. The United States' ocean survey program for FY 1962 is being increased within the limits of ships available for this purpose. I am requesting additional funds

to allow the Coast and Geodetic Survey to extend the operating season of its existing ships, thus making the maximum use of limited ship resources. As already mentioned, funds are included for a new survey ship which will increase our deep-sea survey capability.

6. International Cooperation.

Oceanography is a natural area of opportunity for extensive international cooperation. Indeed, systematic surveys and research in all the oceans of the world represent tasks of such formidable magnitude that international sharing of the work is a necessity.

Our present maps of the oceans are comparable in accuracy and detail to maps of the land areas of the earth in the early part of the 18th century. Precise methods of measuring ocean depths have become available during the last ten years, and these, when combined with new developments in navigation, should make possible for the first time modern maps of the topography of the entire sea floor. An accurate mapping of the oceans will require international cooperation in ship operations and in establishing a world-wide system of navigation. In these endeavors the United States can play a leading part.

This year an Intergovernmental Oceanographic Commission is being established under UNESCO to provide a means whereby interested countries can cooperate in research and in making surveys and maps of the deep sea floor, the ocean waters, and their contained organisms. Membership on the Commission is open to all countries of the UN family that desire to cooperate in oceanography. The United States intends to participate fully in the activities of the Commission.

The United States also will participate in the International Indian Ocean Expedition. Many nations, including the Soviet Union, are cooperating in this expedition under the non-governmental sponsorship of the International Council of Scientific Unions. Over a quarter of the world's people live in the countries surrounding the Indian Ocean. If more can be learned of the Indian Ocean's extensive food resources, these nations can be helped to develop and expand their fishing industries as part of their general economic development.

7. *The Coast Guard.*

At present, the Coast Guard enabling legislation limits the extent to which the Coast Guard can engage in scientific research. Only the International Ice Patrol is authorized to make such studies. I recommend that the statutory limitations restricting the participation by the Coast Guard in oceanographic research be removed. With ocean weather stations, deep sea thermometers, and other data collection devices, our Coast Guard can make a valuable contribution to the oceanographic program.

CONCLUSION

Knowledge and understanding of the oceans promise to assume greater and greater importance in the future. This is not a one-year program—or even a ten-year program. It is the first step in a continuing effort to acquire and apply the information about a part of our world that will ultimately determine conditions of life in the rest of the world. The opportunities are there. A vigorous program will capture those opportunities.

Sincerely,

JOHN F. KENNEDY

APPENDIX 14

Letter to the President of the Senate and to the Speaker of the House Transmitting Reports on Oceanographic Research, March 19, 1964

Dear Mr. —————:

Recognizing the continued interest by the Congress in advancing this Nation's program in oceanography, I am pleased to forward advance copies of two publications of the Federal Council for Science and Technology that set forth Government-wide plans and budget details.

These reports, entitled "National Oceanographic Program, Fiscal Year 1965, Parts I and II," contain an account of oceanic research to meet national goals, in keeping with the long range considerations previously submitted to the Congress. Information is also included concerning proposed funding for research, surveys, new ship and laboratory construction, and concerning program planning and coordination by the Council's Interagency Committee on Oceanography (ICO), to minimize unwitting duplication and program gaps.

The proposed Federal budget in oceanography is $138 million. This is 11% more than Fiscal Year 1964 appropriations, which in turn equalled those for Fiscal Year 1963. This proposed growth is an absolute minimum if the country is to maintain the momentum necessary to achieve those objectives in oceanic research which have been previously enunciated by both President Kennedy and the Congress—to enhance our military defense; to develop marine mineral and fisheries resources; to control pollution; to predict more accurately storms and tides that endanger life and property; to assist state, national and international bodies in wise legislation and regulation of commerce on the sea; and to extend scientific knowledge generally.

I especially should like to call attention to the Government-wide character of this program. Statutory responsibility for the conduct of related sectors are vested in a number of separate agencies. Special measures are thus being continued by the Office of Science and Technology and the Federal Council for Science and Technology, with the assistance of the ICO, to achieve effective interagency planning and coordination.

Sincerely yours,

Lyndon B. Johnson

NOTE: This is the text of identical letters addressed to the Honorable Carl Hayden, President pro tempore of the Senate, and to the Honorable John W. McCormack, Speaker of the House of Representatives.

The letter was made public as part of a White House release concerning the reports published as ICO Pamphlet No. 15, dated March 1964 (50 pp., including Part I: Summary, Fiscal Year 1965, and Part II: The Program and Its Cost).

The release pointed out that the reports were the result of cooperative participation on the part of scientists, engineers, technicians, and administrative officers from numerous universities and industries, and from the State and National Governments. Dr. Donald F. Hornig, Director, Office of Science and Technology, served as Chairman of the Federal Council for Science and Technology, and James H. Wakelin, Jr., Assistant Secretary of the Navy, as Chairman of the Interagency Committee on Oceanography.

APPENDIX 15

Remarks at the Commissioning of the Research Ship *Oceanographer,* July 13, 1966

Secretary and Mrs. Connor, Reverend Harris, Captain Wardwell, my beloved friend Senator Magnuson, Governor Burns of Hawaii, distinguished Members of Congress, guests, ladies and gentlemen:

We meet here today at the beginning of a new age of exploration.

To some, this might mean our adventures in outer space. But I am speaking of exploring an unknown world at our doorstep. It is really our last frontier here on earth. I am speaking of mountain chains that are yet to be discovered, of natural resources that are yet to be tapped, of a vast wilderness that is yet to be charted.

This is the sea around us.

And while our knowledge of the sea is quite primitive, we do know something of its great potential for the betterment of the human race and all mankind.

We know that we can, for instance, greatly improve our weather predictions. We can save thousands of lives and millions of dollars in property each year. We just must start learning more about the sea.

We know that the sea holds a great promise of transforming arid regions of the earth into new, rich, and productive farmlands.

We know that beneath the sea are countless minerals and fuels which can be found and can be exploited.

We know—most important of all—that the sea holds the ultimate answer to food for the exploding population in the world.

Nearly four-fifths of all life on earth actually exists in salt water.

So, using science and technology, we must develop improved ways of taking food from the ocean.

But catching fish is just not enough. It has been said that throughout history we have been simple hunters of the sea. Men must now learn how to farm the sea.

Our scientists are developing a process for turning whole fish into a tasteless but highly nutritious protein concentrate which can be used as a supplement to our daily diet.

In addition, the United States Senate has recently passed a bill for the construction of several pilot plants to begin the commercial development of this fish protein food. The daily output of one of these plants would provide enough high protein supplement for well over half a million people each day.

So, it is toward a goal of understanding all aspects of the sea that we commission the *Oceanographer* today.

Oceanographer is one of the Coast and Geodetic Survey's 14 research ships which will begin to help us to explore the environment around us. Her sister ship, the *Discoverer,* is under construction and also will be commissioned shortly.

In the past decade, our support of marine science and technology has grown from some $21 million to more than $320 million.

The Federal research fleet today totals 115 vessels.

Our progress has been the handiwork, of

course, of many men. These men are in and out of Government. But the Nation owes a very particular debt to those Members of the Congress, men such as our distinguished Senator Magnuson of Washington, who is here today and whose efforts have accomplished so much for oceanography over the last decade.

I want to pay tribute to the Secretary, the Under Secretary, the Assistant Secretaries, all the employees of the Department of Commerce and the Coast and Geodetic Survey, and other Government officials.

But I also want to say that today we must redouble our efforts. In the months ahead, we shall establish our priorities, we shall then set our timetables—and we shall follow them, just as we have followed an orderly and relentless program for the exploration of space. And the distinguished Scientific Adviser to the President, Dr. Hornig, is going to keep seeing that we do this. Because the frontier of the deep challenges our real spirit and we want to see that that challenge from the deep is fully met.

My Science Advisory Committee has recently completed a report on the "Effective Use of the Sea." Through Dr. Hornig I am releasing that report today. I should like to commend it to the attention of all Americans.

I commend it, in particular, to the 100 outstanding high school students who have joined us here today and who have come to the Capital from throughout the States of this Union. I hope that there are among you some of the great oceanographers of tomorrow. You could not choose, in my judgment, a more important or a more challenging career.

I am referring this report from my Science Advisory Committee to the new National Council on Marine Resources and Engineering set up by statute under the leadership of Senator Magnuson. This Council will be headed by our distinguished Vice President; distinguished members of the Cabinet and others will serve on it.

This Council will survey all marine science activities to provide for this Nation a comprehensive program in this field. I will ask them to complete their initial recommendations by the time the new Congress convenes next January.

Truly great accomplishments in oceanography will require the cooperation of all the maritime nations of the world. And so today I send our voice out from this platform calling for such cooperation, requesting it, and urging it.

To the Soviet Union—a major maritime power—I today extend our earnest wish that you may join with us in this great endeavor.

In accordance with these desires I am happy to announce that one of the first long voyages of *Oceanographer* will be a 6-month global expedition in which the scientists from a number of our great nations will participate. It is our intention to invite Great Britain, West Germany, France, the U.S.S.R., India, Malaysia, Australia, New Zealand, Chile, and Peru to participate in the first round-the-world voyage of *Oceanographer*.

We greatly welcome this type of international participation. Because under no circumstances, we believe, must we ever allow the prospects of rich harvests and mineral wealth to create a new form of colonial competition among the maritime nations. We must be careful to avoid a race to grab and to hold the lands under the high seas. We must ensure that the deep seas and the ocean bottoms are, and remain, the legacy of all human beings.

The sea—yes, the great sea—in the words of Longfellow, "divides and yet unites mankind."

So to Captain Wardwell and his distinguished officers and men of *Oceanographer,* we say today: Yours is a most worthwhile mission. May you bring back much for the benefit of all humanity.

We congratulate you on the commissioning of your marvelous new ship. We wish you the best of results, fair winds, and smooth sailing.

And now I look forward with a great deal of personal pleasure to the opportunity to view the ship and some of the developments at first hand.

Thank you very much.

NOTE: The President spoke at 2:10 p.m. at Pier 2, Washington Navy Yard, at the commissioning of the USC & GSS *Oceanographer*. In his opening words he referred to Secretary of Commerce and Mrs. John T. Connor, Rev. Dr. Frederick Brown Harris, chaplain of the Senate, Capt. Arthur L. Wardwell, commander of the *Oceanographer*, Senator Warren G. Magnuson of Washington, and Governor John A. Burns of Hawaii.

APPENDIX 16

Message to the Congress of the United States by President Lyndon B. Johnson on International Decade of Ocean Exploration, 1968

The seas are the world's oldest frontiers. As Longfellow observed, they not only separate—but unite—mankind.

Even in the Age of Space, the sea remains our greatest mystery. But we know that in its sunless depths, a richness is still locked which holds vast promise for the improvement of men's lives—in all nations.

Those ocean roads, which so often have been the path of conquest, can now be turned to the search for enduring peace.

The task of exploring the ocean's depth for its potential wealth—food, minerals, resources—is as vast as the seas themselves. No one nation can undertake that task alone. As we have learned from prior ventures in ocean exploration, cooperation is the only answer.

I have instructed the Secretary of State to consult with other nations on the steps that could be taken to launch an historic and unprecedented adventure—an International Decade of Ocean Exploration for the 1970's.

Together the countries which border the seas can survey the ocean's resources, reaching where man has never probed before.

We hope that those nations will join in this exciting and important work.

Already our marine technology gives us the ability to use the ocean as a new and promising source of information on weather and climate. We can now build and moor electronic buoys in deep water. Unattended, these scientific outposts can transmit to shore data for accurate long-range forecasts.

The benefits will be incalculable. . . .

Source: White House press release, March 8, 1968.

APPENDIX 17

Draft Resolutions Submitted and Principles Supported by the United States in the Ad Hoc Seabed Committee, June 28, 1968

A. U.S. draft resolution containing statement of principles concerning the deep ocean floor

The General Assembly,

Desiring to encourage the exploration, use and development of the deep ocean floor to the fullest extent possible for the benefit and in the interest of all mankind,

Believing that such exploration and use of the deep ocean floor will contribute to international co-operation and understanding,

Convinced that no nation, regardless of geographical location, level of economic development, or technological capability, should be denied the opportunity to participate in the exploration and use of the deep ocean floor,

Conscious of the importance of promoting the general welfare of all peoples, and of furthering scientific study and the conservation of resources,

Reaffirming the traditional freedoms of the high seas under international law,

Recalling its resolution 2340 (**XXII**) of 18 December 1967,

Commends to States for their guidance the following principles concerning the deep ocean floor:

1. No State may claim or exercise sovereignty or sovereign rights over any part of the deep ocean floor. There shall be no discrimination in the availability of the deep ocean floor for exploration and use by all States and their nationals in accordance with international law;

2. There shall be established, as soon as practicable, internationally agreed arrangements governing the exploitation of resources of the deep ocean floor. These arrangements shall reflect the other principles contained in this Statement of Principles concerning the Deep Ocean Floor and shall include provision for:

(a) The orderly development of resources of the deep ocean floor in a manner reflecting the interest of the international community in the development of these resources;

(b) Conditions conducive to the making of investments necessary for the exploration and exploitation of resources of the deep ocean floor;

(c) Dedication as feasible and practicable of a portion of the value of the resources recovered from the deep ocean floor to international community purposes; and

(d) Accommodation among the commercial and other uses of the deep ocean floor and marine environment;

3. Taking into account the Geneva Convention of 1958 on the Continental Shelf, there shall be established, as soon as practicable, an internationally agreed precise boundary for the deep ocean floor—the sea-bed and subsoil beyond that over which coastal States may exercise sovereign rights for the purpose of exploration and exploitation of its natural resources; exploitation of the natural resources of the ocean floor that occurs prior to establishment of the boundary shall be understood not to prejudice its location, regardless of whether the coastal State considers the exploitation to have occurred on its "continental shelf";

4. States and their nationals shall conduct their activities on the deep ocean floor in accordance with international law, including the Charter of the United Nations, and in the interest of maintaining international peace and security and

Source: *MSA*, 3d report (1969), pp. 242–45.

promoting international co-operation, scientific knowledge, and economic development;

5. In order to further international co-operation in the scientific investigation of the deep ocean floor, States shall:

(a) Disseminate, in a timely fashion, plans for and results of national scientific programmes concerning the deep ocean floor;

(b) Encourage their nationals to follow similar practices concerning dissemination of such information;

(c) Encourage co-operative scientific activities regarding the deep ocean floor by personnel of different States;

6. In the exploration and use of the deep ocean floor States and their nationals:

(a) Shall have reasonable regard for the interests of other States and their nationals;

(b) Shall avoid unjustifiable interference with the exercise of the freedom of the high seas by other States and their nationals, or with the conservation of the living resources of the seas, and any interference with fundamental scientific research carried out with the intention of open publication;

(c) Shall adopt appropriate safeguards so as to minimize pollution of the seas and disturbance of the existing biological, chemical and physical processes and balances; each State shall provide timely announcement and any necessary amplifying information of any marine activity or experiment planned by it or its nationals that could harmfully interfere with the activities of any other State or its nationals in the exploration and use of the deep ocean floor. A State which has reason to believe that a marine activity or experiment planned by another State or its nationals could harmfully interfere with its activities or those of its nationals in the exploration and use of the deep ocean floor may request consultation concerning the activity or experiment;

7. States and their nationals shall render all possible assistance to one another in the event of accident, distress or emergency arising out of activities on the deep ocean floor.

June 28, 1968

B. U.S. draft resolution on preventing the emplacement of weapons of mass destruction on the sea-bed and ocean floor

The General Assembly,

Desiring that workable arms limitation measures be achieved that will enhance the peace and security of all nations and bring the world nearer to general and complete disarmament,

Requests the Eighteen-Nation Committee on Disarmament to take up the question of arms limitation on the sea-bed and ocean floor with a view to defining those factors vital to a workable, verifiable and effective international agreement which would prevent the use of this new environment for the emplacement of weapons of mass destruction.

June 28, 1968

C. U.S. draft resolution on the International Decade of Ocean Exploration

The General Assembly,

Recalling its concern for ascertaining practical means to promote international co-operation in the exploration, conservation and use of the sea-bed and the ocean floor, and the subsoil thereof, as manifested in its resolution 2340 (XXII),

Recalling as well that in its resolution 2172 (XXI) it requested that the Secretary-General prepare proposals for ensuring the most effective arrangements for an expanded programme of international co-operation to assist in a better understanding of the marine environment through science, and for initiating and strengthening marine education and training programmes,

Recalling further the proposals made by the Secretary-General in his report (E/4487) pursuant to resolution 2172 (XXI),

Noting that the Bureau and Consultative Council of the Intergovernmental Oceanographic Commission of UNESCO considered the proposed International Decade of Ocean Exploration a useful initiative for broadening and accelerating investigations of the oceans and for strengthening international co-operation,

Noting also the recommendation adopted by the Economic and Social Council on 2 August 1968, inviting the General Assembly to endorse the concept of a coordinated long-term programme of oceanographic research, taking into account such initiatives as the proposal for an International Decade of Ocean Exploration and international programmes already considered, approved and adopted by the Intergovernmental Oceanographic Commission for implementation in co-operation with other specialized agencies,

Aware of the consideration given to the proposal in the *Ad Hoc* Committee to Study the Peaceful Uses of the Sea-Bed and the Ocean Floor, arising from the important contribution which the Decade would make to scientific research and exploration of the sea-bed and deep ocean floor,

1. *Welcomes* and commends to Member States the concept of an International Decade of Ocean Exploration to be undertaken within the framework of a long-term programme of research and exploration under the general aegis of the United Nations;

2. *Invites* interested Member States to formulate proposals for national and international scientific programmes and agreed activities to be undertaken during the Decade with due regard to the interests of developing countries, to transmit these proposals to the Intergovernmental Oceanographic Commission, and to begin such activities as soon as practicable;

3. *Urges* Member States to publish as soon as practicable the results of activities which they will have undertaken within the framework of the Decade and at the same time to communicate these results to the Intergovernmental Oceanographic Commission;

4. *Requests* the Intergovernmental Oceanographic Commission:

(a) To further and co-ordinate, in co-operation with other interested agencies, an expanded, accelerated, long-term and sustained programme of world-wide exploration of the oceans and their resources of which the Decade will be an element, including international agency programmes, expanded international exchange of data from national programmes, and international efforts to strengthen the research capabilities of all interested nations;

(b) To report through appropriate channels to the twenty-fourth session of the General Assembly on the progress made in ocean activities undertaken pursuant to this resolution.

August 26, 1968

D. Draft statement of Agreed Principles proposed for submission to the General Assembly and supported by the United States

(1) There is an area of the sea-bed and ocean floor and the subsoil thereof, underlying the high seas, which lies beyond the limits of national jurisdiction (hereinafter described as "this area");

(2) Taking into account relevant dispositions of international law, there should be agreed a precise boundary for this area;

(3) There should be agreed, as soon as practicable, an international régime governing the exploitation of resources of this area;

(4) No State may claim or exercise sovereign rights over any part of this area, and no part of it is subject to national appropriation by claim of sovereignty, by use or occupation, or by any other means;

(5) Exploration and use of this area shall be carried on for the benefit and in the interests of all mankind, taking into account the special needs of the developing countries;

(6) This area shall be reserved exclusively for peaceful purposes;

(7) Activities in this area shall be conducted in accordance with international law, including the Charter of the United Nations. Activities in this area shall not infringe upon the freedoms of the high seas.

June 28, 1968

APPENDIX 18

Press Release of the Vice President on the Administration's Program in Marine Affairs, October 19, 1969—Details of Five-Point Interim Marine Science Program

1. *Coastal Zone Management*

Legislative proposals will be submitted to the Congress to establish a national policy for the development of coastal areas and to authorize Federal grants, with matching State contributions, that will encourage and facilitate the establishment of State management authorities. Such legislation will assist to insure that rapid coastal development does not destroy limited coastal land and water resources and that all interests in the coastal regions would be assured consideration—for port development, navigation, commercial fishing, mineral exploitation, recreation, conservation, industrial development, housing, power generation and waste disposal.

Grants are anticipated for (1) initial development by States of planning and regulatory mechanisms; and (2) operation of the State management systems that are developed. The latter grants would be made contingent on a State demonstrating a capability to prepare plans that provide for:
- —balanced use of the coastal margin, both land and water, that considers viewpoints of all potential users;
- —access to management-oriented research, including coastal ecology studies;
- —regulatory authority as needed—such as zoning, easement, license, or permit arrangements—to insure that development is consistent with State plans;
- —consideration of the interests of adjacent States;
- —land acquisition and power of eminent domain as necessary for implementation of the plan; and
- —review of proposed Federal, Federal-assisted, State and local projects to insure consistency with plans.

States have responsibility for management of coastal resources but have often lacked regulatory and management capabilities. They have been faced with a diversity of coastal jurisdictions and the absence of ecological information. This program should thus strengthen the States' capabilities, lessen the need for Federal intervention, and facilitate integration of planning, conservation, and development programs among diverse public and private interests.

2. *Establishment of Coastal Laboratories*

Steps will be taken toward establishment of coastal laboratories, supported by the Federal Government, to provide information on resource development, water quality, and environmental factors to assist State authorities and others in coastal management. Existing facilities will be strengthened and consolidated as necessary to provide capabilities to:
- —develop a basic understanding and description of the regionally-differentiated ecology of our 13,000-mile coastline;
- —anticipate and assess the impact on the ecology of alternative land uses, of pollution, and of alterations to the land-water interfaces;

—operate coastal monitoring networks; and

—perform analyses needed for coastal management.

Establishment of these capabilities will be phased with the development of coastal management plans to foster State access to environmental data and research capabilities.

3. *Pilot Technological Study of Lake Restoration*

The feasibility of restoring the Great Lakes with technological as well as regulatory mechanisms will be determined by a pilot study of a lake of manageable size. Existing environmental technology and techniques will be tested, including pollution measuring devices, methods of artificial destratification by aeration, mixing and thermal upwelling techniques, thermal pollution control and enrichment, artificial bottom coating, filtering, harvesting of living plants and animals, and restocking of fishery resources. The program will reinforce current investigations, and bring together additional competence from industry, academic institutions, and Federal laboratories.

4. *International Decade of Ocean Exploration*

Funding will be provided for the U.S. contribution to the International Decade of Ocean Exploration during the 1970s, proposed by the United States and endorsed by the UN General Assembly in December 1968. The United States will propose international emphasis on the following goals:

—preserve the ocean environment by accelerating scientific observations of the natural state of the ocean and its interactions with the coastal margin—to provide a basis for (a) assessing and predicting man-induced and natural modifications of the character of the oceans; (b) identifying damaging or irreversible effects of waste disposal at sea; and (c) comprehending the interaction of various levels of marine life to prevent depletion or extinction of valuable species as a result of man's activities.

—improve environmental forecasting to reduce hazards to life and property and permit more efficient use of marine resources—by improving physical and mathematical models of the ocean and atmosphere which will provide the basis for increased accuracy, timeliness, and geographic precision of environmental forecasts.

—expand seabed assessment activities to permit better management—domestically and internationally—of ocean mineral exploration and exploitation by acquiring needed knowledge of seabed topography, structure, physical and dynamic properties, and resource potential;

—develop an ocean monitoring system to facilitate prediction of oceanographic and atmospheric conditions—through design and deployment of oceanographic data buoys and other remote sensing platforms;

—improve worldwide data exchange through modernizing and standardizing national and international marine data collection, processing, and distribution;

—accelerate Decade planning to increase opportunities for international sharing of responsibilities and costs for ocean exploration, and to assure better use of limited exploration resources.

This U.S. contribution to an expanded program of intergovernmental cooperation reflects four recent developments:

(1) Increased population concentration along the coasts of the United

States and other countries, with attendant threats of harmful degradation of the ocean environment, and increasing demands on the coastal margins and marine resources.

(2) Evolution of technology that is rapidly opening new ocean frontiers.

(3) Recent scientific advances that can improve environmental forecasts if better ocean data is available.

(4) Sharply increasing interests by 100 coastal nations in extracting benefits from marine activities.

The Decade will accelerate needed understanding of the ocean—permitting nations individually to plan investments and collectively to develop arrangements for preserving the ocean environment and managing common ocean resources.

The global character of the oceans and the scope of work to be done make international cost-sharing and data exchange especially attractive. International cooperation in marine affairs is expected to facilitate communication with developing nations, with the Soviet Union, and with others.

The U.S. proposals are compatible with the framework of the expanded and long-term international program of ocean exploration recently developed by UNESCO's Intergovernmental Oceanographic Commission. The extent and nature of the U.S. contribution will depend on the contribution of other nations participating in this program.

5. *Arctic Environmental Research*

Arctic research activities will be intensified, both to permit fuller utilization of this rapidly developing area and to insure that such activities do not inadvertently degrade the Arctic environment.

The program will be directed to (1) the polar icepack including its impact on transportation and global weather and climate; (2) the polar magnetic field and its effects on communication; (3) geological structures underlying the Arctic lands and polar seas both as potential mineral sites and as hazards to construction and resource development; (4) balance of the Arctic eco-system; (5) the presence of permafrost; and (6) slow degradation of liquid and solid wastes under Arctic conditions. Behavior and physiology of man in this environment also will receive increased attention. Initial emphasis will be on strengthening and broadening Arctic research capabilities. Consideration will also be directed to formulating an overall policy framework for Arctic-related activities.

APPENDIX 19

United States Policy for the Seabed—Statement by President Nixon, 1970

THE NATIONS of the world are now facing decisions of momentous importance to man's use of the oceans for decades ahead. At issue is whether the oceans will be used rationally and equitably and for the benefit of mankind or whether they will become an arena of unrestrained exploitation and conflicting jurisdictional claims in which even the most advantaged states will be losers.

The issue arises now—and with urgency—because nations have grown increasingly conscious of the wealth to be exploited from the seabeds and throughout the waters above and because they are also becoming apprehensive about ecological hazards of unregulated use of the oceans and seabeds. The stark fact is that the law of the sea is inadequate to meet the needs of modern technology and the concerns of the international community. If it is not modernized multilaterally, unilateral action and international conflict are inevitable.

This is the time then for all nations to set about resolving the basic issues of the future regime for the oceans—and to resolve it in a way that redounds to the general benefit in the era of intensive exploitation that lies ahead. The United States as a major maritime power and a leader in ocean technology to unlock the riches of the ocean has a special responsibility to move this effort forward.

Therefore, I am today proposing that all nations adopt as soon as possible a treaty under which they would renounce all national claims over the natural resources of the seabed beyond the point where the high seas reach a depth of 200 meters (218.8 yards) and would agree to regard these resources as the common heritage of mankind.

The treaty should establish an international regime for the exploitation of seabed resources beyond this limit. The regime should provide for the collection of substantial mineral royalties to be used for international community purposes, particularly economic assistance to developing countries. It should also establish general rules to prevent unreasonable interference with other uses of the ocean, to protect the ocean from pollution, to assure the integrity of the investment necessary for such exploitation, and to provide for peaceful and compulsory settlement of disputes.

I propose two types of machinery for authorizing exploitation of seabed resources beyond a depth of 200 meters.

First, I propose that coastal nations act as trustees for the international community in an international trusteeship zone comprised of the continental margins beyond a depth of 200 meters off their coasts. In return, each coastal state would receive a share of the international revenues from the zone in which it acts as trustee and could impose additional taxes if these were deemed desirable.

As a second step, agreed international machinery would authorize and regulate exploration and use of seabed resources beyond the continental margins.

The United States will introduce specific proposals at the next meeting of the United Nations Seabeds Committee to carry out these objectives.

Although I hope agreement on such steps can be reached quickly, the negotiation of such a complex treaty may take some time. I do not, however, believe it is either necessary or desirable to try to halt exploration and exploitation of the seabeds beyond a depth of 200 meters during the negotiating process.

Accordingly, I call on other nations to join the United States in an interim policy. I suggest that all permits for exploration and exploitation of the seabeds beyond 200 meters be issued subject to the international regime to be agreed upon. The regime should accordingly include due protection for the integrity of investments made in the interim period. A substantial portion of the revenues derived by a state from exploitation beyond 200 meters during this interim period should be turned over to an appropriate international development agency for assistance to developing countries. I would plan to seek appropriate congressional action to make such funds available as soon as a sufficient number of other states also indicate their willingness to join this interim policy.

I will propose necessary changes in the domestic import and tax laws and regulations of the United States to assure that our own laws and regulations do not discriminate against U.S. nationals operating in the trusteeship zone off our coast or under the authority of the international machinery to be established.

It is equally important to assure unfettered and harmonious use of the oceans as an avenue of commerce and transportation, and as a source of food. For this reason the United States is currently engaged with other states in an effort to obtain a new law of the sea treaty. This treaty would establish a 12-mile limit for territorial seas and provide for free transit through international straits. It would also accommodate the problems of developing countries and other nations regarding the conservation and use of the living resources of the high seas.

I believe that these proposals are essential to the interests of all nations, rich and poor, coastal and landlocked, regardless of their political systems. If they result in international agreements, we can save over two-thirds of the earth's surface from national conflict and rivalry, protect it from pollution, and put it to use for the benefit of all. This would be a fitting achievement for this 25th anniversary year of the United Nations.

APPENDIX 20

Selected International Organizations Active in the Marine Sciences

UNITED NATIONS
General
General Assembly
Economic and Social Council
U.N. Development Program
U.N. Children's Fund
International Atomic Energy Agency
Protein Advisory Group
International Court of Justice
International Law Commission
U.N. Conference on Trade and Development
Specialized Agencies
Food and Agriculture Organization
UNESCO—Intergovernmental Oceanographic Commission
Intergovernmental Maritime Consultative Organization
World Meteorological Organization
World Health Organization
International Civil Aviation Organization
International Telecommunication Union
International Bank for Reconstruction and Development
INTERGOVERNMENTAL ORGANIZATIONS
International Hydrographic Bureau
International Council for the Exploration of the Sea
Permanent International Association of Navigation Congresses
North Atlantic Treaty Organization
Organization for Economic Cooperation and Development
International Commission for the Scientific Exploration of the Mediterranean
Sea
Central Treaty Organization
South Pacific Commission
Indo-Pacific Fisheries Council
Colombo Plan Council for Technical Cooperation in South and Southeast
Asia
Organization of American States
Inter-American Development Bank
Pan American Health Organization
Pan American Institute of Geography and History
NONGOVERNMENTAL ORGANIZATIONS
International Council of Scientific Unions
International Union of Geodesy and Geophysics
International Association for the Physical Sciences of the Ocean
International Union of Geological Sciences
Commission on Marine Geology
International Geographical Union
International Union of Biological Sciences
International Association of Biological Oceanography
Special Committee for the International Biological Program
Scientific Committee on Oceanic Research

Source: *MSA*, 2d report (1968), pp. 191–92.

Scientific Committee on Antarctic Research
International Geophysical Committee
International Union for Conservation of Nature and Natural Resources
Pacific Science Association
Federation of Astronomical and Geophysical Services
    Permanent Service for Mean Sea Level
Scientific Committee on Water Research
Mediterranean Association of Marine Biology and Oceanography
Association of Island Marine Laboratories of the Caribbean
Inter-American Geodetic Society
Union of International Engineering Organizations
Pan American Congress of Naval Engineering and Maritime Transport
International Ship Structure Congress
International Institute of Welding
International Cable Protection Committee
International Association of Lighthouse Authorities
International Association of Ports and Harbors
International Lifeboat Conference
International Commission on Illumination
International Maritime Radio Association
International Radio Consultative Committee
International Chamber of Commerce
International Chamber of Shipping
International Union of Marine Insurance
International Maritime Committee
International Gas Union
Permanent Council of World Petroleum Congress
Offshore Exploration Congress
World Underwater Federation

APPENDIX 21

Convention on the Continental Shelf, 1958

Adopted by the United Nations Conference on the Law of the Sea, April 29, 1958 (U.N. Doc. A/CONF.13/L.55)

The States Parties to this Convention
Have agreed as follows:

Article 1

For the purpose of these articles, the term "continental shelf" is used as referring (*a*) to the seabed and subsoil of the submarine areas adjacent to the coast but outside the area of the territorial sea, to a depth of 200 metres or, beyond that limit, to where the depth of the superjacent waters admits of the exploitation of the natural resources of the said areas; (*b*) to the seabed and subsoil of similar submarine areas adjacent to the coasts of islands.

Article 2

1. The coastal State exercises over the continental shelf sovereign rights for the purpose of exploring it and exploiting its natural resources.
2. The rights referred to in paragraph 1 of this article are exclusive in the sense that if the coastal State does not explore the continental shelf or exploit its natural resources, no one may undertake these activities, or make a claim to the continental shelf, without the express consent of the coastal State.
3. The rights of the coastal State over the continental shelf do not depend on occupation, effective or notional, or on any express proclamation.
4. The natural resources referred to in these articles consist of the mineral and other non-living resources of the seabed and subsoil together with living organisms belonging to sedentary species, that is to say, organisms which, at the harvestable stage, either are immobile on or under the seabed or are unable to move except in constant physical contact with the seabed or the subsoil.

Source: U.N. Doc. A/Conf. 13/L.55

Article 3

The rights of the coastal State over the continental shelf do not affect the legal status of the superjacent waters as high seas, or that of the air space above those waters.

Article 4

Subject to its right to take reasonable measures for the exploration of the continental shelf and the exploitation of its natural resources, the coastal State may not impede the laying or maintenance of submarine cables or pipelines on the continental shelf.

Article 5

1. The exploration of the continental shelf and the exploitation of its natural resources must not result in any unjustifiable interference with navigation, fishing or the conservation of the living resources of the sea, nor result in any interference with fundamental oceanographic or other scientific research carried out with the intention of open publication.
2. Subject to the provisions of paragraphs 1 and 6 of this article, the coastal State is entitled to construct and maintain or operate on the continental shelf installations and other devices necessary for its exploration and the exploitation of its natural resources, and to establish safety zones around such installations and devices and to take in those zones measures necessary for their protection.
3. The safety zones referred to in paragraph 2 of this article may extend to a distance of 500 metres around the installations and other devices which have been erected, measured from each point of their outer edge. Ships of all nationalities must respect these safety zones.
4. Such installations and devices, though under the jurisdiction of the coastal State, do not possess the status of islands. They have no territorial sea of their own, and their presence does not affect the delimitation of the territorial sea of the coastal State.
5. Due notice must be given of the construction of any such installations, and permanent means for giving warning of their presence must be maintained. Any installations which are abandoned or disused must be entirely removed.
6. Neither the installations or devices, nor the safety zones around them, may be established where interference may be caused to the use of recognized sea lanes essential to international navigation.
7. The coastal State is obliged to undertake, in the safety zones, all

appropriate measures for the protection of the living resources of the sea from harmful agents.

8. The consent of the coastal State shall be obtained in respect of any research concerning the continental shelf and undertaken there. Nevertheless, the coastal State shall not normally withhold its consent if the request is submitted by a qualified institution with a view to purely scientific research into the physical or biological characteristics of the continental shelf, subject to the proviso that the coastal State shall have the right, if it so desires, to participate or to be represented in the research, and that in any event the results shall be published.

Article 6

1. Where the same continental shelf is adjacent to the territories of two or more States whose coasts are opposite each other, the boundary of the continental shelf appertaining to such States shall be determined by agreement between them. In the absence of agreement, and unless another boundary line is justified by special circumstances, the boundary is the median line, every point of which is equidistant from the nearest points of the baselines from which the breadth of the territorial sea of each State is measured.

2. Where the same continental shelf is adjacent to the territories of two adjacent States, the boundary of the continental shelf shall be determined by agreement between them. In the absence of agreement, and unless another boundary line is justified by special circumstances, the boundary shall be determined by application of the principle of equidistance from the nearest points of the baselines from which the breadth of the territorial sea of each State is measured.

3. In delimiting the boundaries of the continental shelf, any lines which are drawn in accordance with the principles set out in paragraphs 1 and 2 of this article should be defined with reference to charts and geographical features as they exist at a particular date, and reference should be made to fixed permanent identifiable points on the land.

Article 7

The provisions of these articles shall not prejudice the right of the coastal State to exploit the subsoil by means of tunnelling irrespective of the depth of water above the subsoil.

Article 8

This Convention shall, until 30 October 1958, be open for signature by all States Members of the United Nations or of any of the specialized agencies, and by any other State invited by the General Assembly of the United Nations to become a party to the Convention.

Article 9

This Convention is subject to ratification. The instruments of ratification shall be deposited with the Secretary-General of the United Nations.

Article 10

This Convention shall be open for accession by any States belonging to any of the categories mentioned in article 8. The instruments of accession shall be deposited with the Secretary-General of the United Nations.

Article 11

1. This Convention shall come into force on the thirtieth day following the date of deposit of the twenty-second instrument of ratification or accession with the Secretary-General of the United Nations.
2. For each State ratifying or acceding to the Convention after the deposit of the twenty-second instrument of ratification or accession, the Convention shall enter into force on the thirtieth day after deposit by such State of its instrument of ratification or accession.

Article 12

1. At the time of signature, ratification or accession, any State may make reservations to articles of the Convention other than to articles 1 to 3 inclusive.
2. Any contracting State making a reservation in accordance with the preceding paragraph may at any time withdraw the reservation by a communication to that effect addressed to the Secretary-General of the United Nations.

Article 13

1. After the expiration of a period of five years from the date on which this Convention shall enter into force, a request for the revision of this Convention may be made at any time by any contracting

party by means of a notification in writing addressed to the Secretary-General of the United Nations.

2. The General Assembly of the United Nations shall decide upon the steps, if any, to be taken in respect of such request.

Article 14

The Secretary-General of the United Nations shall inform all States Members of the United Nations and the other States referred to in article 8:

(*a*) Of signatures to this Convention and of the deposit of instruments of ratification or accession, in accordance with articles 8, 9 and 10;

(*b*) Of the date on which this Convention will come into force, in accordance with article 11;

(*c*) Of requests for revision in accordance with article 13;

(*d*) Of reservations to this Convention, in accordance with article 12.

Article 15

The original of this Convention, of which the Chinese, English, French, Russian and Spanish texts are equally authentic, shall be deposited with the Secretary-General of the United Nations, who shall send certified copies thereof to all States referred to in article 8.

IN WITNESS WHEREOF the undersigned plenipotentiaries, being duly authorized thereto by their respective governments, have signed this Convention.

DONE AT GENEVA, this twenty-ninth day of April one thousand nine hundred and fifty-eight.

APPENDIX 22

Nations Which Have Ratified, Acceded to, or Consider Themselves Bound by the Geneva Conventions on the Law of the Sea,[2] 1967

	Convention on the high seas	Convention on the Continental Shelf	Convention on the territorial sea and contiguous zone	Convention on fishing and conservation of living resources of the high seas
Afghanistan	X			
Albania	X [1]	X		
Australia	X	X	X	X
Bulgaria	X [1]	X	X [1]	
Byelorussia	X [1]	X	X [1]	
Cambodia	X	X	X	X
Central African Republic	X			
Colombia		X		X
Czechoslovakia	X [1]	X	X [1]	
Denmark		X		
Dominican Republic	X	X	X	X
Finland	X	X	X	X
France		X [1]		
Guatemala	X	X		
Haiti	X	X	X	X
Hungary	X [1]		X [1]	
Indonesia	X [1]			
Israel	X	X	X	
Italy	X		X	
Jamaica	X	X	X	X
Madagascar	X	X	X [1]	X
Malawi	X	X	X	X
Malaya	X	X	X	X
Malta		X	X	
Mexico	X [1]	X	X [1]	X
Nepal	X			
Netherlands	X	X	X	X
New Zealand		X		
Nigeria	X		X	X
Poland	X [1]	X		
Portugal	X	X	X	X
Rumania	X [1]	X	X [1]	
Senegal	X	X	X	X
Sierra Leone	X		X	X
South Africa	X	X	X	X
Sweden		X		
Switzerland	X	X	X	X
Trinidad and Tobago	X		X	X
Uganda	X	X	X	X
Ukraine	X [1]	X	X [1]	
U.S.S.R.	X [1]	X	X [1]	
United Kingdom	X [1]	X	X [1]	X [1]
United States	X	X	X	X [1]
Upper Volta	X			X
Venezuela	X	X [1]	X [1]	X
Yugoslavia	X	X [1]	X	X

[1] With a reservation or declaration.
[2] Based on information available to the Department of State as of Dec. 8, 1967.

Source: *MSA*, 2d report (1968), pp. 195–96.

APPENDIX 23

Breadth of Territorial Seas and Fishing Jurisdictions Claimed by Selected Countries, 1970

Country	Territorial sea	Fishing limit	Other
Albania	10 miles	12 miles	
Algeria	12 miles	12 miles	
Argentina		200 miles	Sovereignty is claimed over a 200-mile maritime zone but the law specifically provides that freedom of navigation of ships and aircraft in the zone is unaffected. Continental Shelf—including sovereignty over superiacent waters.
Australia	3 miles	12 miles	
Belgium	3 miles	12 miles [1]	
Brazil	12 miles	12 miles	
Bulgaria	12 miles	12 miles	
Burma	12 miles	12 miles	
Cambodia	12 miles	12 miles	
Cameroun	18 miles	18 miles	
Canada	3 miles	12 miles	
Ceylon	6 miles	6 miles	Claims right to establish conservation zones within 100 nautical miles of the territorial sea.
Chile	3 miles	200 miles	
China	3 miles	3 miles	
Colombia	12 miles	12 miles	
Congo (Brazzaville)			
Congo (Kinshasa)	3 miles	3 miles	
Costa Rica	3 miles		"Specialized competence" over living resources to 200 miles.
Cuba	3 miles	3 miles	
Cyprus	12 miles	12 miles	
Dahomey	12 miles	12 miles	100-mile mineral exploration limit.
Denmark	3 miles	12 miles [1]	
Greenland		12 miles	
Faroe Islands		12 miles	
Dominican Republic	6 miles	12 miles	Contiguous zone 6 miles beyond territorial sea for protection of health, fiscal, customs matters, and the conservation of fisheries and other natural resources of the sea.

See footnote at end of table.

Source: *MSA*, 4th report (1970), pp. 281–84.

Country	Territorial sea	Fishing limit	Other
Ecuador	200 miles	200 miles	
El Salvador	200 miles	200 miles	
Ethiopia	12 miles	12 miles	
Federal Republic of Germany.	3 miles	12 miles [1]	
Finland	4 miles	4 miles	
France	3 miles	12 miles	
Gabon	12 miles	12 miles	
Gambia	3 miles	3 miles	
Ghana	12 miles	12 miles	Undefined protective areas may be proclaimed seaward of territorial sea, and up to 100 miles seaward of territorial sea may be proclaimed fishing conservation zone.
Greece	6 miles	6 miles	
Guatemala	12 miles	12 miles	
Guinea	130 miles	130 miles	
Guyana	3 miles	3 miles	
Haiti	6 miles	6 miles	
Honduras	12 miles	12 miles	
Iceland	4 miles	12 miles	
India	12 miles	12 miles	Plus right to establish 100 miles conservation zone.
Indonesia	12 miles	12 miles	Archipelago concept baselines.
Iran	12 miles	12 miles	
Iraq	12 miles	12 miles	
Ireland	3 miles	12 miles [1]	
Israel	6 miles	6 miles	
Italy	6 miles	12 miles [1]	
Ivory Coast	6 miles	12 miles	
Jamaica	12 miles		
Japan	3 miles	3 miles	
Jordan	3 miles	3 miles	
Kenya	12 miles	12 miles	
Korea	3 miles	20 to 200 miles.	Continental Shelf including sovereignty over superjacent waters.
Kuwait	12 miles	12 miles	
Lebanon		6 miles	
Liberia	12 miles	12 miles	
Libya	12 miles	12 miles	
Malagasy Republic	12 miles	12 miles	
Malaysia	12 miles	12 miles	
Maldive Islands	3 miles	6 miles	
Malta	3 miles	3 miles	
Mauritania	12 miles	12 miles	
Mauritius	3 miles	3 miles	
Mexico	12 miles	12 miles	
Morocco	3 miles	12 miles	Exception—6-mile fishing zone for Strait of Gibraltar.
Netherlands	3 miles	12 miles [1]	
New Zealand	3 miles	12 miles	
Nicaragua	3 miles	200 miles	Continental Shelf including sovereignty over superjacent waters.

See footnote at end of table.

Country	Territorial sea	Fishing limit	Other
Nigeria	12 miles....	12 miles....	
Norway	4 miles.....	12 miles....	
Pakistan	12 miles....	12 miles....	Plus right to establish 100-mile conservation zones.
Panama	200 miles...	200 miles...	Continental Shelf including sovereignty over superjacent waters.
Peru	200 miles...	200 miles...	
Philippines			Archipelago concept baselines. Waters between these baselines and the limits described in the Treaty of Paris, Dec. 10, 1898, the United States-Spain Treaty of Nov. 7, 1900, and United States-United Kingdom Treaty of Jan. 2, 1930, are claimed as territorial sea.
Poland	3 miles.....	3 miles.....	
Portugal	No claims..	12 miles [1]...	
Romania	12 miles....	12 miles....	
Saudia Arabia	12 miles....	12 miles....	
Senegal	12 miles....	18 miles....	Fishing zone beyond 12 miles does not apply to those nations which are party to the 1958 Geneva Convention on the Territorial Sea and the Contiguous Zone.
Sierra Leone	12 miles....	12 miles....	
Singapore	3 miles.....	3 miles.....	
Somali Republic	12 miles....	12 miles....	
South Africa	6 miles.....	12 miles....	
Spain	6 miles.....	12 miles [1]...	
Sudan	12 miles....	12 miles....	
Sweden	4 miles.....	12 miles [1]...	
Syria	12 miles....	12 miles....	Contiguous zone—an additional 6-mile area to control security, customs, hygiene, and financial matters.
Tanzania	12 miles....	12 miles....	
Thailand	12 miles....	12 miles....	
Togo	12 miles....	12 miles....	
Trinidad and Tobago	3 miles.....	3 miles.....	
Tunisia	6 miles.....	12 miles....	Fisheries zone follows the 50-meter isobath at specified areas of the coast (maximum 65 miles).
Turkey	6 miles.....	12 miles....	
Ukrainian S.S.R.	12 miles....	12 miles....	
U.S.S.R.	12 miles....	12 miles....	
United Arab Republic	12 miles....	12 miles....	
United Kingdom	3 miles.....	12 miles....	
Overseas areas	3 miles.....	3 miles.....	
United States of America	3 miles.....	12 miles....	

See footnote at end of table.

Country	Territorial sea	Fishing limit	Other
Uruguay	12 miles	200 miles	Sovereignty is claimed over a 200-mile maritime zone but law specifically provides that the freedom of navigation of ships and aircraft beyond 12 miles is unaffected by the claim.
Venezuela	12 miles	12 miles	
Vietnam	3 miles	20 kilometers. (10.8 miles)	
Yemen	12 miles	12 miles	
Yugoslavia	10 miles	10 miles	

[1] Parties to the European Fisheries Convention which provides for the right to establish 3-mile exclusive fishing zone seaward of 3-mile territorial sea plus additional 6-mile fishing zone restricted to the convention nations.

APPENDIX 24

Resolution Adopted by United Nations General Assembly, 2340 (XXII), 1967—Peaceful Uses of the Seabed

Examination of the Question of the Reservation Exclusively for Peaceful Purposes of the Sea-Bed and the Ocean Floor, and the Subsoil Thereof, Underlying the High Seas Beyond the Limits of Present National Jurisdiction, and the Uses of Their Resources in the Interests of Mankind

The General Assembly,

Having considered the item "Examination of the question of the reservation exclusively for peaceful purposes of the seabed and the ocean floor, and the subsoil thereof underlying the high seas beyond the limits of present national jurisdiction, and the uses of their resources in the interests of mankind",

Noting that developing technology is making the seabed and the ocean floor, and the subsoil thereof, accessible and exploitable for scientific, economic, military and other purposes,

Recognizing the common interest of mankind in the seabed and the ocean floor which constitute the major portion of the area of this planet,

Recognizing further that the exploration and use of the seabed and the ocean floor, and the subsoil thereof, as contemplated in the title of the item, should be conducted in accordance with the principles and purposes of the Charter of the United Nations, in the interest of maintaining international peace and security and for the benefit of all mankind,

Mindful of the provisions and practice of the law of the sea relating to this question,

Mindful also of the importance of preserving the seabed and the ocean floor, and the subsoil thereof, as contemplated in the title of the item, from actions and uses which might be detrimental to the common interests of mankind,

Desiring to foster greater international cooperation and coordination in the further peaceful exploration and use of the seabed and the ocean floor, and the subsoil thereof, as contemplated in the title of the item,

Recalling the past and continuing valuable work on questions relating to this matter carried out by the competent organs of the United Nations, the specialized agencies, the International Atomic Energy Agency and other intergovernmental organizations,

Recalling further that surveys are being prepared by the Secretary-General in response to General Assembly resolution 2172 (XXI) of 6 December 1966 and Economic and Social Council resolution 1112 (XL) of 7 March 1966,

1. *Decides* to establish an *ad hoc* Committee to study the peaceful uses of the seabed and the ocean floor beyond the limits of national jurisdiction composed of Argentina, Australia, Austria, Belgium, Brazil, Bulgaria, Canada, Ceylon, Chile, Czechoslovakia, Ecuador, El Salvador, France, Iceland, India, Italy, Japan, Kenya, Liberia, Libya, Malta, Norway, Pakistan, Peru, Poland, Rumania, Senegal, Somalia, Thailand, the Union of Soviet Socialist Republics, the United Arab Republic, the United Kingdom of Great Britain and Northern Ireland, the United Republic of Tanzania, the United States of America, and Yugoslavia, to study the scope and various aspects of this item;

2. *Requests* the *ad hoc* Committee, in cooperation with the Secretary-General, to prepare, for consideration by the General Assembly at its twenty-third session, a study which would include:

   (a) a survey of the past and present activities of the United Nations, the

Source: *MSA*, 2d report (1968), pp. 194–95

specialized agencies, the International Atomic Energy Agency and other inter-governmental bodies with regard to the seabed and the ocean floor, and of existing international agreements concerning these areas;

(b) an account of the scientific, technical, economic, legal, and other aspects of this item;

(c) an indication regarding practical means to promote international cooperation in the exploration, conservation and use of the seabed and the ocean floor, and the subsoil thereof, as contemplated in the title of the item, and of their resources, having regard to the views expressed and suggestions put forward by Member States during the consideration of this item at the twenty-second session of the General Assembly;

3. *Requests* the Secretary-General:

(a) to transmit the text of this resolution to the Governments of all Member States in order to seek their views on the subject;

(b) to transmit to the *ad hoc* Committee the records of the First Committee relating to the discussion of this item;

(c) to render all appropriate assistance to the *ad hoc* Committee, including the submission thereto of the results of the studies being undertaken in pursuance of General Assembly resolution 2172 (XXI) and Economic and Social Council resolution 1112 (XL) and such documentation pertinent to this item as may be provided by the United Nations Educational, Scientific and Cultural Organization and its Inter-governmental Oceanographic Commission, the Inter-governmental Maritime Consultative Organization, the Food and Agriculture Organization of the United Nations, the World Meteorological Organization, the World Health Organization, the International Atomic Energy Agency and other intergovernmental bodies;

4. *Invites* the specialized agencies, the International Atomic Energy Agency, and other intergovernmental bodies to cooperate fully with the *ad hoc* Committee in the implementation of this resolution.

18 December 1967

APPENDIX 25

Resolution Adopted by United Nations General Assembly, 2467A (XXIII), 1968—Establishment of a Standing Seabed Committee

The General Assembly,

Recalling the item entitled "Examination of the question of the reservation exclusively for peaceful purposes of the sea-bed and the ocean floor, and the subsoil thereof, underlying the high seas beyond the limits of present national jurisdiction, and the use of their resources in the interests of mankind",

Having in mind its resolution 2340 (XXII) of 18 December 1967 concerned with the problems arising in the area to which the title of the item refers,

Reaffirming the objectives set forth in that resolution,

Taking note with appreciation of the report prepared by the *Ad Hoc* Committee to Study the Peaceful Uses of the Sea-Bed and the Ocean Floor beyond the Limits of National Jurisdiction, keeping in mind the views expressed in the course of its work and drawing upon its experience,

Recognizing that it is in the interest of mankind as a whole to favour the exploration and use of the sea-bed and the ocean floor and the subsoil thereof, beyond the limits of national jurisdiction, for peaceful purposes,

Considering that it is important to promote international co-operation for the exploration and exploitation of the resources of this area,

Convinced that such exploitation should be carried out for the benefit of mankind as a whole, irrespective of the geographical location of States, taking into account the special interests and needs of the developing countries,

Considering that it is essential to provide, within the United Nations system, a focal point for the elaboration of desirable measures of international co-operation, taking into account alternative actual and potential uses of this area, and for the co-ordination of the activities of international organizations in this regard,

1. *Establishes* a Committee on the Peaceful Uses of the Sea-Bed and the Ocean Floor beyond the Limits of National Jurisdiction, composed of forty-two States;

2. *Instructs* the Committee:

(a) To study the elaboration of the legal principles and norms which would promote international co-operation in the exploration and use of the sea-bed and the ocean floor and the subsoil thereof beyond the limits of national jurisdiction and to ensure the exploitation of their resources for the benefit of mankind, and the economic and other requirements which such a régime should satisfy in order to meet the interests of humanity as a whole;

(b) To study the ways and means of promoting the exploitation and use of the resources of this area, and of international co-operation to that end, taking into account the foreseeable development of technology and the economic implications of such exploitation and bearing in mind the fact that such exploitation should benefit mankind as a whole;

(c) To review the studies carried out in the field of exploration and research in this area and aimed at intensifying international co-operation and stimulating the exchange and the widest possible dissemination of scientific knowledge on the subject;

(d) To examine proposed measures of co-operation to be adopted by the international community in order to prevent the marine pollution which may result from the exploration and exploitation of the resources of this area;

3. *Also calls upon* the Committee to study further, within the context of the title of the item, and taking into account the studies and international negotiations being

Source: *MSA*, 3d report (1969), pp. 234–35.

undertaken in the field of disarmament, the reservations exclusively for peaceful purposes of the sea-bed and the ocean floor without prejudice to the limits which may be agreed upon in this respect;

4. *Requests* the Committee:

(a) To work in close co-operation with the specialized agencies, the International Atomic Energy Agency and the intergovernmental bodies dealing with the problems referred to in the present resolution, so as to avoid any duplication or overlapping of activities;

(b) To make recommendations to the General Assembly on the questions mentioned in paragraphs 2 and 3 above;

(c) In co-operation with the Secretary-General, to submit to the General Assembly reports on its activities at each subsequent session;

5. *Invites* the specialized agencies, the International Atomic Energy Agency and other inter-governmental bodies including the Intergovernmental Oceanographic Commission of the United Nations Educational, Scientific and Cultural Organization to co-operate fully with the Committee in the implementation of the present resolution.

21 December 1968

APPENDIX 26

Resolution Adopted by United Nations General Assembly, 2566 (XXIV), 1969—Promoting Effective Measures for the Prevention and Control of Marine Pollution

The General Assembly,

Recalling its resolution 2414 (XXIII) of 17 December 1968 requesting the Secretary-General to report to the General Assembly at its twenty-fifth session, *inter alia,* on the progress achieved by Member States and organizations concerned to promote the adoption of effective international agreements on the prevention and control of marine pollution as may be necessary,

Recalling also its resolution 2467 B (XXIII) of 21 December 1968 on the prevention of marine pollution which might result from exploration and exploitation of the sea-bed and ocean floor,

Noting that a joint group of experts on the scientific aspects of marine pollution has been established by the Food and Agriculture Organization of the United Nations, the United Nations Educational, Scientific and Cultural Organization, the World Meteorological Organization and the Inter-Governmental Maritime Consultative Organization to give advice to these agencies on this subject,

Taking into account the "Comprehensive outline of the scope of the long-term and expanded programme of oceanic exploration and research", providing for a series of scientific studies which would review the state of the ocean and its resources as regards pollution, and forecast long-term trends to assist Governments individually and collectively to take the steps required to counteract its effects,

Bearing in mind arrangements made by the Food and Agriculture Organization of the United Nations for the holding of a technical conference on marine pollution and its effects on living resources and fishing, to be held at Rome, in December 1970,

Recalling its resolution 2398 (XXIII) of 3 December 1968 on the convening in 1972 of a United Nations Conference on the Human Environment and of the Secretary-General on problems of the human environment which *inter alia,* stresses the problems relating to marine pollution,

Noting the resolution on marine pollution adopted by the sixth Assembly of the Inter-Governmental Maritime Consultative Organization calling for an international conference in 1973 for the purpose of preparing a suitable international agreement for placing restraints on the contamination of the sea, land and air by ships and other vessels or equipment operating in the marine environment,

Considering that in spite of the sustained efforts being made at present many aspects of marine pollution have not yet been dealt with or are not

Source: *MSA*, 4th report (1970), pp. 235–37.

being fully covered, and that additional agreements on this subject may be required,

1. *Requests* the Secretary-General, in co-operation with the specialized agencies and intergovernmental organizations concerned, to complement reports and studies under preparation, with special reference to the forth-coming United Nations Conference on the Human Environment, by:

(*a*) A review of harmful chemical substances, radio-active materials and other noxious agents and waste which may dangerously affect man's health and his economic and cultural activities in the marine environment and coastal areas;

(*b*) A review of national activities and activities of specialized agencies of the United Nations and intergovernmental organizations dealing with prevention and control of marine pollution including suggestions for more comprehensive action and improved co-ordination in this field;

(*c*) Seeking the views of Member States on the desirability and feasibility of an international treaty or treaties on the subject;

2. *Requests* the Secretary-General to report to the Economic and Social Council and the Preparatory Committee for the United Nations Conference on the Human Environment, as appropriate. in the framework of the preparations for the Conference.

13 December 1969

APPENDIX 27

Resolution Adopted by United Nations General Assembly, 2580 (XXIV), 1969—Co-ordination of Maritime Activities

The General Assembly,

Having considered the report of the Enlarged Committee for Programme and Co-ordination,

Noting that the Enlarged Committee was unable in the time available to give thorough consideration to a proposal for more systematic co-ordination of continuing activities of the United Nations system relating to the seas and oceans, is,

Aware of the complexity of the co-ordination of existing international activities with regard to marine science and its applications and that the field of marine science is only one aspect of the existing activities of the United Nations system relating to the seas and oceans,

Noting that use by States of the marine environment is rapidly becoming intensified and diversified,

Noting with appreciation the work done in this field by the organizations in the United Nations system.

Concerned that present international machinery may not permit a prompt, effective and flexible response to existing and emerging needs of States members of the United Nations,

Recognizing that, in order to avoid the overlapping and duplication of programmes and gaps in competence, a full review of the existing activities of United Nations system of organizations relating to the seas and oceans may be urgently required,

1. *Requests* the Economic and Social Council, at its organizational session in January 1970, to consider instructing the Committee for Programme and Co-ordination, after reconstitution, to examine the need for a comprehensive review of existing activities of the United Nations system relating to the seas and oceans in the light of present and emerging needs of Member States, with a view to making the Committee's recommendations available to the Council at is forty-ninth session;

2. *Requests* the Secretary-General to assist the Committee for Programme and Co-ordination in the fulfillment of this task;

3. *Invites* the specialized agencies and the intergovernmental bodies concerned to extend their full co-operation and assistance to the Committee for Programme and Co-ordination.

15 December 1969

Source: *MSA*, 4th report (1970), p. 237.

APPENDIX 28

Resolution Adopted by United Nations General Assembly, 2749 (XXV), 1970—Declaration of Principles Governing the Seabed and the Ocean Floor, and the Subsoil Thereof, Beyond the Limits of National Jurisdiction

The General Assembly,

Recalling its resolutions 2340 (XXII) of 18 December 1967, 2467 (XXIII) of 21 December 1968 and 2574 (XXIV) of 15 December 1969, concerning the area to which the title of the item refers,

Affirming that there is an area of the sea-bed and the ocean floor, and the subsoil thereof, beyond the limits of national jurisdiction, the precise limits of which are yet to be determined,

Recognizing that the existing legal régime of the high seas does not provide substantive rules for regulating the exploration of the aforesaid area and the exploitation of its resources,

Convinced that the area shall be reserved exclusively for peaceful purposes and that the exploration of the area and the exploitation of its resources shall be carried out for the benefit of mankind as a whole,

Believing it essential that an international régime applying to the area and its resources and including appropriate international machinery should be established as soon as possible,

Bearing in mind that the development and use of the area and its resources shall be undertaken in such a manner as to foster the healthy development of the world economy and balanced growth of international trade, and to minimize any adverse economic effects caused by the fluctuation of prices of raw materials resulting from such activities,

Solemnly declares that:

1. The sea-bed and ocean floor, and the subsoil thereof, beyond the limits of national jurisdiction (hereinafter referred to as the area), as well as the resources of the area, are the common heritage of mankind.

2. The area shall not be subject to appropriation by any means by States or persons, natural or juridical, and no State shall claim or exercise sovereignty or sovereign rights over any part thereof.

3. No State or person, natural or juridical, shall claim, exercise or acquire rights with respect to the area or its resources incompatible with the international régime to be established and the principles of this Declaration.

4. All activities regarding the exploration and exploitation of the resources of the area and other related activities shall be governed by the international régime to be established.

Source: *MSA*, 5th report (1971), pp. 105–7.

5. The area shall be open to use exclusively for peaceful purposes by all States, whether coastal or land-locked, without discrimination, in accordance with the international régime to be established.

6. States shall act in the area in accordance with the applicable principles and rules of international law, including the Charter of the United Nations and the Declaration on Principles of International Law concerning Friendly Relations and Co-operation among States in accordance with the Charter of the United Nations, adopted by the General Assembly on 24 October 1970, in the interests of maintaining international peace and security and promoting international co-operation and mutual understanding.

7. The exploration of the area and the exploitation of its resources shall be carried out for the benefit of mankind as a whole, irrespective of the geographical location of States, whether land-locked or coastal, and taking into particular consideration the interests and needs of the developing countries.

8. The area shall be reserved exclusively for peaceful purposes, without prejudice to any measures which have been or may be agreed upon in the context of international negotiations undertaken in the field of disarmament and which may be applicable to a broader area. One or more international agreements shall be concluded as soon as possible in order to implement effectively this principle and to constitute a step towards the exclusion of the sea-bed, the ocean floor and the subsoil thereof from the arms race.

9. On the basis of the principles of this Declaration, an international régime applying to the area and its resources and including appropriate international machinery to give effect to its provisions shall be established by and international treaty of a universal character, generally agreed upon. The régime shall, *inter alia,* provide for the orderly and safe development and rational management of the area and its resources and for expanding opportunities in the use thereof and ensure the equitable sharing by States in the benefits derived therefrom, taking into particular consideration the interests and needs of the developing countries, whether land-locked or coastal.

10. States shall promote international co-operation in scientific research exclusively for peaceful purposes:

(*a*) By participation in international programmes and by encouraging co-operation in scientific research by personnel of different countries;

(*b*) Through effective publication of research programmes and dissemination of the results of research through international channels;

(*c*) By co-operation in measures to strengthen research capabilities of developing countries, including the participation of their nationals in research programmes.

No such activity shall form the legal basis for any claims with respect to any part of the area or its resources.

11. With respect to activities in the area and acting in conformity

with the international régime to be established, States shall take appropriate measures for and shall cooperate in the adoption and implementation of international rules, standards and procedures for, *inter alia:*

(*a*) The prevention of pollution and contamination, and other hazards to the marine environment, including the coastline, and of interference with the ecological balance of the marine environment;

(*b*) The protection and conservation of the natural resources of the area and the prevention of damage to the flora and fauna of the marine environment.

12. In their activities in the area, including those relating to its resources, States shall pay due regard to the rights and legitimate interests of coastal States in the region of such activities, as well as of all other States, which may be affected by such activities. Consultations shall be maintained with the coastal States concerned with respect to activities relating to the exploration of the area and the exploitation of its resources with a view to avoiding infringement of such rights and interests.

13. Nothing herein shall affect:

(*a*) The legal status of the waters superjacent to the area or that of the air space above those waters;

(*b*) The rights of coastal States with respect to measures to prevent, mitigate or eliminate grave and imminent danger to their coastline or related interests from pollution or threat thereof or from other hazardous occurrences resulting from or caused by any activities in the area, subject to the international regime to be established.

14. Every State shall have the responsibility to ensure that activities in the area, including those relating to its resources, whether undertaken by governmental agencies, or non-governmental entities or persons under its jurisdiction, or acting on its behalf, shall be carried out in conformity with the international régime to be established. The same responsibility applies to international organizations and their members for activities undertaken by such organizations or on their behalf. Damage caused by such activities shall entail liability.

15. The parties to any dispute relating to activities in the area and its resources shall resolve such dispute by the measures mentioned in Article 33 of the Charter of the United Nations and such procedures for settling disputes as may be agreed upon in the international régime to be established.

APPENDIX 29

Resolution Adopted by United Nations General Assembly, 2750 (XXV), 1970—Peaceful Uses of the Seabed and Convening of a Conference on the Law of the Sea

Reservation exclusively for peaceful purposes of the sea-bed and the ocean floor, and the subsoil thereof, underlying the high seas beyond the limits of present national jurisdiction, and use of their resources in the interests of mankind, and convening of a conference on the law of the sea

A

The General Assembly,

Reaffirming that the area of the sea-bed and the ocean floor, and the subsoil thereof, beyond the limits of national jurisdiction, and its resources are the common heritage of mankind,

Convinced that the exploration of the area and the exploitation of its resources should be carried out for the benefit of mankind as a whole, taking into account the special interests and needs of the developing countries,

Reaffirming that the development of the area and its resources shall be undertaken in such a manner as to foster the healthy development of the world economy and balanced growth of international trade, and to minimize any adverse economic effects caused by the fluctuation of prices of raw materials resulting from such activities,

1. *Requests* the Secretary-General to co-operate with the United Nations Conference on Trade and Development, specialized agencies and other competent organizations of the United Nations system in order to:

(a) Identify the problems arising from the production of certain minerals from the area beyond the limits of national jurisdiction and examine the impact they will have on the economic well-being of the developing countries, in particular on prices of mineral exports on the world market;

(b) Study these problems in the light of the scale of possible exploitation of the sea-bed, taking into account the world demand for raw materials and the evolution of costs and prices;

(c) Propose effective solutions for dealing with these problems;

2. *Requests* the Secretary-General to submit his report thereon to the Committee on the Peaceful Uses of the Sea-Bed and the Ocean Floor beyond the Limits of National Jurisdiction for consideration during one of its sessions in 1971 and for making its recommendations as appropriate to foster the healthy development of the world economy and balanced growth of international trade, and to minimize any adverse economic effects caused by the fluctuation of prices of raw materials resulting from such activities;

Source: *MSA*, 5th report (1971), pp. 107–12.

3. *Requests* the Secreatry-General, in co-operation with the United Nations Conference on Trade and Development, specialized agencies and other competent organizations of the United Nations system, to keep this matter under constant review so as to submit supplementary information annually or whenever it is necessary and recommend additional measures in the light of economic, scientific and technological developments;

4. *Calls upon* the Committee on the Peaceful Uses of the Sea-Bed and the Ocean Floor beyond the Limits of National Jurisdiction to submit a report on this question to the General Assembly at its twenty-sixth session.

1933rd plenary meeting,
17 December 1970.

B

The General Assembly,

Recalling its resolutions 1028 (XI) of 20 February 1957 and 1105 (XI) of 21 February 1957 concerning the problems of land-locked countries,

Bearing in mind the replies to the inquiries made by the Secretary-General in accordance with paragraph 1 of resolution 2574 A (XXIV) of 15 December 1969, which indicate wide support for the idea of convening a conference relating to the law of the sea, at which the interests and needs of all States, whether land-locked or coastal, could be reconciled,

Noting that many of the present land-locked States Members of the United Nations did not participate in the previous United Nations conferences on the law of the sea,

Reaffirming that the area of the sea-bed and the ocean floor, and their subsoil, lying beyond the limits of national jurisdiction, together with the resources thereof, are the common heritage of mankind,

Convinced that the exploration of the area and the exploitation of its resources must be carried out for the benefit of all mankind, taking into account the special interests and needs of the developing countries, including the particular needs and problems of those which are land-locked,

1. *Requests* the Secretary-General to prepare, in collaboration with the United Nations Conference on Trade and Development and other competent bodies, and up-to-date study of the matters referred to in the memorandum dated 14 January 1958, prepared by the Secretariat, on the question of free access to the sea of land-locked countries and to supplement that document, in the light of the events which have occurred in the meantime, with a report on the special problems of land-locked countries relating to the exploration and exploitation of the resources of the sea-bed and the ocean floor, and the subsoil thereof, beyond the limits of national jurisdiction;

2. Requests the Secretary-General to submit the above-mentioned study to the enlarged Committee on the Peaceful Uses of the Sea-Bed and the Ocean Floor beyond the Limits of National Jurisdiction for consideration at one of its 1971 sessions, so that appropriate measures may be

evolved within the general framework of the law of the sea, to resolve the problems of land-locked countries;

3. *Requests* the Committee to report on this question to the General Assembly at its twenty-sixth session.

1933rd plenary meeting,
17 December 1970.

C

The General Assembly,

Recalling its resolutions 798 (VIII) of 7 December 1953, 1105 (XI) of 21 February 1957 and 2574 A (XXIV) of 15 December 1969,

Recalling further its resolutions 2340 (XXII) of 18 December 1967, 2467 (XXIII) of 21 December 1968 and 2574 (XXIV) of 15 December 1969,

Taking into account the results of the consultations undertaken by the Secretary-General in accordance with paragraph 1 of resolution 2574 A (XXIV), which indicate widespread support for the holding of a comprehensive conference on the law of the sea,

Conscious that the problems of ocean space are closely interrelated and need to be considered as a whole,

Noting that the political and economic realities, scientific development and rapid technological advances of the last decade have accentuated the need for early and progressive development of the law of the sea, in a framework of close international co-operation,

Having regard to the fact that many of the present States Members of the United Nations did not take part in the previous United Nations conferences on the law of the sea,

Convinced that the elaboration of an equitable international régime for the sea-bed and the ocean floor, and the subsoil thereof, beyond the limits of national jurisdiction would facilitate agreement on the questions to be examined at such a conference,

Affirming that such agreements on those questions should seek to accommodate the interests and needs of all States, whether land-locked or coastal, taking into account the special interests and needs of the developing countries, whether land-locked or coastal,

Having considered the report of the Committee on the Peaceful Uses of the Sea-Bed and the Ocean Floor beyond the Limits of National Jurisdiction,

Convinced that a new conference on the law of the sea would have to be carefully prepared to ensure its success and that the preparatory work ought to start as soon as possible after the conclusion of the twenty-fifth session of the General Assembly, drawing on the experience already accumulated in the Committee on the Peaceful Uses of the Sea-Bed and the Ocean Floor beyond the Limits of National Jurisdiction and using fully the opportunity provided by the United Nations Conference on the Human Environment, to be held in 1972, to further its work,

1. *Notes with satisfaction* the progress made so far towards the elaboration of the international régime for the sea-bed and the ocean floor, and the

subsoil thereof, beyond the limits of national jurisdiction through the Declaration of Principles Governing the Sea-Bed and the Ocean Floor, and the Subsoil Thereof, beyond the Limits of National Jurisdiction, adopted by the General Assembly on 17 December 1970;

2. *Decides* to convene in 1973, in accordance with the provisions of paragraph 3 below, a conference on the law of the sea which would deal with the establishment of an equitable international régime—including an international machinery—for the area and the resources of the sea-bed and the ocean floor, and the subsoil thereof, beyond the limits of national jurisdiction, a precise definition of the area, and a broad range of related issues including those concerning the régimes of the high seas, the continental shelf, the territorial sea (including the question of its breadth and the question of international straits) and contiguous zone, fishing and conservation of the living resources of the high seas (including the question of the preferential rights of coastal States), the preservation of the marine environment (including, *inter alia,* the prevention of pollution) and scientific research;

3. *Decides further* to review, at its twenty-sixth and twenty-seventh sessions, the reports of the Committee referred to in paragraph 6 below on the progress of its preparatory work with a view to determining the precise agenda of the conference on the law of the sea, its definitive date, location and duration, and related arrangements; if the General Assembly, at its twenty-seventh session, determines the progress of the preparatory work of the Committee to be insufficient, it may decide to postpone the conference;

4. *Reaffirms* the mandate of the Committee on the Peaceful Uses of the Sea-Bed and the Ocean Floor beyond the Limits of National Jurisdiction set forth in General Assembly resolution 2467 A (XXIII) as supplemented by the present resolution;

5. *Decides* to enlarge the Committee by forty-four members, appointed by the Chairman of the First Committee in consulation with regional groups and taking into account equitable geographical representation thereon;

6. *Instructs* the enlarged Committee on the Peaceful Uses of the Sea-Bed and the Ocean Floor beyond the Limits of National Jurisdiction to hold two sessions in Geneva, in March and in July-August 1971, in order to prepare for the conference on the law of the sea draft treaty articles embodying the international régime—including an international machinery—for the area and the resources of the sea-bed and the ocean floor, and the subsoil thereof, beyond the limits of national jurisdiction, taking into account the equitable sharing by all States in the benefits to be derived therefrom, bearing in mind the special interests and needs of developing countries, whether coastal or land-locked, on the basis of the Declaration of Principles Governing the Sea-Bed and the Ocean Floor, and the Subsoil Thereof, beyond the Limits of National Jurisdiction, and a comprehensive list of subjects and issues relating to the law of the sea referred to in paragraph 2 above, which should be dealt with by the conference, and draft articles on such subjects and issues;

7. *Authorizes* the Committee to establish such subsidiary organs as it deems necessary for the efficient performance of its functions, bearing in

mind the scientific, economic, legal and technical aspects of the issues involved;

8. *Requests* the Committee to prepare, as appropriate, reports to the General Assembly on the progress of its work;

9. *Requests* the Secretary-General to circulate those reports to Member States and to observers to the United Nations for their comments and observations;

10. *Decides* to invite other Member States which are not appointed to the Committee to participate as observers and to be heard on specific points;

11. *Requests* the Secretary-General to render the Committee all the assistance it may require in legal, economic, technical and scientific matters, including the relevant records of the General Assembly and specialized agencies for the efficient performance of its functions.

12. *Decides* that the enlarged Committee, as well as its subsidiary organs, shall have summary records of its proceedings;

13. *Invites* the United Nations Educational, Scientific and Cultural Organization and its Intergovernmental Oceanographic Commission, the Food and Agriculture Organization of the United Nations and its Committee on Fisheries, the World Health Organization, the Inter-Governmental Maritime Consultative Organization, the World Meteorological Organization, the International Atomic Energy Agency and other intergovernmental bodies and specialized agencies concerned to co-operate fully with the enlarged Committee on the Peaceful Uses of the Sea-Bed and the Ocean Floor beyond the Limits of National Jurisdiction in the implementation of the present resolution, in particular by preparing such scientific and technical documentation as the Committee may request.

1933rd plenary meeting,
17 December 1970.

APPENDIX 30

Treaty on the Prohibition of the Emplacement of Nuclear Weapons and Other Weapons of Mass Destruction on the Seabed and the Ocean Floor and in the Subsoil Thereof

The States Parties to this Treaty,

Recognizing the common interest of mankind in the progress of the exploration and use of the seabed and the ocean floor for peaceful purposes,

Considering that the prevention of a nuclear arms race on the seabed and the ocean floor serves the interests of maintaining world peace, reduces international tensions, and strengthens friendly relations among States,

Convinced that this Treaty constitutes a step towards the exclusion of the seabed, the ocean floor and the subsoil thereof from the arms race, and determined to continue negotiations concerning further measures leading to this end,

Convinced that this Treaty constitutes a step towards a Treaty on General and Complete Disarmament under strict and effective international control, and determined to continue negotiations to this end,

Convinced that this Treaty will further the purposes and principles of the Charter of the United Nations, in a manner consistent with the principles of international law and without infringing the freedoms of the high seas,

Having agreed as follows:

Article I

1. The States Parties to this Treaty undertake not to emplant or emplace on the seabed and the ocean floor and in the subsoil thereof beyond the maximum contiguous zone provided for in the 1958 Geneva Convention on the Territorial Sea and the Contiguous Zone any objects with nuclear weapons or any other types of weapons of mass destruction, as well as structures, launching installations or any other facilities specifically designed for storing, testing, or using such weapons.

2. The undertakings of paragraph 1 of this Article shall also apply within the contiguous zone referred to in paragraph 1 of this Article, except that within that zone they shall not apply to the coastal state.

3. The States Parties to this Treaty undertake not to assist, encourage or induce any State to commit actions prohibited by this Treaty and not to participate in any other way in such actions.

Article II

1. For the purpose of this Treaty the outer limit of the contiguous zone referred to in Article I shall be measured in accordance with the

Source: *MSA*, 5th report (1971), pp. 112–14.

provisions of Part I Section II of the 1958 Geneva Convention on the Territorial Sea and the Contiguous Zone and in accordance with international law.

2. Nothing in this Treaty shall be interpreted as supporting or prejudicing the position of any State Party with respect to rights or claims which such State Party may assert, or with respect to recognition or non-recognition of rights or claims asserted by any other State, related to waters off its coasts, or to the seabed and the ocean floor.

Article III

1. In order to promote the objectives and ensure the observance of the provisions of this Treaty, the States Parties to the Treaty shall have the right to verify the activities of other States Parties to the Treaty on the seabed and the ocean floor and in the subsoil thereof beyond the maximum contiguous zone, referred to in Article I, if these activities raise doubts concerning the fulfillment of the obligations assumed under this Treaty, without interfering with such activities or otherwise infringing rights recognized under international law, including the freedoms of the high seas.

2. The right of verification recognized by the States Parties in paragraph 1 of this Article may be exercised by any State Party using its own means or with the assistance of any other State Party.

3. The States Parties to the Treaty undertake to consult and co-operate with a view to removing doubts concerning the fulfillment of the obligations assumed under this Treaty. In the event that consultation and co-operation have not removed the doubts and there is serious question concerning the fulfillment of the obligations assumed under this Treaty, States Parties to this Treaty may, in accordance with the provisions of the Charter of the United Nations, refer the matter to the Security Council.

Article IV

Any State Party to the Treaty may propose amendments to this Treaty. Amendments shall enter into force for each State Party to the Treaty accepting the amendments upon their acceptance by a majority of the States Parties to the Treaty and thereafter for each remaining State Party on the date of acceptance by it.

Article V

Five years after the entry into force of this Treaty, a conference of Parties to the Treaty shall be held in Geneva, Switzerland, in order to review the operation of this Treaty with a view to assuring that the purposes of the Preamble and the provisions of the Treaty are being realized. Such review shall take into account any relevant technological developments. The review conference shall determine in accordance with the views of a majority of those Parties attending whether and when an additional review conference shall be convened.

Article VI

Each Party to this Treaty shall in exercising its national sovereignty have the right to withdraw from this Treaty if it decides that extraordinary events related to the subject matter of this Treaty have jeopardized the supreme interests of its country. It shall give notice of such withdrawal to all other Parties to the Treaty and to the United Nations Security Council three months in advance. Such notice shall include a statement of the extraordinary events it considers to have jeopardized its supreme interests.

Article VII

1. This Treaty shall be open for signature to all States. Any State which does not sign the Treaty before its entry into force in accordance with paragraph 3 of this Article may accede to it at any time.

2. This Treaty shall be subject to ratification by signatory States. Instruments of ratification and of accession shall be deposited with the Governments of the United States, United Kingdom, and Union of Soviet Socialist Republics which are hereby designated the Depositary Governments.

3. This Treaty shall enter into force after the deposit of instruments of ratification by twenty-two Governments, including the Governments designated as Depositary Governments of this Treaty.

4. For States whose instruments of ratification or accession are deposited after the entry into force of this Treaty it shall enter into force on the date of the deposit of their instruments of ratification or accession.

5. The Depositary Governments shall forthwith notify the Governments of all States signatory and acceding to this Treaty of the date of each signature, of the date of deposit of each instrument of ratification or of accession, of the date of the entry into force of this Treaty, and of the receipt of other notices.

6. This Treaty shall be registered by the Depositary Governments pursuant to Article 102 of the Charter of the United Nations.

Article VIII

This Treaty, the English, Russian, French, Spanish and Chinese texts of which are equally authentic, shall be deposited in the archives of the Depositary Governments. Duly certified copies of this Treaty shall be transmitted by the Depositary Governments to the Governments of the States signatory and acceding thereto.

In witness whereof the undersigned, being duly authorized thereto, have signed this Treaty.

Done in _____ at _____ this _____ day of
_____.

APPENDIX 31

Contracts Sponsored by the Marine Sciences Council

Subject	Contractor	Report No.[1]
International legal problems of ocean research.	William T. Burke, Ohio State University.	PB 177724
Law for sea's minerals.......	Louis Henkin, Columbia Law School.	PB 177725
Potential of spacecraft oceanography.	General Electric, Valley Forge, Pa.	PB 177726
Potential of aquaculture.....	American Institute of Biological Science, Washington, D.C.	PB 177767 PB 177768
Encouraging marine resource development. [2]	National Planning Association, Washington, D.C.	PB 178203
Systems analysis of specified trawler operations.	Litton Industries, Beverly Hills, Calif.	PB 178661 PB 178662
Nonmilitary needs for underwater technology.	Southwest Research Institute, San Antonio, Tex.	PB 178687
International law and fishery policy.	Paul W. Dodyk, Columbia Law School.	PB 179427
Multiple use of Chesapeake Bay.	Trident Engineering Associates, Annapolis, Md.	PB 179844
Economic potential of U.S. continental margin.	Economic Associates, Inc., Washington, D.C.	PB 180118
Management of marine data systems, Phase I.	System Development Corp., Santa Monica, Calif.	AD 673992 673993
Evaluation of marine resource statistics.	Surveys & Research Corp., Washington, D.C.	
Outline of marine legal conflicts.	William L. Griffin, Washington, D.C.	
International Indian Ocean Expedition.	Robert G. Snider, State College, Pa.	
Legal aspects of coastal land-sea interface.	Albert Garretson, New York University Law School.	PB 179428
Legal aspects of Great Lakes resources.	.....do....................	PB 186000
Seminar on liability aspects of marine activities.	American Trial Lawyers Association, New York, N.Y.	
Multiple use of the Greater Seattle Harbor.	Management & Economics Research, Inc., Palo Alto, Calif.	PB 183026
Conference on future of fishing industry.	University of Washington, Seattle, Wash.	
Multiple use of Lakes Erie and Superior.	National Planning Association, Washington, D.C.	PB 185163
Multinational investments in marine sciences.	Institute of Politics and Planning, Washington, D.C.	PB 182437
Catalogue of marine research. [3]	Smithsonian Institution, Washington, D.C.	

See footnotes at end of table.

Source: *MSA*, 4th report (1970), pp. 228–29.

Subject	Contractor	Report No.[1]
Science and engineering aspects of Decade of Ocean Exploration.	National Academies of Sciences and Engineering, Washington, D.C.	PB 183679
Gulf of Mexico Research and Environmental Program.	Gulf Universities Research Corp., Houston, Tex.	PB 183680
Economic aspects of selected ocean related activities.	Massachusetts Institute of Technology.	
Collection and analysis of information in support of the gulf environment measurement program.	Florida Institute of Oceanography.	
Federal Planning for U.S. Participation in the International Decade of Ocean Exploration.	National Academies of Sciences and Engineering, Washington, D.C.	
Alternative international seabed regimes governing development of nonliving resources.	The Brookings Institution, Washington, D.C.	
Intergovernmental Relations and the National Interest in the U.S. Coastal Zone.	Harold F. Wise & Associates, Washington, D.C.	PB 184212
Management of Marine Data Systems, Phase II.	System Development Corp., Santa Monica, Calif.	AD 699125 (Vol 1). AD 699126 (Vol 2).

[1] Reports available from the clearinghouse for National Technical Information Service, Springfield, Va. 22151.

[2] Jointly sponsored with National Science Foundation.

[3] For sale by the Superintendent of Documents, U.S. Government Printing Office, Washington, D.C. 20402.

APPENDIX 32

Key Events in the Development of U.S. Marine Affairs Policy

1790		Elements of Coast Guard established in federal government
1807		Coast Survey, first ocean related agency, established
1842		Study of oceans begun by U.S. Navy (publication in 1855 of *The Physical Geography of the Sea*, by Matthew Fontaine Maury)
1872–76		Cruise of HMS *Challenger*, first oceanographic expedition
1899		U.S. Army Corps of Engineers assigned responsibility for control of navigation and pollution of rivers and estuaries
1927		First study of oceanography by National Academy of Sciences, leading to private endowments to three institutions
1945	Sept. 28	Truman Proclamation 2667 extended U.S. sovereignty over submarine resources on continental shelf
1956	Aug. 9	Office of Naval Research (with 3 other Federal agencies) requested National Academy of Sciences study of oceanographic needs and opportunities (NAS Committee on Oceanography appointed Nov. 10, 1957, chaired by Harrison S. Brown)
		Informal Coordinating Committee on Oceanography initiated by Navy
1957	July 1	Beginning of International Geophysical Year (IGY)
	Oct. 4	Soviet Union launched Sputnik
	Nov. 7	Dr. James R. Killian, Jr., appointed by President Eisenhower as Special Assistant for Science and Technology in White House
	Nov. 22	President's Science Advisory Committee (PSAC) upgraded from advisory unit in Office of Defense Mobilization
1958	Jan. 27	S. 3126 introduced by Senator Hubert H. Humphrey regarding federal organization for science and technology
	April 29	First session concluded at UN Conference on Law of the Sea at Geneva with conventions drafted on the continental shelf, territorial sea and contiguous zone, high seas, and fishing and conservation of living resources
	May 28	Dr. George Kistiakowsky replaced Killian

	July 28	National Aeronautics and Space Administration and National Aeronautics and Space Council created by P.L. 85–568
1959	Jan. 1	Navy released report on *Ten Years in Oceanography* (TENOC)
	Feb. 15	NAS Committee on Oceanography released report *Oceanography 1960–1970* recommending doubled funding over next ten years
	Feb. 17	Special Subcommittee on Oceanography appointed in House Merchant Marine and Fisheries Committee (upgraded to Standing Committee February 15, 1961)
	March 13	Federal Council for Science and Technology (FCST) established by Executive Order 10807, upgrading Interdepartmental Committee for Scientific Research and Development; Robert Kreidler appointed executive secretary.
	April 13	H.R. 6298 introduced by Congressman Overton Brooks to amend NSF Act of 1950 to provide categorical grants in oceanography
	May	FCST Subcommittee on Oceanography appointed, upgraded in 1960 to Interagency Committee on Oceanography (ICO) to coordinate programs of 15 agencies engaged in oceanographic research. Assistant Secretary of Navy James H. Wakelin, Jr. appointed chairman
	June 22	Senate Resolution 136 introduced by Senators Warren G. Magnuson, Clair Engle and Henry M. Jackson to strengthen oceanography; unanimously passed by Senate
	Sept. 5	Marine Sciences and Research Act of 1960, S. 2692, introduced by Senator Warren G. Magnuson (with 12 others) to strengthen field of oceanography with coordination through an NSF Division of Marine Sciences; passed by Senate June 25, 1960
1960	April 15	Geographic limitations on operation of U.S. Coast and Geodetic Survey removed by P.L. 86–409 (based on S. 2482 of August 4, 1959) ICO assembled first government-wide budgets in oceanography. Mr. Robert B. Abel appointed ICO executive secretary
	July 1	House Science and Astronautics Committee released report *Ocean Sciences and National Security* urging fourfold increase of funding Intergovernmental Oceanographic Commission (IOC) created in UNESCO
1961	Jan. 20	Inauguration of President John F. Kennedy; Dr. Jerome B. Wiesner appointed President's science advisor
	Feb. 9	S. 901 similar to S. 2692 introduced by Magnuson; passed by Senate July 28
	Feb. 13	Oceanographic Act of 1961, H.R. 4276, intro-

		duced by Congressman George P. Miller to expand and develop aquatic resources by creating Cabinet-level National Oceanographic Council
	Feb. 23	Special message by President Kennedy on Natural Resources, emphasizing oceanography at a presidential level for the first time
	March 29	Executive Communication 734 transmitted to the Congress by President Kennedy, sharply increasing oceanography budget
	July 18	Edward Wenk, Jr., appointed Executive Secretary of FCST
	Oct. 5	P.L. 87–396 extended Coast Guard authority to engage in oceanographic research
		International Indian Ocean Expedition begun
1962	June 9	Office of Science and Technology (OST) created in Executive Office of the President by Reorganization Plan 2—as staff function for President's science advisor, PSAC and FCST
	July 18	Oceanographic Act of 1962, H.R. 12601, introduced by Congressman John Dingell to establish national policy in oceanography, with OST responsibility for coordination
	Sept. 27	House and Senate passed S. 901 with language of H.R. 12601, but pocket vetoed by President Kennedy
1963	Jan. 9	H.R. 13 identical to S. 901 as passed by 87th Congress introduced by Congressman Herbert C. Bonner.
	April 19	Loss of nuclear submarine *Thresher*
	June 12	Oceanographic Act of 1963, H.R. 6997, introduced by Congressman Alton Lennon to overcome objections leading to presidential veto, passed by House August 5
	July 26	FCST released 10-year projections in oceanography, emphasizing uses of sea for the first time
	Sept. 11	Creation of House Select Committee on Government Research, chaired by Carl Elliott
	Nov. 22	President Kennedy assassinated, succeeded by Lyndon B. Johnson
1964	January	Dr. Donald F. Hornig replaced Wiesner
	March 19	FCST/ICO report on oceanography transmitted to Congress for the first time by the president
	April 15	H.R. 10904 to create independent National Oceanographic Agency introduced by Congressman Bob Wilson
	June 10	Continental Shelf Convention went into force (U.S. Senate ratification May 26, 1960)
	July 9	S. 2990 to create National Oceanographic Council at Cabinet level; patterned after Space Council, introduced by Senator Magnuson
1965	Jan. 11	H.R. 2218 introduced by Congressman Lennon, identical to H.R. 6997 of 88th Congress

	Jan. 20	President Lyndon B. Johnson inaugurated
	Feb. 1	S. 944 similar to S. 2990 to create Council introduced by Senator Magnuson
	Feb. 10	S. 1091 to explore and exploit continental shelf through AEC-type agency introduced by Senator E. L. Bartlett
	June 15	H.R. 9064 to establish National Commission on Oceanography introduced by Congressman Paul Rogers
	June 29	S. 2218 to establish contiguous fishery zone of 12 miles introduced by Senators Bartlett, Magnuson and Edward Kennedy
	July 7	S. 2251 to establish Cabinet-level Department of Marine and Atmospheric Affairs introduced by Senator Edmund S. Muskie
	July 13	Environmental Science Services Administration created by Reorganization Plan 2, to consolidate Coast and Geodetic Survey, Weather Bureau and Environmental elements of National Bureau of Standards, headed by Dr. R. M. White
	Aug. 5	S. 944 passed Senate as amended; House passed its version of S. 944, including H.R. 9064
	Aug. 19	S. 2439 Sea Grant legislation introduced by Senator Claiborne Pell
1966	March 7	ECOSOC Resolution 1112 (XL) requested UN Secretary General prepare study of seabed resources
	June 17	Marine Resources and Engineering Development Act of 1966 passed June 2 and signed into law, P.L. 89–454, setting policy and creating interim Cabinet-level National Council on Marine Resources and Engineering Development chaired by Vice President, and Commission on Marine Science, Engineering and Resources
	July 13	President Johnson opposed colonialism on seabed in speech at commissioning of *Oceanographer*, activated Council and released PSAC report on *Effective Use of the Sea* which recommended consolidation of ocean, atmosphere, and geophysical agencies
	July 22	Water Resources Planning Act signed, P.L. 89–80, creating Water Resources Council chaired by the Secretary of the Interior
	July 26	H.R. 16559 to amend P.L. 89–454 to include Sea Grant introduced by Congressman Rogers
	Aug. 13	White House announced appointment of Edward Wenk, Jr., as Executive Secretary of Marine Sciences Council
	Aug. 17	Vice President Hubert H. Humphrey convened first of fourteen meetings of Marine Council
	Oct. 15	Sea Grant enacted, P.L. 89–688. Robert Abel

		subsequently appointed by NSF director as head of Sea Grant Office
	Oct. 15	Department of Transportation created by P.L. 89–670 with transfer of Coast Guard from Treasury
	Nov. 3	P.L. 89–753, Clean Water Restoration Act signed
	Dec. 6	UNGA Resolution 2172 (XXI) passed requesting UN Secretary General to study national efforts in marine sciences and arrangements for cooperation
		89th Congress considered 79 bills related to marine science affairs, compared to 6–8 in three prior congresses
1967	Jan. 9	President Johnson appointed Marine Commission provided for in P.L. 89–454, with Julius A. Stratton as chairman
	March 7	H.R. 6698 on technology assessment introduced by Congressman Emilio Q. Daddario
	March 9	First Annual Marine Affairs report to Congress by President Johnson, with Report of Council announcing nine initiatives on international cooperation, fish protein concentrate, Sea Grant, data systems, estuarine studies, continental shelf surveys, ocean observation and prediction, deep ocean technology and sub-polar research, with increased funding of 13 percent
	March 18	*Torrey Canyon* oil tanker disaster on British Southwest coast
	March 25–30	Marine Council Chairman and Executive Secretary opened discussions on international cooperation with France, United Kingdom, Italy, Netherlands, Germany and Belgium
	July 27	Humphrey opened issue on coastal management in speech at University of Rhode Island
	Aug. 18	Ambassador Arvid Pardo of Malta proposed title to seabed resources beyond national jurisdiction be vested in UN
	August	Marine Council established formal committee structure
	Dec. 18	UN passed Resolution 2340 (XXII) creating an ad hoc Committee on Seabed
1968	Jan. 17	President Johnson urged accelerated international exploration in State of Union Address
	March 8	President Johnson proposed International Decade of Ocean Exploration
	March 11	Second presidential message and Council report on marine affairs released, advancing twelve initiatives, with new emphasis on coastal zone. President Johnson proposed to increase marine science budgets to $516.2 million
	June 7	President issued contingency plan concerned

		with containment, cleanup and liability for oil spills
	June 10	Council's Executive Secretary discussed Decade with officials in U.S.S.R., West Germany, Norway, U.K.
	June 15	National Petroleum Council issued draft report advocating U.S. assert jurisdiction over continental shelf and slope (released March, 1969)
	June 28	U.S. introduced principles for seabed resource development at UN Committee on Seabed
	July	Eighteen Nation Disarmament Committee of UN opened consideration of Seabed Arms Control
	Aug. 3	P.L. 90–454 signed requiring study of nation's estuaries
	Oct. 30	Presidential candidate Richard M. Nixon released statement supporting ocean affairs
	Dec. 15	President Nixon's letter to Congressman Charles A. Mosher released indicating intent to sustain momentum in marine affairs by retention of Council and staff
	Dec. 18	UNGA Resolution 2467 (XXIII) created standing committee on Seabed, urged prevention of pollution, supported International Decade and opened question of international seabed machinery
1969	Jan. 9	Stratton Commission released report *Our Nation and the Sea* recommending consolidation of numerous federal agencies into independent National Oceanic and Atmospheric Agency, and appointment of presidential advisory committee
	Jan. 17	Third presidential message on marine affairs and Council report released, with emphasis on coastal management, the Decade, improved framework of international sea law
	Jan. 20	President Richard M. Nixon inaugurated; Dr. Lee A. Dubridge appointed President's science advisor
	Jan. 28	Santa Barbara oil spill
	Feb. 23	Vice President Spiro T. Agnew requested by the President to have Marine Council review Commission proposals
	Feb. 26	The Vice President convened first of three Council meetings under new administration
	March 18	President Nixon announced U.S. position on seabed disarmament
	April 5	President's Council on Executive Organization appointed, with Roy L. Ash as chairman
	May 16	NAS/NAE released report prepared at Council's request on Decade program, *An Oceanic Quest*
	May 23	Marine Council created Committee on Policy Review at assistant secretary level with Council's Executive Secretary as chairman

	Aug. 8	S. 2802 on coastal zone management assigning responsibility to Marine Council introduced by Senators Magnuson and Philip A. Hart
	Oct. 10	Task group on marine affairs appointed by President Nixon with James Wakelin as chairman; report delivered December 18 and released July 9, 1970
	Oct. 19	5-point program announced by Vice President Agnew in support of oceanography, dealing with coastal management, coastal research, funding for International Decade, Great Lakes pollution and Arctic research
		Subcommittee on Oceanography created in Senate Commerce Committee, Senator Ernest J. Hollings appointed chairman
	Nov. 18	H.R. 14845 introduced on behalf of administration concerned with coastal zone management, assigning responsibility to Interior Department
1970	Jan. 1	National Environment Policy Act signed, P.L. 91–190 creating Council on Environmental Quality in Executive Office of the President and requiring preparation of environmental impact statements
	Jan.	National Estuary Study, prepared under P.L. 90–454, released
	Feb. 1	E. LeRoy Dillon replaced Wenk, as acting Executive Secretary of Marine Council
	May 23	President Nixon proposed policy to UN on legal regime concerning seabed resources, with international trusteeship zone
	June 25	Pacem in Maribus-I convocation on Malta
	June 30	ECOSOC Resolution 1537 (XLIX) called for study of coordination of international machinery in marine affairs
	July 9	President Nixon proposed establishment of National Oceanic and Atmospheric Administration in Department of Commerce by Reorganization Plan No. 4 of 1970
	Oct. 3	Environmental Protection Agency and National Oceanic and Atmospheric Administration created through Reorganization Plans 3 and 4 of 1970
	October	CEQ released report on *Ocean Dumping*
	Dec. 17	UNGA passed Resolution 2749 (XXV) on seabed principles, asserting concept of common heritage and 2750c (XXV) to convene a Law of the Sea Conference in 1973
1971	Jan. 28	President Nixon announced intention to appoint Robert M. White to head NOAA
	Feb. 4	S. 582, amended version of S. 2802 on coastal management, introduced by Senator Ernest J. Hollings, assigning responsibility to Department of Commerce

Feb. 8	Nixon proposed land use policy as framework for coastal management
Feb. 11	Treaty on seabed arms control signed by 62 nations
April 30	Marine Sciences Council terminated
May 26	S. 1963 to accelerate marine technology introduced by Senator Hollings
Aug. 3	U.S. submitted draft convention on territorial sea limits and fisheries to UN
Aug. 15	National Advisory Committee on Oceans and Atmosphere (NACOA) created by P.L. 92–125
Sept. 9	H.R. 9727 to control ocean dumping passed House (similar version passed Senate Nov. 24)
Oct. 19	President Nixon appointed NACOA with William Nierenberg as chairman
Dec. 11	NACOA held initial meeting

APPENDIX 33

Organization Chart of NOAA as of March 1972

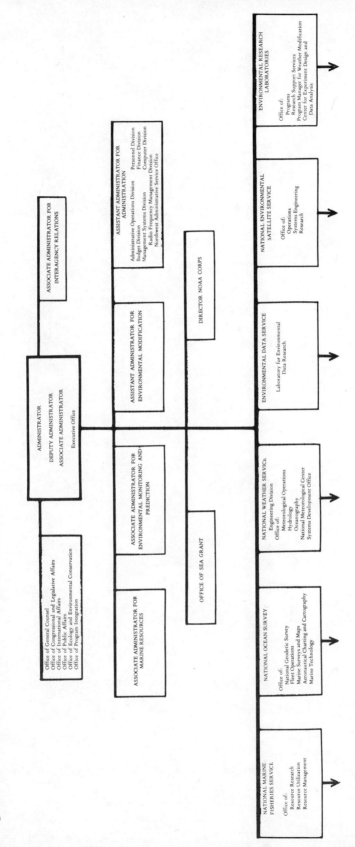

Source: U.S. Dept. of Commerce, NOAA

Notes

INTRODUCTION

1. Columbus O'Donnell Iselin, writing in the *Annual Report of the Smithsonian Institution* (Washington, 1932).

CHAPTER 1

1. Off Alaska, 550,000 sq. mi.; off the Atlantic Coast, 140,000 sq. mi.; off the Gulf Coast, 132,500 sq. mi.; off the Pacific Coast, 26,500 sq. mi.; off Hawaii, Puerto Rico, and Virgin Islands, 13,100 sq. mi. Appendix 1 lists selected countries that border extensive shallow ocean areas (less than 1000 fathom depths).

2. National Council on Marine Resources and Engineering Development, *Marine Science Affairs*, 2d report, 1968 (Washington: GPO, 1968), pp. 21–23. (Hereafter *MSA*, by report no.)

3. *MSA*, 3d report (1969), p. 5.

4. Raymond V. B. Blackman, ed., *Jane's Fighting Ships* (New York: McGraw-Hill Book Company, 1971–72), p. 668.

5. U.S. Coast Guard, *Proceedings of the Marine Safety Council*, 28, no. 12 (Washington: Dept. of Transportation, Dec. 1971).

6. *MSA*, 2d report (1968), p. 209. An inventory of American research ships is given in Appendix G of *MSA*, 4th report (1970).

7. In a study for the U.S. Senate in 1967, the Commerce Committee noted that we should regard "the appearance of any new and vigorous competitor in the shipping field not as a threat but rather as a challenge toward the exertion of the best efforts of all concerned in the direction of a further improvement in the quality, as well as a further reduction in the cost, of the services provided by the maritime industry to the world community." U.S. Senate, Committee on Commerce, *The Soviet Drive for Maritime Power* (Washington: GPO, 1967), p. 33.

8. *MSA*, 3d report (1969), p. 25. Projections for 1975 were 40 million would turn to the sea to swim, 16 million to fish, and 50 million to cruise and sail.

9. *MSA*, 4th report (1970), p. 65.

10. E. B. Potter and Chester W. Nimitz, eds., *Seapower* (Englewood Cliffs, N.J.: Prentice-Hall, 1960), chap. 2.

11. Coincidentally, it was through his responsibility for design of *Aluminaut* for oceanographic research that the author first came in contact with marine sciences.

12. Dynamic station keeping is accomplished by continuous operation of four propellers at whatever speeds are necessitated by sea conditions to prevent horizontal movements. The latter benefited from design of a drilling rig for a scientific try at obtaining a core from the Mohorovicic discontinuity, the interface between the earth's familiar crust and the yet-unseen outer mantle of the earth's core three miles below the crust. "Project Mohole" was begun in 1957 and aborted in 1966.

13. The United States has more small research submarines than all other nations of the world combined; the same is true of offshore drilling rigs.

14. UN Document E/4487.

15. Under "Oceans" and "Oceanography," the *New York Times* Index records the following number of entries: 1956, 7; 1958, 20; 1960, 22; 1962, 24; 1964, 32; 1966, 60; 1968, 79; 1969, 126; 1970, 123. The leveling off after 1969 may be as significant as the sharp increase earlier.

CHAPTER 2

1. U.S. Congress, House, *Ocean Sciences and National Security*. Report of Committee on Science and Astronautics, 86th Congress, 2d sess., July 1960 (Washington: GPO, 1960), pp. 14–15. (Hereafter Congress, *OSNS*.)

2. When interviewed by the author, Dan Markel, a staff specialist in oceanography for the Senate Commerce Committee, and John Drewry, staffing the House Merchant Marine and Fisheries Committee, reported this characterization of the field was jokingly admitted by its leaders as late as 1959.

3. Letter from Chief of Naval Reserve to National Academy of Sciences, cited in U.S. Congress, House, *Study of the Effectiveness of the Committee on Oceanography of the Federal Council for Science and Technology*. Hearings before the Subcommittee on Oceanography, House Committee on Merchant Marine and Fisheries, 87th Congress, 2d sess., February, March, 1962 (Washington: GPO, 1962), p. 23.

4. To strengthen these advisory capabilities during World War I, the National Research Council was founded as a research adjunct to the Academy to play administrative housemother for a proliferating committee structure. During World War II, that advisory capability waas deemed too weak and slow moving to meet the government's needs, and the Office of Scientific Research and Development was created in the Executive Office of the President, having as its own advisory machinery the National Defense Research Committee. When these were metamorphosed after the war into full-time governmental research units of the Office of Naval Research and the National Science Foundation, the NAS/NRC remained as the twin engines of outside advice to all agencies and echelons of the Federal Government. Since 1966, a National Academy of Engineering and an Institute of Medicine were added to engage in similar activities in their respective fields. Negotiations for advice are almost always initiated by the client agency, who formulates the key questions and pays the bill; then follows NAS/NRC

appointment of a committee of specialists to consider the issue, collect the facts, and prepare a report setting forth their findings and recommendations. While some ad hoc panels are dissolved after publication of the report, many committees stay in business to respond to succeeding waves of questions in the same area or to undertake inquiries on their own initiative.

5. P.L. 85–567. This arrangement was changed in 1961 by P.L. 87–26 as follows: the Vice President was substituted for the President as Chairman of the Council and one miscellaneous federal and three public members were removed.

6. U.S. Congress, House, *The National Science Foundation: A General Review of Its First Fifteen Years*. Report of Committee on Science and Astronautics, 89th Congress, 2d sess., January 1966 (Washington: GPO, 1966), p. 32. The Kennedy administration proposed an even further jump —from $360 million to $589 million—for 1964, but Congress granted only $354 million.

7. National Academy of Sciences. Report in 12 parts. (Washington: NAS, 1959.) (Hereafter NAS, *Oceanography*.)

8. Congress, *OSNS*, p. 122. A problem arises as to the scope of activity embraced by NASCO as compared to later tabulations by ICO and the Marine Council that included classified marine research and applied components related to uses of the sea.

9. NAS, *Oceanography*, p. 8.

10. Don Krasher Price, *The Scientific Estate* (Cambridge: Harvard University Press, 1965), pp. 211–12.

11. Navy support for marine sciences does not include their much larger funding for R&D concerned with ship, aircraft, or weapon development. Except for oceanographic vessels, it does not include funds for combat ship design and construction. Also, Maritime Administration funds cover only R&D on ships and not funds for subsidy of ship construction or operation. Thus the total funding for ocean-related activities is far greater than for marine science components alone. Nevertheless, it was the cutting edge of information about the marine environment that lagged. National Council on Marine Resources and Engineering Development, *Marine Science Affairs*, 5th report, 1971 (Washington: GPO, 1968), p. 91. (Hereafter MSA, by report no.)

12. The size of the U.S. merchant fleet, its periodic replenishment and decay, is well documented in the following schedule of tonnage:

1789	123,900	1840	762,800	1890	928,100	1940	3,637,600
1800	667,100	1850	1,439,700	1900	816,800	1950	8,353,000
1810	981,000	1860	2,379,400	1910	782,500	1960	9,610,000
1820	583,700	1870	1,448,800	1920	9,924,700	1970	9,780,000
1830	537,600	1880	1,314,400	1930	6,295,900		

U.S. Congress, *OSNS*, p. 10, and after 1960, *Merchant Fleets of the World*, U.S. Dept. of Commerce, Maritime Administration (Washington: GPO, Dec. 31, 1960, and 1970, same source, Dec. 31, 1970.)

13. Congress, *OSNS*, p. 97.

14. *Congressional Record*, February 17, 1959, pp. 2279–80.

15. U.S. Congress, House, *Frontiers in Oceanic Research*. Hearings before Committee on Science and Astronautics, 86th Congress, 2d sess., April 1960 (Washington: GPO, 1960), pp. 5–6. (Hereafter Congress, *Frontiers*.)

16. Attempts to discharge the Committee through the Legislative Reorganization Act of 1970 were beaten back.

17. U.S. Congress, House, *Oceanography in the U.S.* Hearing before the Subcommittee on Oceanography, Committee on Merchant Marine and Fisheries, 86th Congress, 2d sess., February 1960 (Washington: GPO, 1960).

18. U.S. Congress, Senate Resolution No. 136. 86th Congress, 1st sess., June 22, 1959.

19. U.S. Congress, Senate Bill No. 2482. 86th Congress, 1st sess., August 4, 1959.

20. U.S. Congress, House, *The Office of Science and Technology.* Report prepared by the Legislative Reference Service for the Military Operations Subcommittee of the Committee on Government Operations, March 1967 (Washington: GPO, 1967), pp. 218–19. (Hereafter Congress, *OST*).

21. U.S. Congress, Senate, *Science Program—86th Congress.* Senate Report No. 120, Committee on Government Operations. 86th Congress, 1st sess., March 1959 (Washington: GPO, 1959), p. 117.

22. S. 3126 of 85th Congress; S. 676 of 86th Congress.

23. The TENOC recommendations were consistent with NASCO in identifying needs for study of the oceans and of expansion required to meet these needs in the ten private oceanographic laboratories funded largely by Navy contracts. From a 1959 base of $7.6 million, support was projected to grow roughly $2 million per year, spiced with another $64 million for 18 ships, shore facilities, and piers. Congress, *OSNS*, pp. 127–29.

24. H.R. 9361 of January 6, 1960.

25. H.R. 10412 of February 15, 1960.

26. Congress, *OSNS*.

27. In 1966 I was to discover firsthand by discussions with senior Soviet officials how uncommitted were their efforts to a major thrust in oceanography.

28. Congress, *OSNS*, pp. 174–75.

29. U.S. Congress, Senate, *Marine Science.* Hearings before the Committee on Interstate and Foreign Commerce, 87th Congress, 1st sess., March and May, 1961 (Washington: GPO, 1961).

30. U.S. President (Eisenhower), Science Advisory Committee Report, White House (Washington: GPO, 1958).

31. U.S. President (Truman), Executive Order No. 9912, December 1947, and U.S. President (Eisenhower), Executive Order No. 10807.

32. The precursor ICSRD organization waas explicitly continued as the Standing Committee of the newly created multiagency body over it, to serve as a negotiating agent of the Council at the bureau-chief level.

33. U.S. President (Truman), *Science and Public Policy: A Program for the Nation,* August 27, 1947.

34. ICO members for FY 1965 are as follows: James H. Wakelin, Jr., Chairman, Assistant Secretary of the Navy (Research and Development); H. Arnold Karo, U.S. Coast and Geodetic Survey; Donald L. McKernan, Bureau of Commercial Fisheries; R. D. Schmidtman, U.S. Coast Guard; Ragnar Rollefson, Department of State; Harve J. Carlson, National Science Foundation; Harry G. Hanson, Public Health Service; John N. Wolfe, Atomic Energy Commission; I. E. Wallen, Museum of Natural History, Smithsonian Institution; Edward Wenk, Jr. (Observer), Office of Science and Technology; Robert Fleagle (Observer), Office of Science and Technology; Enoch L. Dillon (Observer), Bureau of the Budget; Athelstan

Spilhaus (Observer), National Academy of Sciences Committee on Oceanography; Robert B. Abel, Department of the Navy (Executive Secretary).

35. Personal interview, October 26, 1970.

36. Theodore C. Sorensen, *Decision-Making in the White House* (New York: Columbia University Press, 1963), and Jerome B. Wiesner, *Where Science and Politics Meet* (New York: McGraw-Hill Book Company, 1965).

37. U.S. President (Kennedy), *Message to the Congress on Natural Resources*, February 23, 1961. A fuller quotation is to be found in Appendix 28.

38. U.S. Congress, Senate, *Coordination of Information on Current Scientific Research and Development Supported by the United States Government*. Report to Committee on Government Operations, 87th Congress, 1st sess., April 1961 (Washington: GPO, 1961). At hearings on this issue Humphrey became so miffed at Department of Defense recalcitrance about filing DOD information centrally that he took the most direct road of gaining cooperation. Working through friends in the House of Representatives, he had the Appropriations Committee inject a rider requiring DOD's participation in a government-wide project registry as a condition to its receiving research and development funds.

39. Paradoxically, however, it was Dingell who eight years later attempted to defeat the creation of NOAA.

40. Richard Fenno, *The President's Cabinet* (Cambridge: Harvard University Press, 1959).

41. George E. Reedy, *Twilight of the Presidency* (New York: World Publishing Co., 1970).

42. *Science* 142 (Nov. 22, 1963): editorial page.

43. Harold Seidman, *Politics, Position, and Power* (New York: Oxford University Press, 1970), p. 215; also *Science* 141 (July 5, 1963):27.

44. Capt. Steven N. Anastasion, USN, "Oceanography and the Government," in *Ocean Sciences* (Annapolis, Md.: U.S. Naval Institute, 1964).

45. That science-related programs crossed agency lines and could not be adequately represented by the structure of budget presentation was recognized by the BOB when they instituted a special analysis in the early 1960s to report scientific categories of oceanography, atmospheric sciences, and water resources research on a government-wide basis. Marine affairs was dropped, however, in FY 1971.

46. U.S. President (Kennedy), *A Message from the President of the United States Transmitting Reorganization Plan No. 2 of 1962 Providing for Certain Reorganizations in the Fields of Science and Technology* in *Weekly Compilation of Presidential Documents* (Washington: GPO, March 29, 1962).

47. The procedure is for the President to submit a reorganization plan that cannot be modified and becomes law unless vetoed by Congress within 90 working days.

48. H.R. 455, introduced July 24, 1963, by Representative Carl Elliott, followed by fourteen colleagues. The committee was authorized by H.R. 504 on September 11, 1963, by unanimous vote.

49. U.S. Congress, House, *Statement by Edward Wenk, Jr., Executive Secretary Federal Council for Science and Technology*, before the Select Committee on Government Research, 88th Congress, 2d sess., November 19, 1963 (Washington: GPO, 1963).

50. U.S. Congress, House, *Interagency Coordination in Research and Development*. Report of the Select Committee on Government Research,

88th Congress, 2d sess., December 1964 (Washington: GPO, 1964), pp. 35–54.

51. Price, *The Scientific Estate*, p. 140.

52. Federal Council for Science and Technology, *Oceanography—The Ten Years Ahead*. ICO Pamphlet no. 10 (Washington: GPO, June 1963).

53. Ibid., p. 7.

54. U.S. President (Johnson), *Letter to the President of the Senate and to the Speaker of the House Transmitting Reports on Oceanographic Research*, March 19, 1964. Public Papers of the Presidents of the United States (Washington: GPO, 1965).

55. U.S. Congress, House, *Creation of the Office of Science and Technology (Reorganization Plan No. 2, 1962)*, 87th Congress, 2d sess., May 15, 1962 (Washington: GPO, 1962). U.S. Congress, Senate, *Organizing for National Security: Science Organization and the President's Office*. Subcommittee on National Policy Machinery, Committee on Government Operations, 87th Congress, 1st sess., 1961.

56. H.R. 10904 of April 15, 1964.

57. Daniel Greenberg, *Politics of Pure Science* (New York: The New American Library, 1967), chap. 9.

58. Price, *The Scientific Estate*, chap. 7.

59. H.R. 921 of January 4, 1965.

60. H.R. 2218 on January 11, 1965; also H.R. 3310 by Thomas M. Pelly and H.R. 3352 by Herbert C. Bonner.

61. S. 1091 of February 10, 1965. "This bill would establish policy to accelerate exploration and development of physical, chemical, geological, and biological resources of the Continental Shelf, to encourage private investment in utilization of its resources, to determine benefits and to disseminate information on resources and to develop an engineering capability to operate on and above the Continental Shelf."

62. H.R. 6457 of March 18, 1965 entitled "National Oceanographic Act of 1965."

63. H.R. 5175 of February 18, 1965, similar to H.R. 11232 by Richard Hanna, May 13, 1964.

64. U.S. Congress, Senate, *Hearings on S. 944*. Report of the Committee on Commerce, 89th Congress, 1st sess., February 19, 1965 (Washington: GPO, 1965), pp. 9–30.

65. U.S. President (Johnson), *Effective Use of the Sea* (Washington: GPO, July 13, 1966).

66. H.R. 5654 of March 2, 1965, introduced by Dante B. Fascell; H.R. 6512 of March 18, 1965, by James G. Fulton; H.R. 7301 of April 8, 1965, by Richard Hanna; H.R. 7798 of May 3, 1965, by J. Oliva Huot.

67. S. 2439 of August 19, 1965, to amend the National Science Foundation Act.

68. An account of Congressional initiatives is given in U.S. Congress, House, *Abridged Chronology of Events Related to Federal Legislation for Oceanography 1956–66*, prepared by Miss Florence Broussard, Legislative Reference Service, as a report of the Committee on Merchant Marine and Fisheries, 89th Congress, 2d sess., January 9, 1967 (Washington: GPO, 1967).

CHAPTER 3

1. Luther J. Carter, "Oceanography: Congress Wants Cabinet Council and Study," *Science* 152 (June 10, 1966):1490.

2. Harold Seidman, *Politics, Position, and Power* (New York: Oxford University Press, 1970), p. 67 and *passim.*

3. Johnson had instructed his 1964 task forces that he wanted to be an activist President, "not a caretaker of past gains" (ibid., p. 67).

4. The mystique of presidential appointment-making was further complicated by the fact that this was one of two extraordinary statutory positions (the other being in the Space Council) where the President makes the appointment and the Vice President is direct boss; later I was to sense the hazards of that predicament.

5. White House press release, August 13, 1966: "President Johnson today announced his intention to appoint Dr. Edward Wenk, Jr., to be Executive Secretary of the National Council on Marine Resources and Engineering Development."

6. Press release, Office of the Vice President, Washington, August 17, 1966.

7. *New York Times*, July 14, 1966, p. 10; July 17, 1966, sec. 3, p. 12.

8. Carter, *Science.*

9. National Council on Marine Resources and Engineering Development, *Marine Science Affairs*, 3d report, 1969 (Washington: GPO, 1969), p. 190. (Hereafter *MSA*, by report no.)

10. Marine Council Members:

Chairman: Hubert H. Humphrey (Aug. 17, 1966 to Jan. 20, 1969); Spiro T. Agnew (Jan. 20, 1969 to April 30, 1971).

Members: Department of State—Dean Rusk (Aug. 17, 1966 to Jan. 20, 1969); William P. Rogers (Jan. 20, 1969 to April 30, 1971). *Department of Defense*—Paul H. Nitze (Aug. 17, 1966 to Sept. 1, 1967); Paul R. Ignatius (Sept. 1, 1967 to Jan. 20, 1969); John H. Chaffee (Jan. 20, 1969 to April 30, 1971). *Department of Interior*—Stewart L. Udall (Aug. 17, 1966 to Jan. 20, 1969); Walter J. Hickel (Jan. 20, 1969 to Nov. 25, 1970); Rogers C. B. Morton (Nov. 25, 1970 to April 30, 1971). *Department of Commerce*—John T. Connor (Aug. 17, 1966 to Feb. 1, 1967); Alexander B. Trowbridge (Feb. 1, 1967 to March 1, 1968); C. R. Smith (March 1, 1968 to Jan. 20, 1969); Maurice H. Stans (Jan. 20, 1969 to April 30, 1971). *Department of Health, Education, and Welfare*—John W. Gardner (Aug. 17, 1966 to March 1, 1968); Wilbur J. Cohen (March 1, 1968 to Jan. 20, 1969); Robert H. Finch (Jan. 20, 1969 to June 23, 1970); Elliot L. Richardson (June 23, 1970 to April 30, 1971). *Department of Transportation*—Alan S. Boyd (Aug. 17, 1966 to Jan. 20, 1969); John A. Volpe (Jan. 20, 1969 to April 30, 1971). *Atomic Energy Commission*—Glenn T. Seaborg (Aug. 17, 1966 to April 30, 1971). *National Science Foundation*—Leland J. Haworth (Aug. 17, 1966 to July 10, 1969); William D. McElroy (July 11, 1969 to April 30, 1971).

Observers: National Aeronautics and Space Administration—James E. Webb (Aug. 17, 1966 to Oct. 7, 1968); Thomas O. Paine (Oct. 7, 1968 to Sept. 15, 1970); George M. Low (Sept. 15, 1970 to April 30, 1971). *Smithsonian Institution*—S. Dillon Ripley (Aug. 17, 1966 to April 30, 1971). *Agency for International Development*—William S. Gaud (Aug. 17, 1966 to Jan. 20, 1969); John A. Hannah (Jan. 20, 1969 to April 30,

1971). *Bureau of the Budget*—Charles L. Schultze (Aug. 17, 1966 to Jan. 29, 1968); Charles J. Zwick (Jan. 29, 1968 to Jan. 20, 1969); Robert P. Mayo (Jan. 20, 1969 to June 30, 1970); George P. Schultz (June 30, 1969 to April 30, 1971). *Council of Economic Advisers*—Gardner Ackley (Aug. 17, 1966 to Feb. 15, 1968); Arthur M. Okun (Feb. 15, 1968 to Jan. 20, 1969); Paul W. McCracken (Jan. 20, 1969 to April 30, 1971). *Office of Science and Technology*—Donald F. Hornig (Aug. 17, 1966 to Jan. 20, 1969); Lee A. DuBridge (Jan. 20, 1969 to Aug. 31, 1970); Edward E. David (Aug. 31, 1970 to April 30, 1971). *Council on Environmental Quality*—Russell E. Train (May 1, 1970 to April 30, 1971). *Environmental Protection Agency*—William D. Ruckelshaus (Feb. 1, 1970 to April 30, 1971). (Within three months of their respective appointments by the President, the Vice President extended formal invitations to Mr. Train and Mr. Ruckelshaus to participate as official observers.)

Executive Secretary: Edward Wenk, Jr. (Aug. 17, 1966 to Jan. 31, 1970); E. L. Dillon (acting, Jan. 31, 1970 to April 30, 1971).

11. Requirements even included such trivia as creating a dignified letterhead, which needed approval from the Secretary of State, who, since 1781, had been guardian of the presidential seal. A zip code had to be designated (we received 20500, the White House number, rather than the less prestigious 20504 of the Executive Office).

12. A list of all studies supported on contract by the Marine Sciences Council is given in Appendix 31.

13. U.S. Congress, House, *National Marine Sciences Program*, Part 1. Hearings before the Subcommittee on Oceanography of the Committee on Merchant Marine and Fisheries, 90th Congress, 1st sess., August 17, 1967 (Washington: GPO, 1968), pp. 3–5.

14. *MSA*, 1st report (1967), pp. iii–v.

15. *MSA*, 2d report (1968), p. v.

16. Ibid., p. iii.

17. *MSA*, 3d report (1969), p. iii.

18. *MSA*, 4th report (1970), p. iii.

19. *MSA*, 1st report (1967), p. v.

20. *MSA*, 2d report (1968), p. iii.

21. Ibid., p. vi.

22. *MSA*, 3d report (1969), p. v.

23. Ibid., p. 200.

24. A friend once compared the bureaucracy to a dinosaur, not so much for its obsolescence as for its physiology—a large, slow-moving body stirred by a relatively small brain. Yet, he added, when it is lumbering along under its own momentum, an alert government official can nudge it toward a new destination. Even though the resulting angular deflection be small, ten years later that dinosaur will be in a far different place than he would have been if unmolested. The metaphor was a source of reassurance when our achievements fell short of goals.

25. Because Council deliberations invariably dealt with budget issues during the preparation stage, they were exempted as executive privilege from the Freedom of Information Act. As noted, the Council's annual reports were unique in their disclosure of unsolved problems, along with details of the President's decisions.

26. *MSA*, 1st report (1967), pp. 101, 102.

27. *MSA*, 2d report (1968), pp. 166–68.

28. *MSA*, 3d report (1969), pp. 201, 202.

29. It was not surprising that the Navy should brush aside an Executive Office assist, because they had long been used to operating with relative indifference to inquiries about their domain even from the Budget Bureau. Although the defense budget represents approximately 50 percent of the federal expenditures, discipline over these budgets had been left to the Office of the Secretary of Defense through an arrangement negotiated by McNamara with Kennedy. The President was persuaded to muzzle the Bureau and to leave the setting of internal priorities and cuts to the Secretary of Defense on grounds that he could build a more proficient group of systems analysts than could be employed by BOB to screen the complex budgetary requests generated within the services. This policy is reflected in the staffing of the Bureau itself; 95 percent of all budget examiners work on civilian affairs that constitute only 60 percent of the budget.

30. *MSA*, 1st report (1967), p. 62.

31. In FY 1972, the first year after the Council was abolished, Sea Grant, which had been transferred to NOAA, was again in trouble because of Bureau cuts on grounds that the program was weak. That time Congress intervened directly.

32. We lent support to the request to extend the Commission's life to a maximum of two years but sought to uncouple our life expectancy from theirs by extending our expiration date to June 30 instead of May 9, 1969. Reluctantly the Bureau concurred. P.L. 89–454 was accordingly amended by P.L. 90–242, signed January 2, 1968.

33. *MSA*, 1st report (1967), p. iii.

34. Compromises expected to rest halfway between House and Senate figures were shaded closer to the lower allocation of the House because Senators often could not match the arguments of better prepared House conferees.

35. Congressional Testimony by Marine Sciences Council on Government-Wide Issues (*MSA*, 4th report [1970], p. 227):

Date	Topic	Committee
10/10/66	Council program and budget for fiscal year 1967.	*House*—Subcommittee on Supplemental Appropriations.[a]
10/17/66	Council program and budget for fiscal year 1967.	*Senate*—Subcommittee on Deficiencies and Supplementals.[a]
3/13/67	Council program and budget for fiscal year 1968.	*House*—Subcommittee on Interior and Related Agencies.[a]
3/17/67	Council program and budget for fiscal year 1968.	*Senate*—Subcommittee on Interior and Related Agencies.[a]
8/17/67	Review of Council activities on first anniversary of Council.	*House*—Subcommittee on Oceanography.[b]
10/11/67	H.R. 13273, to extend deadline for Commission report and lifetime of the Council.	*House*—Subcommittee on Oceanography.[b]
11/28/67	S. 1262, to authorize Corps of Engineers shoreline study.	*Senate*—Subcommittee on Flood Control, Rivers and Harbors.[c]
2/19/68	H.R. 15224, improvements for Coast Guard, research ship.	*House*—Subcommittee on Coast Guard, C&GS, and Navigation.[a]
3/ 7/68	Council program and budget for fiscal year 1969.	*Senate*—Subcommittee on Interior and Related Agencies.[a]

Date	Topic	Committee
3/13/68	Council program and budget for fiscal year 1969.	*House*—Subcommittee on Interior and Related Agencies.[a]
4/ 9/68	H.R. 15490, to increase appropriation for FPC pilot plant.	*House*—Subcommittee on Fisheries and Wildlife Conservation.[b]
5/27/68	H.R. 11584, et al., to establish system of marine sanctuaries.	*House*—Subcommittee on Oceanography.[b]
6/24/68	H.R. 13781, to extend authorization of the Sea Grant Program.	*Senate*—Committee on Commerce.
6/26/68	S. 3030, et al., to enable BCF to proceed with FPC plant.	*House*—Subcommittee on Fisheries and Wildlife Conservation.[b]
7/29/68	H. Con. Res. 803, to express concurrence with objectives of Decade of Ocean Exploration.	*House*—Subcommittee on Oceanography.[b]
3/ 7/69	H.R. 5829, to extend lifetime of Council to June 30, 1970.	*House*—Subcommittee on Oceanography.[b]
4/ 1/69	H.R. 6495, to control oil pollution from ships, and other purposes.	*House*—Subcommittee on Oceanography.
7/ 9/69	Council program and budget for fiscal year 1969.	*Senate*—Subcommittee on the Department of the Interior and related agencies.[a]
7/22/69, 7/28/69	Centralization of Federal science activities.	*House*—Science and Astronautics, Subcommittee on Science, Research, and Development.
7/31/69	Use of marine sources of food to improve nutritional conditions of American citizens.	*Senate*—Select Committee on Nutrition and Human Needs.
9/16/69	National oceanographic program and report of Commission on Marine Science, Engineering, and Resources.	*House*—Subcommittee on Oceanography.[b]

[a] Committee on Appropriations.
[b] Committee on Merchant Marine and Fisheries.
[c] Committee on Public Works.

36. Why that subcommittee had cognizance was something of a puzzle, because all other Executive Office components reported to another subcommittee. We suspected that the clerk of the House had made the choice because of the phrase "Marine Resources" in the Council's title.

37. U.S. Congress, House, *Supplemental Appropriation Bill 1967.* Hearings before Subcommittees of the Committee on Appropriations, 89th Congress, 2d sess., Part 2 (Washington: GPO, 1966), pp. 200–220.

38. U.S. Congress, House, Hearings before Subcommittee on Oceanography, 90th Congress, 1st sess., August 17, 1967, p. 2.

39. *MSA*, 1st report (1967), p. iv.

40. *MSA*, 2d report (1968), p. iv.

41. *MSA*, 3d report (1969), p. iii.

42. *MSA*, 4th report (1970), p. iii.

43. *Washington Post*, December 19, 1964, p. 8; *Science* 146 (Dec. 25,

1964):1659; James A. Crutchfield, Robert W. Kates, and W. R. Derrick Sewell, "Benefit-Cost Analysis and the National Oceanographic Program," *Natural Resources Journal* 7, no. 3 (July 1967):361–75.

44. *Economic Benefits from Oceanographic Research*, National Academy of Sciences, National Research Council Publication 1228 (Washington, 1964).

45. *Science* 153 (July 8, 1966):149.

46. The January 10, 1969 citation read: "To Hubert H. Humphrey, Vice President of the United States, Chairman of the National Council on Marine Resources and Engineering Development from August 1966 to January 1969: Oceanographer of the Nation; 'assistant President,' mediator of science to serve public policy. . . .

"Champion of people and their concerns—who would tame the waters to feed the hungry, refresh the spirit of urban dwellers, foster a bountiful economy, and through acts of cooperation beyond mere words, turn the common interests of nations toward peace.

"Courageous voyager on turbulent bureaucratic seas—who would aid our President turn the energies of a great Government to common purpose, who would steer a determined course through the heavy weather of intragency conflict, and heal the rifts and wounds of competing agencies.

"Nimble spirit—who would employ the creative fruits of science, the muscle of contemporary technology—and the ubiquitous sea—to serve all mankind.

"We were proud to serve with you."

[signed]
Dean Rusk
Donald Hornig
Arthur M. Okun
Leland J. Haworth
Clarence F. Pautzke
Charles E. Bohlen
William S. Gaud
C. R. Smith

S. Dillon Ripley
Glenn T. Seaborg
Thomas O. Paine
Alan S. Boyd
Paul R. Ignatius
Robert A Frosch
Robert M. White
Stewart L. Udall
Edward Wenk, Jr.

47. Richard M. Nixon, "The Sea: Our Last Unexplored Frontier" in *Public Policy for the Seas*, ed. Norman J. Padelford (Cambridge: M.I.T. Press, 1970), pp. 332–38.

48. Members appointed by Vice President Agnew to the Committee for Policy Review included: Herman Pollack—State; Robert A. Frosch—Navy; Russell E. Train—Interior; Myron Tribus—Commerce; James H. Cavanaugh —Health, Education, Welfare; Willard J. Smith—Transportation; Spofford G. English—AEC; Thomas O. Jones—NSF; Homer Newell—NASA; Sidney R. Galler—Smithsonian; Glenn E. Schweitzer—AID; James R. Schlesinger— Bureau of the Budget; Luther T. Wallace—Council of Economic Advisors; John Steinhart—Office of Science and Technology; F. P. Koish—Corps of Engineers; Reuben Johnson—Water Resources Council; Edward Wenk, Jr. —(Chairman) Marine Sciences Council.

49. Section 3(d) of the Marine Sciences Act wisely provided that a council member may designate an alternate "in his unavoidable absence" only if they are policy-level officers "appointed with advice and consent of the Senate." This provision headed off the usual watering down of attendance to the alternates not empowered to speak for their principals and thus able to delay actions.

50. *MSA*, 5th report (1971), p. iii.

CHAPTER 4

1. The President's Council on Recreation and Natural Beauty, *From Sea to Shining Sea: A Report on the American Environment—Our Natural Heritage* (Washington: GPO, 1968).

2. *California and the World Ocean.* Proceedings of the California Governors' Conference, Los Angeles, January/February 1964 (Sacramento: California Office of State Printing, 1964), pp. 11–19.

3. Marine Frontiers Conference at the University of Rhode Island, July 27–28, 1967, *Proceedings* (Kingston, R.I.: England Marine Resources Information Program, 1967), pp. 5–9.

4. Department of the Interior, *National Estuary Study* (Washington: GPO, 1970), 1:8.

5. Roderick Nash, *Wilderness and the American Mind* (New Haven, Conn.: Yale University Press, 1968), p. 40.

6. National Council on Marine Resources and Engineering Development, *Report on the Seminar on Multiple Use of the Coastal Zone, Williamsburg, Virginia, November 13–15, 1968*, PB 184–836 (Springfield, Va.: National Technical Information Service, 1968), p. 15.

7. Commission on Marine Science, Engineering and Resources, *Science and Environment* (Washington: GPO, 1969), vol. 1 of panel reports, pp. III–63.

8. National Council on Marine Resources and Engineering Development, *Marine Science Affairs*, 4th report, 1970 (Washington: GPO, 1970), pp. 17–18. (Hereafter *MSA*, by report no.)

9. Panel on Study of Critical Environmental Problems, *Man's Impact on the Global Environment* (Cambridge: M.I.T. Press, 1970), pp. 126 ff.

10. *National Estuary Study*, 2:56, 57.

11. Council on Environmental Quality, *Ocean Dumping: A National Policy* (Washington: GPO, Oct. 1970), p. iii.

12. P. A. Greve, "Chemical Wastes in the Sea: New Forms of Marine Pollution," *Science* 173 (Sept. 10, 1971):1021–22.

13. *Polluting Spills in U.S. Waters—1970*, U.S. Coast Guard, mimeographed (Washington, D.C., Sept. 1971).

14. Message from President Nixon to the Congress, May 20, 1970, "Marine Pollution from Oil Spills." (Hereafter abbreviated as Nixon message, "Marine Pollution.") ·

15. *Ocean Dumping*, p. 134.

16. Commission on Marine Science, Engineering and Resources, *Our Nation and the Sea: A Plan for National Action* (Washington: GPO, 1969), p. 51.

17. Edward Wenk, Jr., "Coastal Waters and the Nation," address before the American Society of Civil Engineers, New Orleans, February 3, 1969, published in *Civil Engineering*, June 1969.

18. Estuarine zones had been defined by P.L. 89–753 as meaning an environmental system consisting of an estuary and those transitional areas consistently influenced or affected by water from an estuary such as, but not limited to, salt marshes, coastal and intertidal areas, bays, harbors, lagoons, inshore waters, and channels; and the term "estuary" means all or part of the mouth of a navigable or interstate river or stream or other body of water having unimpaired natural connection with open sea and within

which the sea water is measurably diluted with fresh water derived from land drainage.

19. S. 2802 of August 8, 1969, entitled "Coastal Zone Management Act of 1969," introduced by Senators Magnuson and Hart.

20. The texts of the four international Conventions defining national boundaries of the sea and seabed are given in Norman J. Padelford, *Public Policy for the Seas* (Cambridge: M.I.T. Press, 1970).

21. This issue is on the agenda for the 1973 Law of the Sea Conference.

22. P.L. 89–658 (Oct. 14, 1966).

23. See Appendix 21 "Convention on the Continental Shelf."

24. *Our Nation and the Sea*, p. 52. Some controversy now has arisen regarding jurisdiction over offshore islands and submerged lands between them and the coast. State laws prevail in the Great Lakes.

25. *Seminar on Multiple Use of the Coastal Zone*, p. 130.

26. Selected U.S. Milestones in Preserving the Marine Environment (*MSA*, 4th report [1970], pp. 25, 26):

Year	Legislation and governmental initiatives
1899.....	*River and Harbor Act.* Prohibited (1) discharge or deposit of refuse into any navigable waters, except that which flowed from streets and sewers in a liquid state; (2) excavation or filling in navigable waters; (3) construction of piers, dams, bridges, and similar works in harbors and navigable waters without permit from the Secretary of the Army acting through the Chief of Engineers.
1912.....	*Public Health Service Act.* Authorized surveys and studies of water pollution, particularly as it affected human health.
1924.....	*Oil Pollution Act.* Prohibited oil discharges, damaging to aquatic life, harbors and docks and recreation, into the territorial sea and navigable inland waters.
1930.....	*River and Harbor Act.* Authorized the Chief of Engineers under the direction of the Secretary of the Army to make investigations and cooperative studies with States for the purpose of preventing erosion of coastal and Great Lakes shores by waves and currents.
1945.....	*Executive Order 9634.* Provides for establishing fishery conservation zones in areas of the high seas contiguous to the coasts of the United States and allows for establishing marine wildlife sanctuaries as a fishery conservation measure.
1948.....	*First Federal Water Pollution Control Act* with a 5-year expiration date.
1953.....	*Federal Water Pollution Control Act* extended for 3 years. *Outer Continental Shelf Lands Act.* Extended the Secretary of the Army's jurisdiction concerning obstructions in navigable waters to include artificial islands and fixed structures located on the Outer Continental Shelf; authorizes the Secretary of the Interior to require the prevention of pollution in offshore oil or mining operations; the Coast Guard administers the act's safety provisions.
1956.....	*First permanent Federal Water Pollution Control Act.* Extended and strengthened the 1948 law in areas of enforcement and research and initiated grants for construction of waste treatment works.

Year	Legislative and governmental initiatives
1958.....	*River and Harbor Act.* Authorized a comprehensive project to provide for control and erradication of obnoxious aquatic plant growth in navigable waters, their tributaries, and allied waters in 8 States.
	Fish and Wildlife Coordination Act. Requires consultation with the U.S. Fish and Wildlife Service and the responsible State agency whenever the waters of any stream or body of water are controlled or modified.
1961.....	*Federal Water Pollution Control Act Amended.* Further strengthened enforcement authority and increased support for construction of municipal waste treatment works and research; authorized storage of Corps of Engineers and other Federal reservoirs for the regulation of stream flow for the purpose of water quality control.
1965.....	*Water Quality Act, further amending the Federal Water Pollution Control Act.* Established a Federal Water Pollution Control Administration in Department of Health, Education, and Welfare. Required establishment of water qualty standards for all interstate and coastal waters.
1966.....	*Clean Water Restoration Act, further amended Federal Water Pollution Control Act.* Greatly increased authorizations for grants to help build sewage treatment plants, for research, and for grants to State water pollution control programs. Transferred administration of the Oil Pollution Act from the Secretary of the Army to the Secretary of the Interior.
	Reorganization Plan No. 2. Federal Water Pollution Control Administration transferred to Department of the Interior under President's Reorganization Plan No. 2.
	Executive Order No. 11288. Required all Federal agencies to comply with provisions and standards of Federal Water Pollution Control Act and cooperate with the Department of the Interior and State governments in preventing or controlling water pollution.
1969.....	*National Environmental Policy Act.* Enunciated policy to create and maintain conditions under which man and nature can exist in productive harmony, established the Council on Environmental Quality in productive harmony, established the Council on Environmental Quality in Executive Office of the President, provided for annual Presidential environmental quality report, and specified need for interagency cooperation.
1970.....	*Executive Order 11507.* Strengthened requirement for all Federal agencies to comply with the Clean Water Act, the Federal Water Pollution Control Act and the National Environmental Policy Act in prevention, control and abatement of air and water pollution at Federal facilities.

27. Ibid., pp. 32–34.

28. "Radioactive Waste Disposal into Atlantic and Gulf Coastal Waters," NAS/NRC Publication no. 655 (Washington: NAS, 1959).

29. P.L. 89–753, entitled "Clean Water Restoration Act of November 3, 1966," Title II, Section 6(g)(3), A, B, and C.

30. Among other compromises was the excision of the H.R. 25 provision

that Department of Interior must jointly approve estuarine modification previously approved only by the Corps of Engineers.

31. Outdoor Recreation Resources Review Commission, *Shoreline Recreation Resources of the United States.* Study Report No. 4 (Washington: GPO, 1962), p. 30.

32. National Goals Research Staff, *Toward Balanced Growth: Quantity with Quality* (Washington: GPO, 1970), pp. 5–6.

33. John Lear, "Land: Making Room for Tomorrow," *Saturday Review* (March 1971), pp. 45–48. U.S. President (Nixon), "The President's 1971 Environmental Program," in *Weekly Compilation of Presidential Documents* (Washington: GPO, Feb. 15, 1971).

34. Minutes of Marine Science Council, October 27, 1966.

35. Ibid.

36. *MSA*, 1st report (1967), p. iii.

37. Professor Albert Garretson of New York University School of Law.

38. Professors John H. Ryther and J. E. Bardach, through American Institute of Biological Science.

39. Trident Engineering Associates of Annapolis, Maryland.

40. Management and Economic Research, Inc., of Palo Alto, California.

41. National Planning Association of Bethesda, Maryland.

42. Members from: Army Corps of Engineers; Department of the Interior; Department of Health, Education and Welfare; Department of Commerce; Department of Housing and Urban Development; Department of Transportation; Atomic Energy Commission; National Science Foundation; Navy Department; National Aeronautics and Space Administration; Department of State; Smithsonian Institution; Council of Economic Advisors; Office of Science and Technology; Water Resources Council. Observers from: Bureau of the Budget; Marine Sciences Council; Stratton Commission; National Academy of Engineering.

43. *MSA*, 2d report (1968), p. 71.

44. Ibid., p. v.

45. Ibid., p. iv.

46. Ibid., p. 167.

47. The Commissions were multistate-federal planning bodies established by Water Resources Planning Act, P.L. 89–80.

48. This was one of the rare cases when the Budget Bureau shared their internal documents with the Council and afforded the opportunity for rebuttal.

49. *MSA*, 2d report (1968), p. 72.

50. *MSA*, 3d report (1969), p. 74.

51. Ibid., pp. 74–76.

52. *Seminar on Multiple Use of the Coastal Zone*, pp. 10–45.

53. U.S. Congress, Senate, *Beach Erosion Control.* Hearings before the Committee on Public Works, 90th Congress, 1st sess. (Washington: GPO, 1968).

54. U.S. Congress, House, *Department of the Interior and Related Agencies, Appropriations for 1968*, Interior Subcommittee of the Committee on Appropriations, 90th Congress, 1st sess. (Washington: GPO, 1967), pp. 1018–44.

55. P.L. 89–80, entitled "Water Resources Planning Act," July 22, 1965, Title I, Section 101.

56. *From Sea to Shining Sea*, p. 177.

57. *MSA*, 3d report (1969), p. iii.

58. Ibid., p. 79.

59. *National Estuary Study* 5:45.

60. Edward Wenk, Jr., "National Policy for Coastal Management," *Vital Speeches of the Day* 37, no. 6 (Jan. 1, 1971):178–79.

61. Within existing legislation, this is readily possible from the $200 million annual contribution to the Land and Water Conservation Fund from offshore oil revenues, boat fuel taxes, surplus property sales, user fees in recreational areas, etc.

62. COMSER, *Our Nation and the Sea*, pp. 269–70.

63. H.R. 14845, "National Estuarine and Coastal Zone Management Act of 1970," November 18, 1969, and S. 3183 (same title), of November 25, 1969.

64. U.S. President (Nixon), "1971 Environmental Program," February 8, 1971.

65. See CEQ, *Ocean Dumping*, followed by transmittal to Congress of a draft Marine Protection Act in February 1971.

66. U.S. Congress, House, *Coastal Zone Management Conference*, Subcommittee on Oceanography, Committee on Merchant Marine and Fisheries, 91st Congress, 1st sess., October 1969 (Washington: GPO, 1969).

67. H.R. 2492 was introduced by Lennon on January 29, 1971, H.R. 2493 by Lennon on January 29, and H.R. 9229 by Lennon and others on June 17, 1971. Hearings were convened June 22–24, August 3–5, and November 1 and 9, 1971.

68. *MSA*, 4th report (1970), pp. 28, 29.

69. U.S. Congress, Senate, *Creation of a National Program for Coastal and Estuarine Zone Management*, Calendar No. 510, Committee on Commerce, 92d Congress, 1st sess., December 1971 (Washington: GPO, 1971).

70. *New York Times*, July 5, 1971, p. 18.

71. U.S. President (Nixon), "Marine Pollution," May 20, 1970.

72. Edward Wenk, Jr., Testimony before the Department of Interior, Bureau of Land Management, on Proposed Trans-Alaska Pipeline, February 17, 1971. The author called attention to almost complete neglect in the Draft Environmental Impact Statement of adverse effects on the marine environment of transport of oil from Alaska to West Coast ports. When the Final Statement was released in 1972, further comment was submitted on May 1 that it failed to consider alternatives of technology toward prevention of spills in the marine environment, important because environmental effects continued to be largely unknown and containment measures were ineffective.

73. Stewart L. Udall, *The Quiet Crisis* (New York: Avon Books, 1963), p. 94.

74. Richard A. Cooley and Geoffrey Wandesforde-Smith, *Congress and the Environment* (Seattle: University of Washington Press, 1970), p. 208.

75. P.L. 91–190, National Environmental Policy Act of 1969 (Jan. 1, 1970).

76. In early 1965, Congressman Daddario became interested in questions that the author had raised in his capacity of congressional science advisor on imbalance in governmental research between civilian aeronautics and astronautics, which was in the limelight. He sought advice from Charles A. Lindbergh, and asked me to be present. Among Lindbergh's expressed opinions was a dim view of the SST but he then launched on one of his own concerns, the demise of the blue whale. Daddario was so taken with the

urgency of the environmental issue that immediately afterwards he asked me to prepare proposals for an environmental study by his subcommittee. When I reminded him of potential jurisdictional friction with Interior or Public Works committees, or even Merchant Marine and Fisheries, he challenged me to find a topic common to all committees having environmental interests. At that time Richard Carpenter on my staff had completed for Senator Abraham A. Ribicoff a study on fish kills on the Mississippi River from accidental drainage carrying powerful agricultural pesticides. From that special case Carpenter and I developed some generalized criteria whereby unwanted effects could be studied in advance in more detail. The concept of examining possible adverse consequences systematically *before* approval of a new technology was ready-made for Daddario. He approved the theme for a legislative initiative. For want of a brief title, the term "technology assessment" was coined by congressional aide Philip B. Yeager for an acronym when joined with the "Board" that was then conceived as essential to assist the Congress with technology assessment independently of the Executive. Legislative proposals for this general capability for Congress include: H.R. 6698 of March 7, 1967, H.R. 17046 of April 16, 1970, H.R. 3269 of February 2, 1971, H.R. 10243 and H.R. 10246 of July 30, 1971, and S. 2302 of July 19, 1971.

77. U.S. Congress, House, *Managing the Environment*, Committee on Science and Astronautics, 90th Congress, 2d sess. (Washington: GPO, 1968).

78. Cooley and Wandesforde-Smith, eds., *Congress and the Environment*, p. 219.

79. H.R. 13272, "Environmental Quality and Productivity Act of 1969," August 1, 1969.

80. Wenk, testimony on Alaska Pipeline.

81. Legislation in the 92nd Congress concerned with the problem of oil spills focused on offshore drilling regulation (S. 1853, H.R. 4628, H.R. 5617, H.R. 8115, H.R. 12156) and the control of vessel operating and loading activities (S. 2074, H.R. 2522, H.R. 6232, H.R. 9581, H.R. 10051).

82. *Calvert Cliffs' Coordinating Committee, Inc.*, et al., v. *U.S. Atomic Energy Commission and United States of America*, Respondents, *Baltimore Gas and Electric Company, Intervenor*, United States Court of Appeals for the District of Columbia Circuit (No. 24,839) and same petitioners and respondents, Petitions for Review of an Order of the Atomic Energy Commission (No. 24,871), decided July 23, 1971.

83. Executive Order No. 11574 of December 1970. By 1972 the Corps of Engineers had drafted 450 environmental impact statements, over half on coastal initiatives.

84. Council on Environmental Quality, *Environmental Quality*. Report transmitted to Congress August 1970 (Washington: GPO, 1970), pp. 175–78.

85. H.R. 9727, introduced July 13, 1971, and S. 2770, introduced October 28, 1971.

CHAPTER 5

1. National Council on Marine Resources and Engineering Development, *Marine Science Affairs*, 2d report, 1968 (Washington: GPO, 1968), pp. 21–23. (Hereafter *MSA*, by report no.)

2. *MSA*, 4th report (1970), pp. 198–99.

3. During the "cold war" of the 1950s and early 60s, this support was rationalized as an important medium of communication with the U.S.S.R.

4. U.S. Congress, House, *The National Science Foundation: A General Review of Its First Fifteen Years*, Committee on Science and Astronautics, 89th Congress, 2d sess., House Report No. 1219 (Washington: GPO, 1966), pp. 32–33.

5. Proposed by NASCO; see National Academy of Science, National Research Council, *Oceanography, 1960–1970* (Washington: GPO, 1959).

6. For example: International Cooperative Investigations of Tropical Atlantic, and of the Kuroshio and Adjacent Regions (Far East).

7. COLD directors represented the following institutions: Institute of Marine Sciences, University of Miami (Florida); Dept. of Oceanography, Oregon State University; Dept. of Oceanography, University of Washington; Dept. of Oceanography, Texas A&M University; Scripps Institution of Oceanography, University of California (La Jolla); Lamont Geological Observatory of Columbia University; Woods Hole Oceanographic Institution; Graduate School of Oceanography, University of Rhode Island; Dept. of Oceanography, Johns Hopkins University; Institute of Geophysics, University of Hawaii.

8. Full-time graduate student enrollments had jumped from 92 in 1960 to 980 in 1967; Ph.D.'s awarded from 6 to 60; the number of institutions offering degrees from 35 to 63; statistics from the University of Miami reported in *MSA*, 2d report, p. 140, and the ICO, p. 143.

9. These problems in the State Department functions were most recently studied at the request of Secretary of State William Rogers; see *New York Times*, December 9, 1970, "State Department Gives A Revamping Plan. It Seeks a New Breed of Diplomat-Managers."

10. Berkner recognized that science was international in character, provided a neutral forum for a free exchange of views, and could foster intercultural understanding. In noting the dependence of U.S. national security (and thus of free peoples everywhere) on science, he felt U.S. foreign relations should take a more active interest in science. More than that, policy-makers should be aware of the scientific implications of decisions they faced. He therefore called for creation of some scientific locus within the State Department. After the initial response to these findings the State Department lost interest, and a four-year hiatus occurred before the science function was reinstated in 1958 in response to Sputnik. Posts were created for seven science attaches in major U.S. embassies to gather information and foster communication within scientific communities. By 1972, these were increased to nineteen.

11. Eugene B. Skolnikoff, *Science, Technology, and American Foreign Policy* (Cambridge: M.I.T. Press, 1967), pp. 5, 255–73.

12. Responsibility for technical details had been delegated to a panel under the ICO staffed by William Sullivan who reported to McKernan, assisted by panel members from federal agencies and NAS.

13. United Nations General Assembly (hereafter UNGA) Resolution 2172 (XXI), of December 6, 1966. See also discussion in chap. 6.

14. Nations are listed in the five-volume compilation by the Marine Sciences Council, *Marine Science Activities* (Washington: GPO, April 1968).

15. Pollack's post had long been earmarked for some eminent scientist, and while no scientist approached for the position had been willing to take

on the assignment, others unchivalrously pecked away at Pollack's lack of technical qualifications. Yet Pollack as a career officer proved far more influential in State's affairs than his two immediate predecessors, who had scientific credentials.

16. Secretary Rogers was obligated to review his position on continuing CIPME when informed of a further reorganization of Council committees on March 12, 1969. Acting for him on March 29, the Deputy Undersecretary, U. Alexis Johnson, stated Rogers intended to continue the committee, with delineation of authority to block potential Council intrusion.

17. *Note verbale* distributed in UN document A/6695, August 18, 1967: *Malta: Request for the Inclusion of a Supplementary Item in the Agenda of the 22nd Session.*

18. "The White House Staff vs. the Cabinet: Hugh Sidey Interviews Bill Moyers," *Washington Monthly*, February 1969.

19. Harold Seidman, *Politics, Position, and Power* (New York: Oxford University Press, 1970), p. 78.

20. *New York Times*, April 14, 1967, pp. 17–18, presidential statement on Punte del Este.

21. Included were: Howard A. Wilcox (chairman), a former naval scientist, later executive with General Motors and subsequently a private consultant; F. Gilman Blake, senior research scientist for Chevron Research Company; Douglas Brooks, President of Travelers Research Center; Harvey Brooks, Dean at Harvard University, School of Engineering and Applied Science; W. M. Chapman, Director of Research for Van Camp's Seafood Division of Ralston Purina; James Crutchfield, Professor of Resource Economics at the University of Washington, and a member of the Stratton Commission; Robert G. Fleagle, Chairman of the Atmospheric Sciences Department of the University of Washington; Harry Hess, Chairman of the Geology Department at Princeton; Alfred Keil, Head of the Department of Naval Architecture and Marine Engineering at M.I.T.; H. Burr Steinbach, Chairman of the Zoology Department, at the University of Chicago.

22. *New York Times*, November 9, 1967, "U.S. Urges Slow Approach to Issues of Ocean Floor."

23. *MSA*, 2d report (1968), p. iv.

24. U.S. President (Johnson), special Message to the Congress on Conservation: "To Renew a Nation," March 8, 1968.

25. *New York Times*, April 14, 1968, "U.S. Pushes Plan for Ocean Study."

26. National Council on Marine Resources and Engineering Development, *International Decade of Ocean Exploration* (Washington: GPO, May 1968).

27. H.M.S. *Challenger* Expedition, 1885.

28. A hidden Soviet agenda item popped up when Khlestov challenged the buzzing of Soviet merchant and research ships by the U.S. Air Force, contending that such conduct was not conducive to expanded international collaboration in ocean exploration. He further complained about advance notification procedures for calls by Soviet research ships at U.S. ports. My response was to recall our good record of reception of the Soviet oceanographic ships—for example, of the *Mikhail Lomonosov* in San Francisco where Humphrey sent the skipper a wire of welcome. Moreover, we were regularly accepting port calls from Soviet fishing research ships engaged in bilateral research. In contrast, I reminded Khlestov of Soviet denials of

entry to U.S. research ships *Silas Bent* and *Pillsbury*. Khlestov responded that *Silas Bent* was a U.S. Navy ship and, even though unarmed and open to Soviet inspection, was prohibited from entering Soviet waters. I countered by suggesting that the Soviets change their inflexible bureaucratic rules to consider the nonmilitary character of Navy research ships, case by case. Khlestov smiled, nodded that he understood; the verbal scuffle had come out a draw.

29. *New York Times*, June 18, 1968, "Ocean Study Plan Backed in Soviet."

30. Listed in Appendix 20.

31. UNGA Resolution of December 21, 1968, No. 2467A (XXIII). See Appendix 25.

32. Ibid., part B of resolution.

33. Ibid., part D of resolution.

34. Ibid., part C of resolution.

35. *MSA*, 4th report (1970), pp. 195–99.

36. *MSA*, 3rd report (1969), p. 131.

37. *An Oceanic Quest*, NAS-NAE Publication 1709, National Academy of Sciences (Washington, D.C., 1969). The academies identified the following exploration programs to be of particular interest during the next ten years:

a. *Geology and Nonliving Resources:* geological-geophysical surveys of North American Continental Shelves and the eastern Atlantic continental margin; assessment of the mineral resource potential of small ocean basins, such as the Gulf of Mexico, Caribbean, Mediterranean, and the East Indies area; dredging, coring, profiling, and related studies of oceanic ridges and trenches, such as the Mid-Atlantic Ridge and the Peru-Chile Trench; and surveys of selected Pacific sites of manganese nodules and phosphorite deposits.

b. *Fisheries:* assessment of the fisheries production potential of the Gulf of Mexico, Gulf of Alaska, and equatorial eastern and central Pacific; ecological and related studies leading to improved management of fisheries of the northwestern Atlantic; assessment and increased development of fishery resources of the Arabian Sea, offshore southern Chile and Argentina, and in the Indonesian Archipelago; and investigation of the potentially rich euphausiid resources of the Antarctic Ocean.

c. *Biological Studies:* application of recently developed techniques to studies of food chains in the sea; and development of new techniques for measuring biological factors and for modeling ecosystems using computers for areas such as Georges and Grand Banks, the Gulf of Alaska, the Gulf of Mexico, the eastern and central equatorial Pacific, the South Pacific gyre, the western Arabian Sea, and the Antarctic Ocean.

d. *Physics and Environmental Forecasting:* investigation of 1,000 to 3,000 mile, cold and warm anomalies related to "centers of action" in the North Pacific; studies of large-scale, long-term, air-sea interaction, and meso-scale interaction in subtropical upwelling regions; systematic ocean coverage of deep temperature, salinity, and oxygen measurements; and geochemical "benchmark" surveys of selected trace substances on meridional traverses in the Atlantic, Pacific, and Indian oceans.

38. NAS-NAE, *Oceanic Quest*, p. 22.

39. Ibid., p. 87.

40. U.S. Congress, House, *National Marine Sciences Program*, Part 1, hearings before Subcommittee on Oceanography of Committee on Merchant

Marine and Fisheries, 90th Congress, 1st sess. (Washington: GPO, Dec. 1967), pp. 3–57.

41. H. Con. Res. 803, July 26, 1968.

42. Included were the following House Joint Resolutions (cited with their sponsors): 816, James A. Byrne; 817, Frank M. Clark; 818, Thomas N. Downing; 819, Edward A. Garmatz; 820, Richard T. Hanna; 821, Henry Helstoski; 822, Alton Lennon; 823, Paul G. Rogers; 824, Ed Reinecke; 828, Thomas M. Pelly; 829, James B. Utt; 834, L. H. Fountain; 835, H. R. Gross; 837, Don Fuqua; 840, Don H. Clausen. These were placed in the record September 22, 1967.

On October 10, 1967, the following similar resolutions were introduced: 843, Byron G. Rogers; 844, Edwin E. Willis; 850, Robert L. Leggett; 856, Durward G. Hall; 865, Edward J. Gurney; 876, Fernand J. St. Germain. Subsequently other bills of similar or identical content were introduced: October 31, H.J.R. 916, by George Bush; October 25, H. Con. Res. 558, by Jonathan B. Bingham.

43. *Global Ocean Research*, a report of a joint working party of the Advisory Committee on Marine Resources and Research, the Scientific Committee on Oceanic Research, and the World Meteorological Organization, Rome, 1969.

44. The Fifth Session of the Intergovernmental Oceanographic Commission in October 1967 endorsed development of IGOSS, with platforms such as ships, buoys, satellites, coastal towers, etc., to provide environmental data from ocean areas to all countries in a form convenient for use. Two phases were being planned: Phase I (1968–71, approximately) to use existing technology; Phase II (from about 1971) to increasingly incorporate modern developments, especially moored ocean buoys. IGOSS was planned to operate closely with the World Weather Watch.

45. *MSA*, 4th report (1970), pp. 195–96.

46. *MSA*, 5th report (1971), pp. 84–89.

47. Larry L. Booda, "Marine Sciences and ASW Receive Big Budget Increases," *Undersea Technology* 13 (Arlington: Compass Public., March 1972), pp. 30–32.

48. *International Decade of Ocean Exploration*, National Science Foundation, NSF 71-34 (Washington: GPO, 1971).

CHAPTER 6

1. "International Law and the Law of the Sea," in *The Law of the Sea*, ed. Lewis M. Alexander (Columbus: Ohio State University Press, 1967), p. 14.

2. This was expressed in 1967 by a quick embrace by LDC's of an international regime from which proceeds of exploitation would be directed to their benefit. Later, some less developed countries were to reverse their stance until a legal regime was devised to keep the advanced nations from being primary beneficiaries of seabed exploitation because of their available technology. Papers on this issue are listed in *Ocean Affairs Bibliography 1971*, Woodrow Wilson International Center for Scholars (Washington: WWICS, 1971).

3. Full text given in Appendix 11.

4. Santiago negotiations on fishery conservation problems, U.S. Dept. of State, Public Service Div., 1955, pp. 30–32.

5. *Congressional Record*, April 3, 1968, p. S. 3818.

6. John L. Mero, *The Mineral Resources of the Sea* (New York: Elsevier, 1965).

7. For detailed discussion, see U.S. Congress, House, *Exploiting the Resources of the Seabed*, report prepared by George A. Doumani for Committee on Foreign Affairs, 92d Congress, 1st sess., July 1971 (Washington: GPO, 1971), p. 24.

8. Considerations of the territorial sea and contiguous zone date back to 1930 with the Hague Codification Conference sponsored by the League of Nations, although no agreement was reached.

9. William R. Neblett, "The 1958 Conference on the Law of the Sea: What Was Accomplished," in *The Law of the Sea*, p. 37.

10. Robert L. Friedheim, "Factor Analysis as a Tool in Studying the Law of the Sea," in *The Law of the Sea*, chap. 4.

11. G. E. Pearcy, "Geographical Aspects of the Law of the Sea," in *Annals, Association of American Geographers* 49, no. 1 (March 1959):1–23.

12. *The Law of the Sea: The Future of the Sea's Resources*, Proceedings of 2nd Annual Conference, Law of the Sea Institute (Kingston: University of Rhode Island, Feb. 1968).

13. *Christian Science Monitor*, March 9, 1959, p. 13.

14. *Congressional Record*, May 26, 1960, pp. S. 11172–11196; also U.S. Congress, Senate, *Conventions on the Law of the Sea*, Foreign Relations Committee, 86th Congress, 2d sess., January 1960 (Washington: GPO, 1960).

15. Congressman Richard Hanna introduced H.R. 11232 on May 13, 1964; Congressman Alton Lennon, H.R. 5175 on February 18, 1965.

16. Edward Wenk, Jr., speech before the American Society of Civil Engineers, "Engineering for Maritime Exploration and Development," October 22, 1964, published in *Journal of Professional Practice*, ASCE, vol. 91 (Sept. 1965).

17. U.S. President (Johnson), Address at Commissioning of ESSA ship *Oceanographer*, July 13, 1966. Full text given in Appendix 15.

18. *New Dimensions for the United Nations*, 17th Report of The Commission to Study the Organization of Peace (New York: UN Plaza), pp. 44–46.

19. *New York Times*, June 30, 1966, "New Law Urged to Fight Piracy."

20. U.S. Congress, Senate, *The United Nations at Twenty-One*, report by Senator Frank Church before Committee on Foreign Relations, 90th Congress, 1st sess., February 1967 (Washington: GPO, 1967), p. 25.

21. *New York Times*, July 14, 1967, "UN Rule is Urged for Ocean Riches."

22. William T. Burke, "Law and the New Technologies," in *The Law of the Sea*, p. 223.

23. *Note verbale* distributed in UN Document A/6695, August 18, 1967: *Malta: Request for the Inclusion of a Supplementary Item in the Agenda of the 22nd Session*.

24. UN Document, A/C 1/952.

25. Doumani, *Exploiting the Resources of the Seabed*, pp. 53–57.

26. In 1971, when the incumbent government of Malta was swept out of office in a general election, Pardo was replaced. He was later reinstated as Minister Plenipotentiary to the United Nations for Ocean Affairs.

27. National Council on Marine Resources and Engineering Develop-

ment, *Marine Science Affairs*, 1st report, 1967 (Washington: GPO, 1967), p. v. (Hereafter *MSA*, by report no.)

28. The pattern and style of such resolutions, incidentally, reflect all the characteristic strengths and weaknesses of an international body working in an arena of noble idealism, striving to deal with a complex substantive issue while wrestling with nationalism and rivalries among nations of widely varying economic levels and power, yet each having an equal vote. The Resolution opens with a clue as to the bead drawn on exploitation of a new frontier with implications for the LDC's:

"*The General Assembly, Recognizing* the need for a greater knowledge of the oceans and of the opportunities available for the utilization of their resources, living and mineral, *Realizing* that the effective exploitation and development of these resources can raise the economic level of peoples throughout the world and in particular of the developing countries. . . ." Then the Resolution sought to head off anxiety by other international bodies as to UNGA initiatives over this issue by a diplomatic bow of "taking into account with appreciation the activities in the field of resources of the sea" undertaken by the UN, UNESCO, IOC, FAO, its Committee of Fisheries, WMO, the Advisory Committee on the Application of Science and Technology to Development and other "intergovernmental organizations concerned, various governments, universities, scientific and technological institutions, and other interested organizations."

Nevertheless, the Resolution was candid in facing up to the same problem of fragmentation in the international arena that plagued the United States government, with the acknowledgment: "Considering the need to maximize international cooperative efforts for the further development of marine sciences and technology and to avoid duplication or overlapping of efforts in this field. . . ."

Again mindful of the sensitivities of other interests and of protocol, the resolution provided for the Secretary General to submit his surveys and proposals to the General Assembly via the Economic and Social Council, after submission to "the Advisory Committee on the Application of Science and Technology to Development for its comments."

29. *MSA*, 2d report (1968), p. 29. See also Appendix 31.

30. See chap. 5, note 42.

31. U.S. Congress, Senate, *Special Study on United Nations Suboceanic Lands Policy*, Commerce Committee hearings, 91st Congress, 1st sess., September–November 1969 (Washington: GPO, 1969).

32. Doumani, *Exploiting the Resources of the Seabed*, p. 63.

33. Claiborne Pell, *Challenge of the Seven Seas* (New York: William Morrow & Co., Inc., 1966).

34. Senate Resolutions 172 of September 29, 1967, and 186 of November 17, 1967.

35. U.S. Congress, House, *National Marine Sciences Program*, Part 1, Committee on Merchant Marine and Fisheries hearings, 90th Congress, 1st sess., August–December 1967 (Washington: GPO, 1967).

36. *Congressional Record*, December 8, 1967, pp. 35614–615.

37. S. Res. 263, March 5, 1968.

38. *New York Times*, March 5, 1968, "Treaty on Ocean Offered by Pell."

39. *MSA*, 2d report (1968), pp. 30–33.

40. National Petroleum Council, *Petroleum Resources under the Ocean Floor* (Washington: GPO, March 1969), p. 13.

41. *MSA*, 3d report (1969), pp. 242–45.

42. Ibid., p. 53.

43. UN Economic and Social Council Document E/4487.

44. Commission on Marine Science, Engineering, and Resources, *Our Nation and the Sea* (Washington: GPO, 1969), pp. 141–57.

45. See Appendix 19.

46. *MSA*, 5th report (1971), pp. 81, 82.

47. *New York Times*, editorial, "Man's Ocean Heritage," May 27, 1970.

48. Doumani, *Exploiting the Resources of the Seabed*, pp. 66–68.

49. U.S. Congress, Senate, *Outer Continental Shelf*, report by the Special Subcommittee on the Outer Continental Shelf, 91st Congress, 2d sess., December 21, 1970 (Washington: GPO, 1970), pp. 30–32.

50. William T. Burke, "Legal-Political Issues Relating to the 1973 Conference on the Law of the Sea," paper delivered at the Conference on Man and the Oceans, San Francisco, California, December 1, 1971.

51. *New York Times*, August 2, 1970, "U.S. Alters Draft on Seabed Riches."

52. These discussions were the basis for annual conferences at the Law of the Sea Institute, University of Rhode Island.

53. Center for the Study of Democratic Institutions, *Proceedings*, *Pacem in Maribus I and II* (Santa Barbara, Calif., 1971).

54. Edward Wenk, Jr., "Toward Enhanced Management of Maritime Technology," in *Pacem in Maribus-Ocean Enterprises*, ed. E. H. Burnell and P. von Simson, Center for the Study of Democratic Institutions, Occasional Paper, vol. 2, no. 4 (Santa Barbara, Calif., June 1970).

55. Selected international organizations engaged in marine affairs are listed in Appendix 20.

56. Examples of international bodies engaged in fisheries management include: International Commission for the Northwest Atlantic Fisheries (ICNAF), North East Atlantic Fisheries Commission (NEAFC), Inter-American Tropical Tuna Commission (IATTC), International Whaling Commission (IWC), International North Pacific Fisheries Commission (INPFC), International Pacific Halibut Commission (IPHC), International Pacific Salmon Fisheries Commission (IPSFC), Black Sea Fisheries Commission (BSFC).

57. UN: Economic and Social Council Documents E/4449 (of Feb. 21, 1968) and E/4487.

58. This measure somewhat moderated the apprehension of IOC officials that their role in oversight of scientific endeavors would be excised by the UN and, moreover, that rationality of decisions would be seriously distorted by UN politics.

59. *MSA*, 4th report (1970), p. 184.

60. U.S. President (Nixon), United States Policy for the Seabed, May 23, 1970, *The Department of State Bulletin*, LXII, June 15, 1970, pp. 737, 738.

61. *New York Times*, August 30, 1970, "A Seabed Accord is Blocked Again."

62. *MSA*, 5th report (1971), pp. 82, 83.

63. Dr. Frank LaQue, former vice president of International Nickel, quoted in *The Resources of the Seabed* (Ditchley Foundation, Ditchley Park, England, Sept. 1969), p. 16.

64. U.S. Congress, Senate, *Disarmament and Security: A Collection of Documents, 1919–55*. Subcommittee on Disarmament, 84th Congress, 2d sess. (Washington: GPO, 1956).

65. *MSA*, 1st report (1967), p. 103.

66. This point was revealed publicly by former Assistant Secretary of the Navy Robert W. Morse in testimony before Senator Pell on S. Res. 33 in July 1969.

67. U.S. President (Nixon) in *Weekly Compilation of Presidential Documents* (Washington: GPO, March 18, 1969), pp. 227–29.

68. *New York Times*, December 13, 1969, "UN Committee Votes to Refer Seabed Treaty Back to Geneva."

69. E. D. Brown, *Arms Control in Hydrospace* (Washington: Woodrow Wilson International Center for Scholars, 1971).

CHAPTER 7

1. In 1967, five aerospace corporations received more than $1 billion in federal contracts, each thus spending more than five cabinet-level departments.

2. *New York Times*, July 12, 1967.

3. Miller B. Spangler, *New Technology and Marine Resource Development* (New York: Praeger, 1970), pp. 493–96; also Marine Science Council contract study, *Encouraging Marine Resource Development* (see Appendix 31).

4. Major recommendations of Commission on Marine Science, Engineering and Resources are listed in *Our Nation and the Sea* (Washington: GPO, 1969), p. 157. (Hereafter cited as COMSER, *Our Nation and the Sea*.)

5. Initiative 5 on studies of the Chesapeake estuary would also facilitate industry-related interests, but not in the sense of assistance.

6. National Council on Marine Resources and Engineering Development, *Marine Science Affairs*, 1st report, 1967 (Washington: GPO, 1967), p. 17. (Hereafter, *MSA*, by report no.)

7. COMSER, *Our Nation and the Sea*, p. 157.

8. *MSA*, 2d report (1968), p. 158.

9. *MSA*, 4th report (1970), pp. 69–70.

10. *MSA*, 3d report (1969), p. 203.

11. A 1793 law (46 USC 251) in effect requires that vessels over five net tons engaged in our fisheries must be built in U.S. shipyards. A 1966 subsidy law, intended to offset the high costs of construction in the United States, has not had a significant impact on fleet size and at the same time reduces incentives to lower construction costs.

12. See estimates by J. H. Ryther, M. B. Schaefer, W. M. Chapman, and R. L. Edwards and H. W. Graham, cited in M.I.T. contract study for Marine Sciences Council, "The Economics of Fish Protein Concentrate," mimeographed (Washington: GPO, 1970), p. II-9.

13. Francis T. Christy, Jr., and Anthony Scott, *The Common Wealth in Ocean Fisheries* (Baltimore: Johns Hopkins, 1965).

14. Maurice Earl Stansby, et al., *Industrial Fishery Technology* (New York: Reinhold, 1963).

15. Garrett Hardin, "The Tragedy of the Commons," *Science* 162, December 13, 1968, pp. 1243–48.

16. *MSA*, 3d report (1969), pp. 96–97.

17. *MSA*, 3d report (1969), pp. 28–29.

18. *The Future of the Fishing Industry of the United States*, ed. DeWitt Gilbert. University of Washington Publications in Fisheries New Series, vol. 4 (Seattle: University of Washington, College of Fisheries, 1968).

19. *MSA*, 3d report (1969), pp. 95–96.

20. Under a Marine Sciences Council's contract, Litton Industries prepared a report entitled *Systems Analysis of Specified Trawler Operations.* See Appendix 31.

21. John H. Ryther, "Photosynthesis and Fish Production in the Sea," *Science* 166 (Oct. 3, 1969):72.

22. Presidential documents covering marine affairs during the period include:

FISHERIES

1966: (Dec. 29, 1965–Dec. 30, 1966) (1) *Remarks* on Passage of FPC Act, p. 1591.

1967: (Dec. 31, 1966–Dec. 29, 1967) (2) Fish Protein Use, p. 138: *Remarks.*

1968: (Dec. 29, 1967–Dec. 27, 1968) (3) Fishing Interritorial Waters off the U.S., p. 1156, *Remarks;* (4) Fish Inspection, p. 227, *Message to Congress;* (5) Fish Protein Concentrate, p. 257, *Message.*

1969: (Dec. 27, 1968–Dec. 29, 1969) (6) American Fisheries Society, 100th Anniversary Medals, p. 717, *Announcement;* (7) Fisheries Commissions, Recommendations and Actions (EO 11467), p. 638, *Order;* (8) Fishing Operations in the North Atlantic, Convention on Conduct of, p. 563, *President's message.*

1970: (Dec. 29, 1969–Dec. 28, 1970) (9) Fish and Wildlife Coordination Act, p. 1725, *Order;* (10) International Commission for the Northwest Atlantic Fisheries, p. 1582, *Announcement.*

1971: (Dec. 28, 1970–June 14, 1971) (11) Wholesome Fish and Fishery Products Act, p. iii, *Message to Congress.*

23. U.S. President (Johnson), *State of the Union Message* in *Weekly Compilation of Presidential Documents* (Washington: GPO, Jan. 10, 1967).

24. R. L. Jackson, "Effect of Malnutrition on Growth of the Pre-School Child," *Pre-School Child Malnutrition: Primary Deterrent to Human Progress,* NAS-NRC Publication 1282 (Washington: NAS, Sept. 21, 1966).

25. Containing a nutritionally advantageous mix of amino acids not found in cereal sources of protein.

26. Having come from a dairy state, Humphrey was able to lend telling influence to help negotiation with the dairy interests to sidetrack their filing of injunctions against the February action of FDA. As trading currency, the Secretariat was able to offer relaxation of uncoordinated and conflicting standards for milk products imposed by the Department of Agriculture and FDA.

27. *MSA,* 1st report (1967), pp. 54–56.

28. P.L. 89–701 of November 2, 1966.

29. U.S. President (Johnson), *Special Message to the Congress on the Foreign Assistance Programs: "To Build the Peace,"* February 8, 1968.

30. *MSA,* 3d report (1969), p. 102.

31. Ernest F. Hollings, *The Case Against Hunger* (New York: Cowles Book Corp., 1970).

32. *The Economics of Fish Protein Concentrate,* mimeographed study by M.I.T., 1971.

33. *Science* 30 (July 30, 1971):410–12.

34. U.S. Congress, House, Hearings before Subcommittee on Fisheries and Wildlife Conservation, 89th Congress, 2d sess., August 16, 1966.

35. Timothy M. Hammonds and David L. Cole, *Utilization of Protein*

Ingredients in the U.S. Food Industry, mimeographed (New York: Cornell University Department of Agricultural Economics, 1971).

36. V. E. McKelvey and Frank F. H. Wang, *World Subsea Mineral Resources*, text accompanying Geologic Investigations Map I-632, U.S. Geological Survey (Washington: GPO, 1969).

37. U.S. Congress, P.L. 83–31, Title II, Sec. 3–11, 1953.

38. *Oil Pollution: A Report to the President*, by the Secretary of the Interior and the Secretary of Transportation (Washington: GPO, Feb. 1968).

39. Studies on oil spills authorized by the American Petroleum Institute resulted in ten published reports in 1970–71.

40. Offshore exploration for gold was likely to open up if the price should rise by devaluation of U.S. currency above the $35 per ounce.

41. Spangler, *New Technology*, pp. 26–30.

42. Donald N. Taylor, "Worthless Nodules become Valuable," *Ocean Industry* 6 (June 1971):27–28.

43. A contract study for the Marine Sciences Council entitled *Economic Potential of U.S. Continental Margin.* See Appendix 31.

44. Edward Wenk, Jr., see note 16, chap. 6, in this volume.

45. Marine Sciences Council, *U.S. Activities in Spacecraft Oceanography* (Washington: GPO, Oct. 1967).

46. *Underwater Technology Requirements for Nonmilitary Ocean Missions*, report to Marine Sciences Council, Southwest Research Institute, Pb 178687 (Washington: NTIS, 1968).

47. That program ran into trouble in the Congress, however, when old questions of duplication led the House Appropriations Committee to cut the Coast Guard's initial $5 million request. By phone calls, letters, and conferences, the Secretariat supported the Coast Guard in gaining approval, granted in the next budget cycle.

48. *MSA*, 4th report (1970), p. 106.

49. Ibid.

50. Ibid.

51. *MSA*, 1st report (1967), p. 57.

52. Edward Wenk, Jr., speeches, "A New Industry for Maritime Exploration and Development," before Annual Merchant Marine Conference, Galveston, Texas, October 15, 1965; and "Oceanography in Transition," at 21st Conference of Board of Governors and Aerospace Manufacturers Council, Williamsburg, Virginia, May 18, 1967.

53. Robert B. Abel, address at San Diego, September 1971, before the Institute of Electrical and Electronic Engineers.

54. COMSER, *Our Nation and the Sea*, p. 268.

CHAPTER 8

1. Harold Seidman, *Politics, Position, and Power* (New York: Oxford University Press, 1970), p. 3.

2. U.S. Congress, House, *Ocean Sciences and National Security*, 86th Congress, 2d sess., July 1960 (Washington: GPO, 1960), p. 1.

3. Ibid.

4. Yet in its first year of corporate life, in 1971, the new agency's effectiveness was already being sharply questioned by a Congress that had been its major advocate. U.S. Congress, Senate, Hearings before Subcommittee on Oceanography, Seattle, Washington, July 1, 1971, 92d Congress, 1st sess. (Washington: GPO, 1971).

5. See chap. 2, p. 84.

6. H.R. 10904, April 15, 1964, introduced by Congressman Bob Wilson, and S. 2251, July 7, 1965, introduced by Senator Edmund S. Muskie. Actually Lloyd Berkner had recommended this orientation for a Department of Science and Technology in 1958 when testifying before Senator Hubert H. Humphrey.

7. As outlined previously on page 89, the Commission's origins derive from a 1965 proposal by MTS president James Wakelin and advocacy by Congressman Paul Rogers.

8. Otto Klima, Jr., and Gibson M. Wolfe, "The Oceans: Organizing for Action," *Harvard Business Review* 46 (May 1968):98–112.

9. PSAC's report, *Effective Use of the Sea*, seemed unclear in separating operating and coordination functions, suggesting that it may have been a last-minute invention as part of a belated move to head off the Senate's proposal for a Council, mentioned on page 87. This report, prepared by a panel of the President's Science Advisory Committee, was never approved by the parent body, which was steadfastly opposed to accelerating ocean activities.

10. P.L. 89–454, Sec. 5(b). See Appendix 7.

11. In 1966, marine science issues were handled in the Senate by the Committee on Commerce as a whole, so that Senators Warren G. Magnuson and Norris Cotton, its chairman and ranking minority member respectively, were the logical candidates for appointment to the Commission. In the House of Representatives, the Subcommittee on Oceanography of the Committee on Merchant Marine and Fisheries had cognizance. Their chairman and ranking minority member, Alton A. Lennon and Charles A. Moshei, were accordingly appointed. During the Commission's life the House members occasionally attended meetings and their staff were in close communication with Commission staff. The Senate members were not active. The congressional members, however, according to the legislation in P.L. 89–454, were enjoined from taking part "except in an advisory capacity," in the formulation of the findings and recommendations of the Commission.

12. The membership of the Commission on Marine Sciences, Engineering and Resources included:

Chairman:

Julius A. Stratton,
President Emeritus,
Massachusetts Institute of Technology

Vice-Chairman:

Richard A. Geyer,
Head,
Department of Oceanography,
Texas A&M University

David A. Adams,
Commissioner of Fisheries,
North Carolina Department of
Conservation and Development (affiliation at time of appointment)

Leon Jaworski,
Attorney,
Fulbright, Crooker, Freeman,
Bates and Jaworski

John A. Knauss,
Dean,
Graduate School of Oceanography,
University of Rhode Island

John H. Perry, Jr.,
President,
Perry Publications, Inc.

Taylor A. Pryor,
President,
The Oceanic Foundation

Carl A. Auerbach,
Professor of Law,
University of Minnesota

Charles F. Baird,
Under Secretary of the Navy (appointed July 21, 1967, to succeed Robert H. B. Baldwin, former Under Secretary of the Navy)

Jacob Blaustein,
Director,
Standard Oil Company (Indiana)

James A. Crutchfield,
Professor of Economics,
University of Washington

Frank C. DiLuzio,
Assistant Secretary,
Water Pollution Control,
U.S. Department of the Interior (affiliation at time of appointment)

George E. Reedy,
President,
Struthers Research and Development Corp. (affiliation at time of appointment)

George H. Sullivan, M.D.,
Consulting Scientist,
General Electric Reentry Systems

Robert M. White,
Administrator,
Environmental Science Services Administration,
U.S. Department of Commerce

Congressional Advisers: Senators Norris Cotton, Warren G. Magnuson; Representatives Alton A. Lennon, Charles A. Mosher

13. Luther J. Carter, "Oceanography: Will LBJ's New Study Panel Make its Mark?" *Science* 155 (Jan. 20, 1967):306–07.

14. Later Indian-wrestling matches between staff of the two bodies over transfer of Council funds to Commission staff operations almost derailed that treaty.

15. The reports by panels, however, penetrated that discipline, requiring a foreword: "However, it was recognized from the outset that it was neither necessary nor desirable for the several panels to reach total consistency in their proposals or for the proposals to be fully consistent with positions later taken by the Commission as a whole. Although the panels have been guided in their work by the comments of the entire Commission, each panel is solely responsible for its own report. In considering the recommendations advanced by its panels, the Commission adopted some without modification, rephrased or modified others and, in some cases, took no position." Panel Reports of the Commission on Marine Science, Engineering and Resources, 1 (Washington: GPO, 1969), foreword.

16. H.R. 13273 introduced by Congressmen Garmatz, Lennon, Mailliard, and Mosher October 3, 1967, was amended and became P.L. 90–242 on January 2, 1968.

17. Panel Reports of the Commission on Marine Science, Engineering and Resources, 1–3 (Washington: GPO, 1969).

18. P.L. 91–121, Sec. 203 Armed Forces, appropriation authorization, 1970, November 19, 1969.

19. In the forwarding letter by the Commission Chairman to the President and the Congress, these members were exempted from commitment to the final recommendations on organization.

20. Commission on Marine Science, Engineering and Resources, *Our Nation and the Sea: A Plan for National Action* (Washington: GPO, 1969), p. 227. (Hereafter COMSER, *Our Nation and the Sea.*)

21. The only potential major civilian marine agency not included was the large Maritime Administration. Minor elements from other agencies were

omitted: radioactivity-effects research by AEC; biological research and taxonomic cataloguing by the Smithsonian Institution; beach erosion technology by the Corps of Engineers. The proposal also ducked the controversy that would have attended inclusion of the Geological Survey of the Interior Department as proposed by PSAC.

22. COMSER, *Our Nation and the Sea*, p. 245.

23. The Commission's technology panel had originally proposed a far stronger advisory committee to "guide" all the federal activities, in effect giving the program leadership and funding decisions to NACO rather than to NOAA. A parallel case had occurred in 1948. Responding to Vannevar Bush's magnificent scheme to establish a science agency, Congress had then proposed turning over the management of the prospective National Science Foundation completely to the outside National Science Board. President Truman vetoed that arrangement on advice that to give a part-time board authority to operate by making grants was an unconstitutional abdication of presidential powers. Stratton, astutely anticipating a potential clash between NOAA and NACO in a similar effort by outside interests to capture the program, saw to it that the advisory body was placed in perspective. Commissioners who supported the stronger NACO concept privately lobbied the Congress in 1970 to attempt to breathe legislative life into their original concept.

24. Membership of the ad hoc Committee of the Marine Sciences Council to review the Commission report included: Robert A. Frosch—Navy; Herman Pollack and Donald L. McKernan—State; Stanley Cain—Interior; John F. Kincaid—Commerce; William Stewart—HEW; Frank Lehan—Transportation; Spofford G. English—AEC; Daniel Hunt, Jr.—NSF; Sidney R. Galler—Smithsonian; Malcolm H. Merrill—AID; Thomas O. Paine—NASA; John D. Young—Bureau of the Budget; Merton J. Peck—Council of Economic Advisors; Walter Baer—OST; Edward Wenk, Jr.—(Chairman) Marine Sciences Council.

25. U.S. Congress, House, *National Oceanographic Program—1969*, Part 2. Hearings before the Subcommittee on Oceanography of the Committee on Merchant Marine and Fisheries, 91st Congress, 1st sess. (Washington: GPO, 1969), pp. 824–64.

26. This concern was to be reflected in Johnson's unwillingness to receive the Stratton report when it was issued, a procedural courtesy that fell by default to Humphrey.

27. Given the circumstances of an imminent change in administration, the Council Chairman decided not to submit any formal recommendation but to advise President Johnson and the incoming administration that: "This is an imaginative study made by a distinguished Commission, deserving study by all interests, in and out of the Federal Government. I have a strong conviction on the promise of the oceans, on the deep seabed as a legacy for all mankind and on the importance of America's stake therein. The rapid growth of technology makes these problems acute. The next Administration should give Commission recommendations immediate attention, especially as to Federal organization and Coastal Zone management" (minutes of the January 10, 1969, MSC meeting).

28. *New York Times*, January 12, 1969, p. 1, "Ocean Research Urged in Study," and January 18, 1969, p. 30, "Perspectives for Oceanography."

29. *Science* (Jan. 17, 1969):263–65.

30. Norman J. Padelford, ed. *Public Policy for the Seas* (Cambridge: M.I.T. Press, 1970), pp. 332–38.

31. Philip M. Boffey, "Humphrey vs. Nixon: Candidates Sharpen the Science Issues," *Science* 162 (Nov. 1, 1968):549–51.

32. Recommendations for the President prepared by the White House task group for transmittal by Vice President Agnew included:

a. To foster wise and productive use of coastal resources by (1) establishing a Federal grant program to encourage creation of State authorities for coastal area planning and management, (2) establishing coastal zone laboratories, and (3) planning regional port development.

b. To expand ocean exploration of coastal and deep sea resources and weather forecasting services to serve a wide range of public and private interests.

c. To improve the economic position of the U.S. fishing industry to meet foreign competition.

d. To establish national regional laboratories and provide a stronger, more stable base of support for these laboratories, academic research and manpower training.

e. To encourage development of ocean resources with a minimum of international conflict, by U.S. initiatives defining a legal regime for the deep ocean floor.

f. To initiate a long-range Federal contract program in basic marine technology, so as to develop a capability to work in the entire marine environment; e.g., a continental shelf laboratory.

33. Announced April 5, 1969 (Presidential Documents, Week Ending April 11), p. 530: "Named as Chairman of the Council was Roy L. Ash, president of Litton Industries, Inc., Beverly Hills, Calif. Members are Dean George Baker of Harvard University's Graduate School of Business Administration, Boston, Mass.; former Texas Governor John B. Connally, now a member of the Houston law firm of Vinson, Elkins, Weems, and Searls; Frederick R. Kappel, chairman of the executive committee, American Telephone and Telegraph Co., New York; and Richard M. Paget of the New York management consultant firm of Cresap, McCormick, and Paget." The Council was charged with considering: "(1) the organization of the executive branch as a whole in light of today's changing requirements of government; (2) solutions to organizational problems which arise from among the 150-plus departments, offices, agencies, and other separate executive organizational units; and (3) the organizational relationships of the Federal Government to States and cities in carrying out the many domestic programs in which the Federal Government is involved."

34. *Congressional Record* April 21, 1969, 9687–9689; April 22, 1969, 9993–9995; June 12, 1969, 15803–15805.

35. Congressman Wolff introduced into the Congressional Record on April 12, 1967, a copy of a keynote address presented by the author before the IEEE International Convention in New York March 21, 1967, entitled "Marine Sciences—Its Present and Future." The same day he introduced H.R. 8470 that would initiate a feasibility study of a quasi-public corporation for marine sciences research.

36. Edward Wenk, Jr., "Organizing for National Goals in Marine Science," in *Oceans* 1, no. 1 (Jan. 1969):42–49.

37. U.S. Congress, House, National Oceanographic Program, Part 2, *supra.*

38. The criteria for evaluating improvement in marine affairs management were cited as follows: (1) advantages to organizing around the ocean and atmosphere; (2) improvements in government-wide management;

(3) fulfill unmet public need; (4) contribute to improved decision making; (5) stable organization; (6) better performance at same or lower cost; (7) attract resources, good management, and personnel; (8) gains outweigh consequences of dislocation; (9) help the President and Congress to do a better job.

Functions were: international cooperation; coastal management; living resources; military program; nonliving resources; transportation and trade; exploration and surveys; arctic and antarctic; environmental observation and prediction; information management; research and education; engineering and technology.

39. The President's Task Force on Oceanography included James H. Wakelin, Jr., Chairman, Robert O. Briggs, John C. Calhoun, Jr., John P. Craven, Paul M. Fye, Chalmer G. Kirkbride, Edwin A. Link, William A. Nierenberg, Norman J. Padelford, F. Ward Paine, Dixy Lee Ray, Edfred Shannon, C. Monroe Shigley, Athelstan Spilhaus, Elmer P. Wheaton, and George P. Woollard. Consultants to the Task Force were S. Russell Keim and Amor L. Lane.

40. Some months later, the President rejected the Ash Council-Wakelin proposals on the NOAA alternative, and in the subsequent Reorganization Message, the Wakelin task report was released and gracefully acknowledged.

41. *Congressional Record*, March 5, 1970, pp. S3033–39.

42. Hollings had once considered association in Mitchell's firm of bond lawyers and had kept in touch through later years. Mitchell had recently obtained Hollings' support for the controversial nomination of Judge Haynsworth from South Carolina for appointment to the Supreme Court. Hollings had agreed, and fought a tough, fair—though unsuccessful—campaign. Mitchell owed him a favor.

43. Reorganization Plan 4 of July 9, 1970. The President's Message of Transmittal is given in Appendix 10.

44. H. Res. 1210 To Disapprove Reorganization Plan No. 4, September 15, 1970, John E. Moss and John D. Dingell.

45. To build in as much strength as possible, since NOAA was not to be an independent agency, I had urged Will Kriegsman on the White House staff to peg the top position at pay level III, within the range of five executive levels for policy officials ranging from Level I for Cabinet officers down to IV for assistant secretaries and V for special assistants, to assure adequate power for a NOAA head. This rating, higher than that of assistant secretaries in other agencies who had marine responsibilities, would give the incumbent an edge when negotiating interagency issues. Moreover, it was equal in rank to the Under Secretary of Commerce and thus, within the Department's power structure, provided some assurance of access to the Cabinet officer. Behind my recommendation was another more subtle objective—the need for enough independence from the Cabinet officer to avoid close fellowship with special interests that might be inimical to NOAA's future.

46. *MSA*, 1st report (1967), p. 95.

47. *MSA*, 2d report (1968), pp. 157–58.

48. S. 582, introduced February 4, 1971, by Senator Hollings for himself and thirty other senators.

49. S. 1963 introduced by Senators Hollings, Cranston, and Pell on May 26, 1971, and a revised S. 1986 introduced by Senators Hollings, Cranston, Humphrey, Magnuson, Pell, Stevens, and Tunney would amend P.L. 89–454 to "foster a comprehensive, long range, and coordinated na-

tional program in marine science, technology and resource development. . . ."

50. Judy Chase, "NOAA and Oceanographic Research 'Wet NASA' Idea Dries Up," *Science* 173 (July 16, 1971):216–21.

51. Membership in NACOA included Charles F. Baird, International Nickel Company, Inc.; Werner A. Baum, University of Rhode Island; Wayne Burt, Oregon State University; John C. Calhoun, Texas A&M College; William D. Carey, Arthur D. Little Corporation; Dayton H. Clewell, Mobile Oil Corporation and Mobile Research and Development Corporation; John P. Craven, University of Hawaii; Charles L. Drake, Dartmouth College; Thomas A. Fulham, Boston Fish Market Corporation; Joseph J. George, Eastern Airlines; Gilbert M. Grosvenor, National Geographic Society; William J. Hargis, Jr., Virginia Institute of Marine Science; Francis S. Johnson, University of Texas; Ralph A. MacMullan, State of Michigan, Department of Natural Resources; Thomas F. Malone, National Academy of Sciences and University of Connecticut; O. William Moody, Jr., AFL–CIO; Mark Morton, General Electric Company; William A. Nierenberg, Scripps Institution of Oceanography; John J. Royal, Fishermen and Allied Workers Union; Julius A. Stratton, Ford Foundation; Verner E. Suomi, University of Wisconsin; Clement Tillion, Alaska State Legislature; Myron Tribus, Xerox Corporation; Odale D. Waters, Jr., Florida Institute of Technology; Edward Wenk, Jr., University of Washington.

CHAPTER 9

1. Harold Seidman, *Politics, Position, and Power* (New York: Oxford University Press, 1970), p. 11.

2. Ibid., p. 35.

3. U.S. President (Eisenhower), *Farewell Radio and Television Address to the American People*, January 17, 1961.

4. For details, see reports U.S. Congress, House, *The Office of Science and Technology*, 90th Congress, 1st sess. (Washington: GPO, 1967). (Hereafter Congress, *OST*.) U.S. Congress, Senate, *Organizing for National Security: Science Organization and the President's Office*. Report of the Committee on Government Operations, 87th Congress, 1st sess., 1961 (Washington: GPO, 1961).

5. National Academy of Science, *Basic Research and National Goals*. Report to Committee on Science and Astronautics, U.S. Congress, House, 89th Congress, 1st sess., 1965, discussed in chapter by Harvey Brooks, pp. 77–110.

6. U.S. Congress, House, *Toward a Science Policy for the United States*. Report of the Subcommittee on Science, Research, and Development to the Committee on Science and Astronautics, 91st Congress, 2d sess. (Washington: GPO 1970).

7. *New York Times*, October 31, 1971, "Nixon Men Map New Aid to Technology."

8. Alvin Toffler, *Future Shock* (New York: Random House, 1970); Charles Reich, *The Greening of America* (New York: Random House, 1970); Lewis Mumford, *Pentagon of Power* (New York: Harcourt, Brace, Jovanovich, 1970).

9. Edward Wenk, Jr., "SST—Implications of a Political Decision," *Astronautics & Aeronautics* 9 (Oct. 1971):40–49.

10. Martin L. Perl, "The Scientific Advisory System: Some Observations," *Science* (Sept. 24, 1971):1211–15.

11. Although expressed in bland terms, this is the theme of the critique "Toward a Science Policy for the United States," October 15, 1970, released by the House of Representatives Subcommittee on Science, Research and Development.

12. Organization for Economic Cooperation and Development, *Science Growth and Society* (Paris: OECD, 1971).

13. An extensive body of literature has evolved from studies on technology assessment by the NAS, NAE, LRS, NAPA, and George Washington University, sponsored initially by the Science and Astronautics Committee of the House of Representatives. See also note 76, chap. 4, in this volume.

14. Edward Wenk, Jr., "Social Management of Technology," in *Science for Society*, ed. John E. Mock. Proceedings of the National Science Conference, Atlanta, October 1970, pp. 8–31.

15. National Academy of Sciences, *Technology: Processes of Assessment and Choice*, report for U.S. Congress, House, Committee on Science and Astronautics, 91st Congress, 1st sess., 1969; and National Academy of Public Administration, *Technology Assessment System for the Executive Branch*, report for same committee, 91st Congress, 2d sess., 1970 (Washington: GPO, 1969, 1970).

16. Wenk, "Social Management of Technology."

17. U.S. Congress, Senate, *Office of Technology Assessment*. Hearing before the Subcommittee on Computer Services of the Committee on Rules and Administration, United States Senate, 92d Congress, 2d sess., March 2, 1972 (Washington: GPO, 1972).

18. U.S. Congress, House, *National Goals and Policies*, report of the Select Committee on Government Research, 88th Congress, 2d sess., 1964 (Washington: GPO, 1964), part 1, p. 57. See also U.S. Congress, Senate, *The Evolution and Dynamics of National Goals in the United States*, by Franklin P. Huddle, Committee on Interior and Insular Affairs, 92d Congress, 1st sess. (Washington: GPO, 1971).

19. U.S. Congress, House, *Federal Research and Development Program: The Decisionmaking Process*," report by the House Committee on Government Operation, 89th Congress, 2d sess., 1966 (Washington: GPO, 1966), p. 41

20. Congress, *OST*.

21. Harvey Sapolsky, review of *OST* report, in *Bulletin of Atomic Scientists*, March 1968, pp. 46–48.

22. Congress, *OST*, p. 11.

23. Ibid., pp. 1–32 *passim*.

24. Sapolsky, *Bulletin of Atomic Scientists*.

25. Testimony of William D. Carey in U.S. Congress, House, *Presidential Advisory Committees*, hearings before Committee on Government Operations, 91st Congress, 2d sess., March 1970 (Washington: GPO, 1970), p. 162.

26. OST was established by the reorganization route through transfer of two of NSF's dormant functions. According to section 3(a) of the Reorganization Plan of 1962, these functions entailed (1) so much of the Foundation's authority "to develop and encourage the pursuit of a national policy for the promotion of basic research and education in the sciences" as will enable the OST Director "to advise and assist the President in achieving

coordinated Federal policies" in this explicit area; and (2) the Foundation's authority "to evaluate scientific research programs undertaken by agencies of the Federal Government."

Section 3(b) of the Reorganization Plan also provided for the OST Director, in carrying out these functions, to "assist the President as he may request with respect to the coordination of Federal scientific and technological functions." This latter provision for technological functions is consistent with the President's transmittal message where, in four of the five stated purposes, variations on the phrase "science and technology" appear conspicuously. Except when referring to scientific and technical personnel, the NSF enabling legislation makes no reference to the applied developmental or technological extensions of scientific research.

27. Reorganization Plan No. 2 of 1962, transmitted March 29 and effective June 8, 1962, Sec. 3(a) 1 and 2 transferring functions of the National Science Foundation, 42 U.S.C. 1862(a) 1 and 6.

28. Congress, *OST*, pp. 27–29. Similar proposals were made by Philip H. Abelson; see note 42, chap. 2.

29. U.S. Congress, House, *Toward a Science Policy for the United States*, a report by the Subcommittee on Science, Research and Development, Committee on Science and Astronautics (Washington: GOP, 1970).

30. U.S. President (Nixon), statement of July 13, 1969, in *Weekly Compilation of Presidential Documents* (Washington: GPO, July 1969).

31. National Goals Research Staff, *Toward Balanced Growth: Quantity with Quality*. Report to the President, July 4, 1970 (Washington: GPO, 1970).

32. The concept of an annual report on technology set forth by the Congress in *OST*, p. 31–32, had been anticipated by a proposal of William D. Carey for a yearly Presidential report on science, expressed as a strategy for public enlightenment, *Saturday Review* (Nov. 6, 1965):57–58.

33. Seidman, *Politics, Position, and Power*, p. 71.

34. Ibid., p. 190.

35. Harvey Sherman, *It All Depends* (University, Ala.: University of Alabama Press, 1967).

36. Testimony of Edward Wenk, Jr., "Functions and Operations of the Federal Council for Science and Technology," before the Select Committee on Government Research, U.S. Congress, House, 88th Congress, 1st sess., November 1963 (Washington: GPO, 1963).

37. Seidman, *Politics, Position, and Power*, p. 6.

38. *Analysis of the Role of the Vice President* (Berkeley: University of California, Bureau of Public Administration, 1957).

39. P.L. 81–216 signed August 10, 1949.

40. Michael Harwood, *In the Shadow of Presidents* (New York: J. B. Lippincott Co., 1966).

41. Edgar Wiggins Waugh, *Second Consul* (New York: Bobbs Merrill, 1956), p. 87.

42. Irving A. Williams, *The American Vice-Presidency* (New York: Doubleday & Co., 1954).

43. U.S. Congress, Senate, *Proposal to Create an Administrative Vice President*, 84th Congress, 2d sess., 1956 (Washington: GPO, 1956).

44. *New York Times*, July 18, 1968.

45. An approach to public administration proposed by Warren Bennis and criticized by Seidman, *Politics, Position, and Power*, p. 8.

46. Donald Young, *American Roulette* (New York: Holt, Rinehart and Winston, 1965), p. 322.

CHAPTER 10

1. Miller B. Spangler, *New Technology and Marine Resource Development* (New York: Praeger Publishers, 1970), pp. xxix–xxx.

2. Don Krasher Price, *The Scientific Estate* (Cambridge: Harvard University Press, 1965), p. 43.

3. P.L. 85–510 assigned responsibility to the National Science Foundation to initiate and support a program of study, research, and evaluation in the field of weather modification. This assignment was rescinded by P.L. 90–407 which transferred the responsibility for a variety of agency missions to the Executive Office of the President. Recommendations in 1971 by the National Academy of Sciences in *The Atmospheric Sciences and Man's Needs—Priorities for the Future* that administrative coordination and responsibility for government-wide programs relating to weather modification should be assigned to NOAA were not implemented, but P.L. 92–205 now assigns a registry function to the Secretary of Commerce.

4. "Stans Says Economy Needs SST More than Clean Air," *New York Times* (Oct. 27, 1971), 29:1.

5. Judy Chase, "NOAA and Oceanographic Research 'Wet NASA' Idea Dries Up," *Science* 173 (July 16, 1971):216.

6. See UNGA Resolutions: 2172 (XXI) of 6 December 1966; 2340 (XXII) of 18 December 1967; 2413 (XXII) of 17 December 1968; 2414 (XXII) of 17 December 1968; 2467 A-D (XXIII) of 21 December 1968; 2566 (XXIV) of 13 December 1969; 2574 A-D (XXIV) of 15 December 1969; 2602 F (XXIV) of 16 December 1969.

7. UNGA Resolution 2749 (XXV).

8. U.S. institutional initiatives for ocean space are analyzed by Leigh S. Ratiner, "United States Oceans Policy: An Analysis," *Journal of Maritime Law and Commerce* 2 (Jan. 1971): 225–66, and W. Frank Newton, "The New Quest for Atlantis: Proposed Regimes for Seabed Resources," *Judge-Advocate General Journal* 25 (Dec. 1970–Jan. 1971): 79–92. For most comprehensive examples see Draft Ocean Space Treaty by A. Pardo, Ambassador from Malta; Draft Ocean Regime by Elisabeth M. Borgese, Center for the Study of Democratic Institutions; "The Emergence of a Corporate Sovereignty for the Ocean Seas," in *Pacem in Maribus—Ocean Enterprises*, ed. E. H. Burnell and P. von Simson, Center for the Study of Democratic Institutions, Occasional Paper, vol. 2, no. 4 (Santa Barbara, Calif., June 1970), chap. 8. See also Gerard J. Mangone, *The United Nations, International Law, and the Bed of the Seas* (Washington: Woodrow Wilson International Center for Scholars, 1972). The potential role of marine technology in the emergence of transnational enterprises is summarized by Eugene B. Skolnikoff, *International Imperatives of Technology* (Berkeley: Institute of International Studies, University of California, 1972).

9. See for example UNGA Resolution 2580 (XXIV).

10. Edward Wenk, Jr., "Toward Enhanced Management of Maritime Technology," in *Pacem in Maribus—Ocean Enterprises*, ed. E. H. Burnell and P. von Simson.

Index of Names

Abel, Robert B., 72, 83, 121, 135, 147, 533, 556
Abelson, Philip H., 75, 563
Ackley, Gardner, 536
Adams, David A., 201, 556
Adams, John Quincy, 374, 394
Adams, Sherman, 397
Agassiz, Louis, 4
Agnew, Spiro T., 136, 160–63, 200, 352, 353, 399; Chairman, Marine Sciences Council, 156–66, 246, 351–53; staff of, 156, 160, 161; and Marine Commission report, 158–59, 161, 349–57 *passim;* addresses by, 160–61; 5-point marine initiatives announced by, 165, 201, 248, 311; mentioned, 204, 395, 535, 559
Alexander, Lewis M., 549
Alverson, Dayton Lee, 303, 309, 310
Amerasinghe, Hamilton Shirley, 261, 281
Ananichev, K. V., 232, 234
Anastasion, Steven N., 65, 533
Anderson, John B., 155, 352
Arnold, Henry A., 237, 314
Ash, Roy L., 159, 352, 355, 356, 357, 559
Ashley, Thomas L., 88
Auerbach, Carl A., 265, 557

Baer, Walter, 558
Baird, Charles F., 341, 557, 561

Baker, George, 559
Bardach, J. E., 310, 543
Barden, Charles S., Jr., 106
Barrow, Thomas A., 149, 353
Bartlett, E. L., 88, 90, 257, 314
Bartlett, Joseph W., 193
Bauer, Paul S., 53, 54, 71
Baum, Werner A., 561
Bayh, Birch, 395
Beckler, David Z., 68
Behrens, William W., 258
Bennis, Warren, 563
Berkner, Lloyd V., 56, 221, 556
Berman, Al, 164
Beverton, R. H. J., 234
Billingham, John, 164, 328
Bingham, Jonathan B., 549
Black, David S., 194
Blackman, Raymond V. B., 529
Blair, Stanley C., 156
Blake, F. Gilman, 366, 547
Blaustein, Jacob, 265, 270, 353, 557
Boffey, Philip M., 559
Bohlen, Charles E., 193, 539
Bonner, Herbert C., 53, 54, 56, 534
Booda, Larry L., 549
Borgese, Elisabeth Mann, 277, 564
Bowditch, Nathaniel, 4
Boyd, Alan S., 193, 245, 346, 347, 535, 539
Briggs, Robert O., 560
Brockett, E. D., 269

General Index

Note: For cross references to persons, see Index of Names, p. 565